AF841243

Laser Interaction with Matter

Laser Interaction with Matter

Proceedings of the 23rd European Conference, St John's College, Oxford, 19–23 September 1994

Edited by S J Rose, Rutherford Appleton Laboratory, Chilton, Didcot, Oxfordshire, UK

Institute of Physics Conference Series Number 140
Institute of Physics Publishing, Bristol and Philadelphia

Copyright ©1995 by IOP Publishing Ltd and individual contributors. All rights reserved. No part of this publication may be reproduced, stored in a retrieval system or transmitted in any form or by any means, electronic, mechanical, photocopying, recording or otherwise, without the written permission of the publisher, except as stated below. Single photocopies of single articles may be made for private study or research. Illustrations and short extracts from the text of individual contributions may be copied provided that the source is acknowledged, the permission of the authors is obtained and IOP Publishing Ltd is notified. Multiple copying is permitted in accordance with the terms of licences issued by the Copyright Licensing Agency under the terms of its agreement with the Committee of Vice-Chancellors and Principals. Authorization to photocopy items for internal or personal use, or the internal or personal use of specific clients in the USA, is granted by IOP Publishing Ltd to libraries and other users registered with the Copyright Clearance Center (CCC) Transactional Reporting Service, provided that the base fee of $19.50 per copy is paid directly to CCC, 27 Congress Street, Salem, MA 01970, USA.
0305-2346/95 $19.50+.00

CODEN IPHSAC 140 1–476 (1995)

British Library Cataloguing in Publication Data

A catalogue record for this book is available from the British Library.

ISBN 0 7503 0193 7

Library of Congress Cataloging-in-Publication Data are available

Published by Institute of Physics Publishing, wholly owned by The Institute of Physics, London
Institute of Physics Publishing, Techno House, Redcliffe Way, Bristol BS1 6NX, UK
US Editorial Office: Institute of Physics Publishing, The Public Ledger Building, Suite 1035, Independence Square, Philadelphia, PA 19106, USA

Printed in the UK by J W Arrowsmith Ltd, Bristol

Contents

Section 4: Advances in high-power laser technology

Section 5: Radiation physics

Section 6: X-ray lasers

Section 7: Short-pulse interaction with gas targets

Section 8: Coronal interactions

Section 9: Poster presentations

Preface

The 23rd European Conference on Laser Interaction with Matter ECLIM'94 was held at St John's College, Oxford from the 19th to the 23rd of September 1994. This was a particularly interesting time. A recent declassification of formerly secret material by the United States allowed research work concerning the indirect-drive approach to inertial confinement fusion, which had previously not been shown in public, to be presented at the Conference. However, ECLIM'94 was not only concerned with fusion issues, but looked at many other aspects of laser–matter interaction, notably the remarkable new developments in high-power, short-pulse lasers and their applications.

I would like to thank the authors for the careful preparation of their manuscripts and Institute Of Physics Publishing, in particular Kathryn Cantley, for publishing these proceedings so rapidly.

The Organizing Committee is very grateful to Anne-Marie Hamilton and her staff at the Conference Organization Section of the Rutherford Appleton Laboratory for all their work in organizing the Conference.

S J Rose
1995

Organizing Committee

C B Edwards
A M Hamilton (Secretary)
M H Key (Chairman)
S J Rose (Acting Chairman)
P C Thompson
J S Wark
O Willi

Sponsors

Rutherford Appleton Laboratory
University of Oxford
European Union
United States Air Force European Office of Aerospace Research and Development
United States Office of Naval Research, Europe
United States Army Research Development and Standardization Group, UK
Ministry of Defence, UK
Excitech Ltd
Kentech Ltd

Inst. Phys. Conf. Ser. No 140: Section 1
Paper presented at Laser Interact. with Matter Conf., Oxford, 19–23 September 1994

© 1995 IOP Publishing Ltd

Short pulse laser interaction with solid targets

J C Gauthier*

Laboratoire pour l'Utilisation des Lasers Intenses et Laboratoire d'Optique Appliquée (ENSTA), Ecole Polytechnique, 91120 Palaiseau, France

Abstract. The development of lasers delivering ultra-short pulses at the Terawatt power level opens new perspectives for the study of laser-matter interaction at very high intensities. The high intensity laser-produced plasmas look very promising for the generation of sub-picosecond x-rays in the keV range. Several groups have recently begun to explore the physics of laser-heating solid density targets because the availability of such x-ray pulses will have significant impact in many areas requiring time-resolved measurements in plasma and solid state physics.

1. Introduction

The advent of high intensity subpicosecond lasers has opened up new possibilities in the laser-matter interaction field. With the technique of chirped pulse amplification (CPA) [1] it is now possible to build compact solid-state lasers capable of generating powers in excess of 10 TW [2, 3]. With Ti-sapphire amplifiers, sub-50 fs pulses above the TW level are now feasible [4]. The generation of ultra-bright and ultra-fast x-ray pulses is a challenging application of short pulse laser-produced high density and high temperature plasmas [5].

The laser interaction physics of ultrafast lasers with solid targets is very specific [6-12]. Laser energy absorbed within the laser skin depth (≈50Å) gives rise to a thermal plasma of several hundred electron-volts temperature [13,14], approaching solid density. A thermal conduction heat front propagates over a range of order 2000Å towards the bulk of the target [15,16]. Behind the thermal plasma, fast electrons resulting from the non-linear interaction of the incident laser light through the steep density gradient [17-20] eject inner shell electrons of the target plasma, producing Kα radiation [21-22]. Very short x-ray emission from the thermal plasma is expected because electrons cannot transfer a significant fraction of their energy to ions and very weak expansion occurs during the pulse [15]. Kα emission can be [21,23] even shorter because fast electrons last no longer than the duration of the laser pulse. Effects of different electron and ion temperatures, non steady-state population kinetics and non-Maxwellian character of the free electrons in the plasma have to be taken into account [24,25] to understand line formation.

* Work done in collaboration with J.P. Geindre, P. Audebert, A. Rousse, F. Falliès, R. Mancini, A. Mysyrowicz, A. Dos Santos, G. Hamoniaux, J.P. Chambaret, A. Antonetti. LULI and LOA Laboratories are associated with the Centre National de la Recherche Scientifique.

In this review, we will present the state-of-the-art for ultra-short laser pulse interaction with solid targets and focus on laser absorption studies, optical diagnostics of plasma dynamics and x-ray production.

2. Physics of short pulse interaction

There are two major processes which must be considered in describing the interaction of subpicosecond pulses with matter. These are the thermal transport into the target and the absorption of incident laser radiation into the plasma. For the thermal transport, the electron mean free path being larger than the skin depth, non-local thermal transport has to be considered [26]. Here, we will concentrate on laser absorption mechanisms. Laser energy absorption mechanisms in the ultrafast regime depend critically [17,27] on the gradient scale-length L which have been shown by hydrodynamic simulations (see figure 1) to be very much smaller than the laser wavelength λ ($L/\lambda \ll 1$).

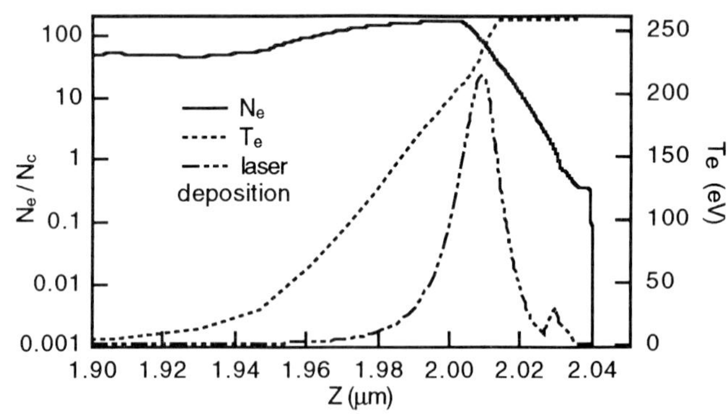

Figure 1: Electron density, temperature and laser energy deposition predicted at the maximum of the laser pulse by the hydrodynamic code FILM. Laser 100fs, 612 nm, 10^{16} W/cm^2.

Laser absorption by a steep-density profile can be calculated using the Drude metallic model for the plasma index of refraction and by solving the Maxwell equations [28-31]. The reflectivity of S- and P-polarised ultra-short light pulses from the plasma has a L dependence that depends strongly on the angle of incidence and collisionality of the material. For S light and $\nu/\omega \sim 1$ the reflectivity increases with L before it begins to decrease in the well-known way. For P light and $\nu/\omega \ll 1$ the angular dependence of the reflectivity presents a minimum which is a strong function of L. For long density scale-length plasmas absorption occurs mainly due to the enhanced field in the critical region, the so-called resonance absorption. For short density scale-length plasmas a new mechanism has been proposed [17], the so-called vacuum heating [27]. The transition from the resonance absorption regime to the vacuum heating regime has been studied by particle-in-cell (PIC) simulations [32] including collisions. It occurs for $L/\lambda \sim 0.01$ where the behaviour of the resonant field becomes complex as the orbits of the resonant electrons extend into the vacuum. In the short pulse regime, wave breaking [33] is the major damping mechanism of the resonant field due to the shortness of the density gradient scale-length. Hydrodynamic codes for short pulse laser interaction are now available [31,34-35] which incorporate an electromagnetic solver to describe the laser absorption in any density gradient configuration and which include a specific treatment of the thermal heat wave propagation in the bulk of the target.

The first measurements of reflectivity (and thus of absorption in the absence of scattering) over a wide range of laser intensities in the subpicosecond regime have been performed by Milchberg and collaborators [8]. Their results are shown in figure 2 and are compared to the results of hydrodynamic simulations using the FILM hydrocode [31]. The absorption maximum around 10^{14} W/cm^2 clearly show the temperature variation of the plasma collisionality. Different models for plasma conductivity [34,36] have been proposed which cover the transition from the degenerate solid to the high temperature classical Spitzer plasma through the non-ideal correlated plasma. High frequency effects on the collisionality have to be incorporated [37,38]. Other absorption studies in the "low" intensity regime have been reported [29,39,40]. Few work has been done on absorption measurements over a wide range of laser energies. Sauerbrey and collaborators [41] have extended the range studied in the early work of Reference 8 up to 10^{17} W/cm^2. A more complete and coherent study over a range extended above 10^{18} W/cm^2 have been performed by the Livermore group [38,42] with a frequency-doubled Ti:sapphire laser and 140 fs pulse duration. Their results show 17 % absorption at 5.10^{13} W/cm^2, a maximum of 30% at 10^{15} W/cm^2 followed by a shallow decrease down to below 10% at 10^{18} W/cm^2. At much higher intensities, in the relativistic regime, where the ponderomotive electron energy is larger that the rest mass of the electron, PIC simulations [43,44] predict strong effects of the ponderomotive laser pressure (hole boring) and copious hot electron generation by the **JxB** force. Ponderomotive effects have been scarcely observed [45,46] through time-resolved Doppler shift measurements of the critical density surface velocity.

It is well established, through x-ray yield measurements [47,49] that the laser absorption is increased if the laser interacts not with the surface of a solid material but with a preformed plasma originating from exposure of the target with an earlier pulse or with prelase from amplified spontaneous emission (ASE) from the laser. The laser intensity contrast problem is crucial in short pulse laser interaction because even with a contrast of 10^{-10}, the laser intensity in the early part of the pulse can be larger than the threshold for plasma formation on most metals. Several techniques have been used to increase the intensity contrast ratio such as transparent target mirrors [50], saturable absorbers [46] and laser frequency doubling [38].

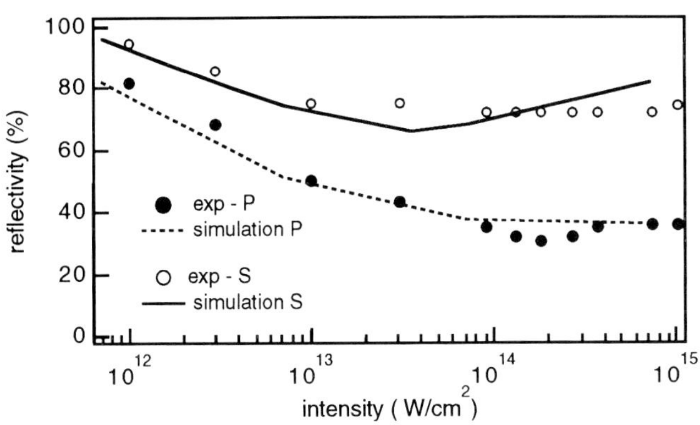

Figure 2: Reflectivity measurements of Milchberg *et.al.* in S and P polarisations compared to hydrocode simulations. Laser 400 fs, 308 nm, incidence 40°.

3. Optical diagnostics of plasma dynamics

To characterise subpicosecond laser-produced plasmas, optical diagnostics have been used. The most simple arrangement involves the time-integrated measurement of the absorption [51] or of the reflectivity of the interacting laser pulse as a function of the incidence angle and of the polarisation angle [8,29,41,52]. More recently, pump/probe techniques have been applied which allow time-resolved studies to be made with a time resolution limited by the duration of the probe pulse. Femtosecond imaging [53] and Schlieren measurements [54] have been used with micron spatial resolution on the position of the expanding plasma surface. Reflection, transmission and scattering measurements of a probe pulse give useful informations on the plasma dynamics. The gradient scale-length and expansion velocities are indirectly deduced by comparison with simulations. These measurements give informations on the *amplitude* variations of the probe pulse induced by the plasma. Femtosecond time-resolved measurements of the *phase* change of a probe pulse by spectral blueshifting [55,56] and second-harmonic generation [57] have been reported. These measurements probe directly the temporal change, with femtosecond resolution, of the plasma refractive index and thus give access to the ionisation dynamics of the plasma during the pump pulse and to its subsequent hydrodynamic expansion.

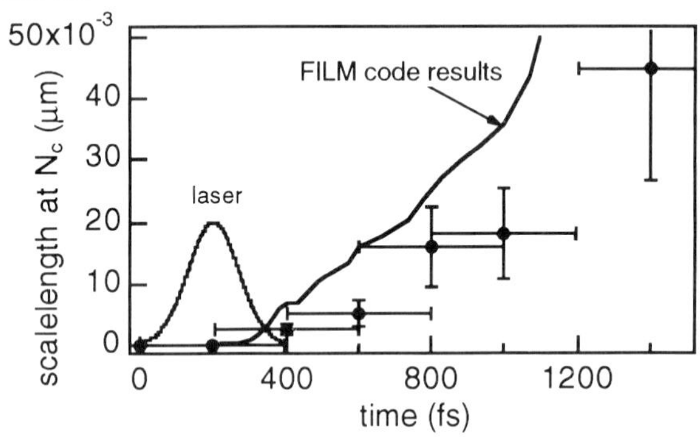

Figure 3: Measured plasma scale-length as a function of time with the double pulse interferometric technique. Laser: 100 fs, 0.62 µm, 10^{15} W/cm^2.

A frequency-domain interferometer for probing the variations of the dielectric constant of a plasma with sub-100 fs temporal resolution and $\lambda/2000$ phase resolution has been recently introduced [58]. It measures the phase shift difference between two probe femtosecond pulses delayed with respect to the plasma-producing pulse. By imaging the plasma on the entrance slit of a spectrograph, spatial resolution along a diameter of the focal spot is provided. The technique has been used to map out the expansion of the critical density surface of a femtosecond laser-produced plasma. Results are given in figure 3 compared to the predictions of hydrodynamic simulations.

4. X-ray diagnostics and conversion efficiency

X-ray spectroscopy has proved to be a powerful diagnostic technique to infer plasma properties in a large variety of laser-produced plasmas involving laser pulses of subnanosecond duration [59]. Experiments with subpicosecond pulses offer a unique opportunity to observe time-dependent atomic kinetics effects since the laser pulse duration is comparable to the atomic processes time-scale [60]. K-shell emission and Li- and He-like satellites have had a rich history of applications in plasma spectroscopy due to the temperature and density dependence of their peak-intensity distribution and broadening properties [59]. Time-integrated spectroscopy of K-shell and satellite emissions of aluminium plasmas have been investigated in great detail [12,24-25,47-48,61-62]. For intensities above 10^{16} W/cm^2, probed densities are in excess of 10^{23} cm^{-3} [63]. Because of time integration, emission from different regions of the target leave their mark in different parts of the spectrum [25]. At intensities approaching 10^{18} W/cm^2, probed densities seem to saturate due to the increasing contribution of the low density expanding plasma and to possible overheating of the observed ion stages. Due to the very high densities, opacity effects have to be taken into account in line modelling [62,64]. Hot electron collisions are equally important in populating doubly-excited levels from which satellite lines originate [25, 24], as depicted in figure 4.

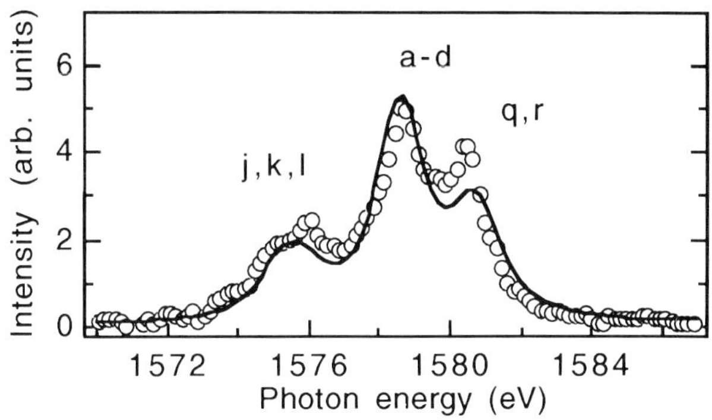

Figure 4: Comparison of experimental spectrum [25] and modelling results with hot electron effects. Laser: 100 fs, 0.62 µm, 10^{17} W/cm^2.

These transient quasi-solid-state plasmas have attracted attention as potential sources for ultrafast, pulsed x-rays [5]. However, a major limitation to their general use has been the relatively low laser-energy to x-ray conversion efficiency observed to date (< 0.1%). This low efficiency is due to the low laser absorption coefficient (see Section 1), rapid conduction into the bulk of the solid and the resulting relatively low plasma temperature. Several techniques have been devised to increase the absorption coefficient by using micro-structured targets which exhibit low reflectivity. Gratings and solid metal clusters (gold black foams) allow to reach a conversion efficiency of the order of 1% for gold. Another technique [66] uses the ASE pedestal to structure the surface state at a fluence close to the target damage threshold.

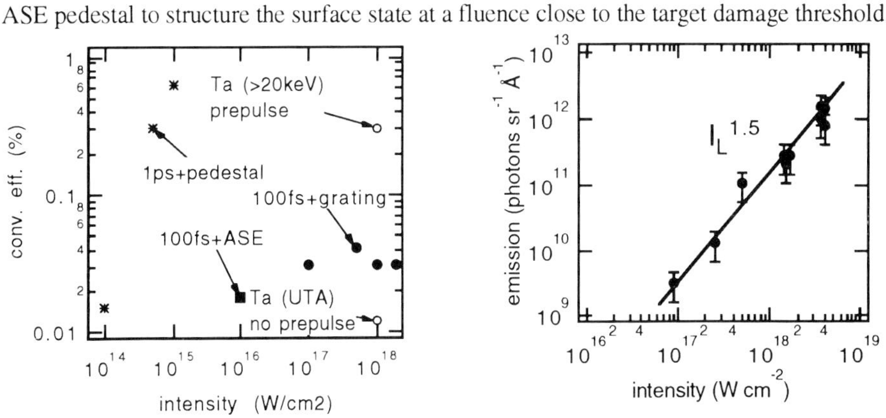

Figure 5: Left, conversion efficiency to x-rays above 1 keV in various target conditions and right, absolute K-shell x-ray yield both as a function of laser intensity.

Both thermal and non-thermal emission have been absolutely calibrated in various experiments. It has been demonstrated [23] that hot-electron-produced Kα line radiation can be scaled up to above 6 keV with a photon yield of 2×10^7 photons/steradian/pulse. This corresponds to a conversion efficiency of 0.015%. Comparable conversions for K-shell thermal plasma line emission have been measured and broadband UTA emission from tantalum with a conversion efficiency of 0.01% have been observed [63]. With a large prepulse, higher x-ray conversion can be obtained (see figure 5, left) to the detriment of pulse duration. Under such conditions, MeV radiation has been detected [67]. The x-ray yield for the supra-keV energy range scales as I^{γ} where γ has been measured to be between 1.5 and 2.5 [63,68] (see figure 5).

The most important feature of short pulse laser-produced plasma incoherent x-rays is their brievety, probably in the subpicosecond range for supra-keV radiation. Up to know, the time duration of thermal x-rays has been found to be less than $\approx$ 2ps, the ultimate resolution of ultrafast streak cameras [5,69-72]. Kα emission is thought to be shorter than thermal emission because it is produced by hot electrons which last no longer than the laser pulse. It should be noted that time-resolved K-shell spectroscopy [47,70] has shown that ions with on or two electrons on their outer shell (e.g. Be-like) also exhibit a very short x-ray emission time, much smaller than laser pulse duration.

5. Conclusions

Recent studies of short pulse laser-matter interaction have shown the very specific nature of the absorption and transport processes in steep density gradients. Reflectivity studies show that the absorption is no larger than 40% at 10^{16} W/cm^2 and that its variations with laser intensity reflects the temperature change of the plasma collisionality. The effects of the quality of the laser pulse (contrast ratio) and of the microstructure of the target surface have been investigated. Among other aspects, these plasmas are of great interest for the production of ultra-short x-ray pulses. Conversion efficiencies are in the 0.1 % range for photons above 1 keV energy. X-ray diagnostic studies have shown that densities well in excess of 10^{23} cm^{-3} can be produced with table-top sized lasers. Regarding applications, the availability of ultra-short pulse x-rays will have significant impact in areas such as time-resolved x-ray diffraction, extended-x-ray-absorption fine structure, probing of molecular dynamics and pumping of x-ray laser plasmas. Many subjects will attract a lot of attention in the near future when higher intensities (>10^{19} W/cm^2) and shorter pulse durations (<50 fs) will be available. Among these, we foresee the study of the laser ponderomotive pressure, the fast ignitor concept in inertial-controlled fusion, absorption by surface waves, giant magnetic fields and polarisation of the x-rays, and high harmonic generation in solids.

References

[1] Maine P *et.al.* 1988 IEEE J. of Quantum Elec. **24** 398
[2] Beaudoin Y *et.al.* 1992 Opt. Lett. **17** 865
[3] Sauteret C *et.al.* 1991 Opt. Lett. **16** 238
[4] Kaypten H C and Murnane M M 1994 "High Field Interaction and Short Wavelength Generation" vol. 16 OSA Technical Digest Series p. 3
[5] Murnane M M *et.al.* Science 1991 **251** 531
[6] Kuhlke D Herpes U and Van der Linde D 1987 Appl. Phys. Letters **50** 1785
[7] Landen O L Campbell M and Perry M D 1987 Opt. Comm. **63** 253
[8] Milchberg M M Freeman R R and Davey S C 1988 Phys. Rev. Lett. **61** 2364
[9] Murnane M M Kaypten H C and Falcone R W 1989 Phys. Rev. Lett. **62** 155
[10] Audebert P *et.al.* 1992 Europhys. Lett. **19** 189
[11] Kieffer J C *et.al.* 1993 Phys. Fluids **B5** 2330
[12] Cobble J A *et.al.* 1989 Phys. Rev. **A39** 454
[13] Rozmus W and Tikhonchuk V T 1990 Phys. Rev. **A42** 7401 and 1992 Phys. Rev. **A46** 7810
[14] Cornolti F Mulser P and Hahn M 1991 Laser Part. Beams **9** 474
[15] Milchberg H M Lyubomirsky I and Durfee C G 1991 Phys. Rev. Lett. **67** 2564

[16] Rosen M, "*Femtosecond to Nanosecond High-Intensity Lasers and Applications.*" 1990 SPIE Proceedings **1229** p.160
[17] Brunel F 1987 Phys. Rev. Letters **59** 52 and 1988 Phys. Fluids **31** 2714
[18] Bonnaud G Gibbon P Kindel J and Williams E 1991 Laser Part. Beams **9** 339
[19] Gamaly E 1993 Physics Fluids **B5** 3765
[20] Gibbon P 1994 Phys. Rev. Lett. **73** 664
[21] Meyerhofer D D *et.al.* 1993 Physics Fluids **B5** 2584
[22] Rousse A *et.al.* 1993 OSA Proceedings on *Short Wavelength V: Physics with Intense laser pulses* Perry M D and Corkum P Eds (Optical Society of America, Washington DC) vol.17, pp.185-189
[23] Rousse A *et.al.* 1994 Phys. Rev. E **50** 2200
[24] Peyrusse O Kieffer J C Coté C Y and Chaker M 1993 J. Phys. B: At. Mol. Opt. Phys. **26** L511
[25] Mancini R C *et.al.* 1994 J. Phys. B: At. Mol. Opt. Phys. **27** 1671
[26] Luciani J F Mora P and Virmont J 1983 Phys. Rev. Letters **51** 1664 and Luciani J F and Mora P 1986 Physics Lett. **A116** 237
[27] Gibbon P and Bell A R 1992 Phys. Rev. Lett. **68** 1535
[28] Milchberg H M and Freeman R R 1989 J. Opt. Soc. Am. B **6** 1351
[29] Fedosejevs R *et.al.* 1990 Applied Physics B **50** 79
[30] Rae S C and Burnett K 1991 Phys. Rev. A**44** 3835
[31] Falliès F 1994 Thèse de Doctorat de l'Université Paris-Sud
[32] Delettrez J 1994 LLE Review **58**, 76
[33] Kruer W L *"The Physics of Laser Plasma Interactions"* 1988 (Addison-Wesley: Redwood City)
[34] Ng A G *et.al.* 1994 Phys. Rev. Lett. **72** 3351
[35] Hüller S (private communication).
[36] Lee Y T and More R M 1984 Phys. Fluids **27** 1273
[37] Ebeling W *et.al.* "*Thermophysical Properties of Hot and Dense Plasmas*" 1991 (Teubner: Stuttgart)
[38] Rozmus W Tikhonchuk V T Cauble R and Shepherd R (this conference)
[39] Riley D Gizzi L A Mackinnon A J Viana S M and Willi O 1993 RAL Annual Report p.16
[40] Kieffer J C et.al. 1989 Phys. Rev. Lett. **62** 760
[41] Sauerbrey R *et.al.* 1994 Phys. Plasmas **1** 1635
[42] More R M, private communication.
[43] Denavit J 1992 Phys. Rev. Lett. **69** 3052
[44] Wilks S 1993 Phys. Fluids **5** 2603
[45] Liu X and Umstadter D 1992 Phys. Rev. Lett. **69** 1935
[46] Kalashnikov M P *et.al.* 1994 Phys. Rev. Letters **73** 260
[47] Willi O *et.al.* 1989 Europhysics Letters **10** 141
[48] Cobble J A Schappert G T Jones L A Taylor A J Kyrala G A and Fulton R D 1991 J. Appl. Phys. **69** 3369
[49] Teubner U Kühnle G and Schäfer F P 1992 Appl. Phys. B **54** 493
[50] Kapteyn H C Murnane M M Szoke A and Falcone R W 1991 Opt. Lett. **16** 490
[51] Landen O L Stearn D G Campbell M 1989 Phys. Rev. Lett. **63** 1475
[52] Wang X Y and Downer M C 1992 Optics Letters **17** 1450
[53] Downer M C Fork R L and Shank C V 1985 J. Opt. Soc. Am. B **2** 595
[54] Benattar R *et.al.* 1992 Optics Communications **88** 376
[55] Wood W M Siders C W and Downer M C 1991 Phys. Rev. Letters **67** 3523
[56] Penetrante B Bardsley J N Wood W M Siders C W Downer M C 1992 J. Opt. Soc. Am. B**9** 2032
[57] Van der Linde D Schultz H Engers T and Schüler H 1992 IEEE J. Quant. Electr. **28** 2388
[58] Geindre J P *et.al.* 1994 Opt. Lett. (in press 1994).
[59] Griem H R 1992 Phys. Fluids **B4** 2346
[60] Edwards J and Rose S 1993 J. Phys. B: At. Mol. Opt. Phys. **26** L523
[61] Riley D. *et.al* 1992 Phys. Rev. Letters **69** 3729
[62] Audebert P *et.al.* 1994 J. Phys. B: At. Mol. Opt. Phys. 27, 3303
[63] Kieffer J C *et.al.* 1994 "High Field Interaction and Short Wavelength Generation" vol. 16 OSA Technical Digest Series p. 69
[64] Apruzesse J P 1993 Phys. Rev. E **47** 2798
[65] Gordon S P Donnelly T Sullivan A Hamster H Falcone R W 1994 Opt. Lett. **19** 7
[66] Audebert P *et.al.* 1991 AIP Conf. Proc. **257** 58
[67] Kmetec J D 1992 IEEE Journal of Quantum Electronics **28** 2382
[68] Teubner U Bergmann J Van Wonterghem B Schäfer F P and Sauerbrey R 1993 Phys. Rev. Lett. **70** 794
[69] Stearns D G Landen O L Campbell E M Scofield J H Phys. Rev. **37** 1684
[70] Kieffer J C *et.al.* 1993 Phys. Fluids **B5** 2330
[71] A. Mens *et.al.* Ultrafast VIII Migus A and Martin J L Eds.1992 (Springer Verlag: Berlin)
[72] Murnane M M Kaypten H C Falcone R W 1990 Appl. Phys. Lett. **56** 1948

Inst. Phys. Conf. Ser. No 140: Section 1
Paper presented at Laser Interact. with Matter Conf., Oxford, 19–23 September 1994
© 1995 IOP Publishing Ltd

Characterization of plasmas produced by picosecond and nanosecond laser pulses

L A Gizzi, T Afshar-Rad[1], M Desselberger[1], A Giulietti, D Giulietti, A J Mackinnon[1], D Riley[1], S M Viana[1], O Willi

Istituto di Fisica Atomica e Molecolare - CNR, via del Giardino, 7 - 56127 Pisa, Italy

[1]The Blackett Laboratory, Imperial College of Science Technology and Medicine, Prince Consort Road, London, SW7 2BZ, UK

Abstract: The results of an extensive experimental investigation devoted to the characterization of laser produced plasmas will be considered in this paper. 600 ps laser pulses were employed to explode thin Al targets and generate long scalelength plasmas for interaction studies relevant to inertial confinement fusion. Thermal transport processes in underdense plasmas were investigated using time resolved X-ray imaging. Transport in micron-sized high density plasmas was also investigated in the picosecond regime, using X-ray spectroscopy. In this case the heat flux was found to be smaller than that predicted by the 1D modelling, with the flux inhibition increasing dramatically at high intensities ($>10^{16}$ W/cm^2). Preliminary results will be presented which suggest that strong self–induced magnetic fields may arise during the interaction which may influence heat transport processes.

1. Introduction

Plasma production by high power laser pulses is of great importance for a wide range of applications as well as basic physics issues. Long scalelength plasmas are presently used as test-beds for laser-plasma interaction experiments related to the physics of inertial confinement fusion[1]. On the other hand the production of hot high density plasmas has strong impact on the development of short pulse X-ray laser sources and basic plasma physics studies including for example, thermal transport processes[2].

Laser interaction with matter was experimentally investigated in the picosecond regime. Solid targets were irradiated using the 12 ps, 0.25 μm laser pulse generated by the SPRITE laser system at the Central Laser Facility (Rutherford Appleton Laboratory, UK). Micron sized, hot dense plasmas were generated and extensively characterized, using a wide range of diagnostic techniques including X-ray spectroscopy, X-ray imaging and calorimetry.

The interaction of high intensity laser light with matter was extensively investigated in the nanosecond regime. Thin Al targets were irradiated at intensities up to 10^{14} W/cm^2 using 600 ps laser pulses produced by the VULCAN laser system at the Central Laser Facility (Rutherford Appleton Laboratory, UK), to generate long scalelength highly uniform plasmas. Two different configurations were adopted. In the so called line-focus geometry, plasmas were produced exploding stripe Al targets. Interaction experiments were performed focusing the delayed beam along the target length. In the cylindrical geometry disk targets were used and interaction experiments were performed focusing the beam along the target symmetry axis. Plasmas were fully characterised[3] in terms of electron temperature and density using time-resolved X-ray imaging and spectroscopy as well as optical interferometry. Advanced numerical techniques were implemented[4] which allowed fully automatic analysis of interferometric patterns. In this paper we focus the attention on some aspects of the investigation which can provide information on electron transport processes in plasmas.

2. Long scalelength plasmas

X-ray imaging in the 1.5 keV photon energy region with ≈200 ps resolution was employed in order to detect the effect of localized heating induced by an interaction laser pulse on a preformed plasma generated in the line-focus geometry. Fig.1 shows a sequence of four time resolved X-ray images of the laser-plasma interaction region. The first frame, temporally located at the peak of the target heating pulse, shows emission from the stripe target during its explosion. The remaining three frames, temporally distributed on the 600 ps interaction pulse, delayed by 2.5 ns with respect to the heating pulse, show X-ray emission from the region of the underdense plasma involved in the interaction. The plot of Fig.2 shows the transverse size of the interaction region as measured from time resolved X-ray images as a function of the interaction intensity in a range of from 10^{13} to 10^{15} W/cm^2 obtained with the same focusing conditions, i.e. 140 μm focal spot. The two sets of data were obtained interacting with plasmas produced from Al and KCl targets respectively.

The observed dependence of the transverse extent of the X-ray emitting region upon the average intensity can can be interpreted in terms of thermal transport processes, once the spatial distribution of the laser intensity in the focal spot has been taken into account. In fact one can reasonably assume that the local heating effect will give rise to X-ray emission in the selected spectral range if the local power is higher than a given reference value. Since the actual intensity distribution is characterized by tails which extend outside the nominal spot size, as the average power of the pulse increases, the effective focal spot size increases.

On the other hand, according to this simple description, as the nominal intensity decreases, the effective transverse size of the focal spot decreases and vanishes when the maximum power approaches the reference value. In this case heat transport processes will play a dominant role in determining the size of the X-ray emitting region, giving rise to a spread of the thermal energy over a volume determined by the characteristic thermal diffusion length. As a simple estimate, we assume[3] a fully ionized background Al plasma at an electron temperature and density of 450 eV and 1.5×10^{20} cm^{-3} respectively. Using a 1D equation[5] of classical heat transport and integrating over the 200 ps gate time of the X-ray imaging system we find a diffusion length of 175 μm which is very close to the lower limit of the measured transverse size given in Fig.2.

In the high intensity limit the interaction beam heavily perturbs the background plasma, giving rise to strong local heating and, consequently, higher thermal conductivity. An accurate calculation of the thermal diffusion length requires, therefore, that the new value of the electron temperature is taken into account. In general, the actual spot size will be a combination of transport processes and effects of the intensity distribution in the focal spot.

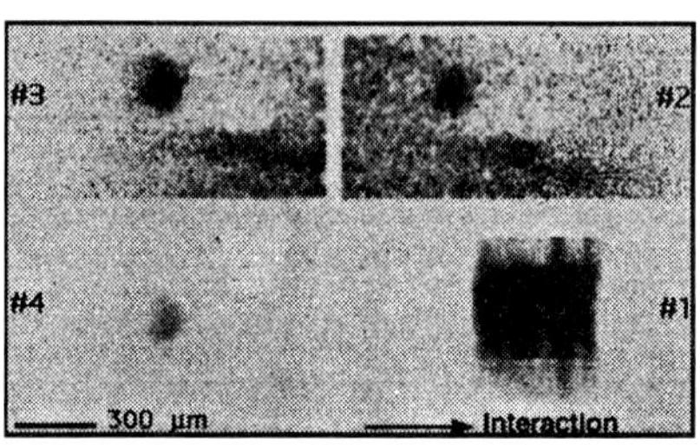

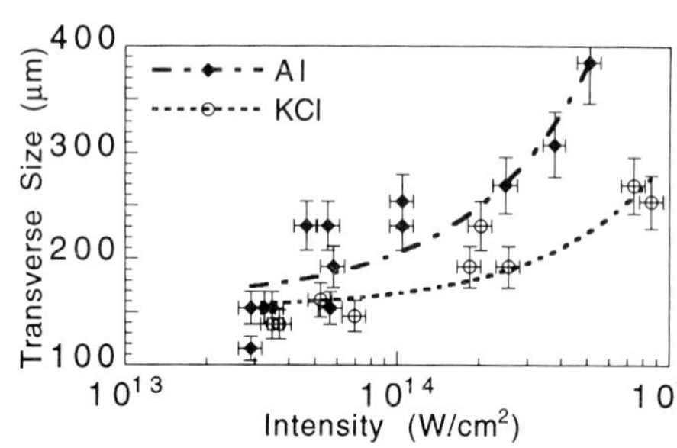

Fig.1.(left) Sequence of time resolved X-ray images of laser interaction with a preformed plasma. The interaction beam intensity 3.5×10^{13} W/cm^2. The first frame was temporally located on the peak of the target heating pulse while the remaining three frames where temporally distributed on the 600 ps interaction pulse, delayed by 2.5 ns with respect to the heating pulse. **Fig.2.(right)** Transverse size of the X-ray emitting region as a function of the nominal intensity of the interaction beam focused on an Al plasma and a KCl plasma.

The observed different behaviour of the Al and the KCl data can be explained taking into account ionization. In fact, at the given background electron temperature, the plasma produced from KCl is not fully ionized. In this case, further ionization will occur during the interaction which will subtract thermal energy from the electrons.Consequently, the transverse size of the X-ray emitting region is expected to be smaller than in the case of the Al plasma. In the limit of high temperature however, both plasmas can be considered fully ionized and differences in the spot size should only come from the dependence of transport processes upon the charge state.

These experimental results show that time-resolved X-ray imaging of the laser plasma interaction region can provide valuable information on basic physics issues such as thermal transport processes. While in this case the thermal diffusion length estimated from these data is consistent with the classical thermal transport theory, transport in micron sized plasmas exhibits a much more complex behaviour.

3. Picosecond interactions

Micron sized plasmas are generated when a high power short laser pulse is focused onto solid targets. Since the hydrodynamics of these plasmas is strictly related to thermal transport, ablation processes can therefore be used to indirectly monitor transport phenomena. Measurements of ablation rate have been performed irradiating layered targets with 12 ps laser pulses, at intensities as high as 3×10^{16} W/cm^2. The propagation of the ablative heat wave in layered targets was investigated in this regime. The target consisted of a 10 μm thick Mylar substrate coated with 2000 Å Al, overcoated with a thin (1000 Å) low Z plastic layer. The history of XUV line emission generated by the signature layers, as the heat front propagated in the target, was recorded using X-ray time resolving spectroscopy. The spectrometer consisted of an X-ray streak-camera fitted with a KBr photocathode, coupled to a spectrometer equipped with a grazing incidence flat field grating.

Assuming [6] that the history of the XUV emission recorded by the spectrometer mainly depends upon the heat propagation and that atomic processes are fast compared to the typical time-scale of these heating mechanisms, one can determine the mass ablation rate. Fig.3 shows the mass ablation rate measured for incident intensities of 3×10^{16} W/cm^2 and 4.3×10^{15} W/cm^2. The incident intensity was converted into absorbed intensity according to the measured[7] absorption coefficient. The results are compared with numerical simulations performed using a 1D hydro-code MEDUSA[8] which treats thermal transport processes according to the classical flux limited description. The calculated dependence of mass ablation rates on the absorbed laser irradiance was investigated for different values of the heat flux limiter. Fig.3 shows the mass ablation rate as a function of laser intensity for two values of the flux limiter. We stress that the dependence upon the flux-limiter predicted by the simulations is a strong indication of the fact that the classical theory of heat conductivity cannot properly describe this regime of interaction, the temperature gradients in the plasma being smaller than the electron mean free path. As shown in Fig.2, this effect increases at higher laser intensities, when larger temperature gradients are established in the plasma.

The mass ablation rate measured at the lower intensity is consistent with the calculated one for a flux limiter of 0.04. This value, though not very different, is smaller than that inferred by the 2D Fokker-Planck simulation performed in an analogous intensity regime[9]. In that study it was found that the heat flux in the supercritical region was consistent with classical theory using a flux limiter of 0.1. However, we cannot exclude that other sources of error could make this experimental result consistent with the 2D FP simulation. In contrast, a large discrepancy is found in the higher intensity regime, where the measured ablation rate is less than half of the calculated one. This would indicate a strong reduction of the heat flux compared with the lower intensity case. It is clear that this discrepancy cannot be simply explained in terms of the classical heat transport theory assumed in the calculation as this conclusion would lead to an unacceptable value of the heat flux, orders of magnitude smaller than the free streaming value. Several effects can be invoked in order to explain this result. Due to the small focal spot size (≈25 μm FWHM), lateral heat flux can play an important role in the subcritical region, where laser-plasma interaction occurs.

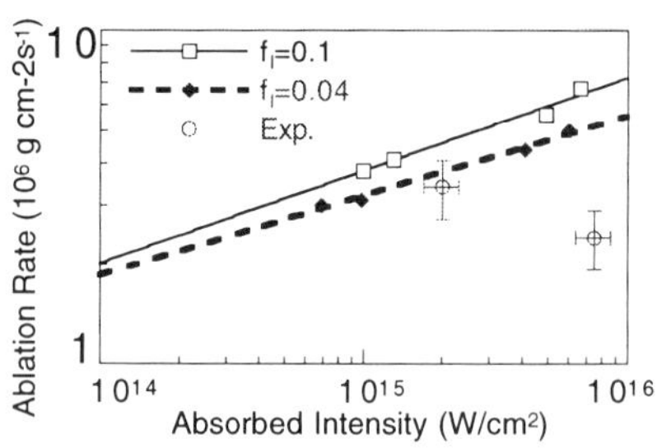

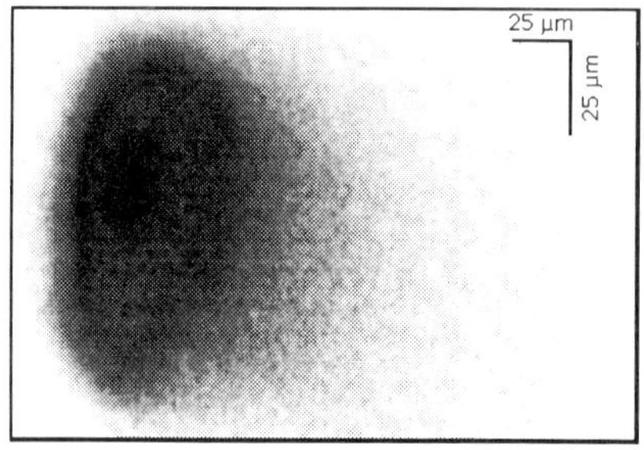

Fig.3. (left) Mass ablation rate as a function of absorbed laser intensity, calculated by MEDUSA for two different values of the flux limiter. The result of the experimental measurement is also plotted in the same graph for comparison. **Fig.4 (right)** X-ray image of the interaction of a 12 ps laser pulse focused to a ≈25 µm focal spot on a solid Al target at an intensity of $\approx 10^{17}$ W/cm^2.

Since this is a purely 2D effect, the 1D calculation presented above would obviously fail to account for it. From the point of view of the experimental results, any lateral flow would result in less thermal energy being transported into the inner layers of the target, effectively reducing the mass ablation rate. A detailed analysis of this aspect can be found elsewhere[6].

On the other hand, it has been shown[10] that strong self-generated toroidal magnetic fields can arise from the $\nabla n_e \times \nabla T_e$ mechanism. If the Larmor radius of the streaming electrons is small compared to their mean–free path, thermal transport can be strongly inhibited. An indication of the presence of magnetic fields is given by the X-ray image of the laser-plasma interaction region shown in Fig.4, obtained in the same experimental condition as above, but at a slightly higher irradiance of $\approx 10^{17}$ W/cm^2. The X-ray emitting region shown in Fig.4 consists of an intense central region, approximately 25 µm in diameter, this being also the FWHM of the laser focal spot. This region is surrounded by a ≈30 µm wide halo similar to that observed in previous experiments [11] on planar targets. This structure has been explained as due to the interaction of hot electrons with the cold material surrounding focal spot. In fact, in presence of electron waves arising from resonance absorption[7], electrons can be accelerated and ejected down the density gradient, away from the target surface. Magnetic fields are established in the plasma and electrons, driven across the target surface, deposit their energy in the cold Al target around the focal spot, giving rise to Kα emission.

More detailed measurements are necessary in order to directly detect magnetic fields. Nevertheless, the results presented here consistently provide evidence that the observed inhibition of thermal transport occurs in a regime where magnetic fields are established in the plasma. From the point of view of the analysis of experimental results, however, it is clear that a complete modelling of interaction regimes like that investigated here requires a self-consistent treatment of non–local heat transport and magnetic field generation in a two dimensional geometry.

[1] Giulietti A *et al.* 1994 this conference
[2] Vu B T V *et al.* 1994 *Phys. Rev. Lett.* **72** 3823-3826
[3] Gizzi L A *et al.* 1994 *Phys. Rev. E* **49** 5628-5643
[4] Gizzi L A *et al.* 1994 CLF- Rutherford Appleton Lab. (UK), Annual *Report* 61-63
[5] Afshar-Rad T *et al.* 1992 *Phys. Rev. Lett.* **68** 942-945
[6] Gizzi L A 1994 *Ph.D thesis*, Imperial College of Science, Technology and Medicine, London.
[7] Riley D *et al.* 1992 *Phys. Rev. E* **48** 4855-4858
[8] Christiansen J P 1974 *Comp. Phys. Comm.* **7**, 271-287
[9] Rickard J *et al.* 1989 *Phys. Rev. Lett.* **62** 2687-2691
[10] Stamper J A, 1975, *Phys. Rev. Lett.* **34** 138-141
[11] Amiranoff F *et al.* 1982 *J. Phys. D: Appl. Phys* **15** 2463-2468

Inst. Phys. Conf. Ser. No 140: Section 1
Paper presented at Laser Interact. with Matter Conf., Oxford, 19–23 September 1994
© *1995 IOP Publishing Ltd*

Experiments on the Interaction of Ultrashort UV-Laser Pulses with Solid Al- and C-Targets

U.Teubner [1], W.Theobald [2], C.Wülker [2], E.Förster [1]

[1] Max-Planck-Arbeitsgruppe Röntgenoptik an der Friedrich-Schiller-Universität Jena, Max-Wien-Platz 1, 07743 Jena, Germany
[2] Max-Planck-Institut für biophysikalische Chemie, Abteilung Laserphysik, Am Faßberg, 37077 Göttingen, Germany

ABSTRACT

We have studied experimentally the interaction of subpicosecond 248 nm laser pulses with solid aluminium and carbon targets at a series of intensities up to several times 10^{17} W/cm^2 and different angles of incidence. Soft X-ray spectra showed that the absorption of the laser pulse and the conversion of the laser energy into X-rays depend strongly on the polarization, the intensity and even more on the angle of incidence.

The absorption of high-intensity subpicosecond laser pulses incident obliquely on a target surface and the conversion into X-ray emission in laser produced plasmas has again recently been of great interest (Kieffer *et al.* 1989), (Fedosejevs *et al.* 1990), (Teubner *et al.* 1993), (Meyerhofer *et al.* 1993). Several interesting features have been found, e.g. that the X-ray yield depends not only on the amount of absorbed energy but on other parameters as well (Teubner *et al.* 1993). Spectra of the emitted radiation, not available for that previous work provide further information on the laser plasma interaction. In this paper we describe the first detailed investigation *with spectral resolution* of laser pulse absorption and X-ray emission produced by high-intensity laser pulses at different intensities and different angles of incidence.

The *experimental setup* was similar to that of Teubner *et al.* 1993. The laser pulses were generated by a combined dye / excimer laser system which was a modified version of that described by Szatmari and Schäfer 1988. Together with a two pass off-axis amplifier a maximum pulse energy of 40 mJ could be obtained. The 248 nm pulses were focused on to a highly polished target of solid aluminium or solid glassy carbon. The target was moved between consecutive shots to always present a fresh surface. The laser radiation was focused to a 1.5 times diffraction limited spot size of 5 µm (FWHM) at normal incidence. The pulse length on the target was 0.7 ps (FWHM). The maximum intensity for normal incidence was $I_{0\perp} = 3 \times 10^{17}$ W/cm^2.

For the angle dependent measurements all intensities on the target surface I_0 were corrected with respect to the angle of incidence α, i.e. $I_0 = I_{0\perp} \times \cos\alpha$. By placing dielectric attenuators in the beam path, the intensity I_0 could be adjusted between 3×10^{14} W/cm^2 and 3×10^{17} W/cm^2. The intensity ratio between the short pulse and the amplified spontanous emission exceeded 10^{10} on the target and so preplasma formation could be avoided. The angle of incidence was varied between 10° and 80°. The polarization of the pumping pulse could be adjusted either to the p or s direction with respect to the plane of incidence and with a degree of polarization exceeding 95 % .

The soft X-ray emission from the plasma was measured in the wavelength range from 15 Å to 150 Å with a transmission grating spectrograph with 2 Å resolution. A series of spectra was measured at 8 different angles of incidence, each at 7 different intensities and for p and s-polarized pump pulses. Thus, it was possible to compare spectra with *only one* changed parameter, and all the others were kept constant.

The dependence of the X-ray emission X on the pump pulse energy *deposited in the target* consists of the dependence of 1) the overall X-ray emission and 2) that of the spectral shape. The overall soft X-ray emissions $X_{p,soft}$ and $X_{s,soft}$ are the integrals over the wavelength for p and s-polarized pump light respectively. For the intensity dependence of $X_{p,soft}$ we assume a simple power law

$$X_{p,soft} = \int_{20Å}^{150Å} X(\lambda)\, d\lambda = C_p(\alpha) \cdot \left(I_A\right)^{\gamma_p(\alpha)} \qquad (1).$$

C_p is a proportional constant describing the conversion efficiency for a constant amount of energy or intensity (if the energy is normalized to the illuminated area and to the pulse duration) deposited in the plasma. A similar law is obtained for s-polarized pump light. For the calculation of the absorbed intensity $I_A = A(\alpha, I_0, \text{polarization}) \times I_0$ we took the absorption values A from our previous measurements (Teubner *et al.* 1993).

Figures 1 and 2 show the intensity dependence of the soft X-ray emission of an aluminium plasma for a *fixed angle of incidence*. While Figure 2 illustrates the changes of the spectral shape, Figure 1 shows the relative strength of the spectra and in particular that the X-ray emission strongly increases with I_A. The exponent γ_p is angular dependent, i.e. for p-polarized pump light $\gamma_p = 1.4 \pm 15\%$ at $\alpha = 10°$ and increases with α to $\gamma_p = 1.7 \pm 15\%$ at $\alpha = 50°$ and even more at larger angles of incidence. Similar, for s-polarized pump light γ_s increases with α too, but faster (for instance $\gamma_s = 1.9 \pm 15\%$ at $\alpha = 50°$).
The electron temperature and density where estimated from measurements of the aluminium Ly-α and He-α line groups, carried out by a von Hamos spectrograph. Together with simulations using the RATION code (Lee *et al.* 1984) and scaling laws for the temperature with I_A (Teubner *et al.* 1992) the electron temperatures of the spectra of Figure 2 are estimated to be between about 30 eV (at 10^{14} W/cm^2) and more than 500 eV (at 10^{17} W/cm^2). The electron density for the spectrum measured at $\alpha = 50°$ with p-polarized pump light was about 2 to 3 times that of the critical density. The spectral shape is similar for both polarizations (Figure 2) and indicates that the transient electron temperature during the time where most of the radiation shown in Figure 2 is emitted is nearly the same in both cases.

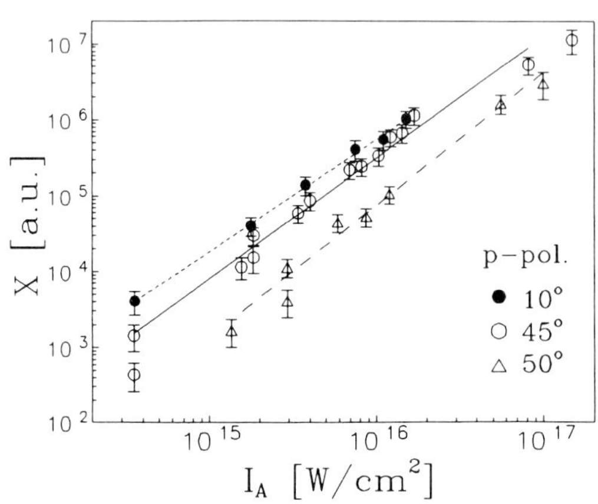

Figure 1. Intensity dependence of the overall X-ray emission $X_{p,soft}$ for p-polarized pump pulses measured at 3 different angles of incidence (α = 10°, 45°, 50°).

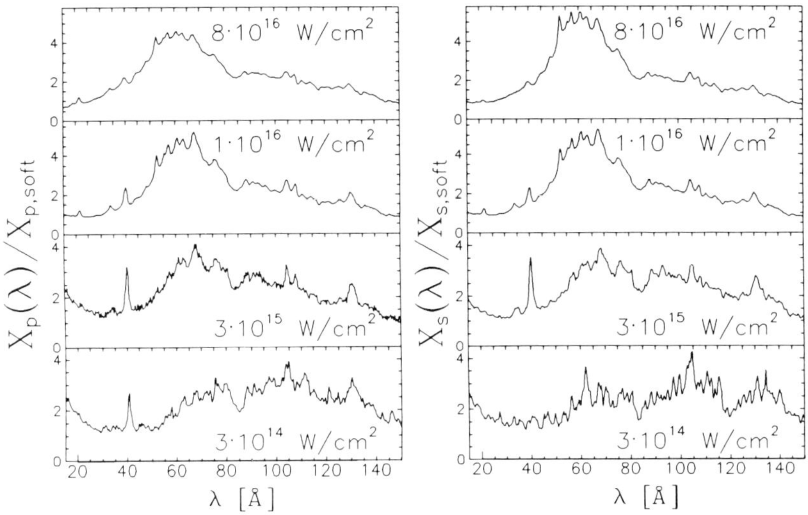

Figure 2. Intensity dependence of the shape of aluminium spectra (left hand side p-polarized pump pulses, right hand side s-polarized pump pulses; all spectra are normalized to their corresponding overall X-ray emission) measured at $\alpha = 45°$. The spectra are convoluted with higher orders.

Figure 1 shows that the proportional constants C_p and C_s are angularly dependent too. This can be seen much more clearly in Figure 3 where the angular dependence of $X_{p,soft}$ and $X_{s,soft}$ is plotted for a *fixed intensity*. In Figure 3(a) and 3(b) the overall X-ray emission is plotted for *fixed intensity on the target surface* and thus its angular dependence contains the angular dependence of the absorption *and* that of the conversion efficiency of the absorbed laser intensity into X-rays. The peak in the $X_{p,soft}(\alpha)$ curve is mainly the result of the collisional absorption (with an angular dependence similar to the absorption calculated with the Fresnel formula) together with resonance absorption. The difference of the shape of the peaks of $X_{p,soft}(\alpha)$ in Figure 3(a) and 3(b) is due to the different intensities and is caused by the different plasma scale-lengths. In contrast to $X_{p,soft}$, $X_{s,soft}(\alpha)$ steadily decreases with α as expected for purely collisional absorption.

A much more interesting topic is to study the angular dependence of the X-ray yield for *constant absorbed intensity* I_A in the plasma (Figures 3(c) and (d)). This provides information about the X-ray conversion efficiency. Although Figure 3(c) does not show more than a rough angular dependence for the soft X-ray yield it may be seen that 1) the overall soft X-ray emission decreases steadily to zero at large α and 2) $X_{p,soft}$ is larger than the corresponding value produced by s-polarized pump light, namely $X_{s,soft}$. Figure 3(d) shows soft X-ray spectra for a constant absorbed intensity. It should be mentioned that the spectral shape of the X-ray emission produced by p and s-polarized pump light is very similar at all angles of incidence. For both polarizations the maximum shifts to longer wavelenghts with increasing α. The angular dependence of the relative strength of the spectra of Figure 3(d) is similar to that shown in Figure 3(c). Comparing Figure 3(d) with Figure 2 shows that changing the angle of incidence α from $\alpha = 10°$ to $\alpha = 60°$ (for I_A = const.) has approximately the same effect on the spectral shape as changing I_A by about 2 orders of magnitude in intensity at constant α !

Carbon spectra measured under the same experimental conditions show in principle a similar dependence on intensity and angle of incidence and therefore will not be discussed here in more detail.

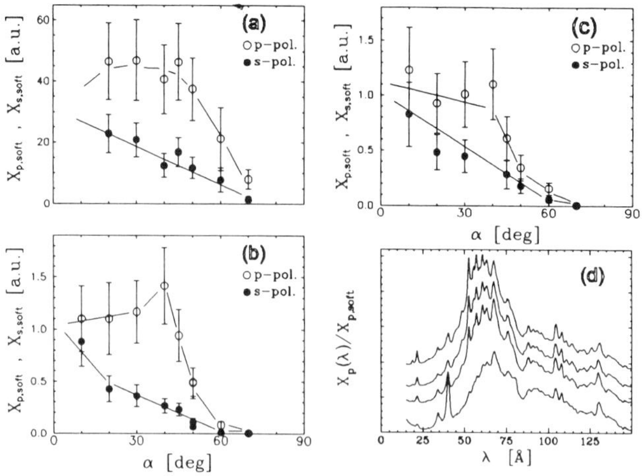

Figure 3. Angular dependence of the X-ray emission of an aluminium plasma. (a) Overall X-ray emission at constant intensity on the target surface ($I_0 = 1 \times 10^{17}$ W/cm^2); (b) like (a) but for $I_0 = 5 \times 10^{15}$ W/cm^2; (c) like (a) but for constant absorbed intensity ($I_A = 4 \times 10^{15}$ W/cm^2). (d) spectra measured at constant absorbed intensity $I_A \approx 10^{16}$ W/cm^2 at different α (from top to bottom: α = 10°, 40°, 50°, 60°). All spectra are normalized to their corresponding overall X-ray emission.

The decrease of the soft X-ray emission at *constant* I_A with increasing α (Figure 3(c)) is the result of the lower conversion efficiency: At normal incidence collisional absorption takes place mainly in the vicinity of the critical density and even more so in the overdense region. With increasing α the absorption volume becomes larger. The result is a lower electron temperature in this volume as can be seen from Figure 3(d) where the emission from the highest ionization stages (i.e. shortest wavelengths) is generally reduced although the amount of absorbed intensity in the plasma is the same. If α is increased the absorption volume and the emission volume both shift to regions with lower electron density where the ionization rate is lower and thus the conversion efficiency into *soft* X-ray radiation reduced. The additional contribution from resonance absorption at intermediate angles always takes place in the vicinity of the critical density. It affects only the total amount of *soft* X-ray radiation but not the electron temperature which is the same for both polarizations (see Figure 2).

In *summary* we have found that the generation of soft X-ray spectra from laser produced plasmas depends strongly on the polarization, the intensity and even more so on the angle of incidence. This can be explained by the different absorption, electron temperature, electron density and scale-length.

References

Fedosejevs, R. et al. 1990 Appl. Phys. B **50**, 79; Phys. Rev. Lett. **64,** 1250
Kieffer, J. C. et al. 1989 Phys. Rev. Lett. **62**, 760; IEEE J. Quant. Electron. QE-**25**, 2640
Lee R. W. et al. 1984 J. Quant. Spectroscop. Radiat. Transfer 32, 91
Meyerhofer, D. D. et al. 1993 Phys. Fluids B **5**, 2584
Teubner, U. et al. 1992 Appl. Phys. Lett. **59**, 2672; Appl.Phys. B **54**, 493
Teubner, U. et al. 1993 Phys. Rev. Lett.**70**, 794
Szatmari, S. and Schäfer, F. P. 1988 Opt. Comm. **68**, 196

Inst. Phys. Conf. Ser. No 140: Section 2
Paper presented at Laser Interact. with Matter Conf., Oxford, 19–23 September 1994

© *1995 IOP Publishing Ltd*

Laser Fusion Investigations on High Power Iodine Laser ISKRA

G.G.Kochemasov

Russian Federal Nuclear Centre - All Russia Scientific Research Institute of the Experimental Physics (VNIIEF), Arzamas-16, Nizhniy Novgorod Region, Russia.

Abstact. A review of ICF investigations will be presented which were performed at VNIIEF (Arzamas-16) on laser facilities Iskra-4 (1 kJ, 0.5 ns, 0.67 mcm) and Iskra-5 (30 kJ, 0.3 ns, 1.315 mcm) with direct-drive and indirect-drive targets including interpretation of the results obtained.

Currently, experimental research in VNIIEF on fundamental laser fusion problems are conducted using high-power iodine lasers Iskra-4 [1] and Iskra-5 [2].

Iskra-4 is a facility mostly intended for studies on the direct-drive physics. Its characteristics include: laser beam energy on the target 0.5-1 kJ for pulse length 0.25-0.6 ns at fundamental frequency. For second harmonic generation the energy on the target is 0.2-0.5 kJ.

The basic conclusion from the studies of direct-drive targets at fundamental frequency under quasiablative conditions is summarised in Fig. 1 [3]. As shown by the figure, the experimental neutron yield ($N_{exp}\approx 10^4\text{-}2\cdot 10^6$) is two or three orders of magnitude lower than the predicted one-dimensional yield for targets having aspect ratio $A_s\geq 50$. For $A_s\leq 20$ this difference is less then order of magnitude. This difference being so strongly dependent on the aspect ratio indicates the main cause for the target to show off-design performance, seems to be gasdynamic instability in the accelerating target, specifically, Rayleigh-Taylor instability. These data were obtained in experiments on glass shells of 80-200 μm diameter filled with DT-gas at 5 to 30 atm pressure. Low-aspect shells were produced by covering a glass sphere by 3μm C_8H_8 layer onto. There was a

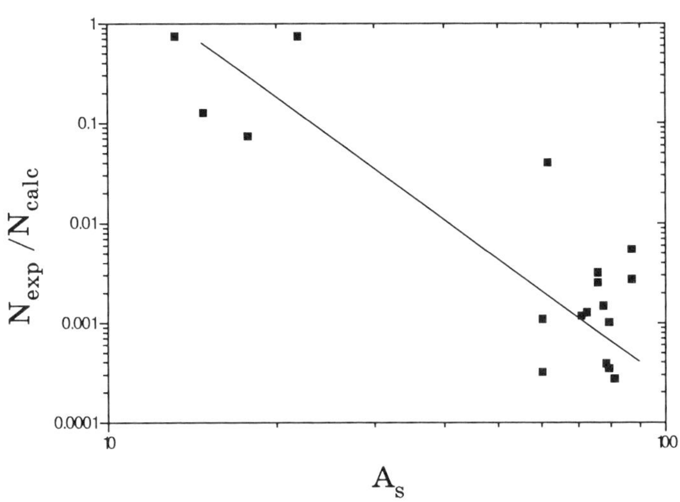

Fig. 1

comparatively small DT compression in the experiments: $30 \leq \delta_V \leq 3000$ (with radial convergence ratio $3 \leq R_0/R_f \leq 7$), with no correlation between N_{exp}/N_{calc} and δ_V observed. As shown by calculations and estimates show, large-scale irradiation nonuniformities that were of to significantly high values $\Delta J/J \sim 1$, could not be responsible for the discrepancy observed for $A_s \geq 50$. However what might definitely contribute to perturbations growth in shell implosion were mid-size nonuniformities ($\Delta x \sim 10$ μm) creating hot spots on the shell surface.

Today we have started to study in more details gasdynamic instabilities in laser-driven foils. Moreover we are looking at possible designs for space-time smoothing of laser beam nonuniformities that would be suitable for iodine laser having narrow amplification bandwidth (plasma phase plates, nonlinear lenses, etc.). We believe, that if there could be found efficient and inexpensive means of beam smoothing, and also ways to reduce the rate of shell damage due to gasdynamic instabilities, direct-drive targets would become again realistically competitive with indirect-drive targets (X-ray targets).

Since 1990 we have been conducting experiments using 12-beam laser facility Iskra-5 (15 kJ energy at the target, 0.25-0.3 ns FWHM, 1.315 μm wavelength, $7 \cdot 10^{-5}$ rad beam divergence).

Initially, the investigations were carried out for inverted-corona targets [4]. Laser radiation was directed into a sphere 2 mm in diameter through 6 holes (two beams in each hole) of 600 μm diameter. The sphere had its inner surface coated with material containing heavy hydrogen isotopes (such as CD_2). As it was evaporated due to laser beams, CD_2 plasma accelerated to implode at the target centre. This would make deuterons heat up to temperatures as high as $T_i \sim (5\text{-}10)$ keV. As a result there is stagnation nonequilibrium hot plasma created in the sphere, which is a unique plasma formation offering special scientific interest. It serves as intense neutron source: the neutron yield achieved in our experiments was about $5 \cdot 10^9$ DD neutrons. Inverted-corona targets also indicated the possibility for the major portion of Iskra-5 laser beams energy to be delivered into the small shell, so that we switched over to indirect-drive, or X-ray, target experiment [5].

For this purpose, the inner surface of a sphere was covered with high-Z material (e.g. Au, Bi), and DT-filled (at 2-20 atm) shell target was placed at the center of the sphere. The target initial aspect ratio ranged between 6 and 45. The indirect-drive target and laser beam injection geometry were designed so that they provided highly symmetric illumination of encapsulated DT. The best experiments were estimated to have illumina-

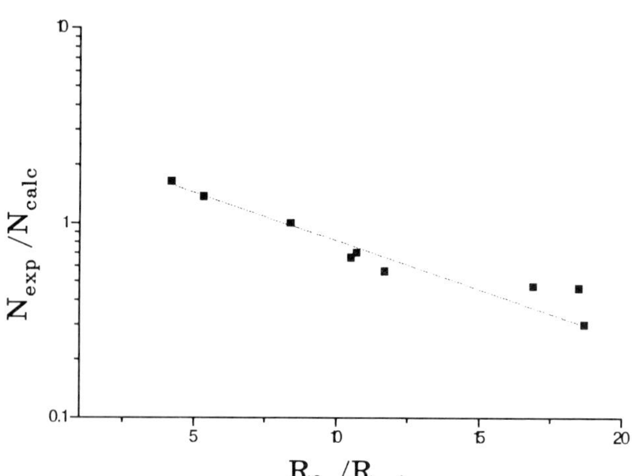

Fig. 2

tion asymmetry within 2 to 3%.

There was a diagnostics package used to study the target performance, which provided the characterisation of laser, X-ray and particle radiation. As shown by time-of-flight measurements of 14.1 MeV neutron spectrum, DT ions had effective temperature as high as 4 keV (for A_s=45-30) at neutron generation time. Total neutrons measured by time-of-flight oscillography were as many as $6\cdot10^9$ and this well agree with integrated measurement techniques involving copper and indium activation. The shell flight velocity was found from the convergence time that was measured by resolving X-rays in time in the photon energy range $h\nu\geq4$ keV. The average convergence velocity estimated in this way was up to $3\cdot10^7$ cm/s.

By varying the parameters of DT-filled shell (DT pressure, wall thickness and materials), we conducted successfully a series of experiments involving different compression levels. Fig. 2 shows experimental to predicted yield ratio, N_{exp}/N_{calc} versus the calculated radial convergence R_0/R_{min} (initial outer ablator radius/DT radius at neutron generation). The value of N_{exp} varied from $5\cdot10^7$ to $6\cdot10^9$. The calculated data as shown in Figs. 1 and 2, were obtained by SNDP radiation gas-dynamic code [6]. As it has been shown by tentative analysis, the increase of N_{calc}/N_{exp} with R_0/R_{min} is almost totally due to turbulent mixing at the DT-gas - shell boundary.

Thus, the experiments performed have demonstrated that Iskra-5 laser facility can be used to generate high-temperature and density plasma and effectively investigate its properties under indirect-drive conditions for DT-filled microballoons.

References

[1] A.I.Zaretskij and et al. *Kvantovaja electronika*, **10**, 756 (1982)

[2] G.A.Kirillov and et al. *Laser and Particle Beams*, **8**, 827 (1990)

[3] S.A.Bel'kov and et al. *ZETF*, **101**, 80 (1992)

[4] A.V.Bessarab and et al. *ZETF*, **102**, 1800 (1992)

[5] F.M.Abzaev and et al. *Pis'ma v ZETF*, **58**, 28 (1993)

[6] S.A.Bel'kov and et al. *Voprosy atomnoj nauki i tehniki*. Ser.: Matematicheskoe modelirovanie fizicheskih processov, vyp.1, 76 (1990)

Inst. Phys. Conf. Ser. No 140: Section 2
Paper presented at Laser Interact. with Matter Conf., Oxford, 19–23 September 1994
© 1995 IOP Publishing Ltd

ZARYA code simulation of physical phenomena occurring in the compression and burn of spherical ICF target

V A Lykov, N G Karlykhanov, V E Chernyakov and A A Chikulaev

Russian Federal Nuclear Centre - Russian Institute of Technical Physics
PO BOX 245, Snezhinsk (Chelyabinsk-70), 456770, Russia

Abstract. The physical models for simulation of spherical ICF targets compression and burn are described. These models are implemented in ZARYA code. The validity of proposed physical models is confirmed by comparison of ZARYA code simulations with known experimental data. Results of ZARYA/ERA code simulations taking into account the energy and momentum transfer by DT-reaction alfa-particles and X-ray radiation are presented.

1. Introduction

Theoretical studies on ICF problem carried out in VNIITF in 1972-93 years were reviewed in reports [1,2]. ZARYA code package was developed in VNIITF for ICF target design [1]. The physical processes pointed out in Ref [3] have been accounted properly in one-dimension ZARYA/ERA code [4,5]. Two-temperature hydrodynamic equations with flux-inhibited electron and ion heat-conductivity, energy and momentum transfer by X-ray, suprathermal electrons and fusion reactions products are based of this code.

ZARYA simulations of experiments data obtained at SOKOL [6], ISKRA-4 [7], PROGRESS [8], KMS-F, JANUS, ARGUS, OMEGA, GEKKO-XII and NOVA laser facilities were performed. These simulations published in Ref [1,6-8] showed a good agreement with experimental data for all this experiments if target compression symmetry was good.

The main goal of theoretical studies on ICF problem in VNIITF performing since 1973 is brought to light on requirements for targets and laser facilities to obtain ignition and high energy gain. Some results of ZARYA code direct-driven targets simulations were published in Ref [1,2,4].

The brief description of theoretical models implemented in ZARYA/ERA code is given bellow, and some results of direct-driven and indirect-driven targets calculation are given for the illustration in this paper only.

2. ZARYA/ERA CODE PHYSICAL MODEL.

Physical models and numerical methods realized in one-dimensional ZARYA/ERA code are discribed in [1,4,5]. The numerical methods of solving of two-temperature hydrodynamic equations with flux-inhibited electron and ion heat-conductivity, energy and momentum transfer by X-ray, suprathermal electrons and fusion reactions products were published in Ref [5]. X-ray transfer can examined with using the effective temperature T_r or multigroup spectral approximations (kinetic, quasi-diffusion or diffusion treatments see Ref [5]). Energy and momentum transfer by charged fusion reactions products and suprathermal electrons can simulated either by one-group approach with the effective free path length [5] or by multigroup spectral kinetic approximation for Landay equations for plasma charged particles with self-consistent electric fields and returning currents [1]. Interpolation formulas are used for the equations of state and radiation opacities calculated by means of atom statistical model, Saha equation and coronal approach. The radiation transfer in lines and continuum jointed with radiative-collisional kinetic model are in realization today. The turbulent mixing model is being implemented in ZARYA/ERA code just now.

Theoretical absorption model used in ZARYA/ERA code was described in detailes Ref [1]. This model takes into account bremsstrahlung, resonant and parametric absorbtion, suprathermal electron generation, steepening of the plasma density profile due to pondermotive forces, refraction and Brillouin scattering in target corona and percularities of laser light focusing on target also. Steepening of plasma density profile in ZARYA/ERA code is taken into account by introducing of "cut off" density ρ_- and desity inhomogeneity scale L_c depending on critical density ρ_c, oscilation electron velocity V_{os} in the field of laser radiation, electron thermal velocity V_{te}, wavelength λ of laser light according to the following relation:

$$\rho_- = \delta \rho_c, \quad L_c/\lambda \simeq 5V_{te}/c\eta, \quad \delta = (1+4\eta)^{-1}, \eta = V_{os}/V_{te}.$$

Factor Ψ considering refraction losses of laser intensity in spherical target corona, is added to the coefficient of bremsstrahlung absorption by the multiplier:

$$\Psi = \left(1 + \zeta_* + (\zeta_c - \zeta_*)^2\right)^{-p}, \quad p = 1 + 0.5\delta, \quad \zeta_* = (1-\delta)^{0.5}.$$

The parameter ζ_c is connected with the focal ratio of focusing optics F, focal shift d from target centre, the size of focal spot d_f and density jump at the radius R_- by the relation:

$$\zeta_c = \left((d + F^2 d_f^2)/R_-^2(1+4F^2)\right)^{0.5}.$$

The interpolation formula for resonance absorption coefficient at critical density jump is as follows:

$$A_p \simeq (0.23 + 0.15(1-\delta))/(1 + 0.4x^{3.5}), \quad x = 17\xi_c\zeta_*^{-1}(L_c/\lambda)^{1/3}.$$

To estimate parametric absorption coefficient the next interpolation formula is used:

$$A_n \simeq 0.15\eta^{5/3}(1 + 2.5\zeta_c^2\zeta_*^{-2})^{-1}.$$

The following interpolation formula is used to estimate SBS of laser light in target corona with effective lenght of L_-:

$$R_B \simeq 0.6y/(1+y), \quad y = 1.2 \times 10^3 \eta^2 L_-/\lambda.$$

Resonant and parametric absorption of laser light with the energy flux density q_c at density jump leads to suprathermal electron generation with the effective temperature T_h equal to:

$$T_h = 5(q_c\lambda^2)^{0.4} T_{ec}^{0.25}, \quad q_c\lambda^2[10^{14} W \cdot \mu m^2 \cdot cm^{-2}], \quad T_{ec}[keV].$$

Validity of laser radiation absorption model proposed was confirmed by comparison of ZARYA/ERA calculations with experimental data obtained at SOKOL, ISKRA-4, PROGRESS, KMS-F, JANUS, ARGUS, OMEGA, GEKKO-XII and NOVA laser facilities. Good agreement with experimental data was obtained for wide range of laser light intensities ($10^{13} \div 10^{16}$ W/cm^2), laser pulse width ($0.04 \div 3$ ns), for various focusing conditions ($F/0.04 \div F/4$), different spot diametres ($d = 50 \div 1000$ μm) and targets[1,6,-11].

3. SOME RESULTS OF ICF SPHERICAL TARGETS SIMULATIONS BY ZARYA CODE.

3.1. Calculation of direct-driven targets.

Results of direct-driven targets simulations by one-dimensional ZARYA/ERA code were published in [1,4] for different target designs. Calculated spectra of X-ray and alfa-particles emitted by target were carried out for glass microbaloon of 14.5 μg and the diameter of $1mm$ covered by beryllium layer of 60 μg from the outer side and DT layer of 3 μg frozen at the inner surface. The calculations were made with the simple model of laser radiation absorbtion for the second harmonic of Nd-laser, with the power about of 25 TW and total energy of 40 kJ.

With using one-group approximation for alfa-particle energy and momemtum transfer and spectral kinetic approach for radiation transfer the gain of $G \simeq 0.9$ was calculated. Notice that in the calculation with T_r approximation for radiation transfer it was obtained $G \simeq 4.5$. The calculation with alfa-particles spectral transfer and T_r approximation for radiation transfer gave $G \simeq 3.6$.

The ZARYA/ERA code calculations of the direct-driven targets gain for the wavelength of $\lambda = 0.25 \div 1.06$ μm and energy of $E = 0.01 \div 10$ MJ were performed with the above-mentioned laser light absorption model. Results of these calculations have been published in [1].

3.2. Calculations of indirect-driven targets.

Some results of indirect-driven targets simulations obtained by ZARYA code were published in Ref. [1,2]. ZARYA code simulations of such targets are being performed as under NOVA test as the NIF facility to be built in LLNL.

ZARYA/ERA code calculations of simple cryogenic spherical targets placed at the center of cylindrical converter have confirmed the possibility to obtain the gain of about $G = 20$ for laser energy of 3 MJ and under condition of ideal symmetry. It was obtained through calculation of the glass shell with the diameter of 4.2 mm, aspect ratio of 20 with the mass of 400 μg DT-ice frozen at the inner shell surface. The hohlraum radiation temperature varies linearly from the value of 50 to 250 eV during time less then 20 ns in this calculation. The calculation give the prediction for fusion energy about of 10 MJ.

According to the tentative results, the X-ray irradiation homogeneity at the target corresponding to a driver with the energy of $E = 1\ MJ$, can be 3 times better to compare with the experimentally achieved one at the operating laser facilities with the energy $E = 40\ kJ$. Despite the fact that such symmetry of X-ray radiation at the target surface can be sufficient for the ignition, it is rather difficult to predict potential decreasing of target gains due to the compression asymmetries thus far.

Acknowledgments.

The authors are indebted to Professor E.Avrorin for continuous interest in this work and to Drs M.Anuchin and V.Popova for help in the translation of this paper.

REFERENCES

[1] Avrorin E N and Lykov V A 1991 Proceedings of the International conference on LASERS'9O (STS PRESS. MLEAN, VA) 811-818

[2] Avrorin E N and Lykov V A 1993 11th International Workshop on LIRPP Abstracts 46, Monterey, CA, 25-29 October 1993

[3] Nuckolls J, Wood L, Thiessen A and Zimmerman G 1972 Nature **239** 139-147

[4] Avrorin E N, Zuyev A I, Karlykhanov H G, Lykov V A and Chernyakov V E 1980 JETP Lett. **32** 7 437-440

[5] Zyev A I 1992 Zh. Vytchislit. Mat. i Mat. Fiz. **32** 82-96

[6] Zysin Yu A, Abramov I A, Volenko V V, et al 1982 Zh. Eksp. Teor. Fiz. **83** 4 1346-1357

[7] Avrorin E N, Eroshenko B A, Zaretcki A I, et al 1984 Zh. Eksp. Teor. Fiz. **87** 2 417-421

[8] Baryulin A E, Bayanov V I, Borodin V G et al 1985 Zh. Eksp. Teor. Fiz. **89** 5 1575-1586

Inst. Phys. Conf. Ser. No 140: Section 2
Paper presented at Laser Interact. with Matter Conf., Oxford, 19–23 September 1994
© *1995 IOP Publishing Ltd*

Implosion experiments of cryogenically cooled targets

T. Yamanaka, K. A. Tanaka, K. Nishihara, T. Norimatsu, H. Shiraga, Y. Kitagawa, N. Miyanaga, H. Azechi, H. Nishimura, M. Nakai, R. Kodama, M. Heya, M. Takagi, Y. Izawa, T. Kanabe, T. Jitsuno, M. Nakatsuka, K. Mima, M. Murakami and S. Nakai

Institute of Laser Engineering, Osaka University, 2-6 Yamada-oka,Suita,Osaka 565, Japan

T. Endo and C. Yamanaka

Institute for Laser Technology, 2-6 Yamada-oka, Suita, Osaka, 565, Japan

Abstract. Cryogenically cooled deuterium targets with plastic ablator are imploded as well as plastic targets with deuterium gas at laser intensities of 4-5x10^{14} W/cm^2 and 1-2x10^{14} W/cm^2. The low intensity shots showed neutron yields normalized by 1-D simulation to be one tenth up to convergence ratio of 30, while those at the high intensity are one hundredth. The observed neutron yields and density radius product ρRs are respectively 10^8 and 8 mg/cm^2 for cryogenic liquid targets. 2-D x-ray images obtained with a 10 ps temporal resolution camera reveal previously undiscovered information of imploding core plasmas.

1. Introduction

A key issue of inertial confinement fusion is to keep spherically symmetric implosion to the final phase with a target design close to the ignition or burn. The spherical symmetric implosion could be degraded by the hydrodynamic instability which is initiated by the surface imperfection of the imploding capsule and the nonuniformity of laser irradiation. We have performed implosion experiments with improved plastic capsules and well power balanced twelve beams at the wavelength of 527 nm in order to investigate the dynamics of stagnation phase using an ultra-high speed 2-D x-ray imaging system with 10 ps temporal resolution. In this paper we report implosion performances of liquid deuterium targets as well as plastic shell targets filled with deuterium gas and Ar tracer.

2. Experimental conditions

Two different spherical targets are mainly used: cryogenic foam shell targets with liquid deuterium (cryogenic target) and plastic shell targets with controlled pressure of deuterium gas (gas fuel target). Both of them have 450-600 μm diameter. The former is employed to study the implosion performance as a candidate for the high gain target. The foam shells are overcoated with 4 μm polyvinylphenol (PVP) of 1.2 g/cm^3 [1]. The surface roughness is 200 nm. The thickness of the foam layer is 10 μm and the mass density is ranging 50-220 mg/cm^3. The filling pressure of deuterium gas is adjusted to have liquid thickness thicker than the foam layer by at most 1 μm [2]. The latter is used to study the stability of stagnation phase. Deuterium gas of 5-50 atm is filled with Ar gas of 0.3 atm as an x-ray tracer using a cryogenic system [3] to avoid explosion of shells and leakage of gas in vacuum chamber. The surface roughness of the plastic shells is 20 nm.

Coherent 530 nm light from the GEKKO XII, 12-beam Nd glass laser system is irradiated onto the target through random phase plates. The energy imbalance of each beam is 2 -5 % rms. The power imbalance of 12 beams is typically 5 % in power balanced shots and is estimated to be 20 % for other shots. The laser energy and intensity of high intensity shots are typically 8 kJ in 2.5 ns with 1 ns rise time and 4-5x10^{14} /cm^2, respectively. Those of the low intensity are typically 4 kJ in 2.5 ns and 1-2x10^{14} /cm^2, respectively. The focusing condition is adjusted to be obtained the best irradiation uniformity and the absorption nonuniformity is estimated 20 % rms without the power balance.

Multi-imaging x-ray streak camera system (MIXS) [4] is introduced to measure the 2-D images at the stagnation phase with 10 ps time and 15 μm spatial resolution.

3. Experimental results and discussion

Typical neutron yields of both liquid and gas fuel targets are 10^8 and 10^7 for high and low intensity shots, respectively. The density radius product ρR of the liquid deuterium layer is 8 mg/cm^2 [5]. The experimentally obtained neutron yields normalized by the HISHO 1-D simulation (normalized neutron yield) are shown in Fig. 1 as a function of convergence ratio of 1-D simulation. Here the convergence ratio is defined as ratio of the initial pellet outer radius to the gas-pusher or -main fuel interface radius at maximum compression in the simulation. The power imbalance of twelve beams is 5 % except for high intensity liquid target shots. The normalized neutron yields of the high intensity shots for both liquid and gas fuel targets are about one hundredth and the yield values coincide with those in the simulation when the shock

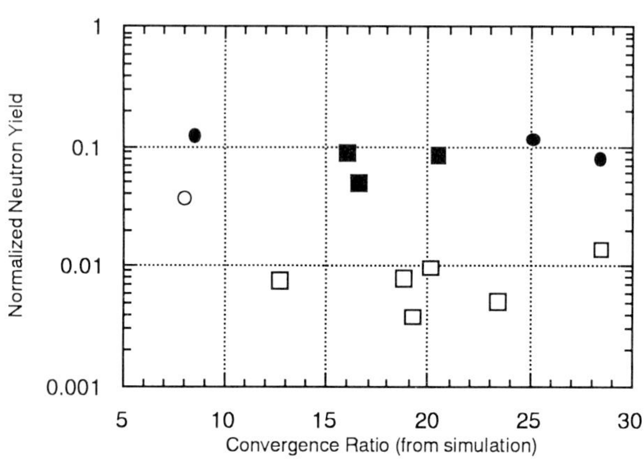

Fig. 1. Normalized neutron yield of liquid and gas fuel targets. Open symbols (○,□) correspond to the high intensity shots and closed symbols (●,■) to the low intensity shots. Squares and circles are for cryogenic and gas fuel targets, respectively.

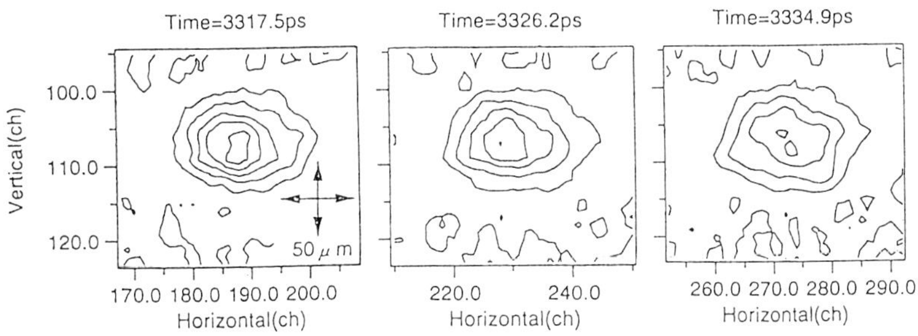

Fig. 2. Isodensity contoures of successively obtained x-ray images at around the time when shock hits the pusher in simulation.

wave reaches at the target center for the second time as described in the previous report [5]. On the other hand for the high intensity shots the normalized neutron yields are about one tenth up to the convergence ratio 30 and the yield values are reproduced at slightly later time in the simulation.

It is important to discuss why the normalized neutron yields are improved in the low intensity power balanced shots. This could be explained as follows. The calculation of irradiation uniformity used actual intensity profile of twelve beams shows that the absorption uniformity in the power balanced shots can be improved only in mode 1 and 2 in the spherical modes, and the effect is rather small if profile controlled beams are not introduced[6]. The simulations predict that the stagnation ratios of low intensity shots are about 3 and lower than 3.8 of the high intensity shots. Here the stagnation ratio is defined as the ratio of core radius at the shock wave colliding with the pusher to that at maximum compression. Therefore the low intensity shots seem to show better implosion performance up to convergence ratio of 30.

The MIXS images with 9 ps frame interval are obtained to the gas fuel targets with Ar gas. The photon energy hν contributed to the image is ranging 2.5-4.9 keV from Ar ions. Typical images are in reference [7]. The shapes of core emission show never observed temporal behaviors by 100 ps resolution framing cameras. The spatial profiles at early time seems to reflect the shock front bouncing back at the target center. At the time when the reflected shock hits the pusher in the simulation the emission at the central part decreases and also the intensity profiles change obviously in 10 ps with low modes ($1 < 6$) as shown in Fig. 2. This fast change of the core shape could result from the fuel-pusher mixing by the Richtmyer-Meshkov and/or Rayleigh-Taylor instabilities. Note the higher mode deformation can not be resolved because of 15 μm spatial resolution .

Some of the above features can be seen in Fig.3. Spatial intensity profiles agree well with the simulation results until the shock wave hits the inner surface of the pusher. After that time the intensity of the center does not increase like the simulation but rather decreases gradually and quenches at around the shock wave reaching at target center again in the simulation. However if we use the fast-rise higher intensity irradiation (0.8 ns rise time, 3×10^{14}W/cm^2) instead of the slow-rise low intensity irradiation (1 ns rise time, 1×10^{14} W/cm^2), the discrepancy between experiments and simulations are takes place later time, typically several tens ps. For both shots the perturbation growth, in-flight aspect ratio and thermal smoothing factor are estimated to understand the mechanisms. A most probable reason is the difference of the thermal smoothing factor. The thermal smoothing factor of the fast-rise high intensity irradiation is estimated to be exp(-0.18m) against exp(-0.08m) for the slow-rise low intensity irradiation, where m is the mode number of spherical modes. The exponents of Rayleigh-Taylor instability growth in acceleration phase are 9.9 and 9.7 at m=100 for the high and low intensity shots, respectively.

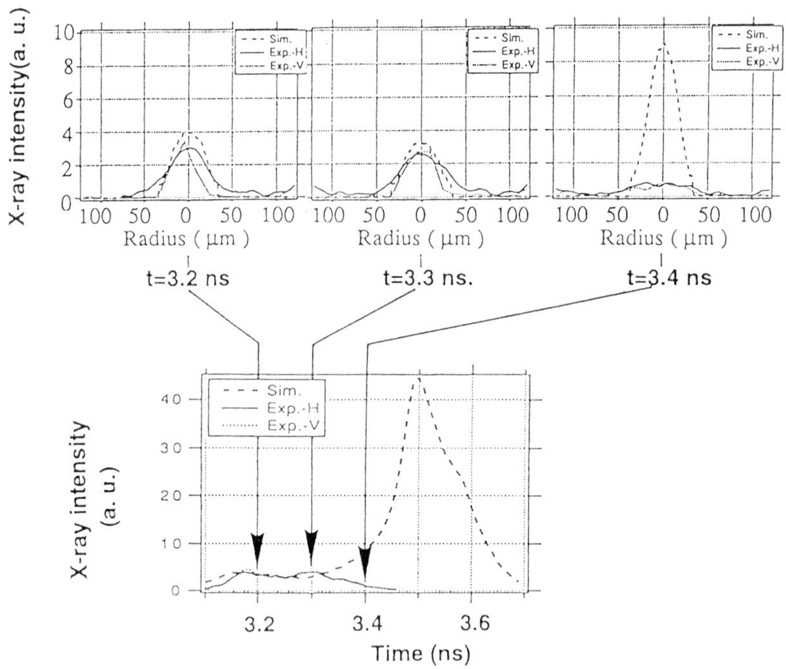

Fig. 3. Spatial and temporal behaviors of the core emission. t=3.3 is the time when shock front collides with pusher in simulastion.

4. Summary

Implosion performances of the deuterium liquid and gas fuel targets with plastic capsules have been studied. The low intensity implosions show higher normalized neutron yields than the high intensity implosions. Detailed behaviors of the compressed core plasma are investigated with an ultra high speed 2-D imaging system. Possible fuel-pusher mixing taking place at the shock wave colliding to the pusher are reported for the first time. Mechanisms why the low intensity implosions show the higher normalized neutron yields should be studied in more ditail under the better irradiation conditions.

References

[1] RICHARD A, et al.1994 Phys. Rev. E **49** 1520-1526
[2] NORIMATSU T, et al.1992 Rev. Sci. Instrum. **63** 3378-3383
[3] TAKAGI M, et al. 1993 J. Vacuum. Sci. Technol. A **11** 2837-2845
[4] SHIRAGA H, et al. 1995 to appear in Rev. Sci. Instrum. **66**
[5] YAMANAKA T, et al. 1993 *Plasma physics and controlled nuclear fusion research* vol.3 (Vienna: IAEA) p 151-158
[6] NISHIHARA K, et al. 1994 Phys. Plasm **1** 1653-1661
[7] NISHIMURA H, et al. *15th international conference on plasma physics and controlled nuclear fusion research* Seville (Spain,26 Sept.-1 Oct. 1994) IAEA-CN-60/B-p-8

Inst. Phys. Conf. Ser. No 140: Section 2
Paper presented at Laser Interact. with Matter Conf., Oxford, 19–23 September 1994

© *1995 IOP Publishing Ltd*

Studies on Radiation Hydrodynamics and Inertial Fusion

J. Meyer-ter-Vehn† , R. Ramis‡ , J. Ramirez‡ and A. Rickert†

† Max-Planck-Institut für Quantenoptik, D-85748 Garching, Germany

‡ E.T.S.I. Aeronauticos, Universidad Politecnica, E-28040 Madrid, Spain

Abstract. Recent work on ICF targets performed in the context of a European initiative for Heavy Ion Inertial Fusion is presented. The code MULTI2D for two-dimensional radiation hydrodynamics has been developed and is applied to a radiation driven target. Another new tool is the opacity code SAPHIR, and results are discussed.

1. Targets for ion beam fusion

Indirect drive is believed to be the only option to achieve sufficient symmetrization for ion beam inertial fusion. Different types of radiatively driven targets have been studied (Meyer-ter-Vehn *et al* 1993). Here we discuss new results with respect to a non-spherical hohlraum configuration with two-sided illumination and localized converter elements. The prototype form of this target is shown in Fig. 1.

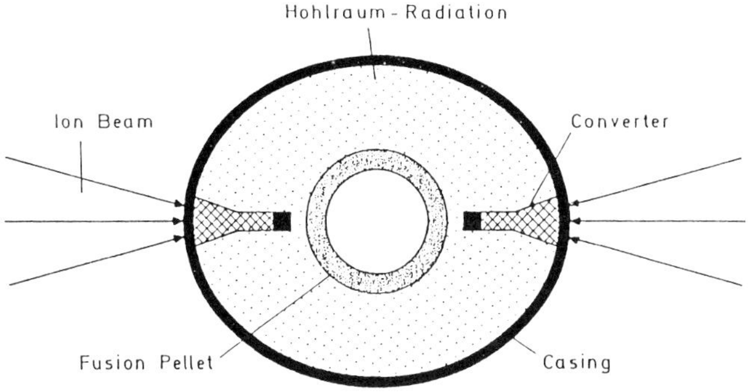

Figure 1. Generic hohlraum target for heavy ion beam fusion with two-sided illumination and localized converter elements (Murakami *et al* 1991)

Murakami and Meyer-ter-Vehn (1991) studied the radiation hydrodynamics of this target by means of 1D simulation and the symmetry of the radiation field in static approximation. Temporal and Atzeni (1992) pointed out that symmetrization is insufficient in this configuration and that at least six converter elements positioned symmetrically in a spherical cavity are required. Kang *et al* (1993) showed that appropriate converter shading would solve the symmetry problem for two-sided illumination.

2. MULTI2D and 2D target simulations

All the symmetry analysis mentioned above was performed in a static approximation, assuming that radiation is exchanged between fixed wall elements of the hohlraum ne-

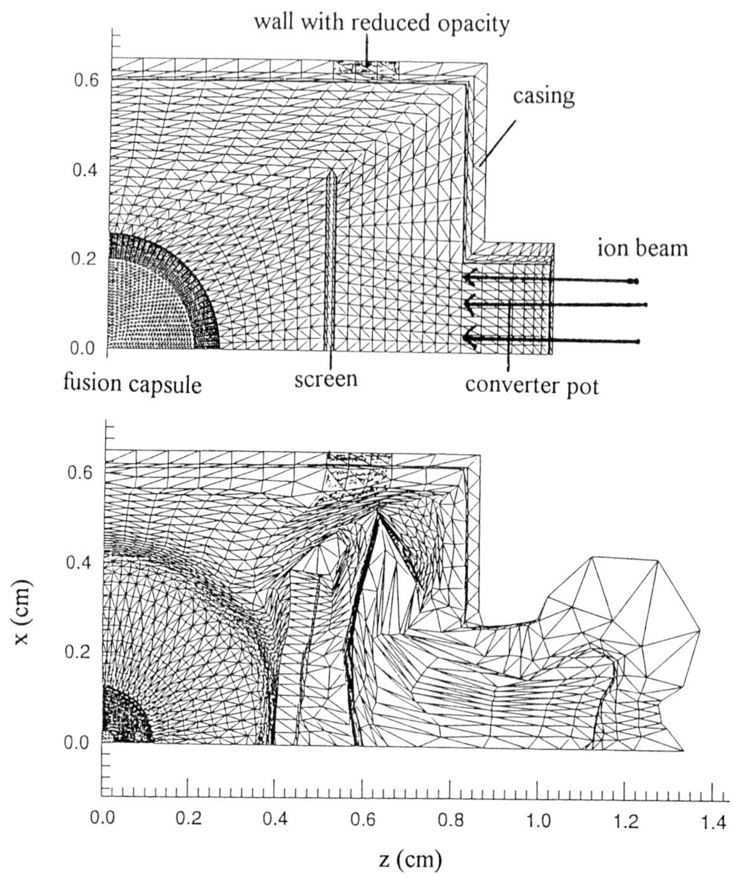

Figure 2. Advanced target design with Lagrangean mesh for MULTI2D simulation. One quadrant of a meridianal cut is shown; the target is axially symmetric about the abscissa, and the ordinate marks the equatorial plane. The upper part (a) displays the initial configuration and the lower part (b) the distorted configuration with 5 MJ beam energy deposited in the converter and the inner capsule ablatively imploded to half its initial radius.

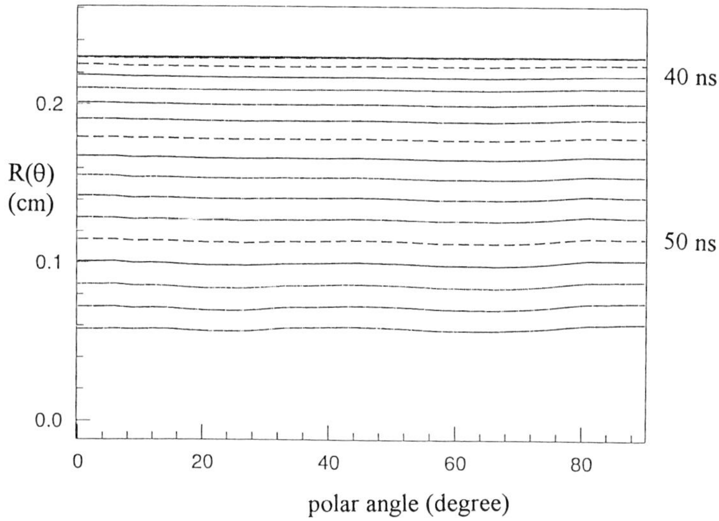

Figure 3. Radius $R(\theta)$ of the ablator/fuel interface at different times of capsule implosion.

glecting hydrodynamics. This approximation is rather crude, and it turned out that advanced tools are required for adequate simulation. Ramis *et al* (1992) have therefore started to develop the code MULTI2D for two-dimensional radiation hydrodynamics. The code treats real transport (not diffusion!) to guarantee sufficient angular resolution of non-isotropic radiation fields and describes optically thin as well as optically thick situations. At present, the code works only with simple equations-of-state and opacities depending on temperature and density in power-law form. The results presented here are therefore only of semi-quantitative nature. Ramis *et al* decribe the code in more detail in a separate contribution to these proceedings.

An advanced version of the Murakami target has been simulated with MULTI2D and is shown in Fig. 2 in terms of the 2D mesh at time $t = 0$ and the time just after beam switch-off. The inner capsule consists of 30 mg plastic ablator, 2.5 mg cryogenic DT fuel, and 0.5 mg DT gas in the centre. The radiation cavity consists of a lead cylinder with converter pots attached at the poles and filled with 35 mg Be; 10 GeV heavy ions with 250 mg/cm^2 range deposit 5 MJ energy into each of the converter. A thin high-Z screen prevents direct irradiation of the capsule. Ho *et al* (1993) proposed this converter design. As we see from the distorted mesh after beam deposition in Fig. 2b, the pot configuration suppresses radial expansion of the converter and insures beam stopping over an extended time. The radiation screen also confines converter flow in axial direction; this enhances conversion into radiation.

The hohlraum is filled with 35 mg/cm^3 low-Z background material (eg He) which becomes optically thin rapidly after a short initialization period. It is needed for numerical reasons to prevent complete collapse of hohlraum cells, but it also helps physically to delay wall ablation and closure between screen and wall which would block radiation flow.

In the lower left corner of Fig. 2b, we see the rather spherical region of ablated material and the imploding core in the centre. Fig. 3 reveals the history of the implosion symmetry in terms of the radius of the ablator/fuel interface plotted versus polar angle for different times. After contraction to one third of the initial radius, the deviation from complete spherical symmetry is a few percent, satisfying ignition requirements. In order to obtain this degree of symmetry, we had to reduce the opacity of the shaded wall elements (see Fig.2a) to tune the l=4 asymmetry which is otherwise the dominating failure mode. Ramirez *et al* discuss the present simulation in more detail in a separate contribution to these proceedings. At present, MULTI2D contains no burn physics and is therefore still inadequate to study ignition.

3. Opacity calculations with SAPHIR

Beside 2D radiation hydrodynamics, reliable information on material functions like equations-of-state and, in particular, opacities is a crucial aspect for quantitative target design. At MPQ, Rickert (1992) is developing the opacity code SAPHIR with emphasis on high-Z materials and including non-LTE plasma situations. The code is based on the collisional radiative model and detailed configuration accounting. It uses relativistic energies calculated from a parameterized potential and accounts, among other features, for dielectronic recombination. Statistical treatment of level spreading in terms of Unresolved Transition Arrays (UTA's) is now being installed to properly account for the width of partially filled configurations.

A benchmark comparison of SAPHIR results with a number of other codes was performed recently at the Third International Opacity Workshop, held at MPQ/Garching in March 1994. Summary plots of the comparison are contained in the final report of the workshop (Rickert *et al* 1994).

References

Ho D D, Lindl J D and Tabak M 1993, preprint UCRL-JC-113066, submitted to *Nuclear Fusion*

Kang K H, Lutz K J, Maruhn J A and Tahir N A 1993 *Il Nuovo Cimento* **106A** 1825

Meyer-ter-Vehn J, Basko M, Ramis R and Rickert A 1993 *Il Nuovo Cimento* **106A** 1883

Murakami M and Meyer-ter-Vehn J 1991, *Nucl. Fusion* **31** 1315

Ramis R and Meyer-ter-Vehn J 1992 *MULTI2D - A Computer Code for Two-Dimensional Radiation Hydrodynamics* Report MPQ174

Rickert A 1993 *Zustand und optische Eigenschaften dichter Plasmen* PhD thesis, TU München, Report MPQ175

Rickert A *et al* 1994 *WorkOpIII Final Report* to appear as MPQ Report

Temporal M and Atzeni S 1993 *Il Nuovo Cimento* **106A** 1429

Inst. Phys. Conf. Ser. No 140: Section 3
Paper presented at Laser Interact. with Matter Conf., Oxford, 19–23 September 1994
© 1995 IOP Publishing Ltd

A review of the Rayleigh-Taylor instability in inertial confinement fusion

O. Willi

Imperial College of Science, Technology and Medicine, London, U.K.

Abstract. The Rayleigh-Taylor (RT) instability places severe contraints on the efficiency of inertial confinement fusion. Small imperfections on the target surface or intensity nonuniformities present in the laser irradiation initiate the RT instability in the ablation region of x-ray or laser driven implosion targets during the acceleration phase. The nonuniformities may feedthrough to the inner pusher-fuel interface and be amplified during the deceleration phase. This reduces the implosion yield and may even cause sufficient fuel-target mix to prevent ignition altogether. Fortunately, the growth of the RT instability in the ablation region is reduced to less than the classically predicted rate primarily due to the effects of ablative flow through the unstable region and finite density gradients. In recent years extensive experimental studies have been carried out to investigate the RT instability in plasmas relevant for inertial confinement fusion. Modulated planar foil targets were driven either with intense soft x-ray or laser radiation. The experimental results were compared to computer simulations and good agreement was obtained. The RT instability induced by laser nonuniformity imprinting was also studied for coherent and smoothed laser beams. Clear target break-up was seen for all conditions investigated. However a promising new scheme involving solid foil targets overcoated with low density foam and a thin gold layer facing the laser beam has been proposed. Initial experimental observations using this target configuration show virtually no imprinting and subsequent RT instability growth. In addition, the first observations have been made of the RT instability in converging geometry. The full implosion process has been modelled with 3-D hydrodynamics code, making a complete study of the RT instability in spherical geometry possible.

1. Introduction

The Rayleigh-Taylor (RT) instability[1] occurs when a heavy fluid is resting on the top of a light fluid in a gravitational field. In inertial confinement fusion, regions where the density and pressure gradients are opposed for both the acceleration and deceleration phases are RT unstable. The hot, low density ablating plasma accelerates a cooler, more dense part of the target and thus this region is susceptible to the RT instability. Any nonuniformities generated in the ablation region may distort the target shell unacceptably or punch through to the inner pusher-fuel interface. Nonuniformities at the inner surface may be amplified during the deceleration phase. This reduces the implosion yield and may even cause sufficient fuel-target mix to prevent ignition altogether.

During recent years a good understanding of the RT instability in x-ray and laser driven planar modulated targets has been obtained. Excellent agreement between simulations and experiments has been achieved in the linear phase of the instability for modulated targets with wavelengths between 30 and 100 μm. However no experimental data has so far been obtained for modulation wavelengths shorter than 20 μm. The RT instability was also investigated in the nonlinear phase. The characteristic bubble and spike formation was clearly observed. An excellent review of these studies is given in reference 2. Recently the RT instability induced by laser beam nonuniformities has been studied. It was clearly observed that planar foil targets break up even if they are driven with spatially and temporally incoherent laser beams [3]. Finally, the first experimental results of the RT instability in converging geometry have been obtained [4].

2. Background

The RT instability occurs in an accelerating fluid system where the density and pressure gradients are anti-parallel. For the incompressible classical case (heavy fluid resting on the top of light fluid in gravitational field), with no ablation, the growth rate γ is

$$\gamma = \sqrt{Akg}$$

where $A = (\rho_{hi} - \rho_{lo}) / (\rho_{hi} + \rho_{lo})$ is the Atwood number, ρ_{hi} is the density of the heavier fluid and ρ_{lo} is the density of the lighter fluid, g is the acceleration, $k=2\pi/\lambda_{pert}$, where λ_{pert} is the modulation wavelength.

In ICF, a thin wall spherical pellet containing the fusion fuel is compressed to high density and temperature by the ablation pressure which is generated by heating the outside of the pellet with intense laser or x-ray radiation. To amplify the ablation pressure upon convergence a high aspect ratio shell ($R_o/\Delta R$) is required. If a shell of density ρ_{sh} and thickness ΔR is accelerated from its initial radius R_o to $R_o/2$ by the ablation pressure then the final pressure P_f is given by

$$P_f \sim \rho_{sh} v_{sh}^2 = P_A \left(\frac{R_0}{\Delta R} \right)$$

Consequently a large aspect ratio is required to obtain a large final pressure. However, it can be shown that the aspect ratio is limited due to the RT instability. In fact, a shell in which the RT growth at the classical rate can only be accelerated over six times its in-flight thickness before surface perturbations grow sufficiently to break up the shell. Typically high gain target designs require aspect ratios in the range between 25 and 50, implying that the growth rate of the RT instability must somehow be reduced to $\gamma/\gamma_{cl}<0.5$.

In the laser driven case, A is close to 1, so that the amplitude a of the perturbation may be expected to grow as $a(t) = a_o \exp \gamma_{cl} t$, where $\gamma_{cl} = \sqrt{kg}$. However, it is fortuate for inertial confinement fusion that ablation reduces the growth rate of the RT instability. This was first discussed by Bodner [5] who obtained

$$\gamma = \sqrt{kg} - kv_a$$

where v_a is the ablation velocity.

A more quantitative study by Takabe, Montierth and Morse [6] used a self-consistent eigenvalue method to obtain

$$\gamma = 0.9\sqrt{kg} - 3kv_a$$

Munro [7] found that further reduction is to be expected in the presence of finite density gradients, so that one might expect

$$\gamma = \alpha\sqrt{kg} - \beta k v_a$$

where $\alpha = (1 + kL)^{-1/2}$, L is the density scalelength at the ablation surface and β is a multiplier thought to be between 2 and 3.

The nonlinear growth phase is expected to commence approximately when $a/\lambda_{pert}\sim 0.1$. The transition from linear to nonlinear growth is a gradual one, but is characterised by the appearance of bubble-and spike structures. The RT instability in the nonlinear phase is expected to obey an $\eta g t^2$ growth law where η typically takes a value of 0.05 for incompressible fluids. A similar value is found in simulations of laser-targets with bubbles of light fluid rising into the heavier fluid and spikes of heavy fluid descending into the lighter one. Measurements have been made during this weakly nonlinear regime showing the growth of higher harmonics (caused by the bubble-and spike structures) with amplitudes up to 50% of the fundamental.

As the bubbles grow in size, they eventually encroach upon and, at the end of the weakly nonlinear regime, coagulate with neighbouring bubbles to form larger structures. During this phase the remaining spikes tend to broaden due to the Kelvin-Helmholz (KH) instability, eventually featuring the characteristic tip 'roll-up'. Repeated KH-type bifurcations and bubble coagulation to ever larger scales (until the fastest-growing mode is excited) soon leads to turbulent mixing between the two fluids. During this entire phase - primary bubble enchroachment to turbulent mixing - RT growth is measured by the growth of the 'mixing region' (typically defined as being the region between the two points at which the concentration of the penetrating fluid is ~1%). The extent of this mixing region is again expected to obey an $\eta g t^2$ growth law, with $\eta\sim 0.05$. The importance to ICF targets of this coagulation and mixing process is that short-wavelength modulations ($\lambda_{pert}\sim 5$ μm) might provide effective sources of longer wavelength disturbances which are more likely to cause shell break-up. The most dangerous modes have $\Delta R<\lambda_{pert}<3\Delta R$, where ΔR is the shell thickness. Such short wavelength modes may have very fast growth rates in the linear regime, so that they could very rapidly reach the nonlinear and coagulation stages.

3. Experimental techniques

The RT instability in laser produced plasmas is generally studied either with side-on or face-on time resolved x-ray radiography. A laser produced x-ray source is used to backlight the RT unstable foil target. The target is imaged by a pinhole camera with a typical spatial resolution of about 10 μm. The image is recorded either with an x-ray streak camera or with a framing camera. In addition, a soft x-ray imaging technique using a multilayered mirror has recently been developed to investigate the RT instability in laser produced plasmas [3,8]. This technique is capable of a spatial resolution of about 1 micron which is an order of magnitude better than is obtained with the pinhole camera assembly. Furthermore, due to the higher target material opacity at the probing energy of about 250 eV this technique yields are far higher sensitivity compared to the harder x-ray backlighter technique, allowing imprinting and the early phase of the RT instability to be studied.

4. Imprinting and induced RT growth

The growth and saturation of the imprinting process was investigated on foam-foil and bare foil targets using a coherent, RPP or ISI smoothed laser beam [3]. Face-on transmission radiography with a probe wavelength of about 250eV was used to observe the mass perturbations across the target. The transmitted signal was recorded either with a soft x-ray streak camera or a four frame gated microchannel plate intensifier. These diagnostics also allowed the early phase of the Rayleigh-Taylor instability to be studied. The observations clearly show that significant target break-up occurs on thin foils at early times caused by the imprinting process with a perturbation wavelength similar to the laser structure. Later in time, the longer wavelength modes become more pronounced. Figure 1 shows two gated soft x-ray images which were recorded at about 1 and 2ns after the beginning of the laser pulse. The target was a 7.5 μm thick CH foil which was irradiated with a RPP smoothed laser beam. In contrast, virtually no target break-up was observed when a foil/foam target overcoated with a thin layer of high Z material facing the laser beam was used [3,9]. These measurements are consistent with hydrodynamic simulations [10].

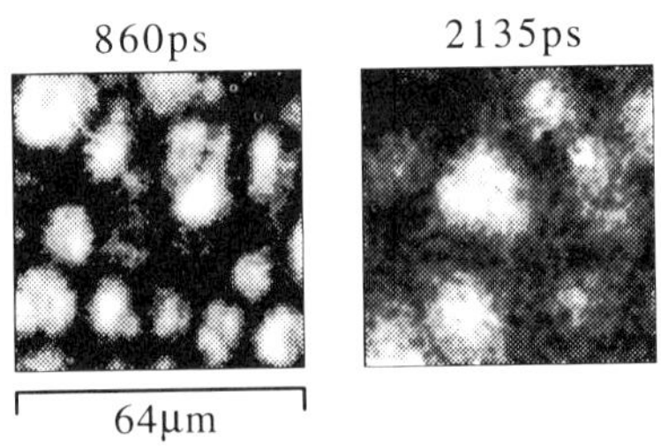

Figure1: Two soft x-ray images showing laser imprinting and RT growth induced by a Random Phase Plate speckle pattern.

5. Planar x-ray or laser driven target experiments

The RT instability of planar targets with perturbations between 30 and 100 μm driven by an intense source of x-ray radiation is now quite well understood. In general smaller growth rates for x-ray than for laser drive have been observed which has been attributed to higher ablative stabilisation as the mass ablation rate is higher for x-ray drive than laser drive at the same incident flux. The most complete experimental studies have been performed on NOVA. The x-ray drive used in these experiments was produced inside a cylindrical gold hohlraum with a typical radiation temperature up to 190 eV. The modulated foil target was mounted across a hole in the wall with the perturbations facing inwards. Single mode targets consisting of fluorosilicone (FS) or bromine-doped (CHBr) plastic foils were investigated in the linear regime. The experimental observations were modelled with the hydrodynamic code LASNEX. Good agreement was obtained showing that the growth rate is smaller for the CHBr foils than for FS targets caused by the larger ablation velocity and density scale length [11]. For CHBr foils with a modulation wavelength of 50μm, γ/γ_{cl} was smaller than 0.5. The nonlinear regime was studied with targets with larger initial modulation amplitudes [12]. The observations showed growth of higher harmonics leading to bubble and spike formation. In addition to the single mode experiments mode coupling was investigated with two and eight mode targets [13]. It was observed that the modes grow independently in the linear regime but become coupled in the nonlinear regime, leading to the appearance of $k_i \pm k_j$ modes. Again the observations are in agreement with simulations.

The first direct drive experiments of the RT instability using face-on radiography were performed in the early 1980s [14] and were followed by a series of experiments at NRL [15]. These early experiments were performed at low incident irradiance and without beam smoothing. The first RT instability experiments with beam smoothing used large initial amplitudes leading to nonlinear growth and saturation during the observations [16]. In recent experiments higher harmonics' growth and saturation were observed with growth rates of about 70% of classical [17]. All these experiments so far used targets with perturbation wavelengths > 20 μm. The growth rates for target modulations with wavelengths < 20 μm are predicted to be small by the Takabe-Morse formula. Experimental verification of this short wavelength regime is however urgently required.

6. Experiments in converging geometry

The first experimental observations of the RT instability in converging geometry have been obtained by researchers from the Los Alamos National Laboratory. A cylindrical polystyrene target with modulations on the outer target surface with peak to value amplitude of 9 μm was placed in a hohlraum. The hohlraum was heated by eight NOVA laser beams. A layer of chlorinated polystyrene was placed on the inside surface of the cylinder and was backlit with x-rays from a laser irradiated Ag target. Feedthrough and growth of the RT instability in the deceleration phase was clearly observed [4].

7. 3-D RT simulations

3-D simulations of the Rayleigh-Taylor instability in the deceleration phase of ICF implosions have previously been carried out with the 3-D hydrocode PLATO [18,19]. In order to model the complete implosion, PLATO has been substantially modified with the addition of classical electron thermal transport, laser energy deposition, multi material capability and a semi Lagrangian sliding mesh. The effects of the initial laser inhomogeneity on the shape of the final compressed core are currently being investigated by varying the laser perturbation wavelength and amplitude, the target aspect ratio and the gas fill pressure. Results indicate significant feedthrough of the ablation surface instability to the pusher-fuel interface for an l=6 perturbation imposed on a moderate aspect ratio target ($R_0/\Delta R$=30) [20].

8. Conclusions

A good understanding of the RT instability in planar geometry in the long wavelength regime has been obtained in recent years. The effect of laser imprinting on target break-up has been investigated and a detailed computational interpretation is presently in progress. For x-ray drive, there is a good agreement between experiments and simulations for perturbation wavelengths between 30 and 100 μm. It has been shown that $\gamma/\gamma_{cl}<0.5$ with ablation pressures close to fusion-like conditions. For direct drive, the growth rates are slightly higher with $\gamma/\gamma_{cl}\sim0.7$, however present experiments address the issue of reducing the growth rates to an acceptable level by enhancing the stabilising process. The first measurements of the feedthrough RT instability in convergent geometry have been made using cylindrical targets. The Rayleigh-Taylor instability for a complete implosion process has been studied with a 3-D hydrodynamics code which modelled the acceleration and the deceleration phase.

Acknowledgments

The author would like to thank many of his colleagues for supplying viewgraphs for the presentation. In particular, J. Kilkenny, B. Remington, J. Dahlburg, W. Hsing, H. Azechi and B. Jones are acknowledged.

References

[1] Taylor, G. 1950 Proc. R. Soc. London Ser. A **301**, 192; Lord Rayleigh 1883 Proc. London Math. Soc. **14**, 170
[2] Kilkenny et al. 1994 Phys. Plasmas**1**, 1379
[3] Taylor, R., et al. see this proceedings
[4] Hsing, W. 1994 private communication
[5] Bodner, S. 1974 Phys. Rev. Lett. **33**, 761
[6] Takabe, H., Montierth, L., Morse, R.L. 1983 Phys. Fluids **26**, 2299; Takabe, H., Mima, K., Montierth, L., Morse, R.L. 1985 Phys. Fluids **28**, 3676
[7] Munro, D.H. 1988 Phys. Rev. A. **38**, 1433
[8] Desselberger, M., Afshar-rad, T., Khattak, F., Viana, S., Willi, O. 1991 Appl. Optics **30**, 2285; Willi, O., Afshar-rad, T., Desselberger, M., Dunne, M., Edwards, J., Khattak, F., Taylor, R., 1992 Review of Scientific Instrum. **63**, 4818
[9] Dunne, M. et al. see this proceedings
[10] Jones, M. et al. see this proceedings
[11] Remington, B.A., et al 1993 Phys. Fluids B **5**, 2589
[12] Remington, B.A., et al 1994 Phys. Plasmas **1**
[13] Remington, B.A., et al 1994 Phys. Rev. Lett. **73**, 545
[14] Cole, A., et al. 1982 Nature **299**, 329
[15] Grun, J., et al. 1984 Phys. Rev. Lett. **53**, 1352
[16] Desselberger, M., Willi, O., Savage M., Lamb, M. 1990 Phys. Rev. Lett. **65**, 2997; Desselberger, M., Willi, O.1993 Phys. Fluids B **5**, 896
[17] Glendinning, S.G., et al. 1992 Phys. Rev. Lett. **69**, 1201
[18] Town, R.P.J., Bell, A.R. 1991 Phys. Rev. Lett. **67**, 1863
[19] Town, R.P.J., Jones, B.J., Findlay, J.D., Bell, A.R. 1994 Laser And Particle Beams, **12**, 163
[20] Jones, B., Town, R.P.J., Bell, A.R. see this proceedings

Inst. Phys. Conf. Ser. No 140: Section 3
Paper presented at Laser Interact. with Matter Conf., Oxford, 19–23 September 1994
© *1995 IOP Publishing Ltd*

Turbulent Mixing in Laser Driven Implosion

H. Takabe, H. Azechi, N. Miyanaga, H. Nishimura, H. Shiraga, Y. Isayama, A. Sunahara, K. Mima, and S. Nakai

Institute of Laser Engineering, Osaka University, Suita Osaka 565 JAPAN

Abstract. We have employed a turbulent mixing model to evaluate a degradation in implosion performance. A mixing model is coupled with a one-dimensional implosion code. The turbulent mixing in the stagnation phase is studied by assuming a fraction of turbulent energy compared to the implosion kinetic energy. We have obtained a systematic agreement with the experimental neutron yield, when we assume that 10-20 % of the implosion kinetic energy is converted to the turbulent kinetic energy in the acceleration.

1. Introduction

In laser driven implosion, hydrodynamic stabilify is the most important key issue. Many implosion experiments have been carried out with high power laser system such as GekkoXII; however, the experimental results have not been explained without taking account of hydrodynamic instability regardless whether it is direct [1,2] or indirect [3] schemes. Long-wavelength nonuniformity induces a deformation from implosion sphericity, while short-wavelength nonuniformity becomes a source of fluid turbulence with a help of instability and its nonlinear process. In most of the experiments carried out so far, we have used beam profiles not uniform enough to eliminate the generation of short-scale turbulent fluid motion. In order to analyze the experimental results, therefore, it is essential to model a capsule mixing by the turbulent convection.

Relating to the direct-drive glass-micro-balloon implosion experiment, it has been concluded that a systematic disagreement between the experimental neutron yields and the conventional one-dimensional implosion code results stem from the Rayleigh-Taylor instability in the stagnation phase. However, we can enumerate the following three scenarios for explaining the disagreement.

(1) If hot electrons or tail component of heated electron preheat the glass shell in the acceleration phase and the density of the glass reduces drastically compared to the conventional code result, the glass pusher can not reflect the shock wave coming from the DT plasmas, consequently no stagnation is observed.

(2) If the hydrodynamic instability grows substantially in the acceleration phase, the shell-breakup takes place in this phase. Then, averaged density of the glass shell reduced due to mixing and dynamical pressure (ρu^2) may reduce and no stagnation occurs due to the same reason as case(1).

(3) Even if the shell does not break up in the acceleration phase, the hydrodynamic instability in the stagnation phase prevents the heating and compression in the stagnation phase.

Actually, the above three may occurs in the same time, while in the present paper we focusing our attention to the mixing in the stagnation phase. In the gas-filled GMB implosion, glass plasma penetrates into the DT fuel, consequently neutron yield reduces due to cooling of the fuel. We have obtained a systematic agreement with the experimental neutron yield, when we assume that 10-20 % of the implosion kinetic energy is converted to the turbulent kinetic energy in the acceleration phase. It is also found that the same relative fraction of the turbulent energy can explain neutron yield, core temperature, and density-radius product for the high density compression experiment with CD-shell targets as reported in Ref.[4].

2.Turbulent Mixing in GMB target implosion

The turbulent mixing model has been applied to a single shock tube problem [5,6] and also applied to the laser implosion [7]. The numerical constants relating to the turbulent diffusion coefficients are adjusted so that the experimental results is well reproduced in the shock tube experiments[6]. Although a trial to apply the mixing model to the laser implosion has been repented is a short articles[7], it is not clear how the model works to explain some experimental results.

We have applied the turbulent mixing model to the case of high neutron yield experiments reported in Ref.[1]. In the experiment, so-called LHART-type targets characterized with large diameter (~ 1mm), thin shell (~1μm) glass micro balloon are used to contain deuterium tritium (DT) mixture fuel gas of about 10 atmosphere pressure. Most of target irradiations have been done with about 10kjoule, green light (λ=0.53μm) of Gekko XII laser system. The neutron yield of about 10^{10}~10^{13} has been observed.

Implosion dynamics has been compared with conventional one-dimensional simulation. It is reported that a systematic agreement is obtained for the implosion timing by adjusting an electron heat flux limiter [8]. Implosion parameters such as neutron yield, compressed ion temperature and density, and so on have also been compared with one-dimensional simulation for many shots. It is reported that the neutron yields are not well reproduced[1] The same kind of comparison on neutron yield has been carried out in several institutes and the same kind of disagreement has been reported. The disagreement becomes remarkable with the increase of importance of the stagnation dynamics in neutron production. It has been phenomenologically concluded in Ref.[1] that because of high level nonuniformity in irradiated laser intensity profile, uneven ablation pressure play a role in generating perturbation source of the Rayleigh-Taylor instability. Since the Rayleigh-Taylor instability grows explosively in the stagnation phase [9], we may not observe the neutrons generated in this phase in the conventional one-dimensional codes.

In the present report, we have included the turbulent mixing in the stagnation phase. For the case of highest neutron shot (#3826), the details of radius-time diagram of trajectories of each fluid element (numerical grid of Lagrangian description) is plotted in Fig.1 (a). It is typical that a shock wave collides at the center and a reflected shock wave in generated and propagates outward. Then, the shock wave collides at the fuel-pusher interface (25-th flow line from the center) then reflected again toward the center. After same time the pusher (glass material)tend to be decelerated due to the pressure of the fuel. These dynamics is called "stagnation dynamics" and shows the same evolution as in the shock tube experiments [for example, 5]. In such situation, the contact surface between DT fuel and glass (SiO_2) pusher is unstable for Richtmyer-Meshkov and Rayleigh-Taylor instabilities.

In Fig.1-(b), the flow-diagram is shown for the case when the turbulent mixing model is coupled. In this simulation, the turbulent mixing starts to be solved at the time ($t=t_s$) when the shock wave collides at the center ($t\approx 2.1$ nsec) and the turbulence energy of the value of $\xi(=2k/u^2_0$ at $t=t_s)=20\%$ is assumed to be generated in the acceleration phase ($0<t<t_s$). It is clearly seen in Fig.1-(b) that the glass pusher can not be decelerated by the reflected shock wave and freely penetrates toward the center. Since the material can diffuse among Lagrangian zones, this flow diagram does not tell how the glass penetrates into the fuel region.

In Figs.2-(a) and (b), the equi-contours of glass concentration are plotted for the cases without and with the turbulent mixing model, respectively. In Fig.2-(a), the trajectory of decelerating contact surface is recognized. However, by including the mixing effect, the glass material is not decelerated by the fuel pressure and the material font characterized with sharp front of the glass concentration penetrates toward the center. As the results, the fuel is cooled down and the neutron production which is very sensitive to the temperature of the fuel is suppressed. In the case, the neutron yield without the mixing was $Y_N=2x10^{13}$, while it reduced to $Y_N=1x10^{13}$ by including the mixing. It is noted that the corresponding experimental neutron yield is $Y_N=1x10^{13}$. What we can conclude in such study is that about 20% of the kinetic energy given by the rocket effect in the acceleration phase may be converted to the energy of disordered fluid motion due to uneven ablation pressure acceleration and/or hydrodynamic instability. It is found that in this case the fraction of the turbulence energy grows up to almost 10% of the total kinetic and thermal energy in the fuel at the maximum compression (t=2.4 nsec).

The same kind of simulations have been carried out for a variety of experimental shots. In Fig.3, the ratio of the neutron yields with the turbulent mixing model (Y_N*) and those without the turbulent mixing model [Y_N (o)] is plotted as a function of the radius convergence ratio (R_o/R_f). The radial convergence ratio is obtained from a conventional one-dimensional simulation without the mixing model which yields the fusion neutrons Y_N (o). The radius R_o is the initial radius of a glass-fuel interface and R_f is the minimum value of the radius of the interface. In the simulation, we have varying the radius convergence ratio by varying the initial pressure of the DT gas without raring the parameters of the glass shell and the irradiated laser. In Fig.3, the marks ×, ◇, ○, ⊞, □ and △ represent the cases when the levels of turbulence are assumed to be ξ=0, 2, 6, 10, 20 and 40%, respectively. Here, $\xi=2k/u_o^2$ is the ratio of the turbulence energy k and the kinetic energy of implosion $1/2u_o^2$ at the time where the shock wave generated in the gas fuel collides at the center.

In Fig.3, the solid circles are the the experimental neutron yields divided by these obtained with the conventional 1-D code without the mixing model[1]. As reported in Ref.[1], with the increase of the contribution of the stagnation dynamics to the neutron yields, the radial convergence ratio increases due to the compression by this dynamics and the discrepancy between the experimental yields and the corresponding 1-D simulation ones increases. By comparing the neutron yields obtained in the experiment and in the 1-D simulation with the mixing model, we can conclude that the turbulent energy of 2-20% out of the kinetic energy of implosion is generated before the stagnation phase. In the stagnation phase, the turbulence energy is seen to be enhance by the instability to finally terminate the predominant production of the neutrons due to the cooling of the fuel through the penetration of pusher material (glass) by the turbulent diffusion process.

3. Summary and Conclusion

In order to treat the effect of small scale turbulence in laser driven implosion, we have installed the mixing model into the one dimensional implosion code ILESTA-1D and used the code to analyze the experimental results obtained by imploding DT gas-filled glass micro balloon with Gekko XII laser system at Osaka. The k-ε type mixing model is introduce to couple with the one-fluid, two-temperature fluid equations.

We have focused on the turbulent mixing phenomena in the final compression called stagnation phase and studied how high level of the turbulence energy has been generated in the acceleration phase. As the results, it is found that about 2-20% out of the kinetic energy for implosion goes to the energy of turbulent mention in this phase.

References

[1] Takabe H et al 1988 Physics of Fluids 31 2884

[2] Azechi H et al 1991 Laser and Particle Beams 9 193

[3] Nishimura H et al 1994 to be published in Physics of Plasma

[4] Takabe H et al 1994 Proc. 15-th Int. Conf. on Plasma Physics and Controlled Nuclear Fusion Research, Seville,26 September-1 October 1994.

[5] Andronov V A 1977 Sov. Phys. JETP **44** 424

[6] Gauthier S and Bonnet M 1990 Phys. Fluids **A2** 1685

[7] Andronov V A et al 1979 JETP Lett. **29** 56

[8] Azechi H et al 1989 Appl. Phys. Lett. **5 5** 945
[9] Hattori F et al 1986 Phys. Fluids **2 9** 1917

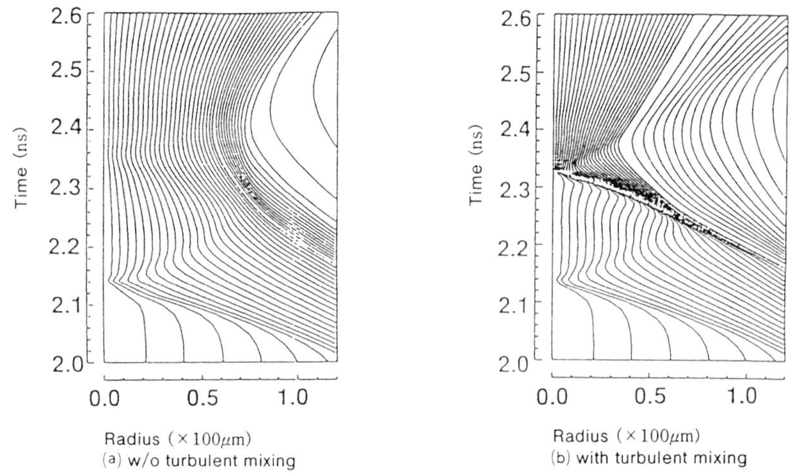

Fig. 1 Flow diagram of gas-filled GMB implosion.

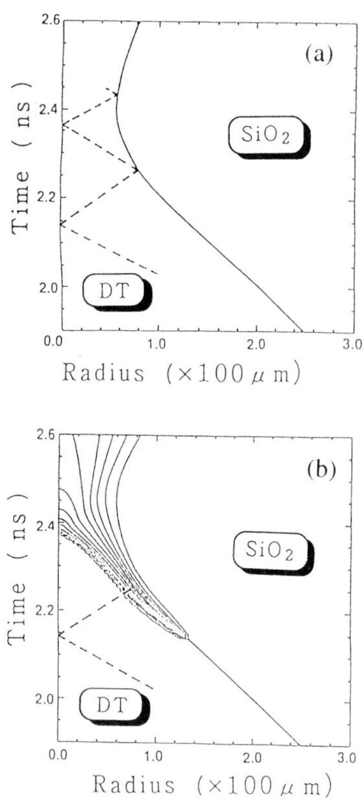

Fig. 2 Equi-contour of glass concentration for cases without and with mixing.

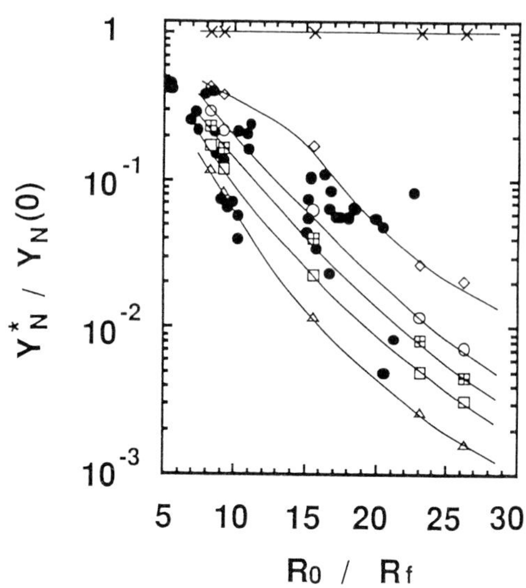

Fig.3 Reduced neutron yield by turbulent mixing.

Inst. Phys. Conf. Ser. No 140: Section 3
Paper presented at Laser Interact. with Matter Conf., Oxford, 19–23 September 1994
© *1995 IOP Publishing Ltd*

NUMERICAL SIMULATIONS AND EXPERIMENTAL STUDY OF RICHTMYER-MESHKOV INSTABILITY.[1]

I.V.Popov, V.F.Tishkin, V.V.Nikishin, N.V.Zmitrenko, A.P.Favorsky
Institute of Mathematical Modeling, Moscow 125047,Miusskaya sq., 4-A, Russia
S.G.Zaytsev, A.N.Aleshin, E.V.Lazareva
The Krzhizhanovsky Power Engineering Institute, Moscow 117071,Leninsky pr. 19, Russia
V.B. Rozanov, I.G.Lebo, S.Yu. Gus'kov
P.N. Lebedev Physical Institute of Academy of Scienties, Moscow 117924, Leninsky pr.53, Russia

Abstract. The paper presents the results of numerical simulations and experimental study of Richtmyer-Meshkov instability (RMI) with relation to the problem of ICF-target. The possibility to use shock-tubes (ST) in the experimental study of RMI in ICF-targets is shown. The proposed numerical simulations are effective for the solution of the considered problem of the ICF-target.

1. Introduction

The conditions for the occurrence and development of the RMI are observed in the laser targets (LT). In combination with the Rayleigh-Taylor instability (RTI) the above mentioned RMI leads to the destruction and mixing of the shell and the target thermonuclear fuel. This may result in the decrease or stopping of the thermonuclear reaction.

The direct experimental study of RMI is difficult, since the processes are very fast ($1 \div 2ns$), the geometrical dimensions of the target are small, and the shock wave (SW) is strong, and so on.

The RMI has been conventionally studied with the ST, but the gas dynamic parameters attained in these experiments are different then those in the LT.

We have developed an Euler numerical method. This method allows us to theoretically investigate the RMI with the SW parameters as well as with real laser experiment

[1]This work is partly supported by Russian fund for fundamental researches grant N 94-01-01144-A

parameters. The developed technique has been checked with the experimental data obtained from the SW. Then taking the initial data from the 1-D and 2-D Langangian numerical calculations for the LT, we have simulated the target shell and DT-gas behavior as applied to the Euler technique.

The numerical simulations have indicated that the experimental research of RMI with ST could be applied for ICF.

2. Experimental studies with shock tubes.

The experiments were carried out with $7.2 \times 7.2cm^2$ cross section ST.A plane SW moves from a light gas to a heavy one and has a diffraction at the contact boundary (CB),which was perturbated by a sinusoidal-like law [1]. The Mach number was changed within an interval $2.5 \div 3.5$ and an initial pressure was $0.5atm$. Experiments were fulfilled for 2-D and 3-D perturbations with different wavelengthes.

As a result we show some data for 2-D case and $\lambda = 24$ mm in the moment $t = 100$ mks in the fig.1. This case corresponds SW transition from Ar to Xe ($M = 3.6 \pm 0.15$, initial amplitude of perturbation $a_0 = 10mm$).

For 3-D perturbation we have taken $\lambda_{3D} \cong \sqrt{2} \cdot \lambda_{2D}$ to equal increments of the modes. The corresponding Topler pictures are shown in fig.2 (3-D, $Ar - Xe$, $M = 2.5 \pm 0.15$, $\lambda_x = \lambda_y = 36mm$ $a_0 = 10mm$). The comparison of 2-D and 3-D perturbations shows, that in the 3-D case the heavy gas "spike" penetrates considerably deeper into a light gas, than in the 2-D case.

3. Numarical simulation of shock-tubes experiments.

Mathematical modelling of RMI was fulfilled on the base of model of ideal gas dynamic.

The quasy-monotonous conservative finite-difference schemes of a high order of an accuracy were used to solve numerically the gas dynamic equations. Difference schemes convergence [2] was checked by the means of calculations with condenced grids. In fig.3 the density isolines are shown for the grids of 20×250, 40×500 sizes ($Ar - Xe$, $\lambda = 24mm$, $a_0 = 10mm$, $M = 3,5$). In fig.4 a growth of amplitude of contact boundary perturbation is shown for the same grids.

The comparison of numerical and experimental results shows, that numerical method gives a good description both the sizes of a "spike" and the shape of an arizing "mushroom".

In fig.5 the calculated density isolines for the experimental results on described above (see fig.1) are shown.

In fig.6. the results of 3D calculations ($\Lambda_x = \Lambda_y = 36\,mm$, $M = 2.5$, $A_r - X_e$, $a_0 = 10mm$) are presented. The more fast amplitude growth occurs both in 3D calculation and experiments in comparison with 2D case.

In fig.7 an amplitude growth is shown for 2D and 3D cases. The wavelengthes were different so, that the linear increments to be equal.

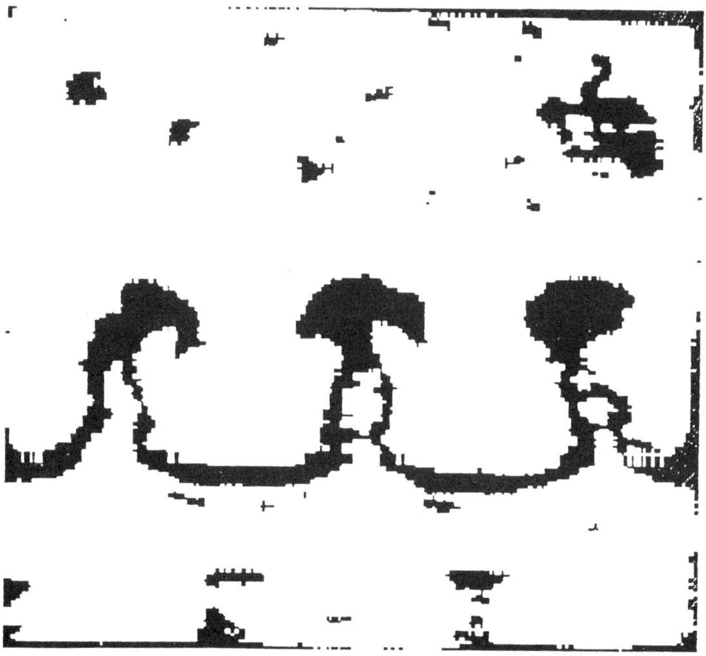

Fig. 1

Fig. 2

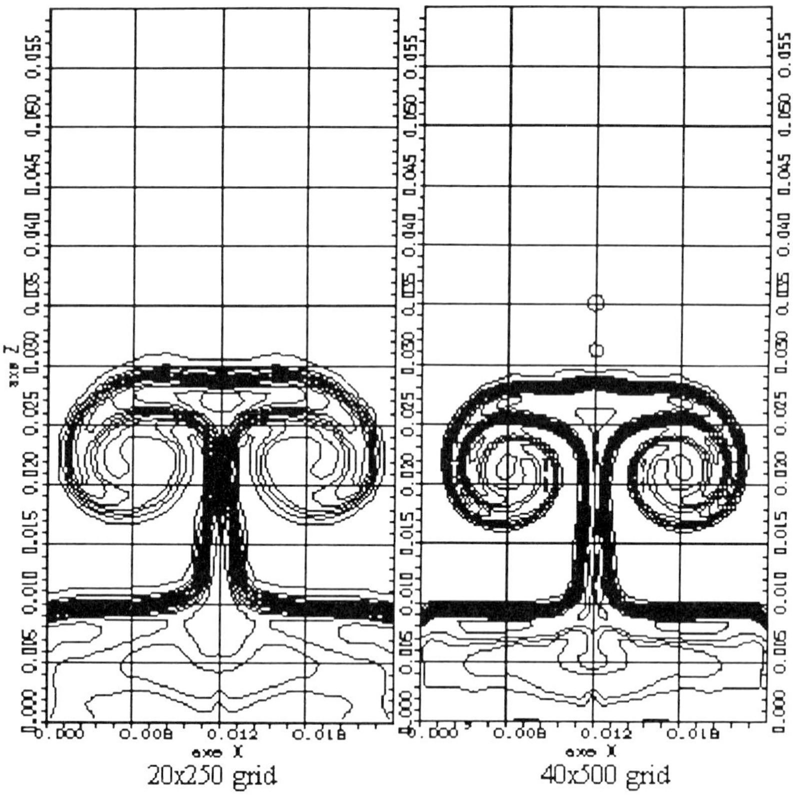
20x250 grid
40x500 grid

Fig. 3

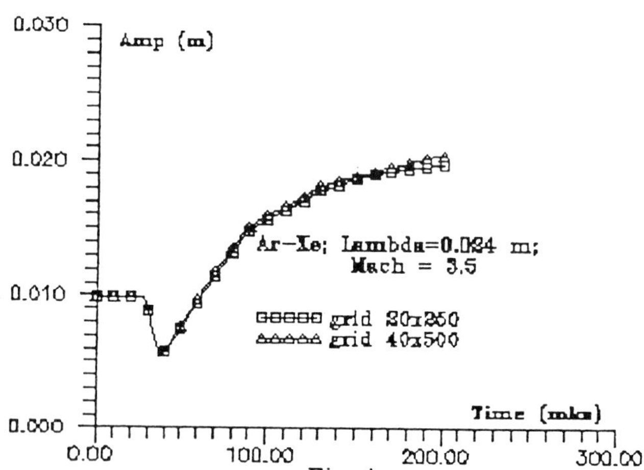
Amp (m)
Ar-Xe; Lambda=0.024 m;
Mach = 3.5
grid 20x250
grid 40x500
Time (mks)

Fig. 4

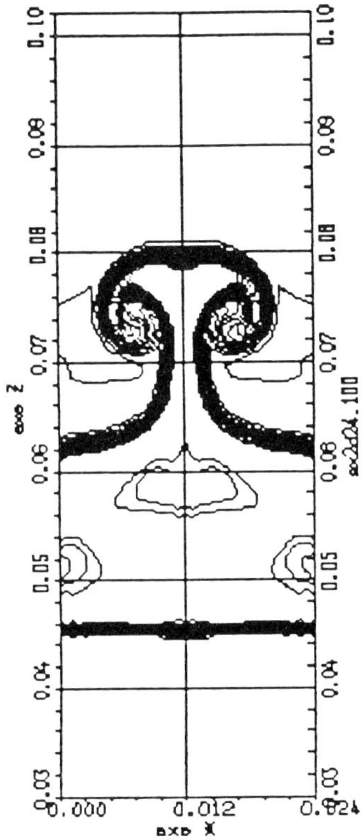

Fig. 5

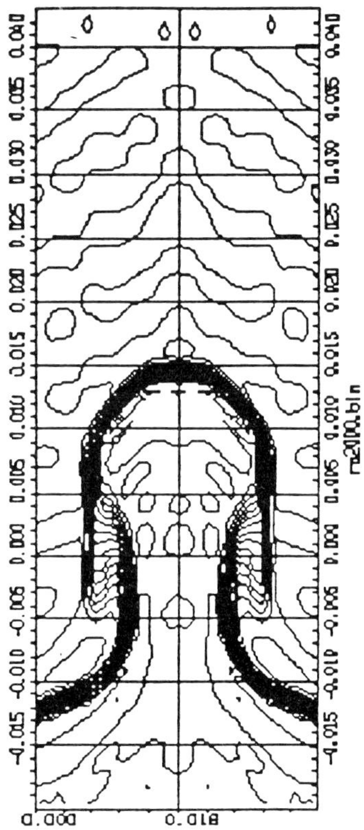

Fig. 6

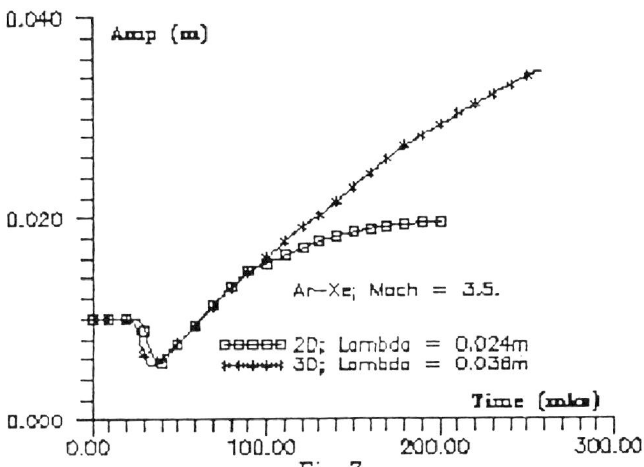

Fig. 7

4. Theoretical explanation of "mushroom - like" structures.

Let us assume, that the "spike" destruction of RMI perturbation occurs due to the Kelvin-Helmgoltz instability (KHI) development.

The growth rate of KHI is given as $\gamma = qvf$, where $\Lambda = 2\pi q^{-1}$ is a wavelength, velocity v is close to "spike" velosity v_0 and factor $f = \frac{\sqrt{\alpha \cdot \delta}}{1+\alpha \cdot \delta}$ takes into account density relation $\delta = \frac{\rho_l}{\rho_h}$ and geometric coefficient α. For 2D calculations of $Ar - Xe$ shock wave transition we can take $f \cong 0.39$ [3] and $q \cong 0.2mm^{-1}$. This value of q corresponds to $\Lambda \simeq 2\pi \cdot d \simeq 2(a+b)$, where $a+b$ is a total size ("spike" and " bubble") of RMI perturbation and d is the effective thickness of a "spike", when KHI does not develope. So, we obtain $\gamma \simeq 0.012\,mm^{-1}$ in a good agreement with calculated magnitude $\gamma \simeq 0.01\,mm^{-1}$.

We can generalize this approach for different cases. We suppose, that KHI starts to develop from the moment, when $ka = 1$ ($\lambda = \frac{2\pi}{k}$ is a wavelength of RMI), at further moments $a+b = a + \frac{1}{k}$, the desfruction moment corresponds to the condition $\zeta \cong d$, when ζ is an amplitude of KHI and d is a width of RMI "spike". Also let us neglect the compressibility and heat conductivity. The KHI wavelength Λ must be determine as $\Lambda = 2\sqrt{\lambda^2/4 + (a+b)^2} = \lambda \cdot \sqrt{1 + \left(\frac{ka+1}{\pi}\right)^2}$ in 2D case and as $\Lambda = \frac{2}{3}\sqrt{\frac{2\lambda^2}{4} + (a+b)^2} = \frac{\lambda\sqrt{2}}{3} \cdot \sqrt{1 + \left(\frac{ka+1}{\pi\sqrt{2}}\right)^2}$ in 3D case. Under these assumption we can calculate the growth factor $\Gamma = \int_{t_0}^{t} \gamma dt = \int_1^{ka} \frac{\gamma}{kv_0} d(ka)$, so, that $\zeta = \zeta_0 \cdot e^{\Gamma}$

The condition $\xi \simeq d$ leads us to equations

$$B_2 = y^{\frac{1}{2 \cdot \bar{f}}} \cdot \left(y + \sqrt{1+y^2}\right), \quad ka = \pi y - 1 \qquad (2D)$$

$$B_3 = z^{\frac{1}{3 \cdot \bar{f}}} \cdot \left(z + \sqrt{1+z^2}\right), \quad ka = \pi\sqrt{2}z - 1 \qquad (3D)$$

Here $\bar{f}$ is an average value of f factor. Calculations show, that the typical values of ka are

$$ka = 2.2\ (Ar - Xe, 2D,\ \ \bar{f} = 0.35,\ \ B_2 = 2.57)$$

$$ka = 2.9 \div 5.1\ (Ar - Xe, 3D,\ \ \bar{f} = 0.4,\ \ B_3 = 2 \div 4)$$

$$ka = 2.6\ (\mathrm{He} - Xe, 2D,\ \ \bar{f} = 0.15,\ \ B_2 = 4.08)$$

$$ka = 3.3 \div 6.5\ (\mathrm{He} - Xe, 3D,\ \ \bar{f} = 0.2,\ \ B_3 = 2 \div 9)$$

Therefore we have found, that destruction criterium depends only on $\bar{f}$ value and so it will not be essential difference between a ST and LT for the reason of close values of δ. In LT δ for DT - fuel and pusher is close to 0.3 like in the case of Ar-Xe in ST.

5. Simulation of instabilities evolution for the laser target conditions.

It is important to understand what an information from ST experiments can be used for laser target design. Otherwize it is interesting to find out the possibilities of the traditional laser target 2D Lagrangian codes to describe an instabilities evolution.

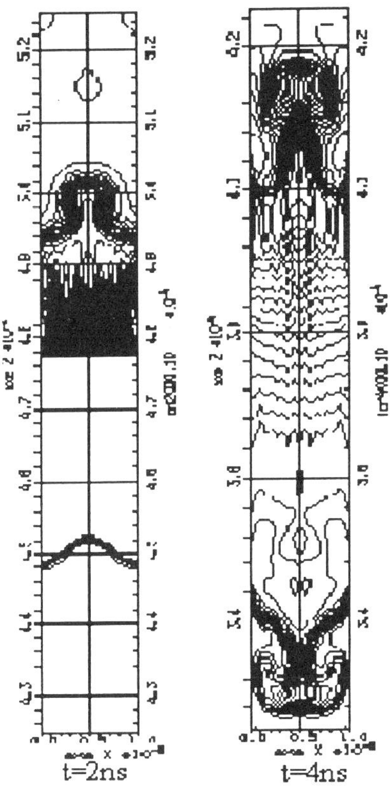
t=2ns
t=4ns

Fig. 8

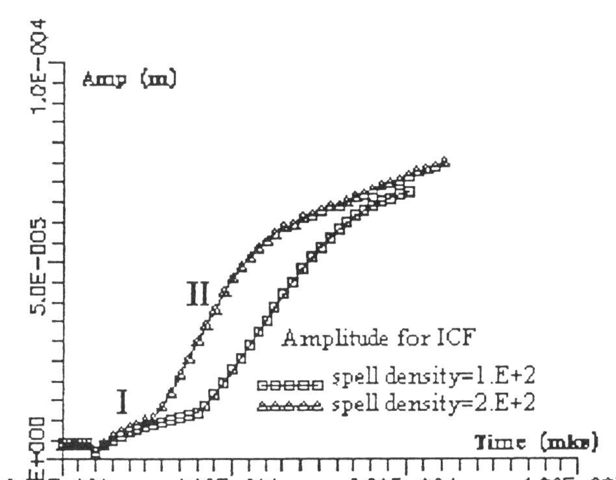
Amp (m)
Time (mks)
Amplitude for ICF
spell density=1.E+2
spell density=2.E+2
I
II
1.0E-004
5.0E-005
0.00E+000
0.00E+000
4.00E-004
8.00E-004
1.20E-003

Fig. 9

For this purpose we have simulated the process of a pusher deceleration due to a reflected from a center a first SW. The initial date for such problem were taken from calculation fulfilled by the means of 2D Lagrangian code ATLANT [3,4]. It is considered a DT-fuel ($\rho_f = 0.05\, g/ccm$), CH_2– pusher ($\rho_p = 0.1$ g/ccm, $\frac{\rho_p}{\rho_f} = 2$, thickness is $100\, \mu m$), surrounding by a plasma corona (CH_2, $\rho = 3\, mg/ccm$). The SW have moved from DT-fuel to pusher with a relative velocity of $500\, \mu m\, /ns$.

The density isolines are shown at fig.8 for the moment 2 and 4 ns and the wavelength of the perturbation $\lambda = 10 \mu m$. We can see the perturbation change when SW passes from the pusher to the corona. In the fig.9 the dependences of perturbation amplitudes at the fuel-pusher boundary upon a time are shown. These pictures show, that in a perturbation development the second stage appears and this stage is connected with an interaction of a rarefaction wave with primary contact boundary.

The fig.9 also shows the amplitude dependences for different Atwood numbers ($\rho_p\, /\, \rho_f = 2$ and 4). So, we can conclude, that the ideas, developed for ST can be used and for LT. Only, we have to take into account an influence of a heat conductivity.

6. Conclusion.

The main results are the next.
1. The evolution of a perturbation development from a linear stage through "mushroom" structure up to mixing stage is studied in the ST experiments.
2. The essential difference between 2D and 3D perturbation growth is observed both in experiments and numerical simulations. 3D perturbations develope to greater values of amplitude than 2D ones.
3. The probably cause of "mushroom - like" structure origin is the KHI development due to a penetration of a heavy "spike" into a light surrounding medium.
4. The Large set of 2D and 3D calculations was carried out. The good agreement with physical experiments was shown.
5. On the base of created methods it is shown, that the nature and the properties of the hydrodynamical instabilities development in ST and LT may be the same.

References

[1] Zaytsev S., *et al "Richtmyer - Meshkov instability for two and three dimensional interfaces". 4th international workshop on the physics of compessible turbulent mixing. 29 march-1 april 1993, Cambridge, England.*

[2] Vyaznikov K., *et al 1989 Mathematical modeling in Russian, V1, N 5, p.95 - 120.*

[3] Rozanov V., *et al 1992 P.N.Lebedev Phys. Inst., preprint N°16.*

[4] Gamaly E., *et al 1988 Quantum Electronics (in Russian) V15, N°8,* p.1622 - 1632

[5] Lebo I., Popov I., *et al 1994 Journal of Russian Laser Research V15, N°2,* p.136 - 143

Inst. Phys. Conf. Ser. No 140: Section 3
Paper presented at Laser Interact. with Matter Conf., Oxford, 19–23 September 1994
© *1995 IOP Publishing Ltd*

Measurement and simulation of a plasma collision at the rear of laser-irradiated low-Z foils

O. Rancu, C. Chenais-Popovics, J.C. Gauthier, H. Kawagoshi*,
Laboratoire pour l'Utilisation des Lasers Intenses, École Polytechnique, 91128 Palaiseau Cédex, France
P. Renaudin, H. Pépin,
INRS Énergie-Matériaux, Université du Québec, C.P. 1020, J3X1S2, Varennes, Québec, Canada
M. Dirksmöller, I. Uschmann, E. Förster
Max-Planck "Röntgenoptik" Arbeitsgruppe, Friedrich-Schiller Universität, Max-Wien Platz 1, D07740 Jena, Germany
O. Larroche, O. Peyrusse
Centre d'Etudes de Limeil-Valenton, B.P. 27, 94190 Villeneuve-Saint-Georges, France
* on leave from PNC, Tokai, Japan

Abstract. We have studied the collision of aluminium exploding foils by space-resolved spectroscopy in the range 5-7 Å and by time and space-resolved X-ray monochromatic imaging. Interpenetration of the colliding plasmas could be visualized by replacing one of the foils by a magnesium foil. Time and space evolution of the colliding plasmas has been measured and compared to 1D multifluid code simulations post-processed by an atomic-physics package which calculates the observed x-ray emission.

1. Introduction

The hydrodynamics of the collision of two hot plasmas is of interest for the understanding of different phases of inertial fusion, specially in the indirect drive scheme where collisions can occur inside the hohlraum and deteriorate the symmetry of the imploding microballoon.[1, 2] Plasma collisions have also been proposed to create a large scale plasma for laser-plasma interaction or X-ray laser studies.[3] We present here an experiment in which two laser-irradiated thin foils are accelerated and collide in the middle of the inter-target region. The created plasma is probed by X-ray diagnostics and simulated by a one-dimension multifluid eulerian code post-processed by an atomic-physics package. The aim of the experiment is to map out the spatial distribution of the ionic species and to measure the electronic and ionic temperatures during the plasma interpenetration and stagnation.

2. Experimental arrangement

The colliding plasmas are produced by focusing two synchronised beams of the LULI laser facility at 0.53 µm laser wavelength onto two 0.8 µm thick aluminium foils (atomic number Z=13). Magnesium foils (Z=12) of 1 µm thickness have also been used to visualize the interpenetration of the two plasmas during the collision. The foil thickness has been optimized by experiments and hydrocode simulations to be equal to the ablation depth for the laser intensities used in the collision experiment. The target distance has been varied between 200 and 900 µm and the laser intensity between 5 10^{13} and 10^{14} W/cm^2. Random phase plates have been used to obtain 250 µm diameter focal spots. The main diagnostics were time-resolved monochromatic imaging and a space-resolving ADP flat crystal spectrograph. One pinhole camera was space-resolving the inter-target region and another one was imaging the focal spot of one of the laser beams.

Two monochromatic images were recorded at 0.6635 nm, the wavelength of the $1s^2$-$1s3p$ (Heβ) aluminium line by using two toroidal quartz $(1\ 0\ \bar{1}\ \bar{1})$ crystals coupled with two gold cathodes of a Kentech 120 ps frame-duration x-ray camera. The monochromaticity achieved ($\Delta\lambda/\lambda = 8.5\ 10^{-3}$) allowed to follow the interpenetration of the colliding plasmas when one of the aluminium foils was replaced by magnesium. Indeed, the aluminium plasma produced on one foil can be imaged with only a weak continuum from the magnesium foil. The monochromatic imaging crystals provided 10 μm spatial resolution with an x-ray film detector and 35 μm with the framing camera.[4]

3. Experimental results

The space and time evolution of the emission of the Heβ line measured for 450 and 600 μm inter-target distances and for a laser intensity of 5 and 8 10^{13} W/cm^2 respectively has been recorded. A maximum of emission occurs between 500 and 700 ps after the maximum of the laser pulse and is localised on around 100 μm along the inter-target axis and 300 μm across. Figure 1 shows the monochromatic images obtained before and during the collision phase for the 450 μm distance and 5 10^{13} W/cm^2 case. Five emission zones are observed and appear clearly on the 1100 ps frame of figure 1 : the plasmas produced at the front and the rear of each foil, and a collision region, which spreads laterally in the middle of the inter-target region. Pinhole images filtered for energies below 500 eV with an aluminized mylar foil show a much wider lateral spread than the Heβ emission, i.e. around 500 μm wide, twice the focal spot diameter. The monochromatic images and the spectra obtained with inter-target distances smaller than 300 μm show that the time-integrated Heβ is emitted from the whole inter-target region.

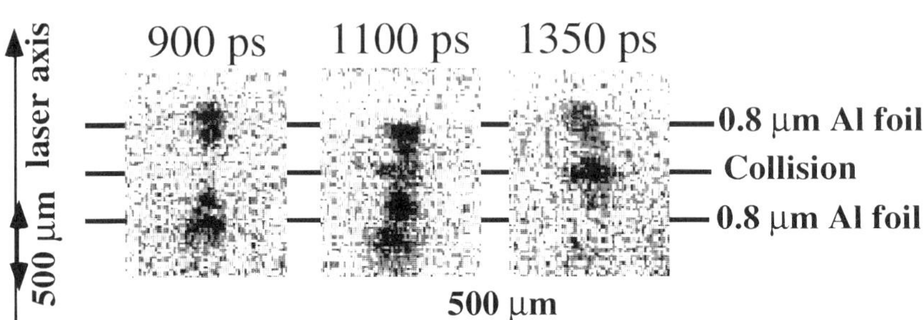

Figure1: Monochromatic images of Heβ emission at different times and for a inter-target distance of 450 μm; laser intensity 5 10^{13} W/cm^2. Times are measured from the foot of the pulse, i.e. 600 ps before the maximum.

Figure 2 shows the comparison of time-integrated monochromatic images obtained with Al/Al and Al/Mg targets. It shows clearly that no interpenetration of the plasmas is observed by X-ray imaging in the experimental conditions achieved here, i.e. 450 μm inter-target distance and 5 10^{13} W/cm^2 laser intensity.

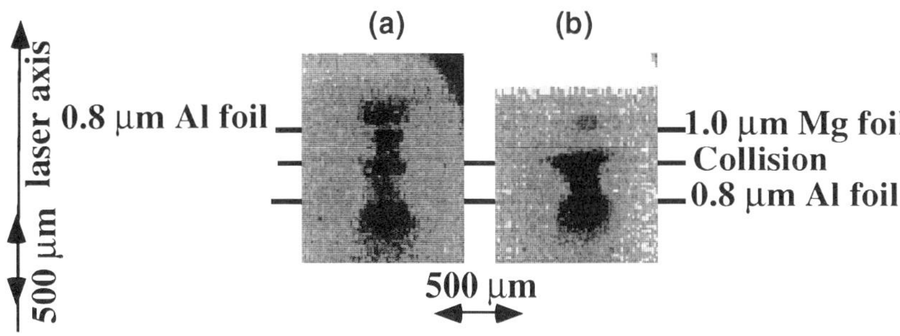

Figure 2: Monochromatic images of Al/Al and Al/Mg collisions at the wavelength of the Al Heβ emission for 5 10^{13} W/cm^2 laser intensity ; (a): 0.8 μm Al foils, ; (b): 0.8 μm Al /1 μm Mg foils.

The analysis of the time-integrated, space-resolved spectra obtained for the same conditions as figure 1 is shown on figure 3 (0.8 μm Al foils, at a distance of 450 μm irradiated with 5 10^{13} W/cm^2 laser intensity). An experimental spectrum traced in the collision region is compared to the best-fit spectrum calculated with the code RATION run for a steady-state homogeneous plasma.[5] The electronic temperature obtained for the best fit is 360 eV. The electronic density has been taken from the code ($n_e = 10^{21}$ cm^{-3}) , and the fit is not very sensitive to its value. This measurement is only indicative as the hydrodynamics of the collision varies on a time scale of 100 ps, comparable to the time of evolution of the ionic populations responsible for the X-ray emission. One of the more striking features on the experimental spectrum is that the Heβ and Heγ lines have a comparable intensity which shows that the plasma is largely out of equilibrium and that reabsorption is not negligible.

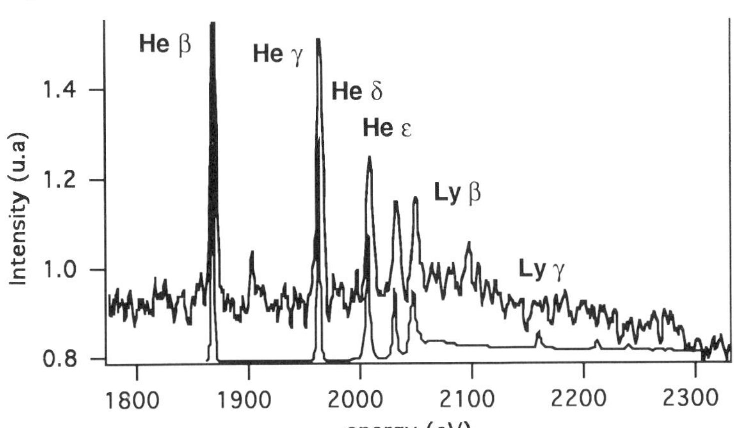

Figure 3: Comparison of an experimental time-integrated spectrum of aluminium measured in the collision region with a spectrum calculated with RATION code.

4. Comparison of data to simulations results.

The experiment has been simulated with a multispecies eulerian monodimensional fluid code.[6] The model uses standard FCT modules improved for the description of the plasma-vacuum interface.[7] A perfect gas equation of state is assumed, and the average charge state is fixed (Z = 10). Transport coefficients are described by a flux-limited Spitzer model. A consistent treatment of ionic transport is performed in the case of interpenetrating fluids. The input of the code is taken from the standard monodimensional lagrangian fluid hydrocode FILM.[8] Calculations have been performed for 5 10^{13} and 10^{14} W/cm^2, 0.53 μm laser wavelength and 600 ps laser pulse duration. Two atomic packages have been used to post-process the multifluid code: the time dependent version of the code RATION (called FLY)[5] , and the code TRANSPEC which additionally takes into account non-homogeneous radiative transfer.[9] Figure 4 shows the simulated emission of the Heβ line along the inter-target axis as a function of time. It has been calculated by the multifluid code post-processed with TRANSPEC in the same conditions as those of figure 1. One can see that an overall agreement is obtained with an increase of the Heβ line during a few hundreds of picoseconds at a time corresponding to the measurements (1100 ps after the beginning of the laser pulse). This increase is associated with an increase of the electronic density and electronic and ionic temperatures at this time in the centre of the inter-target region. The code shows that no interpenetration is expected with these conditions where the velocities of the colliding plasmas are rather low. Calculations performed at higher laser intensities and larger distances between the foils exhibit more interpenetration.

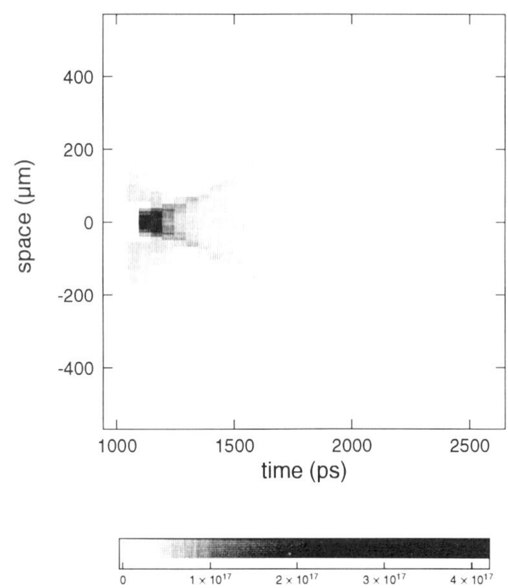

Figure 4: Calculated Al Heβ emission as a function of time and distance along the 450 μm inter-target space for a laser intensity of 5. 10^{13} W/cm^2.

5. Conclusion and future work.

The experimental data and the simulations show that for a distance larger than 300 μm the x-ray emission in the inter-target region is very inhomogeneous. For 450 and 600 μm distances an increase of density and temperatures occurs in the centre of the inter-target region on a short time scale. A radial spread of the x-ray emission is observed, mainly in the soft X-ray region, meaning that a low temperature plasma surrounds the hot collision core. The multifluid code predicts a very strong increase of the ionic temperature and more interpenetration for higher laser intensity. Future experiments will be focused on these two points.

Acknowledgements.

We would like to thank the Institut de Physique Nucléaire for target preparation and CDSI-Institut d'Optique for film densitometry, both in Paris Sud University. The technical staff of the LULI facility is gratefully acknowledged, in particular, R. Konig and C. Montheil.

References.

[1] R.L. Kauffman et al., ECLIM'94 Conference Proceedings, to be published.
[2] M. Cray, ECLIM'94 Conference Proceedings, to be published.
[3] B. Yaakobi, D. Shvarts, T. Boehly, P. Audebert, R. Epstein, B. Boswell, M.C. Richardson and J.M. Soures, IEEE Trans. Plasma Science, **16**, 505 (1988).
[4] M. Dirksmöller, I. Uschmann, E. Förster, O. Rancu, P. Renaudin et C. Chenais-Popovics, submitted to Optics Commun.
[5] R.W. Lee, B.L. Whitten and R.E. Strout, J. Quant. Spect. Radiat. Transf., **32**, 91 (1985)
[6] O. Larroche, private communication.
[7] J.P. Boris, NRL Memorandum Report 3237 (1976); J.P. Boris and D.L. Book, J. Comput. Phys. **12**, 67 (1976).
[8] J.C. Gauthier, J.P. Geindre, N. Grandjouan and J. Virmont, J. Phys. D, **16,** 321 (1983).
[9] O. Peyrusse, Phys. Fluids B **4**, 2007 (1992).

Inst. Phys. Conf. Ser. No 140: Section 4
Paper presented at Laser Interact. with Matter Conf., Oxford, 19–23 September 1994

© *1995 IOP Publishing Ltd*

Advances in high power laser technology

S. Nakai, M. Nakatsuka, N. Miyanaga

Institute of Laser Engineering, Osaka University
2-6 Yamada-oka Suita, Osaka 565 Japan

Abstract. The goal of inertial fusion energy (IFE) development is to prove that fusion energy can be available to society as electric power source. Recent progress of laser fusion research and development enable us to examine technical and economical feasibility, and to plan the realistic strategy and program to the commercial power plant. The most important key issue for IFE is driver technologies. The progress of high power laser through out the world and it related technologies have been reviewed in the view point for the advanced implosion and for ignition and burn experiments.

1. Introduction

The requirements to the fusion driver are specified depending on the developmental phase of demonstrations toward the commercial power plant. The goals of the major demonstration are the following

(1) Ignition demonstration (ID) demonstrate ignition and thermonuclear burn in an ICF target and determine minimum driver conditions necessary
(2) High gain demonstration (HGD) demonstrate sufficiently high driver efficiency-gain product, $\eta G>10$,
(3) Engineering demonstration (ED) demonstrate high pulse rate operations in an integrated system and chose best reactor design,
(4) Commercial demonstration (CD) demonstrate safe, environmentally benign, reliable, economic near commercial operation.

Table 1 Summarizes the key technical issues associated with each of these demonstrations.[1]

2. Requirement to fusion driver

The specifications of driver for ICF are 1) energy, 2) intensity, 3) pulse shape, 4) wavelength (or particle range), 5) spatial uniformity, 6) efficiency, 7) repetitive operation, 8) cost (capital and operational, including life and maintenability), 9) reliability or availability.

The energy gain scaling of implosion on the driver energy, which is optimized regarding to the intensity, pulse shape, wavelength, is shown in Fig. 1.[2] The simulation code to predict the implosion at higher energy region is examined and reinforced with the available implosion data by presently operational facilities.

The sensitivity of gain on the nonuniformity of fuel pellet and irradiation intensity give us the required experimental condition such as pellet sphericity and surface roughness, low and

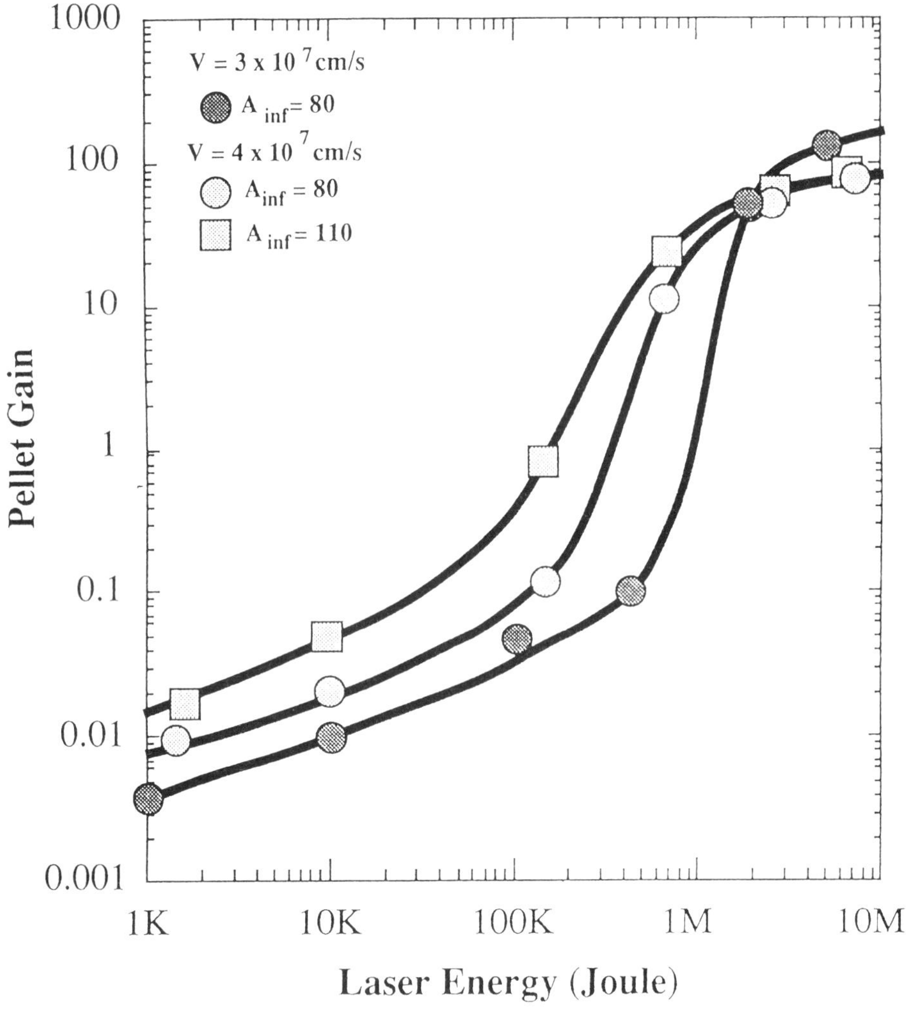

Fig. 1 Pellet gain as a function of laser energy with the implosion velocity (V) and the inflight-aspect-ratio (A_{inf}) as a parameter.

Table 1 Technical issues associated with the various required demonstrations for each power plant system

	Required demonstrations			
Power plant systems	Ignition demonstration	High gain demonstration	Engineering demonstrations	Commercial demonstrations
Target: Physics	$\rho R .0.3 g/cm^2$ More α particle self heating than from driver Hot spot formation Gain ≈ 1	$\rho R > 0.3 g/cm^2$ Propagating burn from hot spot through cold fuel Gain ≈ 30~100		
Fabrication	Cryogenic layering Sphericity and uniformity <1% Surface finish <100nm		Mass production, delivery rate 1-10 Hz Batch mode operation	Continuous production Long term reliability Inventories
Drivers	Single shot operation Illumination uniformity <1% Beam power balance Pulse shaping $E_d \approx$ 300kJ-2MJ	Single shot operation Efficiency $\eta \approx$ 5-30% (coupled with Gain as $\eta G > 10$) $E_d \approx$ 1-10MJ	Pulse rate ≈ 1-10Hz Beam pointing Burst mode operation	Reliable, safe, continuous, long term operation Maintainability
Reaction chamber	Single shot operation Final optics survivability	Single shot operation Containment and survivability	Pulsed operation≈ 1-10Hz First wall protection Blanket operation Target injection and tracking Final optics or magnets survivability	Continuous, safe, reliable, long term operation
Remainder of Plant (ROP)			T_2 extraction and materials management Thermal energy production and transport	Closed power and materials recycling loops Safe, low environmental impact, reliable, economical operation

Table 2 Major operating or planned driver related facilities

Country	System Name	Location	Energy / Duration	No.of Beams	Wave length	Comments
Glass lasers						
USA	NOVA OMEGA OMEGA UG NIF	LLNL, Livermore LLE, U. Rochester (to be selected)	50kJ/1ns 3kJ/1ns 30kJ/3ns 1.8MJ/5ns	10 24 60 244	0.35μm 0.35μm 0.35μm 0.35μm	Precision completed Under construction Design complete
Japan	GEKKO XII KONGOH	ILE,Osaka U	15kJ/1ns 300kJ/3ns	12 92	0.35μm 0.35μm	Precision completed Design complete
China	SHENGUANG SG-II SG-III	SIOFM,Shanghai	1.6kJ/1ns 2.4kJ/1ns 20kJ/1ns	2 8 32	1.06μm 0.35μm 0.35μm	 Under construction Plan
Russia	DELFIN PROGRESS	Lebedev Inst. Vavilov Inst.	3kJ/1ns 1kJ/1ns	108 6	1.06μm 1.06μm	
Italy	ABC	ENEA, Frascati	0.2kJ/2ns	2	0.53μm	
France	PHEBUS Mega-J LULI	CEL-V Ecole Poly.Tech.	14kJ/2.5ns 1.8kJ/15ns 0.5kJ/0.6ns	2 288 6	0.53μm 0.35μm 1.05μm	 Design underway 2, 3, 4 ω
UK	VULCAN HELEN	RAL, Chilton AWRE	3kJ/1ns 1.3kJ/0.2ns	6 3	0.53μm 0.53μm	
KrF lasers						
USA	NIKE MERCURY	NRL LANL	5kJ/4ns 1kJ/5ns	56 1	0.25μm 0.25μm	Under construction
Japan	ASHURA Super-ASHURA	ETL, Tukuba	0.7kJ/15ns 8kJ/22ns	6 12	0.25μm 0.25μm	22J/10ps/0.25μm Under construction
UK	SPRITE TITANIA	RAL, Chilton	90J/60ns 0.36kJ/0.5ns	6 1	0.25μm 0.268μm	 2.5J/0.15ps/0.25μm
Iodine lasers						
Russia	ISKRA-5	VNIIEP,Arzamas	30kJ/0.25ns	12	1.3μm	
FRG	ASTERIX	MPQ,Garching	0.66kJ/0.4ns	1	0.44μm	

high mode irradiation nonuniformity, and the excess driver energy to overcome the gain reduction due to various 2D effects. Fig. 2 (a) and (b) show the normalized gain reduction around ignition and high gain region respectively. For more quantitative and credible prediction on the sensitivity, various physics models of instability of different mode at difference phase of implosion are investigated experimentally and theoretically, the results of which are introduced into simulation code to reinforce it.

For the ignition demonstration (ID), the first five of the above listed specifications are critical. As shown in Fig. 1, laser energy of 0.1~1MJ is required for the ignition experiment. It should be noticed that the energy depend strongly on the implosion mode, that is defined by intensity and pulse shape for a optimized target, spatial uniformity of the irradiation intensity on the fuel pellet as shown in Fig.2. There are a lot of design proposals for the implosion scheme such as isobaric, isochoric, volume ignition and fast igniter concept. Therefore, the driver for next generation should have the versatile flexibility in pulse shape and spatial uniformity control.

For the high gain demonstration (HGD), sufficient driver efficiency is needed such that the product of efficiency and target gain is greater than about 10..High pulse rate operation is required for the engineering demonstration. Cost and reliability are important for all the demonstrations but strictly critical for commercial demonstration.

it is worth emphasizing that while the solid state laser has demonstrated all the characteristics necessary for ID stage, no driver has demonstrated the characteristics necessary for the HGD and power plant. Since the ID stage is rapidly being approached, it is, therefore, essential that a vigorous driver development program be pursued in parallel with the construction of the ignition facilities, at a level well above the existing one.

3. High power laser facility

3.1. Present and plan of high power laser

The exsisting and planned laser systems are summarized in the Table 2. Solid state lasers are mostly used in the fusion and related experiments. The LLE, Rochester is now installing the new facility, OMEGA Upgrade which can deliver 30kJ output at 0.35μm wavelength in 60 beams. The final level of nonuniformity is expected to be less than 1%. The beamline project as a prototype for the NIF(National Ignition Facility) system having minimumly 192 beamlets has been tested at LLNL, which can deliver 6kJ output from a segment amplifier of 30×30cm aperture. Shenguang upgrade has been also constructing at SIOFM, Shanghai which is 8 beam system of an output energy of 2.4kJ/0.35μm.

Excimer laser has some advantages as the reactor driver, such as short wavelength, high repetition rate, higher efficiency, lower cost and its ideal uniformity of irradiation. NIKE system, NRL, U.S.A. has a primary characteristic of lower beam nonuniformity less than 0.4% which will be performed soon. ASHURA is now under construction for a first multibeam implosion system at 0.25 μm at ETL, Japan, which has 12 beams in 61 cm aperture final amplifier for the output of 8kJ/22ns. TITANIA is an upgrade system of SPRITE which will be used in highfield, short pulse interaction physics experiments at RAL, UK.

Iodine lasers of 1.3μm wavelength with the frequency conversion optics are also used at MPQ, Garching, and Arzamas, Russia.

3.2 Advanced staging and system design of high power glass laser

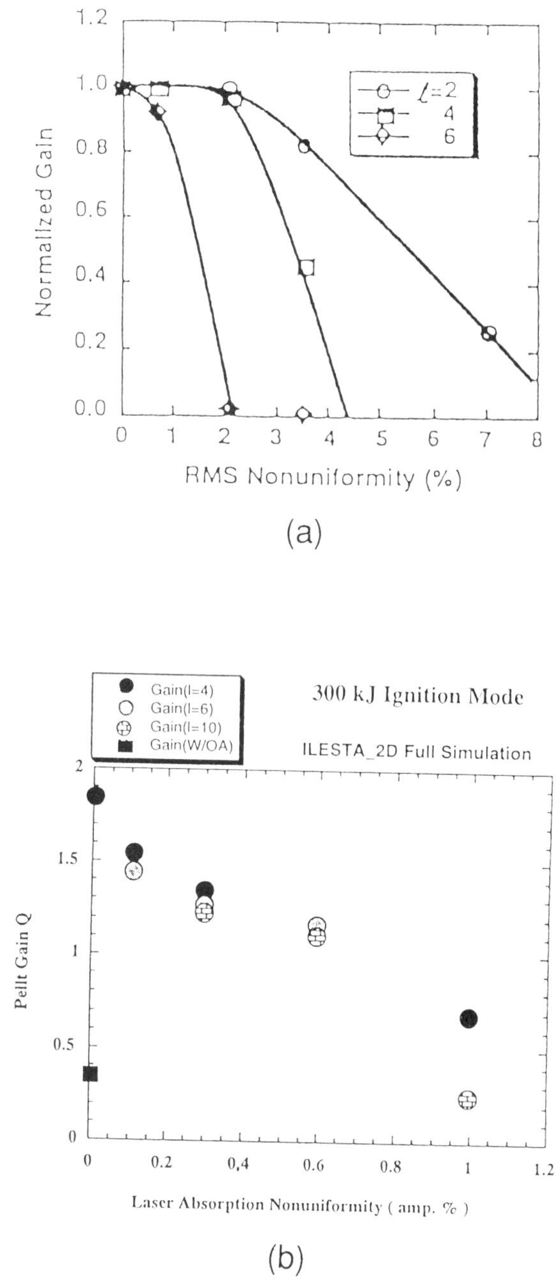

Fig. 2 Normalized pellet gain as a function of the laser irradiation nonuniformity with spherical mode number ℓ as a parameter at high gain region (a) and at ignition (b).

The improvement of energy extraction efficiency of glass laser system is crucial item in designing the next generation facility for ignition and burn experiments in near future. The architecture of most matured glass laser system uses MOPA(master oscillator and power amplifier) staging with multibeam arrangement. Regenerative amplification is very promissing to get higher efficiency of energy extraction in which smaller number of amplifiers and optical elements are used at high fluence laser beam. Cost reduction is also expected by using components of same specification in multibeam segmented amplifier lining.

LLNL has developed large aperture Pockels cell, mensioned below, as a fast optical switch by which the laser energy is extracted from the regenerative amplifier cavity with a spatial filter used as the incident injecter of a seed pulse. The experimental prototype(Fig.3.(a)) in so called Beamline Project, has been succeeded to realize an efficient amplification up to $6kJ/30\times30cm^2$ by an arrayed square amplifier with good beam pattern which will be adapted at NIF installation. CEL-V, France is designing new scheme multipass amplification by an arrayed amplifier without using a large aperture Pockels cell and polarizer, which is called L-turning arrangement shown in Fig. 3. (b).

ILE, Osaka is designing a Cassegrainian type three pass amplifier system for GEKKO upgrade, 600kJ laser. The injection spatial filter shown in Fig. 3. (c) can relay a beam image to output point of a booster amplifier at third pass which can avoid vacuum component in a multipass cavity.

3.3 New optical technologies

The system cost raises to \$300M to \$1000M with the increasing output energy which is necessary for ignition and burn experiment. The advanced technology is required to reduce the cost by an efficient performance of laser and a mass production of low cost material and components with easy finishing process.

i) new laser material

Phosphate glass is a most widely used material for a high power glass laser which has a low nonlinear optical index. LHG80 is a new material (similar as LG760) having an emission cross-section of 20% larger than LHG8 which is adopted at the precision-GEKKO at Osaka.

ii) new optical material

Sol-gel method is very useful and still being improved as an antireflection surface and also high reflectance film, which was used routinely at CEL-V, France and LLNL. This technology can be applied for large size optic materials including a fragile and deliquescent crystal such as KDP.

iii) nonlinear optical crystal

KDP(potasium dihydrate phosphate) is only crystal for a frequency converter of a large size beam of several tens centimeter in diameter. The growth technology of a large size crystal of 40cm class has been established at ILE, Osaka. Recently, a rapid growth rate over 20 mm a day was demonstated for KDP and KD*P crystals at Moskow university. This method gave also high damage threshold of the crystal which was tested at LLNL and ILE.

Partially deuterated KDP has a frequency conversion responsibility for a wide band laser light which was also tested at LLNL and ILE, Osaka. A 15.7% deuterated KDP, for example, gives 2 nm band width, in contrast with 0.6 nm for an ordinary KDP at Type-I arrangement.

The polishing technology of KDP crystal is established in Japan to decrease scattering loss for window-free frequency converter, which will be applied to be installed at Omega upgrade at LLE, Rochester.

iv) light weight mirror

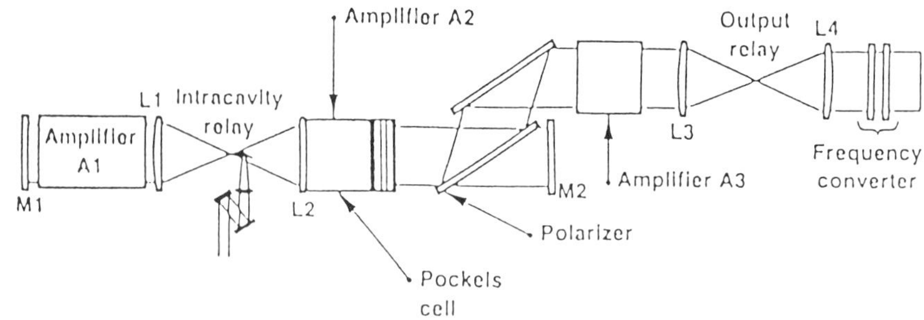

(a)

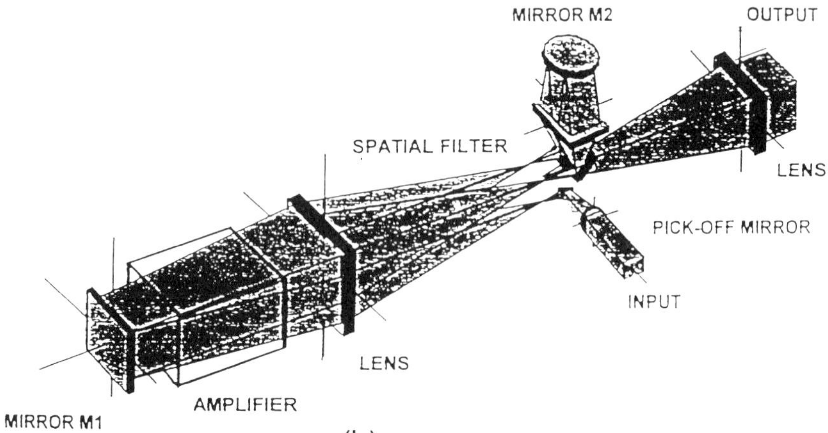

(b)

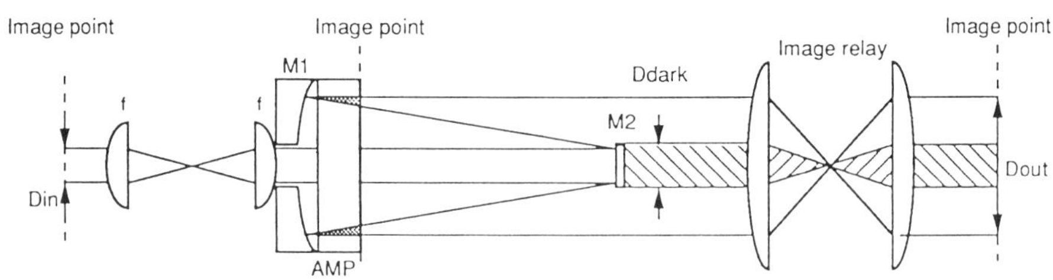

(c)

Fig.3 System arrangments of next generation booster amplifier at a) Beamline prototype in LLNL, b) L-turning scheme in Mega-J laser CEL-V c) GEKKO upgrade in ILE.

Porous fused quartz plate has been developed for a light mirror which has 30% weight of usual fused quartz. The size of such mirror is scaled up to 600 mm diameter.

v) mold optics

CEL-V, France has developed a technology of fabricating a large size optics by molding method which can form any designed surface such as flat, spherical and aspherical ones with using a shaped mother plate. The technology is promissing to decrease the cost of shaped optics of large apperture by mass-production

vi) UV-coating

The damage threshold of UV-coating has been increased up to 7-8 J/cm^2 at 1 ns pulse width for 0.35-0.25μm. The coating material with a low absorption coefficient at working wavelength was selected, and the precise coating thickness control enabled larger number multi-layer coating for higher performance.

vii) new optical devices

LLNL completed development of 27 cm square, large aperture Pockels cell as a beam switch at multipass regenerative amplifier system for NIF driver.

LLE proposed new phase plates, so called Distributed Phase Plate(DPP), of which the principle is diffraction by grating in place of usual binary phased Random Phase Plate(RPP). The energy efficiency of 15% up over the RPP is expected.

LLE also developed liquid crystal optical elements such as polarizer, soft aperture and isolator.

viii) adaptive optics

The adoptive optics is very promising to compensate the phase aberration generated by optical elements through a laser system. The NIF driver will use the adoptive optics which has been tested in the prototype beamline for NIF system at LLNL. SIOFM, China used the adoptive mirror having 19 actuators in the Shenguang laser to compensate the wave front aberration due to a long rod laser glass. At ILE, Osaka, the adoptive mirror of 200-360 mm diameter have been tested for wave front control in a line focus optics.

ix)high power laser diode

Arrayed laser diode is greatly developed as a pump sourse for solid state laser resulting in higher pump efficiency and lower thermal loading. The power density of emission energy is 3kW/cm^2 with stacked array. The final problem is only its higher cost which could be overcome by mass production with increasing the application user's fields.

4. Improvement of irradiation uniformity

In the direct drive ICF research, the several intensive works are being carried out in the world for minimizing the irradiation nonuniformity. The laser irradiation uniformity depends on the beam intensity distribution, their geometrical arrangement on the target surface (single beam factor and geometrical factor defined in reference [3]) and the beam-to-beam power balance. It is recognized that the rms nonuniformity ($\sigma_{rms}= (\Sigma\sigma_\ell{}^2)^{1/2}$, where σ_ℓ is the nonuniformity of the ℓ-th spherical harmonic mode) required for the fuel ignition would be a level of 1% or less.

4.1 Required beam number and envelope profile control of focused beam pattern

The beam number required to achieve the above mentioned nonuniformity is quoted to be 60 or more because the larger beam number relaxes the profile control of the intensity envelope of focused beam pattern, the power balance and the uniformity degradation due to the shrinkage of the critical surface during implosion.

To reduce the lower mode nonuniformity in the existing laser systems with relatively small beam number, the envelope profile control is quite essential. Whereas, an elegant solution has been found to suppress the lower mode nonuniformity. The single beam factors at lower l-th mode can be greatly reduced as the intensity distribution on the target surface is closer to cos2q or cos4q, where q is an angle measured from the beam axis [4]. The envelope profile control has been demonstrated by optimizing both of the element size of a random phase plate (RPP, developed at ILE, Japan [5]) and the focusing condition [6]. The intensity envelope on the target plane leaves from a sharp distribution (square of sinc function) and becomes flatter when the beam is defocused from the far-field position. On the other hand, the multi-lens array (MLA, proposed by SIOFM, China) [7, 8] is useful not only to control the envelope shape without significant energy loss but also to reduce the long wavelength nonuniformity that is rather unavoidable for RPP's because of the correlation in the distribution of phase shifter. Furthermore the design of kinoform phase plate (KPP, proposed by LLNL, USA) has been recently advanced [9], and one can easily obtain an arbitrary intensity distribution on the target plane if the fabrication technology of KPP is successfully developed.

4.2 Beam smoothing

The beam smoothing is based on the techniques to divide a nonuniform beam into a huge number of beamlets. The RPP's are being widely used because of their optical simplicity and easiness in the fabrication. However the random interference of coherent multi beamlets results in the fully developed speckle pattern that increases the higher mode nonuniformity (almost white spectrum) in spite of the substantial reduction of lower and medium mode nonuniformities (see Fig. 4 (a)). Thus the problem results in the study of speckle suppression. For this purpose the several concepts have been proposed and some of them are being supplied in the experiments on the fuel implosion and the hydrodynamic stability. The techniques reviewed in this section are ISI (induced spatial incoherence, NRL, USA) [10], echelon-free ISI (NRL) [11], PCL (partially coherent light, ILE) [12] and SSD (smoothing by spectral dispersion, LLE, USA) [13].

ISI is the first proposed methods for the speckle reduction. The spatial incoherence of a broad band laser is induced by an echelon pair installed just before the focusing optics that gives each beamlet different time delay larger than the coherence time. Then the beamlets are superposed at the far-field position creating a smooth pattern as a result of the time integration. To apply this method to the multi-beam high power laser system it needs large echelons.

The concept of the echelon-free ISI or PCL is similar to that of ISI, but the spatial incoherence is introduced at a laser oscillator itself using the amplified spontaneous emission (ASE [14, 15]) or at a front-end system using an optical fiber [16, 17]. In the echelon-free ISI scheme being adopted in Nike KrF laser system in NRL, a uniform far-field pattern with an appropriate profile shaped beforehand at the front-end is relayed faithfully onto the target plane. Nike laser has demonstrated an ultra-uniform focused beam pattern of which intensity variation approaches 1% level in 4 ns. This uniformity was taken with one beam, thus the experiments on hydrodynamic stability being planed at NRL would be done with a further uniform illumination improved by superposition of multibeams. While, the PCL has been developed for the existing glass laser systems in which the tolerance of beam divergence is not enough large for the beam smoothing. Therefore the angular dispersion of spectrum is introduced to smooth out the medium mode nonuniformity caused by the residual spatial coherence. The PCL in Gekko XII laser has a 0.6-nm band width and a 32-times diffraction-limited divergence (both measured at the fundamental, w). The angular dispersion is 239 μrad/nm at the KDP crystal that is adjusted to the index matching condition along the

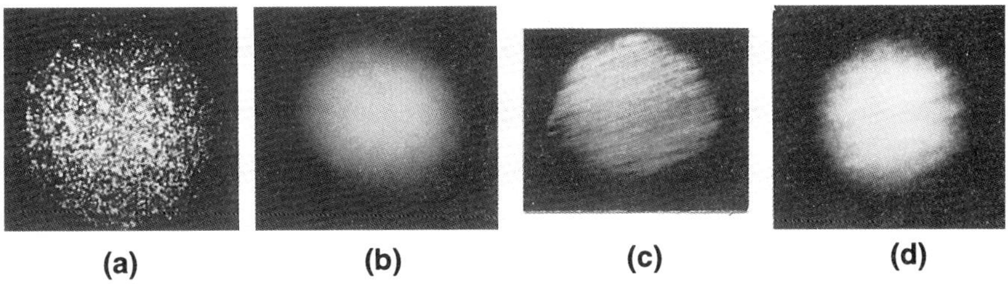

Fig. 4 Focused beam patterns measured with different smoothing techniques in GEKKO XII. (a): coherent laser , (b): PCL with the angular dispersion of spectrum, (c): 1-D SSD, (d): 2-DSSD. In all cases RPP's were used.

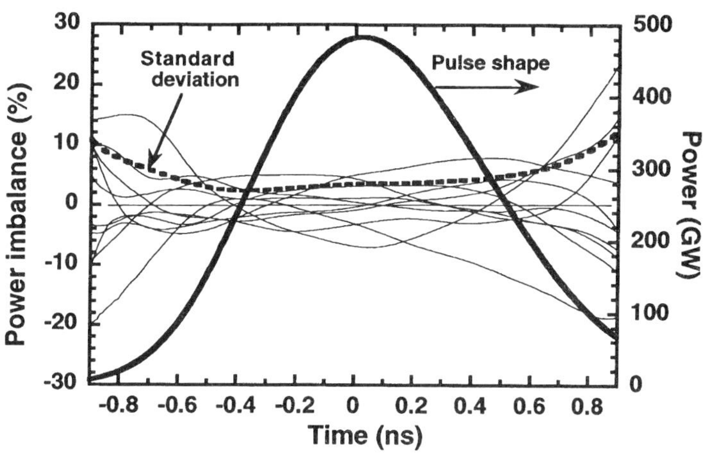

Fig. 5 Power balance control demonstrated with Precision GEKKO XII.

extraordinary axis of type II KDP. Because the maximum efficiency of the second harmonic conversion was about 50%, the advances in the growing technology of new nonlinear crystal such as KTP and LBO are necessary. The smoothing time of the focused beam pattern by which the standard deviation of time averaged intensity distribution reached 10% was measured to be ~40 ps, and the intensity fluctuation of time integrated pattern was as low as 3% for a ~2-ns pulse (see Fig. 4 (b)). The irradiation nonuniformity was estimated for a spherical target using measured pattern. The mode structure shows the substantial reduction by a factor of 40 (σ_ℓ =0.05%) at higher mode ($l \geq 40$) comparing with that for the coherent laser with RPP. The rms nonuniformity (σ_{rms}) was 3.8%.

The laser beam being used in SSD scheme has the temporal and spatial color cycles maintaining the high efficiency of harmonic conversion, and the effect of SSD on the implosion performance has been studied by changing the band width. The focused beam pattern of the usual SSD shows a streaked feature of RPP speckle because the smoothing occurs one-dimensionally along the spectral dispersion (see Fig. 4 (c)). Therefore the higher mode nonuniformity is larger than that has been achieved with PCL by a factor of two or more, and the reduction of medium mode nonuniformity is not significant in comparison with PCL when the same band width and the same angular dispersion are introduced. A few approaches for two-dimensional (2-D) SSD have been proposed, but they are still under the development. Figure 4 (d) shows an example of 2-D SSD that was demonstrated with a spatial color cycle of which direction was perpendicular to that of angular dispersion [18].

4.3 Pulse shaping

The pulse shaping is important for controlling the entropy of imploding shell and for stabilizing implosion. Advanced pulse shaping systems have been prepared for the Beamlet laser in LLNL (rack-mounted integrated optics), Omega up-grade (a foot pulse and a main pulse will be generated separately) and Precision Gekko XII (optical fiber couplers for PCL). As well as the Gaussian pulse, the square pulses with a square foot pulse (LLNL) and a separated short prepulse (ILE) are being supplied routinely respectively for the medium isentrope implosion and the high isentrope implosion.

4.4 Power balance control

The power imbalance appears to be another significant source of lower mode nonuniformity as the beam uniformity has been substantially improved. However, the universal method and the generalized algorithm are still unclear to date because of the individuality of laser systems, although the principal concepts being adopted at ILE, LLNL and LLE are quite similar.

The procedure of power balance control for Gekko XII is as follows[18, 19]: the similarity of ω pulse shape that can be achieved by the amplifier gain control; the power density balance on the KDP crystal that is adjusted by a variable linear loss; the similarity of pulse shape after the second harmonic conversion; and finally the power balance of 2ω pulse on the target plane. An initial test result is summarized in Fig. 5 that shows a 2ω power balance ranging from 3-5% rms (at around pulse peak) to 10% rms (at one hundredth of peak intensity).

On the Precision Nova, the power gain from the small signal region through the saturated amplification region is controlled by adjusting both of the splitting ratio (input to main amplifier chain) and the gain of amplifier being installed in the lower reaches (discharge timing of flash lamps). The achieved power imbalance for a shaped (two-step) pulse with a contrast ratio of 3-4 was 10% rms and 5% rms at foot pulse and main pulse, respectively [20].

5. Reactor driver

For the high gain demonstration (HGD), sufficient driver efficiency is needed such that the product of efficiency and target gain is greater than 10. For the engineering demonstration (ED), high pulse rate operation is required. For the commercial reactor, the driver cost, and reliability or availability are essential requirements to be competitive to other energy sources. Recent design studies of commercial laser fusion power plant have considered and examined the feasibility of KrF gas laser (Sombrero, Prometheus-L, Sirius-PB), Heavy ion beam accelerator (Hylife-II, Osirus, Prometheus-H), Light ion beam (Libra), and LD pump solid state laser (KOYO). Since the ID stage is rapidly being approached, it is essential that a vigorous driver development program be pursued, in parallel with the construction of ignition facilities, at the level above the existing one.

6. Summary

Fusion ignition and burn is well in scope physically and technically by constructing advanced glass laser of 300kJ~MJ output energy. Future R&D toward reactor driver which could satisfy high efficiency (10%), high repetitive operation (~10 Hz) with low cost for commercialization seems to be feasible.

Because of the separability of power plant systems, it is possible to develop the various subsystems and drivers needed in different countries. If an effective international collaboration on IFE were developed, then its development could proceed efficiently, and all countries would derive benefit from it. It should be noticed that advance drivers for ID, HGD, ED and CD, and pellet implosion, which produce ultra-high density of matter and intense neutron source, it self is opening and could open new field of science and also industrial technologies.

Acknowledge

This paper has been written with the collaboration of world major ICF laboratories in collecting newest data on drivers. We would like to acknowrage the kind collaborations of Dr.s D. Banner of IAEA, R. McCrory and J. Soures of LLE, University of Rochester, S. Bodner of NRL, M. Campbell and W. Hogan of LLNL, Y. Owadano of ELT in Tsukuba, X. Deng of NLHPLP in Shanghai, N. Basov of LPI Moscow, R. Sigel of MPQ, Garching, A. Caruso of ENEA, J. Coutant of CEL-V, Pesme of Ecol Poly.Tech., M. Key and C. B. Edwards of RAL.

References

[1] Energy from Inertial Fusion, IAEA Review Working, to be published in 1995.

[2] Nakai S et al. 1994 IAEA 15th International Conference on Plasma Physics and Controled Thermonuclear Fusion Research CN-60/B-I-1.

[3] Skupsky S and Lee K 1983 J. Appl. Phys. 54 3662-3671.

[4] Murakami M et al. 1993 J. Appl. Phys., **74** 802-808.

[5] Kato Y et al. 1984 Phys. Rev. Lett. **53** 1057-1060.

[6] Miyanaga N et al. 1992 Laser Interaction and Related Plasma Phenomena (Plenum Press, New York) Vol.**10** 251-280.

[7] Deng X et al. 1986 Appl. Opt. **25** 377-381.

[8] Nishi N et al. 1993 SPIE Vol. **1870** 105-111.

[9] Dixit S N et al. 1994 Opt. Lett. **19** 417-419.

[10] Lehmberg R H and Obenschain S P 1983 Opt. Commun. **46** 27-31.

[11] Lehmberg R H et al. 1987 J. Appl. Phys. **62** 2680-2701; 1987 Fusion Technol. **11** 532-541.

[12] Nakano H et al. 1993 J. Appl. Phys. **73** 2122-2131.

[13] Skupsky S et al. 1989 J. Appl. Phys. **66** 3456-3462.

[14] Nakano H et al. 1990 Opt. Commun. **78** 123-127.

[15] Lehmberg R H et al. 1993 SPIE **1870** 163-174.

[16] Véon D et al. 1988 Opt. Commun. **65** 42-46.

[17] Nakano H et al. 1993 Appl. Phys. Lett. **63** 580-582.

[18] Miyanaga N et al.1994 IAEA-CN-60/B-P-11.

[19] Nishihara K et al.1994 Phys. Plasmas **1** 1653-1661.

[20] Caird J.A. et al. 1993 ICF Quarterly Report, Lawerence Livermore Nationa Laboratory, Vol. **4**, No.1 p.10.

Inst. Phys. Conf. Ser. No 140: Section 4
Paper presented at Laser Interact. with Matter Conf., Oxford, 19–23 September 1994

© 1995 US Government

National Ignition Facility Design, Performance, and Cost

W. J. Hogan, J. A. Paisner, W. H. Lowdermilk

Lawrence Livermore National Laboratory, P.O. Box 808, Livermore, CA 94550

M. S. Sorem

Los Alamos National Laboratory, P.O. Box 1663, Los Alamos, NM 87185

J. D. Boyes

Sandia National Laboratory, P.O. Box 5800, Albuquerque, NM 87185

S. A. Kumpin

University of Rochester, 250 East River Road, Rochester, NY 14623

Abstract. A conceptual design for the National Ignition Facility (NIF) has been completed and its cost has been estimated by an multilaboratory team. To maximize the performance/cost ratio a compact, segmented amplifier is used in a multipass architecture. Many recent optical and laser technology developments have been incorporated into the final design. The Beamlet project has successfully demonstrated the new concept. The mission of ICF Program using the NIF is to achieve ignition and gain in the laboratory. The facility will be used for defense applications such as weapons physics and weapons effects experiments, and for civilian applications such as inertial fusion energy development and fundamental studies of matter at high energy density. This work was performed under the auspices of the U.S. Department of Energy by Lawrence Livermore National Laboratory under contract W-7405-Eng-48.

1. Introduction

In accordance with the 1990 recommendations of the National Academy of Sciences[1], the U.S. Secretary of Energy, in January, 1993, requested a conceptual design of a NIF which uses a neodymium glass laser operating at the third harmonic (0.35 μm). A Memorandum of Agreement established a project organization including technical teams from the Lawrence Livermore National Laboratory, Los Alamos National Laboratory, Sandia National Laboratory, and the Laboratory for Laser Energetics at the University of Rochester. A 7000 page conceptual design report[2] (CDR) was completed in April, 1994.

2. NIF design criteria

The baseline ignition target, shown in Figure 1, requires a laser system that delivers a 1.8 MJ (at 0.35 μm) shaped pulse with a 50:1 contrast ratio. Table 1 includes a subset of the many primary criteria and functional requirements for the facility. The study assumed that the NIF would be located at a generic DOE site. Symmetrical implosion of the capsule requires two-sided target irradiation with two cones per side; each having an

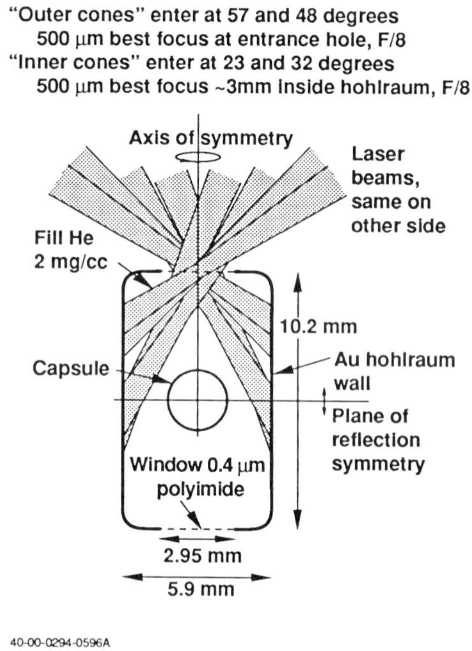

Fig. 1. Generic NIF ignition target.

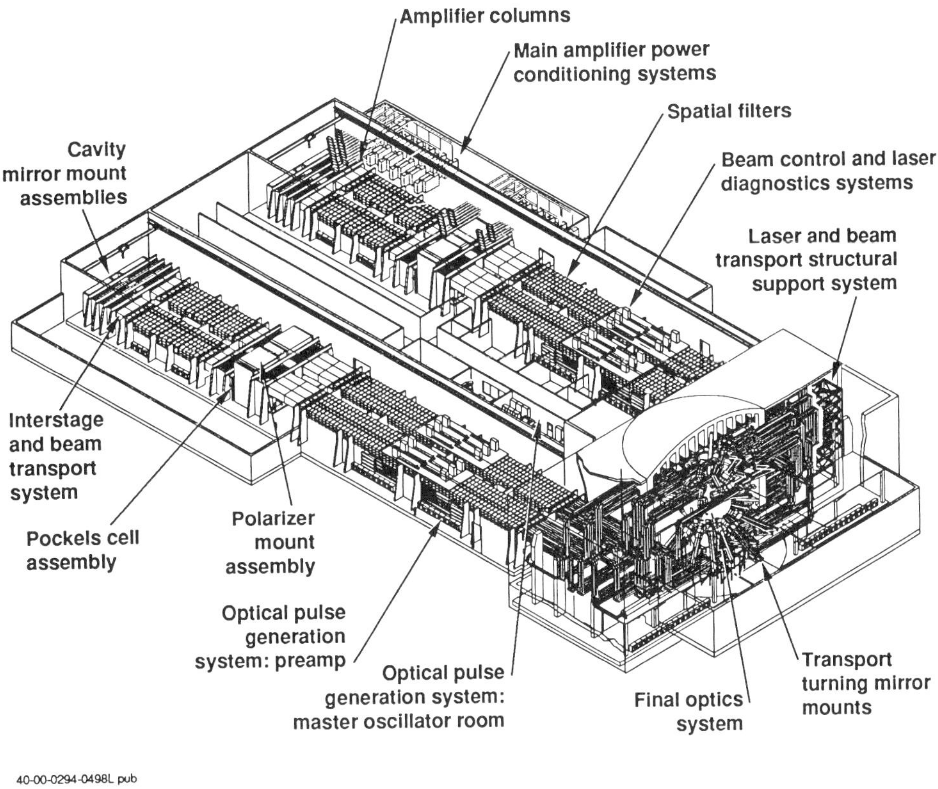

Fig. 2. The National Ignition Facility overview. The NIF will be a low hazard, non nuclear facility.

Table 1. Primary criteria for the National Ignition Facility.

Laser pulse energy	1.8MJ
Laser pulse peak power	500 TW
Laser pulse wavelength	0.35 μm
Beamlet power balance	<8% rms over 2 ns
Beamlet pointing accuracy	<50 μm
ICF target compatibility	Cryogenic and non-cryogenic
Annual number of shots with fusion yield	100 with yield 1 kJ—100 kJ 35 with yield 100 kJ—5 MJ 10 with yield 5 MJ—20 MJ
Maximum credible DT fusion yeild	45 MJ
Classification level of experiments	Classified and unclassified

outer cone to inner cone laser power ratio of two to one, and at least eight-fold azimuthal irradiation symmetry. The cone angles, nominally at 53° (outer) and 27° (inner), and laser power ratio are chosen to maintain time-dependent symmetry of the x-ray drive seen by the imploding capsule. To avoid laser-plasma instabilities, such as filamentation and stimulated scattering, the baseline indirect-drive target hohlraum requires laser spatial beam smoothing using phase plates, and laser temporal smoothing using a combination of four beams at different center wavelengths, each separated by 3.3 Å (3.3 x 10^{-4} μm). This separation was set by the requirement on the motion of the kinoform-induced speckle pattern at the target focus. As a consequence of these design rules, the laser system must deliver at least 192 beams to the target chamber. The overall design of the NIF in the indirect drive configuration is shown in Figure 2. The U-shaped building configuration shown in Figure 2 provides for the future addition of a second large target area to accommodate special requirements of other communities, such as for weapons effect tests, with minimal interruptions to system operations in the main target area.

3. Laser system design

The neodymium glass laser system must provide at least 632 TW/3.3 MJ in a 5.1-ns pulse at 1.053 μm to account for modest beam transport losses, and energy and power conversion efficiencies of 60% and 85%, respectively. A schematic of one NIF beamline is shown in Figure 3. It uses a four-pass architecture with a large aperture optical switch consisting of a plasma electrode Pockels cell and polarizer combination. The full aperture Pockels cell switch is an enabling technology for this laser design. The laser chain used in the CDR as the baseline for estimating NIF system cost and performance has a hard aperture of 40 cm x 40 cm. The amplifiers contain 19 neodymium-doped glass laser slabs arranged in a 9-5-5 configuration as shown in Figure 3. Each Brewster-angled slab is 3.4-cm thick. At the design point each laser chain should generate 3.9 TW/20.5 kJ of 1.053 μm light. Because only 162 beams are required to achieve the energy and power requirements, the 192-beam system has a laser design margin of greater than 15%. Since many target designers believe that targets may be ignited with from 0.9 to 1.3 MJ of laser energy, the overall design margin for achieving ignition is approximately a factor of two.

The NIF laser design incorporates 4-high x 12-wide arrays of laser beams as shown in Figure 2. The design is very compact compared to previous laser fusion systems, increasing overall electrical and optical efficiency while simultaneously reducing system size and cost. The optical pulse generation system provides individually controlled input pulses from one of four tunable fiber oscillators and an integrated optics network located in the master oscillator room. The outputs from the master oscillator room are delivered on single-mode polarization preserving fibers to each of 192 preamplifiers. These stand-alone packages, located beneath the transport spatial filters, provide individual power balance capability for each of the 192 beams. The output beams from the preamplifiers are injected into the far-field pinholes of the transport spatial filters, passed through the boost-amplifier stages, the optical switch assemblies, and are then captured inside the

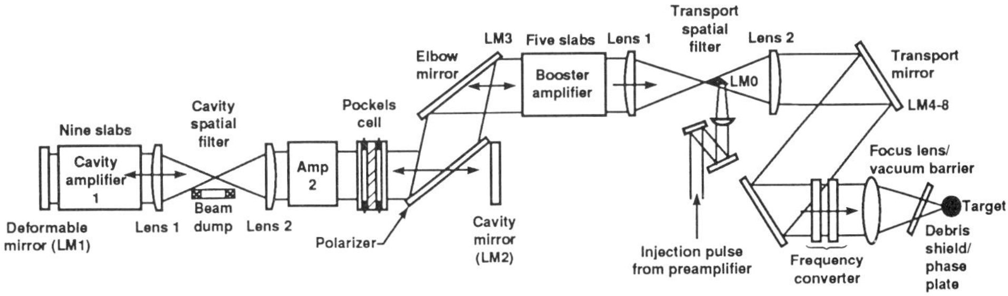

40-00-0394-0789Apb01

Fig. 3. A schematic of one beamline of the NIF laser from pulse injection to final focus on target.

11SAJ/gbh

Fig 5.6-2

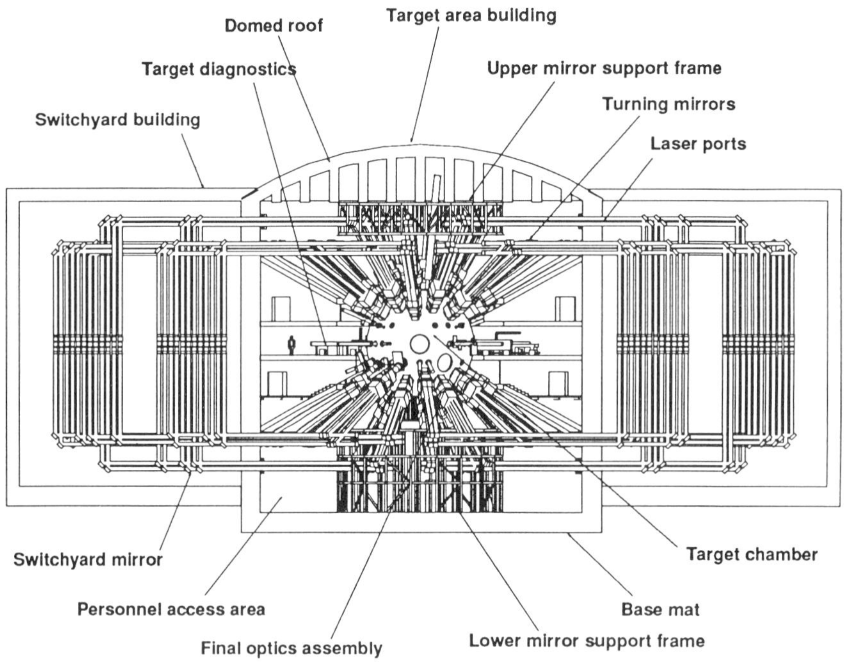

Fig. 4. A cutaway view of the NIF target area showing major subsystems.

23VPK/lwh

multipass cavities. The flashlamps located in the amplifier enclosures that uniformly pump the glass laser slabs are energized with approximately 260 MJ of electrical energy from a modular bank of thin film, metallized dielectric capacitors. After four passes through the cavity amplifiers, the pulses are switched out of the multi-pass, further amplified by the boost stage, and then transported to the target chamber. The laser arrangement allows for top and bottom access to the amplifiers and the optical switch arrays. The pulsed power is transmitted to the amplifiers overhead with large, 30-cm-diameter, coaxial conductors. The space below the amplifiers allows access for assembly and maintenance of any 4-high amplifier column.

The Beamlet Project was conceived to provide an integrated near full scale demonstration of a complete beamline. The Beamlet consists of a 2 x 2 amplifier cavity of which one beamline with an aperture of 35 cm x 35 cm is activated. Tests in April 1994 demonstrated successful multipass operation and switchout. Tests in September with a reduced aperture (30 cm x 30 cm) frequency conversion crystal system (with a 84% fill factor) produced 3ω energies up to 5.9 kJ. This resulted in a mean fluence of 8.0 J/cm^2, exceeding the NIF 3ω design fluence of 7.4 J/cm^2 and, therefore, demonstrating that the NIF will be able to meet its design requirement. In December, tests will be run with a full aperture crystal system.

The NIF is currently designed to achieve about 700 full-system performance shots/year. It is expected that through continued engineering design the system shot rate will increase substantially. This is consistent with experience on all previous ICF glass laser systems, including Nova, which has had its shot rate increase by a factor of six since operations began in 1985.

4. Target area design

A cutaway view of the switchyard and target area is shown in Figure 4. The 192 laser beams are optically relayed via the transport spatial filters in 48 2 x 2 groupings to the final optics assemblies. The (48) final optics assemblies are positioned on the 53° outer cone (16 assemblies) and 27° inner cone (8 assemblies) at the top and on the bottom of the target chamber. At the chamber each 2 x 2 grouping is converted to 0.35 μm by a Type I (KDP)/Type II (KD*P) crystal array in the final optics assembly. The final optics assemblies mount to the exterior of the chamber, and also provide 2 x 2 lens arrays for focusing the light onto the target and 2 x 2 debris shield/kinoform phase plate arrays for protecting the lenses from target shrapnel. Each beam in every 2 x 2 grouping can be operated at a different center wavelength to provide the requisite laser temporal beam smoothing. The final optics assemblies are offset from the nominal cone angles by ±4 degrees to provide isolation between opposing beamlines.

The target chamber is housed in a reinforced-concrete building with three separate operational areas. The upper and lower polar regions of the target chamber house the final optics and turning mirrors in a Class 1000 clean room. Personnel access to these areas will be limited. The cantilevered floor sections of the building provide a separation of the clean-room enclosures at the polar regions from the equatorial target diagnostics area. This horizontal, planar architecture simplifies the design of the access structures required to service the optical components and target diagnostics.

The NIF baseline target chamber is a 10-cm-thick by 10-m internal-diameter spherical aluminum shell. The aluminum wall provides the vacuum barrier and mounting surfaces for the first wall panels, which protect the aluminum from soft x-rays and shrapnel. The unconverted laser light hitting the opposite wall is absorbed by other panels. The exterior of the chamber is encased in 40 cm of concrete to provide neutron shielding. The chamber is supported vertically by a hollow concrete pedestal and horizontally by radial joints connected to the cantilevered floor. The target area building, chamber, and auxiliary systems are designed to handle a maximum credible yield of 45 MJ and routine operations as shown in Table 1.

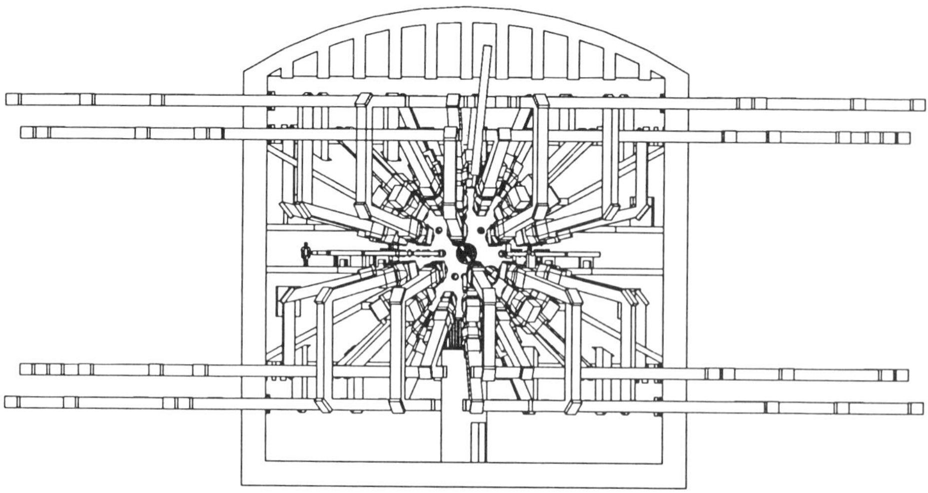

40-00-0694-2750 pub

Fig. 5. Implementation of direct drive requires that 24 of the 48 beams be re-positioned. This can be easily accomplished using the NIF optical system design.

17JAP/dsm

Table 2. Summary of NIF costs for 192 beam system.

	Base costs	Contingency	Total	Total
		($M FY94)		($M escalated)
TEC	586	121	707	842
OPC	199	N/A	199	231
TPC	785	121	906	1073
Annual operating costs	57	N/A	57	N/A

Recent engineering analyses and target physics calculations show that the baseline design can be easily modified, as illustrated in Figure 5, to incorporate a direct-drive ignition capability. To reconfigure for direct drive, 24 of the 48 final optics assemblies (along with the associated transport hardware) must be moved into new locations around the equator. This option can be preserved at minimal cost by adding 24 holes to the target chamber and the corresponding holes in the concrete structures for beam transport paths. Alternatively, 24 additional final optics assemblies and transport mirrors can be in place at all times and used when half the beams are switched into the alternate transport paths. Costs for including these options have not yet been developed and, therefore, the current cost estimate does not include them.

5. NIF project schedule and cost

The NIF Total Project Cost (TPC) is the sum of the Total Estimated Cost (TEC) and the Other Project Cost (OPC). The TEC is funded by Plant and Capital Equipment (PACE) funds and the OPC is funded by Operating Expense (OPEX) funds. Division of costs between TEC and OPC is provided in DOE guidelines. TEC activities include, for example, Title I and II design, and Title III engineering; building construction; procurement; assembly and installation of all special equipment; and sufficient spares to pass the acceptance test procedures. OPC activities include, for example, conceptual design; advanced conceptual design; NEPA documentation; vendor facilitization and pilot production; vendor component qualification/reliability/lifetime testing; operational readiness reviews; startup costs; and operational spares.

The costs shown in Table 2 were derived from a "bottom-up" estimate based on a detailed work breakdown structure. The Martin Marietta Energy Systems, Inc. Automated Estimating System (AES) was adopted by the Project as its cost estimating tool. Contingency information provided by each estimator for every Level 3 item was used in a separate probabilistic contingency analysis performed by Bechtel Corporation using their Microrac code. The results of that analysis were entered into the AES. A detailed, integrated project schedule was developed and this was combined with the cost information in the AES to estimate escalation and calculate the budget profiles required. The NIF design and construction project would last from 1996 through 2002. The annual operating costs for the facility, shown in Table 2, were estimated based on Nova experience. It does not include, per DOE guidance, the annual ICF Program costs (currently at approximately $175 M/year). The engineering design was sufficiently detailed to generate costs, typically, at level 5, and, in some cases, at level 6 or 7. of the WBS. Approximately 70% of the costs (in dollars) were derived from catalog prices, vendor estimates, or engineering drawings. The costs in Table 2 have been validated by the DOE and by an Independent Cost Estimator (ICE) team.

6. Summary

The National Ignition Facility design is the product of the efforts of a multi-laboratory team, representing a more than twenty year experience base at the Lawrence Livermore National Laboratory, Los Alamos National Laboratory, Sandia National Laboratory, and the Laboratory for Laser Energetics at the University of Rochester. Using the world's most powerful laser to ignite and burn ICF targets, the NIF will produce conditions in matter similar to those found at the center of the sun and other stars. New, well characterized, high energy-density regimes will be routinely accessible in the laboratory for the first time. The NIF will impact and extend scientific and technical fields such as controlled thermonuclear fusion, astrophysics and space science, plasma physics, hydrodynamics, atomic and radiation physics, material science, nonlinear optics, advanced coherent and incoherent x-ray sources, and computational physics. The importance and uniqueness of the NIF to these wide-ranging fields of science and technology have been recently reviewed in a series of workshops.[3] If authorized in FY1996, the NIF could begin operations in FY2003.

7. References

[1] Committee for a Second Review of DOE's ICF Program, *Final Report—Second Review of the Department of Energy's Inertial Confinement Fusion Program*, National Academy of Sciences, Commission on Physical Sciences, Mathematics, and Applications, National Research Council (National Academy Press, Washington, D.C., September 1990).

[2] *National Ignition Facility Conceptual Design Report*, Lawrence Livermore National Laboratory, Livermore, CA, UCRL-PROP-117093, (1994).

[3] *Utility of the National Ignition Facility*, Lawrence Livermore National Laboratory, Livermore, CA, NIF-LLNL-94-240, UCRL-117384, (1994).

Inst. Phys. Conf. Ser. No 140: Section 4
Paper presented at Laser Interact. with Matter Conf., Oxford, 19–23 September 1994
© *1995 IOP Publishing Ltd*

Uniformity Performance of Laser Irradiation on Gekko XII

N. Miyanaga, M. Nakatsuka, T. Kanabe, T. Jitsuno, K. Tsubakimoto*, N. Nishi, S. Matsuoka, A. Ando, S. Amano, S. Urushihara, N. Morio, T. Kawasaki, K. Suzuki, S. Matsuo, H. Asahara, K. Kawakami, M. Murakami, H. Azechi, K. Nishihara, and S. Nakai

Institute of Laser Engineering, Osaka University,
2-6 Yamada-oka, Suita, Osaka 565, Japan

*Institute for Laser Technology
2-6 Yamada-oka, Suita, Osaka 565, Japan

Abstract The laser irradiation uniformity that has been performed by Precision Gekko XII is discussed. This project includes the following three key technologies. An optimum design of a random phase plate enables us to minimize the lower-mode nonuniformity. The beam smoothing techniques such as the partially coherent light with angular dispersion of spectrum and the two-dimensional SSD can substantially reduce the higher-mode nonuniformity. Thus, it is expected that the absorption nonuniformity of 1-2% will be achieved because of the improved power balance. Furthermore the pulse shaping of partially coherent light is also reported.

1. Introduction

Uniform laser irradiation is one of critical issues in the laser fusion research. For this purpose, Precision GEKKO XII has been completed for the technology development such as the beam focusing control for an optimum intensity envelope on the target, the beam smoothing, the pulse shape control, and the power balance control among beams. The envelope profile control is essential to reduce lower mode nonuniformity (σ_ℓ, ℓ=6, 12 dominantly, where ℓ is the spherical harmonic mode number) which originates in the decahedron arrangement of twelve beams. While the beam smoothing is needed to suppress medium and higher mode ($20 \le \ell < 40$ and $\ell \ge 40$, respectively) nonuniformities that are caused by the random interference (speckle) of multi beamlets being divided by a random phase plate, RPP [1]. Substantial developments of the envelope profile control and the beam smoothing enable us to achieve an irradiation nonuniformity ($\sigma_{rms} = (\Sigma\ \sigma_\ell^{\ 2})^{1/2}$) of 1% level. Then the power imbalance also should be improved to the same level to avoid the increase of the lower mode nonuniformity.

2. Envelope profile control of focused beam pattern and beam smoothing

The lower mode nonuniformity can be minimized by controlling the intensity envelope on the target surface, $I(\theta)$, where θ is an angle measured from the beam axis [2]. The

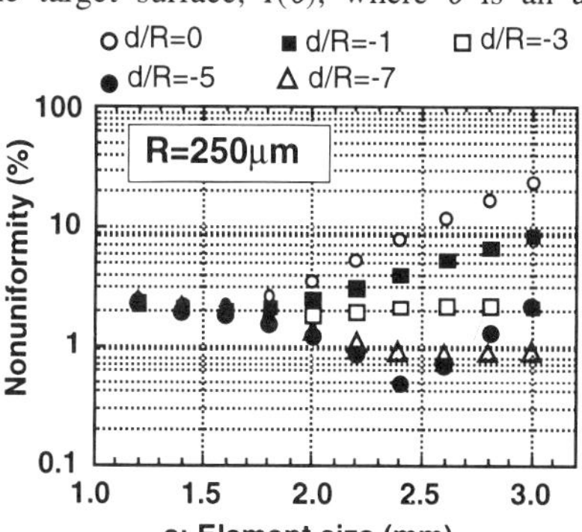

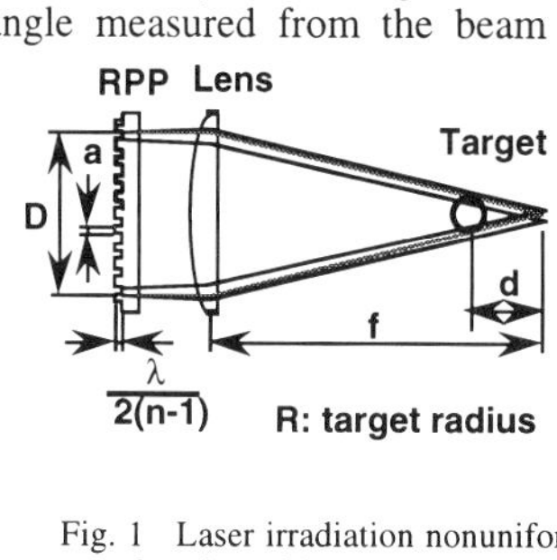

Fig. 1 Laser irradiation nonuniformity (σ_{rms}) as a function of hexagonal RPP element size calculated for the incoherent laser with different focusing conditions. The element size is defined by the diameter of circumscribed circle.

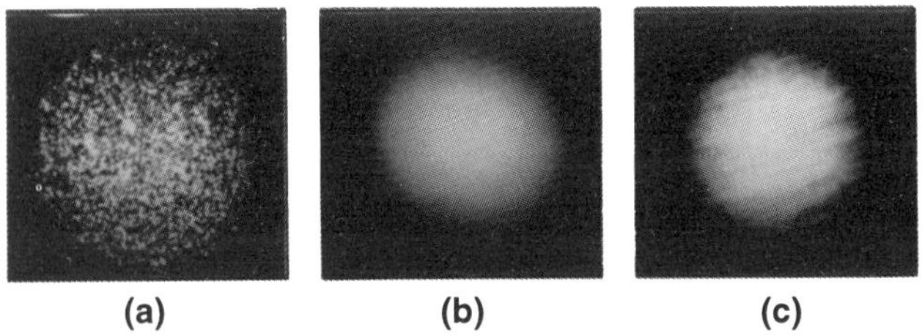

Fig. 2 Focused beam patterns measured at a tangential plane of spherical target of 500-μm diameter. (a): coherent laser with an old RPP of 2-mm square element, (b): angularly dispersed PCL with the old RPP, (c): SSD with a new RPP of 1.85-mm hexagonal element.

envelope profile can be controlled by optimizing the RPP's [3] or the aspherical multi-lens array [4]. Figure 1 shows the rms irradiation nonuniformities calculated for the incoherent laser with RPP, where *d* and *R* are the focal position and the target radius, respectively. There are two candidates. One of them is an easier way to use the far-field illumination with an RPP of smaller (1.6-1.8 mm) element. The other is the quasi far-field illumination with an RPP of relatively large (2.4 mm) element, where $I(\theta)$ almost obeys $\cos^2\theta$ distribution that is greatly effective to reduce the single beam factor [5] of irradiation nonuniformity. The present choice is 1.85-mm element because the latter case requires a quite uniform near-field pattern. The error of RPP coating thickness was clarified by means of Zernike interference technique to be 1-2% for 0.527-μm laser (2ω).

RPP's divide a nonuniform beam into a huge number of small beamlets, then these beamlets randomly interfere with each other creating a fully developed, fixed speckle pattern with relatively uniform intensity envelope (see Fig. 2(a)). To suppress the small scale imprint due to this speckle, the partially coherent light (PCL, a concept similar to echelon-free ISI [6]) has been developed for the high-power Nd:glass laser as shown in Fig. 2(b) [7]. The PCL has the 0.6-nm band width, the 32 times diffraction-limited beam divergence (both measured at the fundamental, ω) and the angular dispersion of spectrum of 239 μrad/nm (at the inlet of KDP frequency doubling crystal). The smoothing time in which the single beam non uniformity reached 10% was measured to be ~40 ps. Therefore, the PCL is suitable for a foot pulse, although it degrades the efficiency of harmonic conversion (40-50%). While the smoothing by spectral dispersion (SSD) [8] having 0.6 nm band width (at ω) with 10 GHz frequency modulation is promising for a main pulse. Moreover a two-dimensional (2-D) SSD (Fig. 2(c)) has been demonstrated to reduce medium mode non-uniformities caused by a streaked feature along the dispersion direction. Here we adopted the spatial color cycle of which direction was perpendicular to that of angular dispersion of spectrum.

The laser intensities on the target surface were estimated from these patterns by assuming the equivalent twelve beams. In the case of coherent laser with RPP the mode structure of nonuniformity showed a nearly white spectrum of ~2% per mode. By the introduction of PCL the substantial reduction of intensity perturbation was observed at higher modes ($\sigma_\ell \approx 0.05\%$ for $\ell > 40$). For 2-D SSD (Fig. 2(c)) the medium and higher mode nonuniformities were somewhat larger than those of PCL, but the σ_{rms} with the thermal smoothing was almost the same as that of PCL. The estimated nonuniformities (σ_{rms}) were 19.7% (Fig. 2(a)), 3.8% (Fig. 2(b)) and 4.7% (Fig. 2(c)), respectively.

3. Pulse shaping for partially coherent light

In the experimental study of implosion stability, it is required to control the laser pulse shape flexibly, because one can quantitatively estimate the entropy of the imploding shell. The system we have developed for the fiber-generated PCL (Fig. 3(a)) has a capability to deliver an arbitrary pulse shape (such as tailored pulse, square pulse and picket fence) with a fast rise time of order of 100 ps. Before the introduction of spatial incoherence using the step-index multi-mode fiber [3, 9, 10], a broad-band laser pulse was temporally shaped using optical fiber couplers. First, a single compressed pulse of ~15 ps was stretched by a grating (G_2 in Fig. 3). Next it was split and then stacked to be a 200-ps flat top pulse. This 200-ps

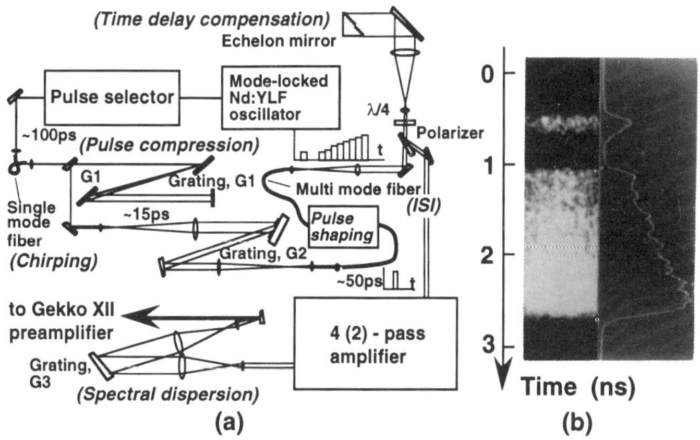

Fig. 3 (a): An optical fiber system for supplying PCL with flexible pulse shaping. (b): An example of shaped pulse measured at the front-end output.

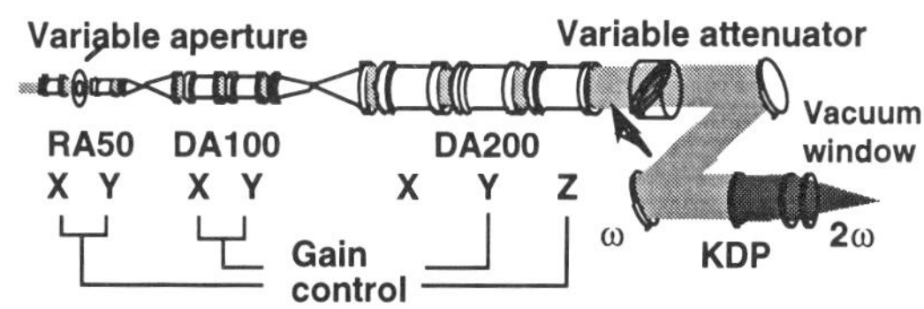

Fig. 4 Schematic arrangement of main amplifier chain of Gekko XII. The power balance is achieved by controlling the amplifier gains, the added linear loss and the beam diameter.

pulse was split into 16 channels using a wave-guide type fiber coupler. Finally, they were combined again with different attenuations and different time delays of 200-ps step. Figure 3(b) shows an example of pulse shape at the output of this front-end system that was supplied for the high isentrope compression. The time delay (~300 ps) across the beam diameter caused by the grating (G_3 in Fig. 3) being used for the angular dispersion of spectrum was compensated using an echelon mirror of 15 steps with 33-ps time separation.

4. Power balance control

Figure 4 shows the principle of power balance control [11]. To achieve the 2ω power balance, the power density balance of ω beam incident onto the KDP crystal is essential. First, we concentrate to obtain the similar pulse shapes of ω beams by controlling the time dependent power gain. Here, we adjust mainly the pumping voltage of rod amplifiers (RA50X and Y with the same pumping) under the balanced gains of disk amplifiers (DA100 and DA200). Then ω power density balance can be realized by installing an additional variable attenuator (partial reflection mirror with adjustable tilt angle) at the outlet of amplifier chain. The variable loss compensates not only the output energy imbalance but also the reflectivity imbalance of mirrors being set after the amplifiers. The expected imbalance of ω power density was ~1% level ensuring the similarity of 2ω pulse with an accuracy better than ~2%. The final 2ω power balance on a target is achieved by adjusting diameters of variable apertures being set between the rod amplifiers, thus both the imbalances of harmonic conversion efficiency and the transmission of focusing optics among beams are compensated. Finally, the intensity balance on the target surface is adjusted by slightly changing the focusing condition, d/R. This procedure of power balance control was confirmed by improving the accuracy (~1%) of energy monitors of ω and 2ω beams and by introducing 4 GHz oscilloscopes for the pulse shape measurement.

Figure 5 shows a preliminary test result of the power balance control of 2ω Gaussian pulse. The observed power imbalance was 10% (rms) even at the foot region (one hundredth of the peak intensity), and reached 3-5% (rms) at around the pulse peak.

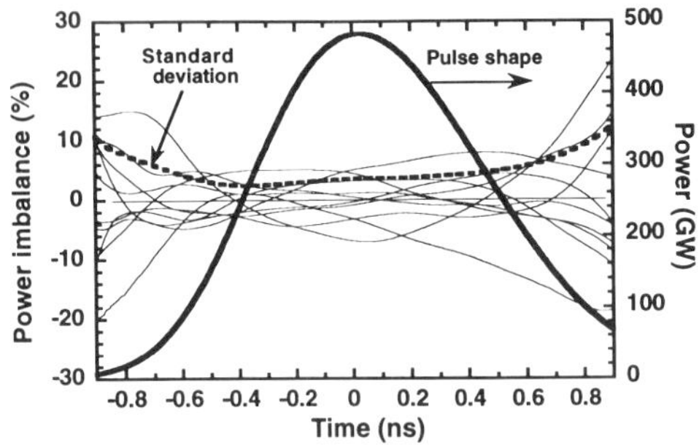

Fig. 5 An initial test result of power balance control of 2ω pulses. The power imbalances of different beams and their standard deviation are plotted as a function of time. Here variable attenuators have not been used yet.

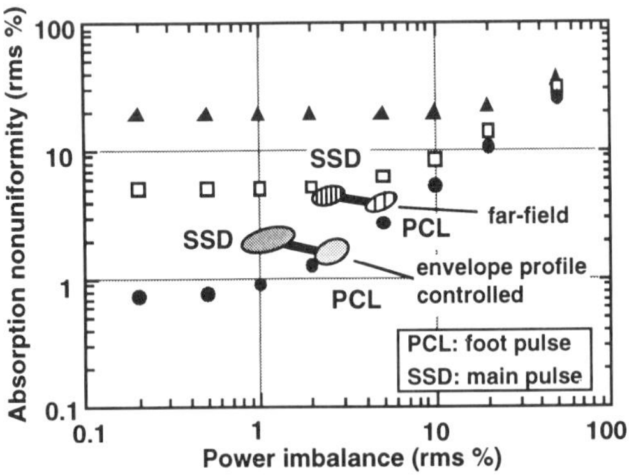

Fig. 6 Absorption nonuniformities depending on the power imbalance for the typical laser conditions. ▲: the coherent laser with old RPP, □: the angularly dispersed PCL with old RPP, ●: the angularly dispersed PCL with an RPP of 2.4-mm hexagonal element. Shaded regions show the near-term and future goals of Precision Gekko XII.

5. Overall absorption uniformity and summary

The absorption nonuniformities for different laser conditions were calculated by means of ray tracing [12]. A target simulated was a plastic shell of 600-μm diameter and 6-μm wall thickness irradiated by 6-kJ, nearly trapezoidal pulse having a full width at half maximum of 2 ns (rise time is 1 ns). Here we used the time integrated patterns (Fig. 2(a) and (b)) and a calculated pattern for the PCL with the envelope profile control, and took into account the fixed power imbalance. The absorption nonuniformities were evaluated in the imprint phase as plotted in Fig. 6. Expected absorption nonuniformities for PCL foot pulse and SSD main pulse are marked in this figure for two types of RPP designed both for the far-field irradiation and for the profile-controlled irradiation. The expected overall irradiation (or estimated absorption) nonuniformity was in a few % level, and the further improvement is needed to overcome 1% level. The several technology developments suggest the feasibility of ignition demonstration with a 300-500 kJ driver.

References

[1] Kato Y *et al.* 1984 Phys. Rev. Lett. **53** 1057-1060.
[2] Murakami M *et al.* 1993 J. Appl. Phys. **74** 802-808.
[3] Miyanaga N *et al.* 1992 *Laser Interaction and Related Plasma Phenomena* (Plenum Press, New York) Vol.10 251-280.
[4] Nishi N *et al.* 1993 SPIE Vol. **1870** 105-111.
[5] Skupsky S and Lee K 1983 J. Appl. Phys. 54 3662-3671.
[6] Lehmberg R H *et al.* 1987 J. Appl. Phys. **62** 2680-2701.
[7] Nakano H *et al.* 1993 J. Appl. Phys. **73** 2122-2131.
[8] Skupsky S *et al.* 1989 J. Appl. Phys. **66** 3456-3462.
[9] Véron D *et al.* 1988 Opt. Commun. **65** 42-46.
[10] Nakano H *et al.* 1993 Appl. Phys. Lett. **63** 580-582.
[11] Miyanaga N 1993 IAEA-CN-60/B-P-11.
[12] Nishihara K *et al. 1994* Phys. Plasmas **1** 1653-1661.

Inst. Phys. Conf. Ser. No 140: Section 4
Paper presented at Laser Interact. with Matter Conf., Oxford, 19–23 September 1994

© *1995 IOP Publishing Ltd*

On the generation of ultrabroad bandwidth light for inertial confinement fusion

G S McDonald†, G H C New†, L L Losev‡, A P Lutsenko‡ and M J Shaw§

† Laser Optics & Spectroscopy Group, The Blackett Laboratory, Imperial College of Science, Technology and Medicine, Prince Consort Road, London SW7 2BZ, UK

‡ P.N. Lebedev Physical Institute, Leninsky Prospekt 53, 117924 Moscow

§ Rutherford Appleton Laboratory, Chilton, Didcot, Oxfordshire OX11 0QX, UK

Abstract. Symmetric bichromatic pumping in stimulated Raman scattering is predicted to lead to the generation of a multifrequency beam consisting of nearly 50 Raman lines of comparable amplitude. Such a beam may have application in laser fusion. We show that ultrabroad bandwidth may be generated in a time much less than the dephasing time of the nonlinear medium and that two distinct regimes of multifrequency generation exist - the coherent and incoherent regimes. In the coherent regime we have discovered that a large number of long-lived soliton pulse trains are spontaneously generated at the distinct Raman frequencies.

1. Introduction

In nearly all previous studies of stimulated Raman scattering (SRS), higher order Stokes and anti-Stokes waves are assumed to be of negligible amplitude. Remarkably, SRS with symmetric bichromatic pumping (identical input pump and Stokes field envelopes) has only very recently been examined in detail [1, 2]. Modelling of this regime has led to the prediction that multifrequency beams consisting of nearly 50 distinct Raman waves of comparable amplitude may be generated. Such broadband multifrequency beams may have application in inertial confinement fusion (ICF) where high gain targets require collisional absorption to be the dominant process in the laser-target coupling. Experiments have shown that reducing the coherence of the incident light can suppress laser-driven plasma instabilities [3,4]. However, a simple increase in the bandwidth of the incident light can also increase thresholds and lower growth rates for such instabilities [5-7]. This approach, as an alternative to or in combination with the use of incoherent sources, has therefore been suggested for ICF [8, 9].

2. The model

To model broadband SRS, the total electric field is expanded in terms of constituent plane waves whose carrier frequencies are given by $\omega_n = \omega_o + n\omega_R$ $(n = 0, \pm 1, \pm 2, \ldots)$ where ω_o and ω_R are the pump frequency and Stokes shift, respectively. To study the essential physics of the broadband generation process, we recast the governing equations into dimensionless form. For the propagation of the nth normalised electric field envelope, A_n, and the dynamics of the polarisation wave, P, one finds [2, 10]

$$\frac{\partial A_n}{\partial Z} = \frac{\omega_n}{2\omega_0}[P^* A_{n+1} e^{-i\gamma_{n+1}Z} - P A_{n-1} e^{i\gamma_n Z}] \qquad (1)$$

$$(\frac{T_2}{t_p})\frac{\partial P}{\partial \tau} = -P + \sum_j A_j A^*_{j-1} e^{-i\gamma_j Z} \qquad (2)$$

$Z = gI_0z$ is the gain-length product, g is the Raman gain coefficient, I_0 is the peak input intensity, τ is local time (in units of input pulse width t_p), and T_2 is the medium dephasing time. Dispersion gives rise to a set of finite values of normalised mistuning, γ_n, which can be parametrised by a single value, $\gamma_1 = (k_1 + k_{-1} - 2k_0)/gI_0$. The input fields are assumed to drive a resonant transition and, for pumping with Gaussian pulses, are taken as $A_0(\tau) = A_{-1}(\tau) = \exp(-\tau^2)$. We also consider here square input pulses, defined as $A_0 = A_{-1} = 1$ for $0 \le \tau \le 1$ and $A_0 = A_{-1} = 0$ for $\tau < 0$ and $\tau > 1$. Results are presented for rotational SRS in H_2 gas pumped by the second harmonic of a Nd:YAG laser, $\omega_R/(2\pi c) = 587\text{cm}^{-1}$ and $\omega_o/(2\pi c) = 18900\text{cm}^{-1}$. However, we expect that our overall conclusions will have much wider applicability.

3. Incoherent and coherent regimes

Towards the steady-state limit, $T_2/t_p \to 0$, distinct temporal regions become decoupled (incoherent) and are, ultimately, independent. One finds that the generated Raman waves can be modulated with rapidly-varying envelopes which are a direct consequence of the input pulse shapes. In this limit, square input pulses lead to square output pulses. In dispersionless propagation, and for large Z, a broad temporal region exists where the system reaches a stationary state, corresponding to saturation of the Raman conversion processes [1, 2]. Generated bandwidth shows a strong dependency on input amplitude and its peak value is found to coincide with the pump frequency, $32\omega_R$ [1]. The incorporation of finite dispersion greatly increases the complexity of the temporal profiles of the interacting waves, imposing phase effects that cause rapid oscillations in the time domain. However, in this case, the conversion processes remain Z-dependent and unsaturated, which can result in bandwidth greater than the pump frequency.

A finite value of T_2/t_p introduces memory effects and allows the polarisation wave to grow even while the magnitude of the polarisation source term falls to zero. For dispersionless propagation we have found that the effect of transiency, alone, can be to increase the bandwidth generated. The presence of both finite coherence time and finite dispersion can result in highly complex patterns in the time domain - see Fig. 1(a). The profiles of the interacting waves exhibit irregular and rapidly varying oscillations. However, the combined effect results in a smooth, high amplitude polarisation wave which can mediate efficient Raman conversion processes, not saturating at low Z, and

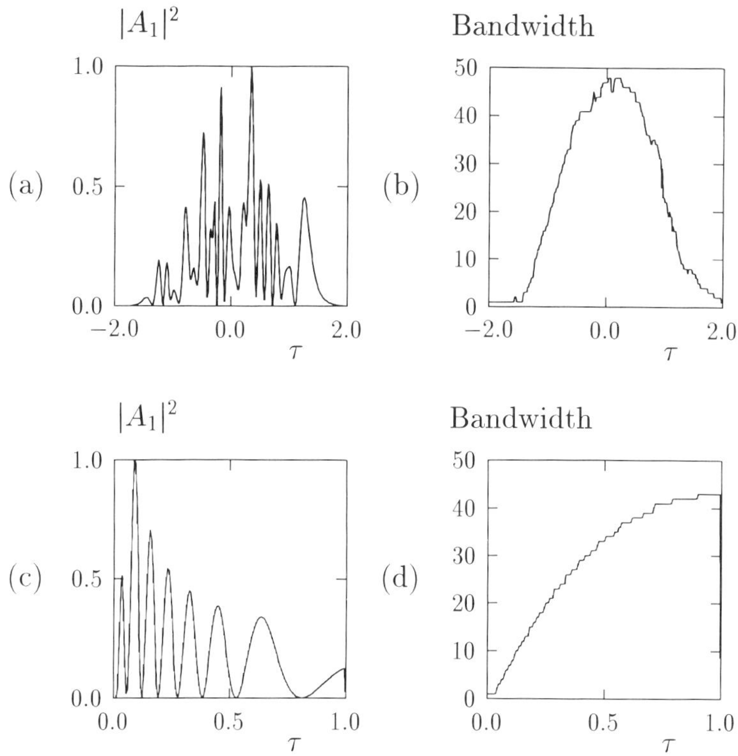

Figure 1. Characteristics of the generated Raman spectra: the normalised intensity profile at the anti-Stokes frequency ($n = +1$) and the generated bandwidth (in units of the Stokes shift). (a) and (b) show results for Gaussian input pulses in the incoherent regime with $T_2/t_p = 0.25, Z = 250$ and $\gamma_1 \approx 2 \times 10^{-3}$ ($gI_0 = 1\text{cm}^{-1}$, 1 atmosphere of H_2 gas) while (c) and (d) are for the coherent regime with $T_2/t_p = 4, Z = 250$ and $\gamma_1 = 0$.

enabling the generation of a peak bandwidth of nearly $50\omega_R$ - see Fig. 1(b). An optimal (dispersive) phase mismatch between the interacting waves and the collectively-written polarisation grating is shown to increase output bandwidth by 50%.

Considering cases where $T_2 > t_p$ (the coherent regime), we find, in contrast to some of the highly complex patterns found in the incoherent regime, that well-defined pulse trains dominate both the Stokes and anti-Stokes orders. For square input pulses, the complexity of the generated waveforms is further reduced - see Fig. 1(c) and (d). In fully transient SRS, pulse trains appear for both Gaussian and square input pulses; they are intrinsic to the nonlinear dynamics of the system and are not a simple consequence of input pulse shape. These *spontaneous* structures are robust Raman soliton pulse trains, resulting from the collective self-organisation of the interacting waves. Note that, in units of τ, T_2 is 4 here and thus, contrary to previous expectations [9], we find that ultrabroad bandwidth may be switched on in a time much less than T_2. In applications, bandwidth switch on time can be a crucial parameter [11].

4. Concluding remarks

An important consideration is the distribution of *energy* across the orders. In [2] we outlined the dependency of energy bandwidth on T_2/t_p for calculations with finite dispersion ($\gamma_1 \approx 2 \times 10^{-3}$). It was shown that, in this case, there is an overall increase in energy bandwidth with decreasing T_2/t_p and that, at the same time, the bandwidth switch on time is lowered. Considering target instabilities, if the instability growth rate is smaller than ω_R then each frequency can drive its own instability. Alternatively, if this growth rate is much larger that ω_R then the discrete spectrum can be considered as continuous [9, 11]. In either regime, the use of broadband spectra can be highly advantageous. For applications in ICF, the incoherent regime appears optimal for $\gamma_1 \approx 2 \times 10^{-3}$, allowing the *rapid* generation of *larger* bandwidth and also enhancing the temporal bandwidth of each wave. However, as $\gamma_1 \rightarrow 0$ and for small T_2/t_p we have found that saturation of the conversion processes occurs at moderate values of gain-length product. For $\gamma_1 = 0$ and large Z we can report that energy bandwidth is optimised in the coherent regime. In this case the global solution is strongly "locked" to a set of soliton pulse trains. This latter consideration may be an important feature if these results are to be applied to other Raman media, such as N_2 [11], where the Stokes shift can be much smaller.

Acknowledgements

This research was supported in part by UK SERC grant GR/J04746. Two of us (L L and A L) wish to acknowledge financial support under a Royal Society Joint Project on Raman lasers.

References

[1] Losev L L and Lutsenko A P 1993 *Kvantovaya Electron.* (Moscow) **20** 1054

[2] McDonald G S, New G H C, Losev L L, Lutsenko A P and Shaw M J 1994 *Opt. Lett.* **19** 1400

[3] Obenschain S P, Pawley C J, Mostovych A N, Stamper J A, Gardner J H, Schmitt A J and Bodner S E 1989 *Phys. Rev. Lett.* **62** 768

[4] Coe S E, Afsar-rad T, Desselberger M, Khattack F, Willi O, Giulietti A, Lin Z Q, Yu W and Danson C 1989 *Europhys. Lett.* **10** 31

[5] Thomson J J and Karush J I 1974 *Phys. Fluids* **17** 1608

[6] Thomson J J 1975 *Nucl. Fusion* **15** 237

[7] Guzdar P N, Liu C S and Lehmberg R H 1991 *Phys. Fluids B* **3** 2882

[8] Eimerl D, Hargrove R S and Paisner J A 1981 *Phys. Rev. Lett.* **46** 651

[9] Eimerl D, Kruer W L and Campbell E M 1992 *Comments Plasma Phys.* **15** 85

[10] Hickman A P and Bischel W K 1988 *Phys. Rev. A* **37** 2516

[11] Eimerl D, Milam D and Yu J 1993 *Phys. Rev. Lett.* **70** 2738

Inst. Phys. Conf. Ser. No 140: Section 4
Paper presented at Laser Interact. with Matter Conf., Oxford, 19–23 September 1994

© *1995 IOP Publishing Ltd*

About testing experiment with targets for gain 1 based on using of the powerful pulsed lasers pumped by γ-radiation of the underground nuclear explosion

E K Bonyushkin, R A Il'kaev, A I Pavlovski, A P Morovov
All-Russian Scientific Research Institute For Experimental Physics, 607200, Nizcni Novgorod region, Arzamas-16, Russia

N G Basov, S Yu Gus'kov, V B Rozanov
P N Lebedev Physics Institute, Leninski prospect 53, 117924, Moscow, Russia

N V Zmitrenko
Institute For Mathematical Modelling, Miusskaya sq. 4-A, 125047, Moscow, Russia

Abstract. Scheme of the experiment on laser target ignition by using the radiation of a powerful excimer laser pumped by γ-radiation of an underground nuclear explosion is proposed. As a target for this experiment the direct driven target of a new type, which is able to provide a uniform deposition of laser energy, with the quantity of laser beams exceeding (6 - 8), is considered.

1. Introduction

In this paper we should like to pay attention to the possibility to accomplish a not so expensive experiment on an ignition of a laser driven thermonuclear target. This experiment may be prepared during a short period of (2 - 3) years. We speak about the experiment with using a short wavelength radiation ($\lambda = 0.353\,\mu m$) of a powerful excimer laser pumped by a γ-radiation pulse of an underground nuclear explosion [1, 2]. A relatively small underground nuclear explosion of $0.5\,ktn$ is needed to pump such a laser system with output energy of $0.1 - 1\,MJ$. The "ignition" experiment may be accomplished both with direct and indirect driven target. In this report we give physics basis for the direct implosion experiment, which needs less amount of laser energy, compared to the indirect implosion experiment.

2. Laser system

The proposal on a laser driver for this experiment has been prepared in All-Russia Scientific Research Institute For Experimental Physics (Arzamas-16). The concept of a powerful excimer laser for the "ignition" experiment is based on a combination of two major elements. The first is the scheme of a single-pass longitudinal pumping by the γ-pulse in the running wave regime. It is needed to ensure the short laser pulse duration ($t < 10\,ns$) under the conditions of a small inversion time of an excimer laser. Another important element is the cone amplifier-former which provides highly directed laser radiation. This arrangement is a combination of the right circular cone amplifier, which has the main part of the active medium volume, and a single pass master oscillator which is a low power laser with small volume of active medium. The apex of the cone amplifier is directed to the pumping source. The single pass master oscillator is placed in the cone amplifier apex. The main and auxiliary volumes are connected by a diaphragm with the diameter, which is much less than the cone length. The single pass master oscillator provides the saturation of the amplification in the apex part of the cone.

The wave front close to a spherical shape is formed under the propagation of a laser radiation along the cone amplifier, and divergence of laser radiation is approximately the ratio of the diaphragm diameter to the cone length. So, the concept of the cone excimer running wave laser allows one to solve simultaneously the problems of a divergence and contrast under laser pumping by γ-pulse of a nuclear explosion.

Researches of the cone excimer running wave laser were carried out in experiment at the laser energy of about $1\,kJ$ [2]. The cone was $15\,m$ long, clear aperture at its base being $0.8\,m$. The active medium presents the mixture of Ne, Xe and NF gases with $1000 : 26.2 : 6.2$ volume ratio at $5\,atm$ pressure. The pump pulse duration was a few nanoseconds. The experimental results close to the theoretical model were obtained: output energy - $700\,J$, laser radiation power - $1.4\,10^{11}\,W$, laser efficiency -3%, output radiation contrast - 10^4, divergency - 10^{-4} . The larger experiments were also made: the output energy - $6\,kJ$ at the laser pulse duration of a few nanoseconds and at the laser efficiency - 3.8%.

The numerical simulations reveal that the single independent arm of the cone excimer laser system pumped by γ-radiation of a nuclear explosion with energy of $(70-90)\,kJ$ may be realized. The estimates of the cost and the time of building of such type laser with the energy of $0.1-0.5\,MJ$ is $(4-6)\,M\$$ and (2 -3) years, respectively.

So, the excimer laser pumped by γ-radiation is the perspective laser for the "ignition" test experiment.

3. Target

The main demand to ICF target for experiment with the laser having small quantity of beams is the ability to provide uniform deposition of the driver energy into the target. One of the types of such a target may be indirect driven capsule which makes conversion of the laser radiation into X-rays.

We should like to propose an other variant of the experiment by using a new type direct driven target which is able to provide the uniform distribution of the absorbed laser energy. The concept of this target [3] was proposed in P.N.Lebedev Physics Institute and is theoretically developed now under colloborative research of Lebedev Institute and Institute for Mathematical Modelling.

The proposed approach to the absorbed laser energy distribution smoothing for a direct target is based on the initiation of a high speed electron conductivity processes at the absorption of the laser radiation into a low density foam absorber of light elements. When the laser radiation having the intensity above $10^{15}\,W/cm^2$ is absorbed by extensive plasma with the density less than critical density, the plasma may be heated up to temperature of $(2-4)\,keV$. The electron heat conductivity waves which propagates in a plasma with such high temperature and low, undercritical density are able to smooth the nonuniformities of plasma heating by the laser beams during the time, which is less than the laser pulse duration. Under these conditions during the "smoothing" time the wave of electron heat conductivity remains a super sonic wave, which goes ahead a hydrodynamic wave. The analysis shows that the electron heat conductivity is able to make the uniform temperature distribution before the hydrodynamic perturbations arise in the target driven by more than (6 - 8) laser beams.

We called this target as the Laser Green House bearing in mind the principle of the absorber operation. The general design of this target presents two co-axial shells. The external shell has holes for input of the laser beams into the target. The internal shell contains the thermonuclear fuel. The shells include the layers of a heavy element material to absorb the thermal plasma radiation. The part of the target, which is placed between these shells is the absorber of the laser beams. The absorber material presents a low density foam matter of light elements (for example, foam plastic) with density close to the laser-produced plasma critical density.

1D code "DIANA" [4] and 2D code "ATLANT" [5] have been used for numerical simulations . The target for the excimer laser ($\lambda = 0.353\,\mu m$) pulse with the energy of $200\,kJ$ and pulse duration $3\,ns$ has the following parameters: the shells present the combination of the copper and plastic layers, the quantity of the external shell holes, 6; the internal shell radius, $0.09\,cm$; the absorber thickness, $0.05\,cm$; the DT fuel and internal shell masses , $20\,\mu g$ and $300\,\mu g$, respectively; the absorber density close to $5\,10^{-3}\,g/cm^3$. The 1D numerical simulations of this target give implosion velocity, about $4\,10^7\,cm/s$, the hydrodynamic efficiency, 0.15, thermonuclear gain, 4. The preliminary 2D numerical simulation with show effective operation of the "smoothing" absorber.

4. Conclusion

The relatively not expensive test experiment on ignition of the direct target driven by the excimer laser pumped by the γ-pulse of a relatively small underground nuclear explosion may become an important step of the ICF activity. This experiment may be accomplished within the frame work of international collaboration under International Atomic Energy Agency monitoring, under the condition that all ecological requirements will be met.

References

[1] Jacob J H et al 1986 Applied Physics Letters **48** 318-320

[2] Alekhin B V et al 1992 Abstracts of Conference on Physics of Nuclear Induced Plasmas and Problems of Nuclear Pumped Lasers (Obninsk, Russia) 354

[3] Gus'kov S Yu and Rozanov V B 1993 Report on seminar in Centre d'Etudes de Limeil-Valenton (Limeil-Valenton, France)

[4] Samarski A A et al 1983 Problems of Nuclear Science and Engineering **2** 38-42 (in Russian)

[5] Gamaly E G et al 1988 Quantum Electronics **15** 1622-1632 (in Russian)

Inst. Phys. Conf. Ser. No 140: Section 5
Paper presented at Laser Interact. with Matter Conf., Oxford, 19–23 September 1994

© *1995 IOP Publishing Ltd*

Evaluation of a foam buffer target design for use in ICF

M Dunne, M Borghesi, A Iwase, M W Jones, R Taylor and O Willi

Imperial College of Science, Technology and Medicine, London, SW7 2BZ, UK

Abstract. Experimental observations are presented demonstrating that the use of an x-ray preformed low density foam layer on the surface of a directly-driven foil is effective in smoothing spatial nonuniformities present in a coherent drive laser. Observations have been made on the thermal smoothing in the preformed foam layer, the acceleration history of the hybrid target and on the levels of parametric instability growth, with encouraging results.

1. Introduction

A prime contender for a viable Inertial Confinement Fusion (ICF) driver is direct illumination by a large number of high power optical laser beams. Such a driver has the benefits of geometrical simplicity and potentially high overall energy efficiency. However, the wavefront of a high power laser beam suffers from severe and unpredictable intensity nonuniformities. This leads to nonuniform energy deposition and the seeding of a number of deleterious instabilities in the underdense and overdense plasmas. Premature target breakup can ensue, resulting in a drastic reduction in the achievable fusion yield. To date, laser-based techniques have proved inadequate in overcoming the imprinting of intensity nonuniformities in the early-time 'start-up' phase of the laser-target interaction [1,2]. This imprinting arises due to the combination of two effects. Firstly, in all smoothing schemes the instantaneous laser intensity profile remains grossly nonuniform, with effective smoothing only occurring over a large number of coherence times [2]. This effect is then exacerbated by the proximity of the absorption region to the ablation surface in the initial plasma, so that there is an insufficient distance over which transverse *thermal* smoothing of the initial nonuniform shock can occur before it reaches the ablation surface. The target response to the shock generated by these residual nonuniformities results in mass flow perturbations across the target, seeding a number of hydrodynamic instabilities. The magnitude of this imprinting can far outweigh that induced by the surface roughness of the target and acts to severely limit the maximum allowed aspect ratio of the fusion capsule.

Computational analysis of this outstanding problem led us to propose a solution in which the optical drive pulse is preceeded by a short x-ray burst, incident onto a thin, low density foam layer overcoated on the solid target surface [3]. The x-ray heating occurs through the passage of a supersonic ionization wave, thus preventing any shock waves reaching the fuel interface prior to the thermal front induced by the main drive laser. The supersonic production of super-critical, homogenous plasmas from x-ray irradiated foam layers has already been demonstrated [4] and the thermal smoothing properties of x-ray preformed solid wires well established [1,5]. The initial distance between the critical and ablation surfaces is approximately set by the thickness of the overcoated foam layer, with an insignificant amount of underdense plasma created prior to the arrival of the optical interaction. The foam buffer is introduced to influence only the initial (t<500ps) start-up phase of the interaction. The subsequent implosion physics should remain essentially unchanged from a conventional direct drive interaction and so will not be discussed here. The use of such a hybrid target design has applications in any situation where the integrity of a laser accelerated foil has to be maintained.

This paper now aims to present initial experimental evidence which establishes the validity of the proposed smoothing scheme. Planar targets are used throughout to be representative of a segment of the shell in an implosion target. Key parameters to be investigated include the level thermal smoothing occurring in the foam layer, the effect this layer has on the hydrodynamic response of the target (and thus the implosion efficiency) and any change in parametric instability generation in this preformed plasma buffer. Details of a fully quantitative experiment, in which a face-on radiographic system was used to measure the levels of areal density nonuniformity growth, will be published elsewhere.

2. Experimental details

Data will be presented from an experiment performed on the VULCAN glass laser system of the Central Laser Facility. A second harmonic 6-beam cluster configuration was used to irradiate the target, with a seventh, axial beam used both to drive and act as a backscatter probe. Irradiances on target were in the range (1 to 5)x10^{14} Wcm^{-2}. Coherent beams were used throughout to present the most taxing irradiation conditions for the smoothing scheme.

The targets used consisted of triacrylate ($C_{15}H_{20}O_6$) foam cylinders [6], 50mg/cc in density, held onto a 12μm solid mylar ($C_{10}H_4O_8$) foil. The plastic layer represents the 'payload' in an implosion pellet and ensured that a significant fraction of the target always remained at or above solid density during the acceleration measurements, for direct comparison with simple planar foil targets. The thickness of the foam layer was varied from 50 to 200μm. Cell sizes in the foam were less than 1μm in diameter, with a very high degree of homogeneity throughout the foam. This was confirmed through the use of off-line, point source x-radiography and scanning electron microscopy. These targets were then overcoated with 400Å bismuth on the laser side. This overcoat acted to produce a burst of x-rays during the leading edge of the drive pulse which supersonically preformed the foam target, as discussed below. A time-resolved XUV diode monitored the absolute levels of soft-x-ray emission from the front of the overcoated targets. The irradiation and target geometry are shown schematically in Figure 1(a). The simplicity of such a target design should allow a relatively easy conversion to spherical geometry. In the context of an ICF pellet, the target design would be essentially unchanged from a conventional direct-drive pellet, with the simple addition of a 50μm thick foam layer on the outer side of the pellet, overcoated with a thin ($\rho x \approx 4 \times 10^{-5}$ g/cm^2) layer of high-Z material. The ability to produce a supersonic x-ray ionization wave in this 'overcoat' geometry was confirmed using streaked XUV absorption images as used in [4], with an example image displayed in Figure 1(b). The preforming x-ray wave propagates well ahead of the shock front at all times in the interaction. The cell closure time in the foam (10-20ps) is far shorter than the delay before the arrival of the shock front (>100ps), allowing ample time for the cellular foam to homogenise.

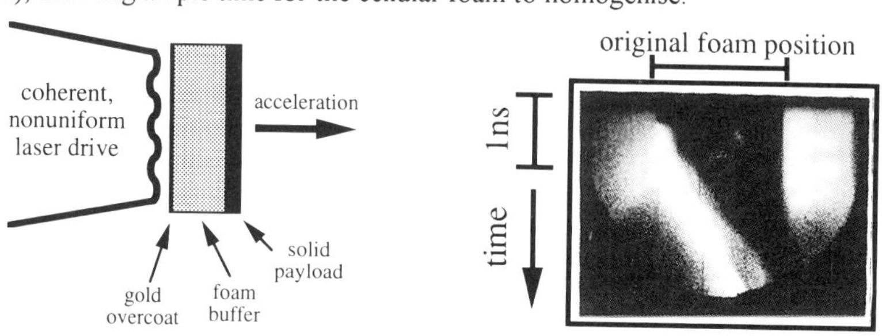

Fig 1 (a) Schematic of the target geometry (b) Side-on XUV image of a Au-coated foam target. The foam was 200μm long for these studies and shows a radiatively driven wave propagating ahead of the laser thermal front

3. Thermal smoothing results

The thermal smoothing properties of the foam buffer layer are demonstrated in figure 2. Here, an XUV imaging system [7] (operating at 50±5Å and at a magnification of 100x) was used in a 2D side-on configuration to demonstrate the formation of a uniform heat front as the preheated foam layer is traversed. Results from the bismuth-coated foams are compared to foam targets where no bismuth layer was present. The foam layer consists of a series of solid density walls, separated by the pore size of the material. Without a precursor x-ray ionization wave, the critical and ablation surfaces are coincident at the beginning of the interaction, allowing no scope for thermal smoothing. This is reflected in the data. The target self-emission, imaged at 90° to the laser interaction, shows intense filaments throughout the uncoated foam layer. These filaments are substantially smoother for the gold-coated foam targets. In absorption, a grossly nonuniform profile is obtained for the uncoated foam, whereas the absorption front for the preheated foam is smooth to within the resolution of the imaging system. The thermal smoothing results from the hot, long-scalelength plasma which exists as a result of the x-ray preforming process. The side-on system used is able to detect nonuniformities with scalelengths greater than 5μm and transmission differences of better than 2%. For these measurements the greatest contribution to the relative transmission arises from areal density nonuniformities. The diagnostic therefore provides an indication of the relative degree of target breakup that would result from each target design. Indeed, the face-on measurements (to be published) show a dramatic reduction in areal density nonuniformity when the full, preheated foam targets are used in comparison to conventional solid foils, or foils overcoated with an *uncoated* foam. The reduction in the level of target nonuniformity in the *preheated* foam arises as a result of the pore closure induced by the x-ray wave well before the arrival of the laser thermal front. Throughout the optical interaction, then, the laser front encounters a homogenous, long scalelength, supercritical plasma in which the critical surface is separated from the solid (ablation) surface by a distance on the order of the thickness of the foam overcoat. Simulation of the laser interaction with these hybrid foam targets has resulted in good agreement in the dynamics of the thermal and radiation wave propagation. These results, along with details on the simulation of the thermal smoothing will be published elsewhere. Generic simulations of the response of foam buffered targets can be found in [3].

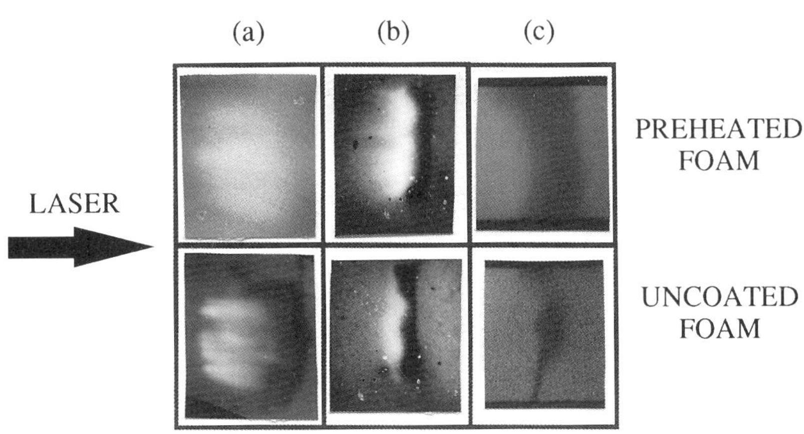

Fig 2 Demonstration of the thermal smoothing capabilities of an x-ray preheated foam layer (top row) compared to bare, uncoated targets (bottom row) (a) time-integrated self emission, (b) gated images showing self emission 100-200ps after the start of the interaction, (c) gated absorption images 500ps after the start.

4. Hydrodynamic efficiency

Such effective thermal smoothing is useless if it compromises the efficiency with which the target can be accelerated. Side-on streaked XUV imaging, as used in [8], demonstrated that the presence of a foam overcoat does not affect the degree to which a solid 'payload' can be accelerated. The efficiency of the acceleration of a foam-buffered target is almost identical to that of a simple, directly heated solid planar foil even though the hybrid targets which were studied contained 17% more mass. These observations are in good agreement with detailed hydrocode simulations of the interaction [3]. Note that in a real ICF pellet design there will be no significant difference in the total mass of a foam coated target compared to a conventional target. For the same areal mass density, the hydrocode simulations show that the kinetic energy of the payload remains unchanged when a foam overcoat is present. This efficiency can be understood from the fact that in the interaction of the laser with a solid foil, the distance between the critical and ablation surfaces reaches a steady-state value on the order of 20-50μm in the first few 100ps of irradiation. The presence of the foam buffer, then, introduces this steady-state distance from the start of the interaction. For timescales of interest to ICF (>10ns), the difference in energy transfer in this initial transient stage becomes irrelevant.

The observed retention of hydrodynamic efficiency provides an indication that little energy is lost to parametric processes occurring in the preformed plasma. This was confirmed using calibrated, time-resolved diodes, which demonstrated that less than 5% of the incident laser energy was lost to B-SBS. This is in comparison to 3% losses that resulted from interactions with a plain, solid CH foil. Levels of B-SRS were also shown to remain essentially unchanged. The backscattered SBS was imaged onto a 2D gated optical imager and shown to be concentrated in intense filaments, as expected in the underdense plasma.

6. Summary

In summary, we have demonstrated that an x-ray preformed foam layer is capable of thermally smoothing laser nonuniformities in a coherent beam, without compromising the hydrodynamic performance of the target. Parametric instability generation has also been shown to be insignificant under the current conditions. Fully quantitative measurements will be published in the near future. These results appear to be encouraging for the prospects of direct-drive ICF, showing drastic reductions in the mechanism that is the primary limiting factor on the viability of high aspect ratio pellets. It now remains to convert these proof-of-principle experiments to a realistic spherical implosion.

Thanks are due to W. Nazarov, J. Falconer and G. Lyall for supplying the foam targets, B. Evans for the XUV mirror coating, and to the staff of the CLF for help during the experiments. The initial investigation into this smoothing scheme was performed on the Los Alamos TRIDENT laser. Details of these experiments will be published at a later date.

References

[1] M. Desselberger *et al.* 1992 Phys. Rev. Lett. **68** 1539
[2] M. H. Emery *et al.* 1991 Phys. Fluids B **3** 2640
[3] M. Desselberger, M. Jones, J. Edwards, M. Dunne and O. Willi, submitted to Phys. Rev. Lett.; see also M. Jones *et al.*, these proceedings
[4] T. Afshar-rad *et al.* 1994 Phys. Rev. Lett. **73** 74
[5] M. Dunne *et al.* 1994 Annual Report to the Laser Committee RAL-94-042 48-49 (unpublished)
[6] J. Falconer *et al.* 1990 J. Vac. Sci. Technol. **A8** 968
[7] O. Willi *et al.* 1992 Rev. Sci. Instr. **63** 4818
[8] J. Edwards, M. Dunne, R. Taylor, O. Willi, C. Back and S. Rose 1993 Phys. Rev. Lett. **71** 3477

Inst. Phys. Conf. Ser. No 140: Section 5
Paper presented at Laser Interact. with Matter Conf., Oxford, 19–23 September 1994

© *1995 IOP Publishing Ltd*

NUMERICAL SIMULATION OF LASER-IRRADIATED CAVITIES

R Ramis

E.T.S.I. Aeronáuticos, Universidad Politécnica de Madrid, Spain

J Meyer-ter-Vehn

Max-Plank-Institut für Quantenoptik, Garching, Germany

Abstract. The two-dimensional radiation-hydrodynamics code **Multi2D** has been used to simulate the behavior of high-Z hollow-shells, whose interior is irradiated by laser beams penetrating through holes. The investigated issues are: plasma blow-off from the walls, soft X-rays transport, plasma collision at the center, and hole closure. The results are compared with experiments.

1. Experiments

We apply our code **Multi2D** [1] to the study of the hohlraum experiments described in [2]. Figure 1 summarizes the geometry used. Axisymmetric cavities with a diameter of $2.2mm$, formed by $15\mu m$ thick gold walls, are irradiated by 4/5 laser beams ($500J$ each and $800ps$ HWHM), entering through hole EG, and inciding over the conical section CB. The soft X-rays generated there, heat indirectly the rest of the wall. After some time, cavity thermalization is expected. A diagnostic hole, drilled through the cylindrical section CD allows observation of the interior (radiated power, spectrum, ...) By other hand, optical diagnostics permit to observe the emergence of the ablation-driven shock wave at the rear of the plate AB. By using plates with variable thickness, the wave speed can be determined. The conical shield EF is introduced for technical reasons, to impede the laser defocused light to perturb the diagnostics.

2. Model

The hydro-code solves radiation transport with angular resolution. Photon propagation is considered over a discrete number of directions (72) that cover the entire 4π solid angle. For each direction, the radiation intensity is obtained by sweeping through the computational grid. The algorithm ensures appropriate accuracy in both, optically thin and optically thick limits. This treatment of radiation goes beyond the diffusion approximation, frequently implemented in other codes. The opacities are taken from [3] as functions of density and temperature (no frequency dependency is considered at the moment) The fluid motion is solved using a fixed (eulerian) unstructured grid formed

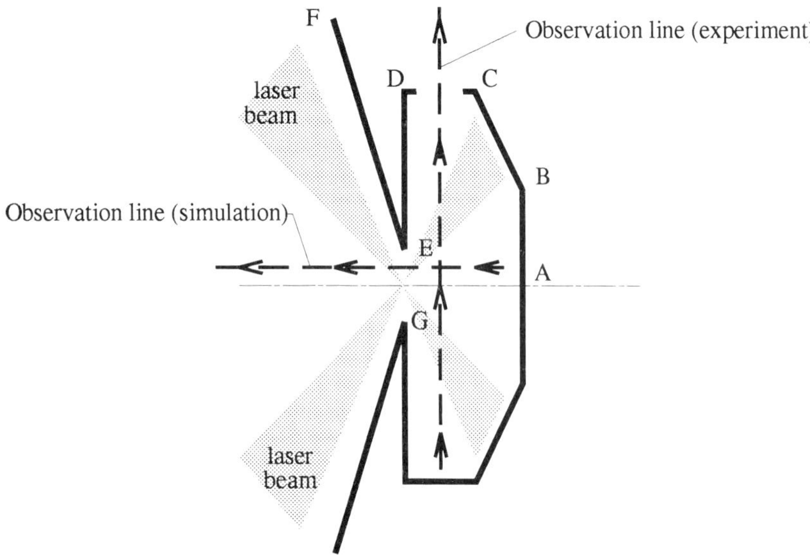

Figure 1

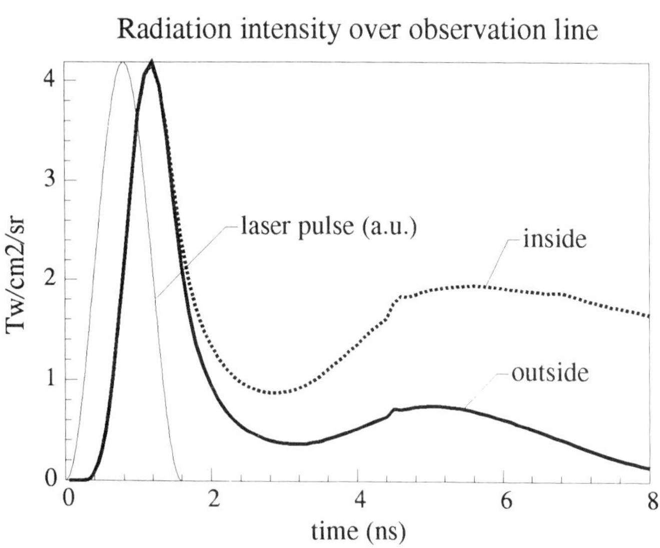

Figure 3

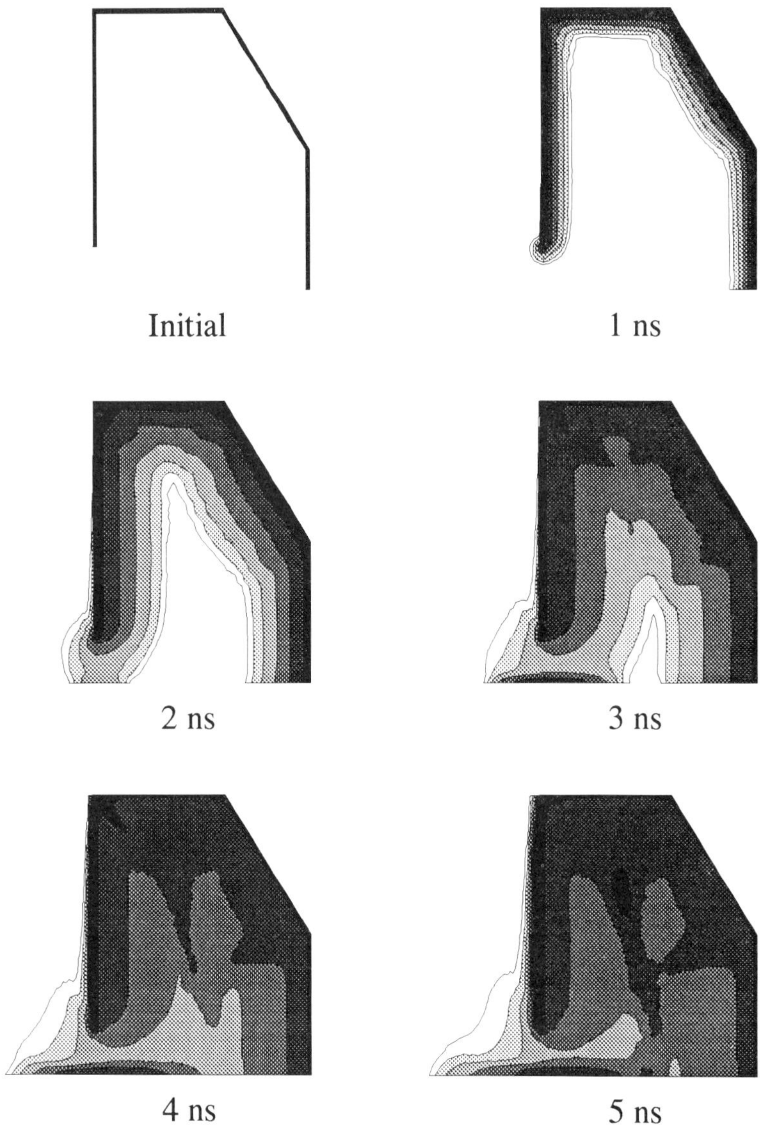

Figure 2

by triangular elements. We assume an ideal gas equation of state, with coefficients chosen to fit the SESAME tables [4] at typical working conditions ($100eV$, $1g/cm^2$) The laser beams are simulated by taking an averaged conical beam, absorbed only at the critical surface.

3. Results

Figure 2 shows isodensity lines for 1,3,10,30,100,300,and $1000mg/cm^2$.At $1ns$ the plasma blow off is practically one-dimensional, except at the border of the entrance hole, where the lines take a characteristic spiral form. The plasma cloud is thicker at the directly irradiated section *BC*, than at the rest of the wall. At $3ns$, plasma collision become visible at the symmetry axis of the entrance hole, forming a spike-like structure that grows with time. Its border is a cylindrical diverging shock wave, practically isothermal due to the strong dependence of the radiated heat with temperature. Collisions are visible near the corners between wall sections. At $5ns$, plasma coming from opposite walls forms a disk-like structure at the center of the cavity. We can characterize the degree of thermalization by comparing the temperatures corresponding to equilibrium with the radiation intensity emitted by each wall element. At $1ns$ we get: 105, 163, 113, and $121eV$ for *AB*, *BC*, *CD*, and *DE* respectively. The differences decrease with time, and, after $2ns$, a uniform temperature ($\pm3eV$) is reached. In experiments, the measured quantity is the intensity of radiation through the diagnostic hole. This value is not available from the simulation, because it implies three dimensional geometry. We consider instead radiation through the entrance hole (its size and irradiation conditions are very similar), with the observation line separated $100\mu m$ from the symmetry axis. In figure 3, the outcoming intensity is plotted as a function of time. There is a delay of $400ps$ between the main radiation pulse and the laser pulse. A less intense second radiation pulse, generated by plasma collisions, has a maximum at $5ns$. From this curve, a mean brightest temperature of $89eV$ results (using intensity as a weight function) This value agree with the experimental value $108\pm6eV$. A substantial part of the difference is due to the fact that in simulation we are looking to wall segment *AB*, that is about $8eV$ cooler than the experimentally observed *BC*. Finally, we observe for the ablation-driven shock wave in AB, speeds of about $6\text{-}7km/s$. This value separates from the experimental one of $9.0km/s$ due to the rather crude approximation used for the equation of state.

4. Conclusions

A reasonable agreement is found between experiments and **Multi2D** simulations. The structure of the cavity filling by plasma can be quantitatively described. This phenomena is important to understand measured values of radiation intensity: the doted curve in figure 3, computed *without* taking into account reabsorption by this plasma, gives too large values. There is some evidence of the spike-like structures in pinhole photographies, but more detailed studies are needed to confirm its existence.

References

[1] Ramis R and Meyer-ter-Vehn J 1992 *Max-Plank Institut für Quantenoptik Report* MPQ-174
[2] Löwer T et all 1993 *Submitted to Phys. Rew. Letters.*
[3] Tsakiris G D and Eidmann K 1987 *J. Quant. Spectrosc. Radiat. Transfer* **38** 353-368
[4] Bennet B I et all 1978 *Los Alamos report* LA-7130

Inst. Phys. Conf. Ser. No 140: Section 5
Paper presented at Laser Interact. with Matter Conf., Oxford, 19–23 September 1994

© 1995 IOP Publishing Ltd

A study of laser plasmas as X-ray sources in the 1-10 keV spectral region

S. Bastiani, D. Giulietti†, A. Giulietti, L.A. Gizzi, T. Ceccotti and A. Macchi

Istituto di Fisica Atomica e Molecolare, Via del Giardino 7, 56127 Pisa, Italy
† Dipartimento di Fisica, Università di Pisa, Piazza Torricelli 2

Abstract. X-ray emission from plasmas produced by focusing the beam of a Nd laser on massive Al or Cu targets was studied. The temporal evolution of the emission was investigated by using an X-ray streak-camera. The spectral analysis of the emission was performed with both low and high resolution. The X-ray conversion efficiency, in the investigated spectral region, was measured to be 1% for Cu and slightly lower for Al. Two main spectral components were observed in the emission. In the case of Cu target, the first component was found to be centred at ≈ 1.2 keV and the second, whose intensity was measured to be 10^{-3} of the previous component, at ≈ 7 keV. From line intensity ratios, obtained from high resolution spectra, we could estimate temperature and density of the emitting region.

1. Introduction

Laser produced plasmas represent versatile and relatively inexpensive XUV sources for application in different fields of research and technology [1].

On the other hand, X-ray radiation gives valuable information on plasma parameters such as temperature, density and opacity [2].

The study presented here refers to plasmas produced by focusing a Nd laser on targets of Cu and Al [3]. The laser was monomode transversal but multimode longitudinal and delivered up to 3 J in 3 ns (FWHM) or 20 J in 20 ns at 1.064 μm wavelength.

2. Source characterisation

The intensity of X-ray radiation strongly depends on the laser focusing configuration on target [4]. For this reason X-ray flux was maximised by scanning the target position along the beam axis, in the proximity of the nominal focal position. The maximum of the X-ray emission was found with the target displaced roughly 0.5 mm away from the laser with respect to the waist of the beam, wich was focused with an f/8 lens.

The spectral window of the X-ray detector was determined by the spectral sensitivity of the PIN-diode combined with the transmittivity of the Al filters placed in front of the PIN. It was included between 0.5 and 10 keV, with a width depending on the different filters used for the Al or Cu targets.

Fig. 1 shows the behaviour of X-ray yield versus nominal laser intensity, for the position of maximum emission, and for a marginal position. In the first case experimental data points are fitted by a curve more than quadratic, in the second by a straight line. As we can see, in the first case the experimental points are, shot by shot, more scattered than in the second case. All this seems to indicate that the plasma influences the laser focusing and the interaction conditions by a non-linear process. It is interesting to observe that measurements on second harmonic emission [5], versus target position and laser intensity, showed similar behaviour.

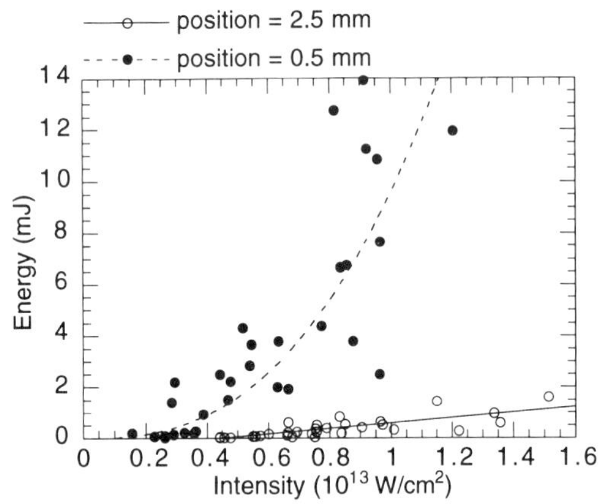

Fig. 1 X-ray energy versus laser intensity for two different positions of Cu target.

The conversion efficiency of laser radiation into X-ray energy was ≈ 0.3 % for aluminium and ≈ 1 % for the copper. In the case of aluminium the emission was mainly due to recombination processes in the 1÷3 keV range, while in the case of copper the main contribution was given by line emission in the 1÷1.5 keV range.

The temporal evolution of the emission was studied using an X-ray streak-camera. The duration of the emission was appreciably shorter than the laser pulse duration and resulted 1.5÷2 ns for both targets. The emission was *spiky* and the duration of a single spike resulted comparable to the duration of the spikes present in the laser pulse, which result from mode beating in the laser oscillator cavity.

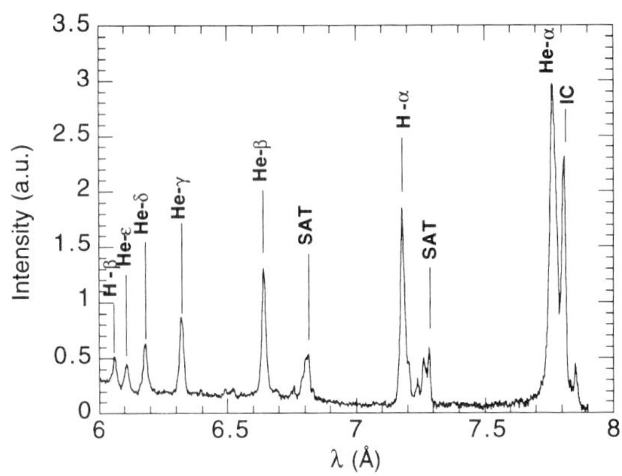

Fig. 2 Time integrated K-shell aluminium spectrum.

The electron temperature and density of the aluminium plasma were determined by comparing K-shell intensity ratio with the predictions of a steady-state, atomic physics code (RATION [6]) and were found to be approximately ≈ 300 eV and ≈ 10^{21} cm^{-3} respectively. Such values were compared to that obtained with 1D hydrodynamics code (MEDUSA [7]). The values of temperature and density given by the simulations resulted higher than the experimental ones. This is due to the fact that the actual expansion is not 1D for the whole interaction time, as shown by the X-ray images of the plasma we obtained with a pin-hole camera.

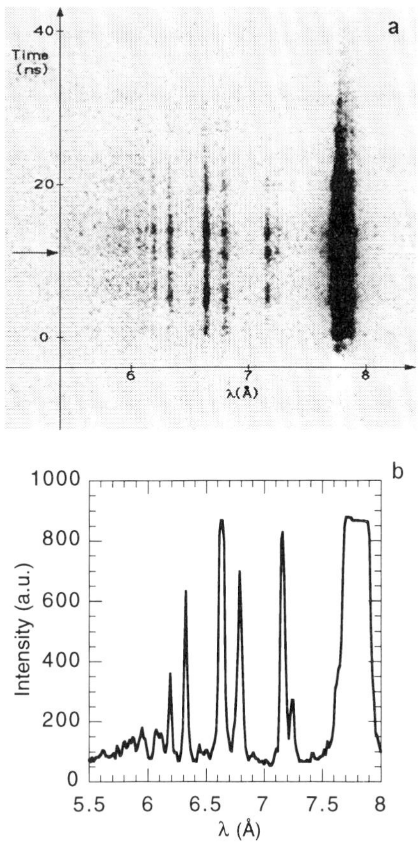

Fig. 3 a) Time-resolved X-ray spectrum of K-shell emission from an aluminium plasma produced with a 20 ns laser pulse. b) Horizontal lineout (taken at the time indicated by the arrow in a)) showing identification of spectral lines.

Fig. 3 shows a typical time resolved spectrum of K-shell emission from an aluminium plasma produced with a 20 ns laser pulse, at a nominal intensity of $1.5 \cdot 10^{13}$ W/cm^2. The He-α line is merged with the intercombination line. The line is saturated, its intensity being much stronger than that of the other lines. The electron temperature deduced by spectra like these was found to be lower (≈ 200 eV) than in the case of plasmas produced by 3 ns laser pulses, as shown by the greater intensity of He-like lines with respect to the H-like lines. Incidentally we notice that the plasma produced in this regime can be considered a quasi-monochromatic X-ray source in the investigated spectral range, due to the high intensity of the He-α line.

3. Suprathermal emission

A weak component of the X-ray emission at higher photon energy was observed for both Al and Cu plasmas. Such a component was detected in the target position which maximise the X-ray emission and resulted (for the Cu target) of the order of the 10^{-3} of the overall X-ray energy. The high energy component was spectrally located at ≈ 7 keV. In Fig. 4 we report the PIN output voltage versus the thickness of the aluminium filter, for the Cu plasma. In a) the Cu target is in the position of maximum X-ray emission, while in b) it is away from such a position.

As one can see, while in the first case the data we reported on semilogarithmic scale can be fitted by two straight-lines in the second case they lie on a single straight-line. In fact the intensity of the X-ray radiation detected by the PIN is, in the case of emission of two spectral components λ_1 and λ_2 ($\lambda_1 > \lambda_2$),

$$I = I_1 \exp(-\rho\mu_1 x) + I_2 \exp(-\rho\mu_2 x)$$

$\mu_{1,2}$ being the mass absorption coefficients of the filter for the two wavelengths and ρ its density. As the thickness $x \rightarrow 0$, $I \rightarrow I_1 + I_2 \approx I_1$, provided that one of the two spectral components largely overcomes the other ($I_1 >> I_2$), as in our condition. For $\rho\mu x >> 1$, being generally $\mu_1 >> \mu_2$, $I \rightarrow I_2 \exp(-\rho\mu_2 x)$. Therefore from the slope of the straight lines, and from the mass absorption coefficients of the filters, one can estimate the wavelength of X-ray radiation, while from the ratio of the intercepts of the two straight lines with the ordinate axis, the ratio of the intensity of two spectral components can be inferred.

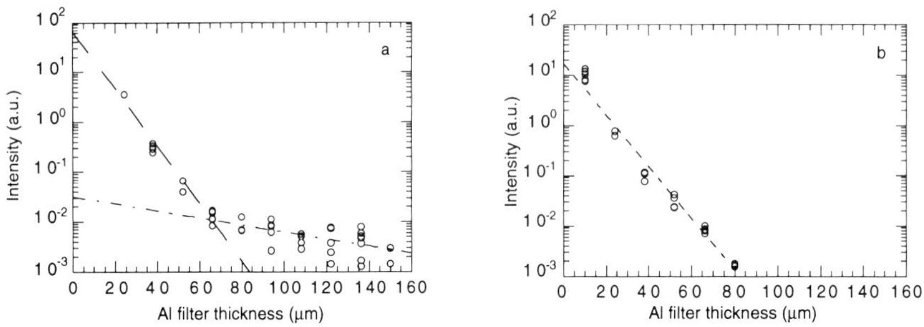

Fig. 4 X-ray intensity versus aluminium filter thickness a) in the position of maximum emission (2.5 mm) and b) far from previous position (0.5 mm).

The high energy component is presumably due to fast electrons produced by electron waves [8]. Among all the instabilities that can generate such waves in a plasma, two plasmon decay seems to be, in our experimental conditions, the most probable process. In fact the energy and the intensity of the most energetic X-rays is consistent with the amplitudes and phase velocities of waves produced by such an instability [9].

References

[1] Sigel R 1994 *Laser-induced radiation hydrodynamics: X-ray generation and application* XLV SUSSP, St. Andrews

[2] Gizzi L A et al 1994 *Phys. Rev. E* **49** 5628

[3] Bastiani S 1994 *Studio dello spettro e dell'evoluzione temporale dell'emissione X di plasmi prodotti da laser* (Degree Thesis, Università di Pisa)

[4] Amiranoff F 1979 *Etude des mechanismes d'interaction laser-matiere par mesure du rayonnement continu* (Doctorial Thesis, 3^0 level)

[5] Biancalana V et al 1993 *Europhys. Lett.* **22** 175

[6] Lee R W et al 1984 *J. Quant. Spectr. Radiat. Transfer* **32** 91

[7] Christiansen J P et al 1974 *Comp. Phys. Comm.* **7** 271

[8] Lamoureux M et al 1984 *Phys. Rev. A* **30** 429

[9] Gizzi L A et al 1992 *Laser and Particle Beams* **10** 65

Inst. Phys. Conf. Ser. No 140: Section 5
Paper presented at Laser Interact. with Matter Conf., Oxford, 19–23 September 1994
© 1995 IOP Publishing Ltd

Velocity and Temperature Profiles in Laser-Created Plasmas

O. Renner†, T. Mißalla‡, M. Kopecký†, E. Krouský† and E. Förster‡

† Institute of Physics, Academy of Sciences of the Czech Republic, CZ - 18040 Prague, Czech Republic

‡ X-ray Optics Group of Max-Planck-Gesellschaft at the Friedrich-Schiller-Universität Jena, D-07743 Jena, Germany

Abstract. High-resolution x-ray spectra of chlorine He-like resonance line group emitted from low radiance plasma have been measured at two observation angles by using a vertical dispersion variant of the Johann spectrometer. The results of computerized analysis of the 1-D spatially resolved spectra were used to evaluate the distribution of the main plasma parameters. The electron temperature gradient $7.5 \cdot 10^4 \frac{eV}{cm}$ computed by modeling the measured spectra with the collisional-radiative package RATION and the blowoff velocity gradients $(0.9 - 1.6) \cdot 10^9\, s^{-1}$ determined from the Doppler shifts of different spectral lines are consistent with the results of a simplified hydrodynamic simulation.

1. Introduction

Spectroscopic data combining high spectral and spatial resolution provide important information about laser-produced plasmas; the knowledge of spatial distribution of macroscopic characteristics such as temperature, density, but also blowoff velocity is essential to investigate energy transport, uniformity of implosion, radiation transfer and other properties.

Large velocity gradients accompanying the plasma expansion are of crucial importance for the success of different x-ray laser schemes [1]. In recombination pumped lasers, the differential Doppler shift of spectral lines seen by different parts of the expanding plasma can solve the problem of reabsorption increasing the population of the lower lasing level. Contrary to this, in resonant photopumping schemes, where the strong x-ray source pumps the lasant, the effect of counter-streaming plasmas may offset the resonance defect and thus increase the achieved gain.

Due to the Doppler effect, the positions, shapes and intensities of the emitted spectral lines strongly depend on the velocity profiles and angles of observance with respect to the direction of expansion. In the past, the precision of the velocity measurements has been limited by relatively low resolution of the Bragg crystal spectrometers used or by the lack of spatial resolution, see e.g. [2, 3].

Recently, well resolved emission line profiles and positions of the hydrogenic and helium-like ions of aluminium have been reported [4, 5]. The spectra emitted from the high radiance x-ray source have been detected at different observation angles using a novel high resolution double crystal spectrometer (DCV) [6] and interpreted by modeling the radiative transfer in the plasma with large velocity gradients. Here we present the results of the high-resolution velocity and temperature profile measurements at moderate parameter plasmas using a high luminosity vertical dispersion variant of the Johann spectrometer (JV). Obviously, these papers and the experimental method used are complementary.

2. Experimental

The experiment was performed using the Nd-glass laser system at the Max Planck Institute for Quantum Optics in Garching. The laser pulses containing $3-9\,J$ of $0.53\,\mu m$ light in a pulse length of $2\,ns$ were incident normally on to massive targets of KCl or NaCl. The focal spot of diameter $60\,\mu m$ corresponded to incident intensities $(0.5-1.5)\cdot 10^{14}\frac{W}{cm^2}$.

The scheme of experimental setup is shown in Fig. 1. The dispersive element was a quartz crystal (11.0) cylindrically bent to a radius of $76.8\,mm$. In the JV, both the source and the detector are placed at the intersection of the Rowland circle with the central ray which strikes the crystal at the Bragg angle θ_o corresponding to the wavelength λ_o. The rays with slightly shorter wavelengths λ are dispersed in the vertical direction at an angle $\pm\phi$ given by relation $\lambda = \lambda_o \cos(\phi)$, i.e. two sets of symmetrical spectra are recorded simultaneously [7].

The JV has the same extremely high dispersion as the DCV but its collection efficiency is much higher. The results of ray tracing procedure suggest that the instrument provides the spectral resolution 13 500, the spatial resolution at the level of $10\,\mu m$ and

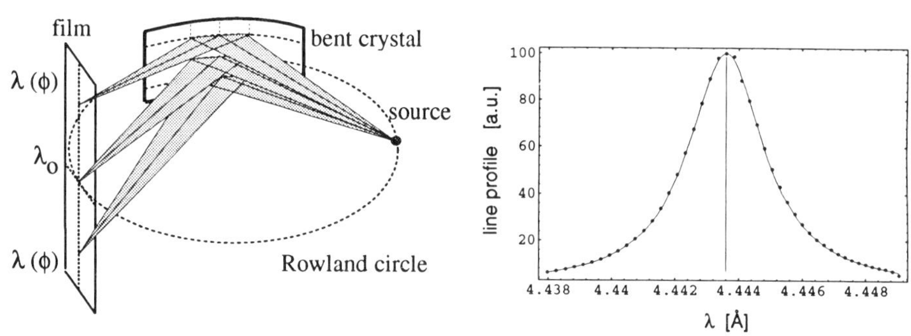

Figure 1. A schematic diagram of the experiment (left) and a comparison of the spectral line profile transmitted through the spectrometer (points) with the theoretical one (solid line).There is a negligible shift in the position of the line maximum, its width (FWHM) is increased by 2%.

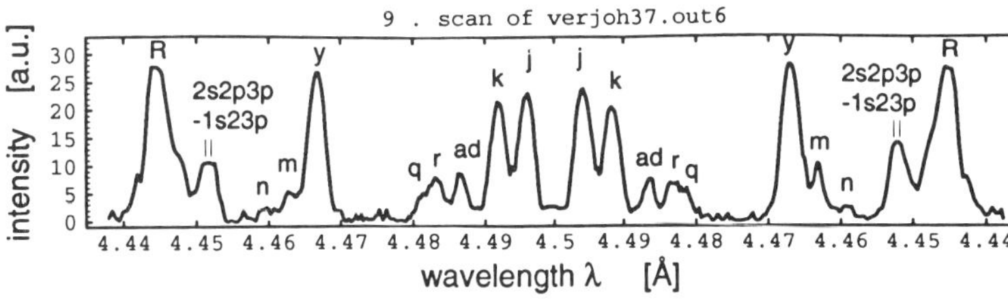

Figure 2. The spectrum of the chlorine He_α group scanned in a track close to the maximum intensity of the recorded spectral lines.

the luminosity up to three orders of magnitude higher than the DCV. The distortion of the registered spectral lines is negligible, as can be also seen in Fig. 1.

The spectra were recorded in single shots on Agfa Structurix D7 x-ray film and converted to an intensity scale using an unpublished calibration curve. The procedure of the target alignment limits the precision of λ_o setting to $0.8\,m\mathring{A}$, the relative positions of spectral details registered in one shot may be determined with precision $2 \cdot 10^{-5}$.

3. Results and discussion

The spectra obtained for the chlorine He-like resonance, intercombination and accompanying satellite lines were scanned using a photodensitometer slit $8 * 100\,\mu m^2$ (spatial * dispersion direction) and then corrected for non-linear dispersion and finite source size.

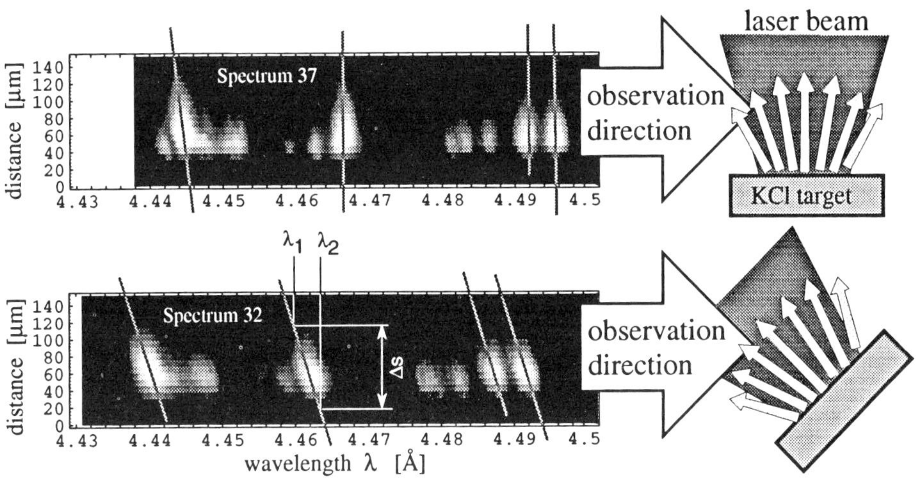

Figure 3. Spatially resolved spectra recorded at different observation angles were used to determine the expansion velocities and gradients.

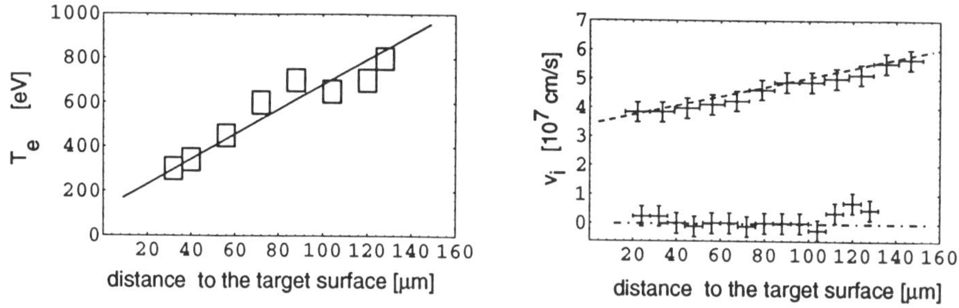

Figure 4. Electron temperature T_e and ion velocity v_i distribution in the plasma. Velocities have been computed from the motional Doppler shift of the intercombination line viewed parallel (— · —) and 45° (- - - -) to the target surface, error bars are shown.

The spectrum shown in Fig. 2 was taken along a line of the sight at 88° with respect to the laser beam. The total exposed length of the film was 50 mm and the well resolved satellite structure demonstrates the resolving power of the instrument. By fitting such spectra with the results of modeling based on collisional-radiative code RATION [8], the spatial distribution of electron temperature ($T_e = 250 - 800\, eV$) and density ($n_e = (0.5 - 1.0) \cdot 10^{22}\, cm^{-3}$) was derived.

The comparison of the perpendicularly viewed (Fig. 3, top) and the motionally Doppler-shifted spectrum taken at 45° (bottom) provides the plasma expansion velocities and gradients. The values obtained depend on the choice of the spectral line evaluated. By assuming that the expansion velocity vector is perpendicular to the target surface, we have found the maximum velocities 6.1, 5.8, 4.6 and 4.2 $\cdot\, 10^7 \frac{cm}{s}$ and gradients 1.6, 1.5, 0.96 and 0.90 $\cdot\, 10^9\, s^{-1}$ for the resonance (R), intercombination (y) and j,k satellite lines, respectively. The spread of these parameters may be qualitatively explained by recognizing that these lines originate from different regions of the plasma.

The tilt of the optically thick resonance line observed in the top spectrum is a consequence of opacity; the spectrum is formed mainly in the plasma region facing the spectrometer. The maximum velocity component parallel to the target surface was thereby estimated as $9.5 \cdot 10^6 \frac{cm}{s}$. The profile of optically thin intercombination and satellite lines is formed by the emission from the full plasma volume; the surface velocity component broadens the line profiles symmetrically and therefore no tilt is observed.

The electron temperature and velocity profiles shown in Fig. 4 were compared with the results of a simplified hydrodynamic simulation using the code FILM [9]. The experimentally determined temperature gradient of $7.5 \cdot 10^4 \frac{eV}{cm}$ is comparable with the plasma parameters formed $0.4 - 2.2\, ns$ after the end of the laser pulse, the observed velocity gradients of $(0.9 - 1.6) \cdot 10^9\, s^{-1}$ are somewhat smaller than the value $3.7 \cdot 10^9\, s^{-1}$ computed for the end of the pulse. The validity of this comparison is limited by the experimental conditions used (the spectrometer views the plasma across the regions with considerably different sets of macroscopic parameters and integrates over time) and by approximations in the modeling.

4. Conclusions

We have shown that the vertical dispersion variant of the Johann spectrometer may provide spectroscopic data with high spectral and spatial resolution and luminosity. The spectra of $Cl\ He_{\alpha}$ group produced by a plasma of modest parameters have been measured at two lines of sight, the 1-D spatial distribution of electron temperature and blowoff velocity has been determined with precision probably not previously available. A more accurate interpretation of experimental data requires better knowledge of the experimental conditions; the use of spatially localized emitters and time-resolving detectors is highly desirable. If these conditions were properly defined, this technique is capable of providing high-quality data suitable to test the results of detailed hydrodynamic simulations.

Acknowledgments

This program is supported by Grant Agencies of the Czech Republic and of the Academy of Sciences under Grants No. 202/93/1023 and 110427. Part of this work was carried out during the stay of O. R. in Jena under the COST Go West Scheme, contract No. ERB-CIPA-CT-93-0632. The authors gratefully acknowledge fruitful discussions with C. Chenais-Popovics, the help of I. Uschmann and the staff of the Max Planck Institute for Quantum Optics in performing the experiment.

References

[1] Pert G J 1994 Laser and Particle Beams 12 209-222

[2] Moreno J C, Goldsmith S and Griem H R 1992 J. Opt. Soc. Am. B9 339-343

[3] Fiedorowicz H, Denus S, Jeziak K, Kolanowski M, Parys P, Pawlowicz W and Wolowski J 1988 J.Phys. 49 NC-1 369-371

[4] Wark J S, He H, Renner O, Kopecký M, Förster E and Mißalla T 1994 J. Quant. Spectrosc. Radiat. Transfer 51 397-406

[5] Wark J S, Djaoui A, Rose S J, He H, Renner O, Mißalla T and Förster E 1994 Phys. Rev. Lett. 72 1826-9

[6] He H, Wark J S, Förster E, Uschmann I, Renner O, Kopecký M and Blyth W 1993 Rev. Sci. Instr. 64 26-30

[7] Renner O, Kopecký M, Krouský E, Förster E, Mißalla T and Wark J S 1994 Laser and Particle Beams 12 (in print)

[8] Lee R W, Whitten L and Stout R E 1984 J. Quant. Spectrosc. Radiat. Transfer 32 91-101

[9] Chenais-Popovics C, private communication

Inst. Phys. Conf. Ser. No 140: Section 6
Paper presented at Laser Interact. with Matter Conf., Oxford, 19–23 September 1994
© 1995 IOP Publishing Ltd

X-ray lasers

E E Fill

Max-Planck-Institut fur Quantenoptik, Hans-Kopfermann-str. 1,
D-85748 Garching, Germany

Abstract. In recent years there has been considerable progress in the development of soft X-ray lasers. Improvements with respect to their efficiency, coherence and brightness have been achieved and progress in their application has been made. More recently, encouraging results towards table-top X-ray lasers have been obtained.

1. Introduction

The challenge of extend the laser principle to short wavelengths meetws two main obstacles: The cubic frequency dependence of the ratio of spontaneous to stimulated emission rate coefficients leads to a pump intensity requirement proportional to $\nu^3 \Delta\nu$, where ν and $\Delta\nu$ are the frequency and line width of the lasing transition. Secondly, high-reflectivity mirrors are not available in the X-ray region and so high Q resonators cannot be used. Instead, one has to resort to an amplified spontaneous emission (ASE) mode of operation, which requires high gain.

All this means that typical pumping intensities for X-ray lasers exceed 10^{13} watt/cm^2, making large laser fusion installations the natural choice as the pump source.

Nevertheless, a review on X-ray lasers of 1991 [1] already lists some 50 X-ray laser lines spanning the region of 36 Å up to about 300 Å. By mid-1994 an update of this compilation would yield almost 100 X-ray laser lines, the majority of the new ones appearing around 200 Å. Prominent X-ray laser lines are listed in table 1.

2. Pumping mechanisms

As is obvious from the table, the most successful pumping mechanisms for X-ray lasers are electron collisional pumping and recombination pumping. Electron collisional pumping of neon-like and nickel-like ions results in robust high-gain lasers, which lase in a wide range of parameters. Their shortcoming is a low efficiency (since they operate on $\Delta n = 0$ transitions) and the high pump intensity they require.

Recombination lasers, operating on $\Delta n = 1$ or 2 transitions, have a higher gain efficiency (defined as *gain-length product / pump laser energy*) but are more sensitive to the choice of experimental parameters. They require a narrow line focus or a fibre target (to obtain a sufficiently high rate of adiabatic cooling), a short pulse duration and an intensity matched to the ionisation stage of the parent ion [2]. Furthermore, certain aspects of these lasers are not well understood and they have not been scaled to large gain-length products.

Table 1: Prominent X-ray laser lines

λ [Å]	Lasing ion	Electron. conf.	Pumping	Ref.	Comment
35.6	Au^{52+}	Ni-like	el. coll.	3	shortest laser wavelength
43.2	W^{46+}	Ni-like	el. coll.	4	above carbon K-edge
44.8	Ta^{45+}	Ni-like	el. coll.	4	below carbon K-edge
69	Cu^{18+}	Na-like	recombination	5	
81	F^{8+}	H-like	recombination	6	
154	Al^{10+}	Li-like	recombination	7	
155	Y^{29+}	Ne-like	el. coll.	8	brightest X-ray laser
182	C^{5+}	H-like	recombination	9,10	
206	Se^{24+}	Ne-like	el. coll.	11	
209	Se^{24+}	Ne-like	el. coll.	11	
232	Ge^{22+}	Ne-like	el. coll.	12	most popular X-ray laser
236	Ge^{22+}	Ne-like	el. coll	12	

Various other pumping mechanisms have been suggested, such as photo-pumping [13,14], charge transfer pumping [15] and pumping by multiphoton excitation [16]. In spite of progress achieved, e.g. in photo-pumping, none of these has resulted in a laser exhibiting a gain-length product greater than two.

For a detailed discussion of the various pumping mechanisms see the X-ray laser book by Elton [17].

3. Current issues of X-ray laser research

3.1 Gain saturation

Saturation of an ASE laser, required for efficient operation, can only be achieved if the gain-length product is high, with a numerical value around 20. The saturation intensity of an X-ray laser is typically about 10^{10} W/cm^2. Saturation has been achieved for a germanium laser by using a half-cavity arrangement [18].

3.2 Coherence

Longitudinal coherence is not an issue in X-ray lasers: Gain narrowing reduces the relative line width of the emission to a value around 10^{-4}, yielding a coherence length $\lambda^2/\Delta\lambda$ of about 10^4 λ. The resulting longitudinal coherence lengths of >100 μm are sufficient for applications.

An adequate transverse coherence length is much more difficult to achieve since an ASE laser does not usually emit spatially coherent light. The coherence function of a selenium X-ray laser was measured by means of a slit array, the diffraction pattern of which was analyzed [19]. The beam obtained with a 300 μm wide line focus showed a transverse coherence length of less than 50 μm. A significantly higher coherence was achieved by narrowing the line focus of the pump laser. With a 100 μm wide line focus the coherent region was found to be about 200 μm. The obvious explanation of this result is that in an X-ray laser medium with a lower Fresnel number the generation of high-order modes is prevented, yielding a beam of higher spatial coherence.

In the experiment just discussed the transverse coherence length was considerably smaller than the beam diameter. Furthermore, it was observed that the coherence did not improve when the length of the gain medium was increased from 3 to 7.5 cm. This observation is explained by recent work analytically solving the wave equation with a parabolic gain profile [20]. It is shown that small "hose-like" fluctuations of the gain region lead to energy exchange between propagating modes, preventing the low-order modes from being singled out.

A considerable improvement of the spatial coherence was achieved for a germanium laser using an arrangement similar to an oscillator-amplifier chain in a conventional laser [21]. With a distance of more than one meter between the "injector" and the amplifier only low-order modes reach the amplifier, which results in a more coherent beam. A Young´s double-slit experiment showed fringes of high contrast up to a slit separation of 200 μm. Nevertheless, much remains to be done in order to achieve an X-ray laser beam with a high coherence over its full diameter.

3.3 Polarization

For some applications, especially in atomic physics experiments, a polarized X-ray beam is needed. A degree of polarization of 98 % has been achieved with the injector-amplifier arrangement [22]. For this purpose the beam emitted by the "injector" was reflected twice from a polarizing X-ray mirror with a 45° angle of incidence. Two similar mirrors were used as analyzers for vertical and horizontal polarization. It was verified that the high degree of polarization was maintained even after passing the amplifier, which was operated at a gain-length product of 5.

3.4 Anomalies

Although much research has been done on modelling X-ray lasers several experimental observations are still only poorly understood. A prominent anomaly occurs in lithium-like recombination lasers: Simple arguments as well as sophisticated computer codes predict considerably smaller gain for the 3d-5f line than for the 3d-4f line, the main reason for this being the smaller oscillator strength of the 3d-5f transition. In experiments, however, it is generally observed that the 3d-5f transition exhibits gain similar to or even higher than 3d-4f, a discrepancy unexplained to date.

A different anomaly affects collisionally excited lasers: The relative magnitude of the gain on fine-structure components of the 3p-3s gain transition considerably differs from that predicted. Specifically, a 3p ($J = 0$) - 3s ($J = 1$) line (termed $J = 0 - 1$ line) was predicted to have the highest gain, but showed much smaller gain experimentally than two of the $J = 2 - 1$ lines. This problem is close to being solved since recently it was found that collisional X-ray lasers lase predominantly on the $J = 0 - 1$ line if a prepulse several ns before the main pulse is applied to the target [23,24].

In experiments at Max-Planck-Institut fur Quantenoptik [25] the intensity of a prepulse 5 ns ahead of the main pulse was systematically varied. For neon-like zinc a prepulse intensity of around 1 % of the main pulse intensity was optimum to obtain the highest gain on the $J = 0 - 1$ line at 212 Å. An explanation for this observation is derived from numerical simulations of a laser plasma using the LASNEX computer code [24]. These calculations suggest that the effect of a prepulse is to reduce the steep electron density gradient in the expanding plasma, which otherwise would result in deflection of the generated X-ray beam away from the target. Since the J = 0 - 1 line is generated at relatively high electron density, it is most seriously affected by a refractive index gradient.

This explanation is corroborated by recent work at ILE Osaka in which a curved target was used to avoid beam refraction out of the gain region [26,27]. In these experiments, in which a series of 100 ps pulses with a separation of a few 100 ps was applied to the target, the J = 0 - 1 line dominated the spectrum and was emitted with high brightness and narrow divergence.

3.5 Brightness

If one compares the brightness of X-ray lasers to that of synchrotrons, of a laser plasma and of harmonics of high-power laser radiation, it turns out that X-ray lasers are by far the brightest soft X-ray sources, reaching a power of 10 MW and a brightness of 10^{23} photons / s, mm^2, $mrad^2$, 0.01 % bandwidth [8].

The superiority of X-ray lasers can also be appreciated by comparing the brightness temperature of a laser-plasma (≈200 eV) to that of an X-ray laser: The first collisionally pumped X-ray lasers had brightness temperatures of around 40 keV. To date, the most powerful X-ray lasers have MeV brightness temperatures.

3.6 Applications

X-ray laser applications were discussed long before X-ray lasers came on the scene. The applications predicted to be the most likely ones were in microscopy, holography and lithography. More recently, applications in atomic physics research, materials science and plasma diagnostics have been stressed. This is reflected in the proceedings of a workshop on X-ray laser applications, held in San Francisco in 1992 [28]. At that meeting scientists from different areas discussed the implications of X-ray lasers on their field of research. The result was an impressive list of their potential impact in various disciplines.

As examples of such applications, we discuss X-ray microscopy and laser plasma diagnostics. X-ray microscopy is a well-established technique, usually carried out with synchrotron radiation [29]. Its advantage with respect to electron microscopy is the ability to image biological objects in vitro, as being alive. The use of an X-ray laser instead of a synchrotron is motivated by radiation damage caused by the high dose of radiation required to obtain good contrast. With the long exposure times needed using synchrotron radiation, fine details of the specimen tend to get destroyed and therefore cannot be resolved in the microscope.

With X-ray laser radiation a similar dose is required on the sample, but due to the short pulse duration there is not enough time for the structure to move, it being "inertially confined". Another argument for the use of X-ray lasers is the hope that table-top systems will eventually be developed, making them available to more laboratories than synchrotron radiation.

X-ray microscopy with an X-ray laser was demonstrated at Livermore [30]. By using a tantalum laser with a wavelength of 44.8 Å a resolution of 500 Å was obtained.

A different application of X-ray lasers is as a diagnostic for laser-produced plasmas. With X-rays one can probe a laser plasma at very high densities. Various techniques, such as absorption, Moiré deflectometry and interferometry can be applied.

An example of such an experiment, using the yttrium laser at 155 Å and a Mach-Zehnder X-ray interferometer, is described in these proceedings [31].

4. Table-top X-ray lasers

It is obvious that applications of X-ray lasers will never become widespread unless they are made available to a greater number of laboratories. Thus, instead of using a laser fusion installation as the pump source, an independent and relatively small pump laser must be developed or a non-laser pump of some kind must be found.

Recently, considerable progress in this direction has been made at several laboratories. The greater part of it was achieved by means of new ultra-short pulse (ps or fs) lasers, such as Ti:sapphire lasers, short-pulse KrF lasers and lasers with chirped pulse amplification (CPA-lasers).

- At Princeton University a compact soft X-ray laser, operating on the 3-2 transition of hydrogenic carbon ($\lambda = 182$ Å) is being developed [32].
- At MIT progress has been made towards the demonstration of gain in low-Z nickel-like ions. With a train of pulses containing less than 1 J per pulse being used for pumping a gain-length product of 3 was obtained at 204 Å in nickel-like niobium [33].
- At DRAL the Vulcan laser was converted into CPA-Vulcan, which emits more than 30 J in a pulse of 2 ps duration. High gain on the 182 Å line of hydrogenic carbon was reported [34].
- At RIKEN optical field ionization [35,36] was used to generate an over-ionized, strongly recombining plasma, achieving transient gain on the Lyman-α line of hydrogenic lithium at 135 Å [37].
- At the University of Lund gain on the 3d-5f transition in lithium-like nitrogen (λ = 512 Å) was measured in an optical-field-ionized plasma [38].
- At Stanford a 10 Hz XUV laser operating at 418 Å in optical-field ionized Pd-like Xe was demonstrated [39].
- Finally, an exciting result was obtained at Colorado State University, where for the first time high gain on a transition pumped by a capillary discharge was demonstrated [40]. A gain-length product of 6 was reported on the 3p-3s $J = 0 - 1$ transition in neon-like argon at a wavelength of 468 Å. This is the only non-laser-pumped XUV laser to date.

5. Summary and outlook

Collisionaly excited X-ray lasers are already well developed. They will be further improved, especially with respect to their coherence, brightness and efficiency. However, these lasers have almost reached their limits in achieving shorter wavelengths. Recombination lasers, being less developed, have the potential to be scaled to shorter wavelengths with less pump power. Other pumping methods may also eventually contribute to the realization of short-wavelength lasers. Better X-ray optics will be developed and X-ray laser applications will continue to emerge.

Table-top systems are presently making fast progress especially in the XUV. New such lasers will be discovered with wavelengths approaching the real X-ray region. In the more distant future exciting things such as X-ray laser nonlinear optics, tunable X-ray lasers and inner-shell X-ray lasers may become a reality.

Acknowledgment: The author is indebted to many of his colleagues of the X-ray laser community for giving information on their current research, and for making unpublished papers available.

References

[1] Skinner C H 1991 *Phys. Fluids B* **3** 2420-2429
[2] Pert G J and Ramsden S A 1974 *Optics Commun.* **11,** 270-273
[3] MacGowan B J et al. 1990 *X-Ray Lasers 1990*, Bristol: IOP, ed. Tallents G J 221-230
[4] MacGowan B J et al. 1990 *Phys. Rev. Lett.* **65**, 420-423
[5] Steingruber J and Fill E E 1994 *Appl. Phys. B* **58**, 29-34
[6] Grande M et al. 1990 *Optics Commun.* **74**, 309-312
[7] Jaeglé P et al. 1987 *J. Opt. Soc. Am B* **4**,563-574
[8] Matthews D et al. 1987 *J. Opt. Soc. Am B* **4,** 575-587
[9] Suckewer S et al. 1985 *Phys. Rev. Lett.* **55**, 1753-1756
[10] Chenais-Popovics C et al. 1987 *Phys. Rev. Lett.* **59**, 2161-2164
[11] Matthews D L et al. 1985 *Phys. Rev. Lett.* **54**, 110-113
[12] Lee T N, McLean E A and Elton R C 1987 *Phys. Rev. Lett.* **59**, 1185-1188
[13] Vinogradov A V, Sobelman I I and Yukov E A 1975 *Sov. J. Quant. Electron.* **5**, 59-63
[14] Chichkov B N and Fill E E 1990 *Phys. Rev. A* **42**, 599-610
[15] Vinogradov A V and Sobelman I I 1973 *Sov. Phys. JETP* **36**, 1115-119
[16] Clark C W et al. 1986 *J. Opt Soc. Am. B* **3**, 371-378
[17] Elton R C 1990 *X-Ray Lasers* (New York; Academic)
[18] Jaeglé P et al. 1992 *X-Ray Lasers 1992,* Bristol: IOP, ed. Fill E E, 1-7 and Lewis C L S et al. 1992 ibid. 23-30
[19] Trebes J E 1992 *Phys. Rev. Lett.* **68**, 588-591
[20] Amendt P, London R A and Strauss M, to be published
[21] Cairns G et al. 1994 *Appl. Phys. B* **58,** 51-56 and Cairns G et al. 1994, *4th Intern. Coll. on X-Ray Lasers,* Williamsburg, May 16-20, 1994
[22] Rus B et al. 1994 *4th Intern. Coll. on X-Ray Lasers,* Williamsburg, May 16-20, 1994
[23] Nilsen J, Moreno J C, MacGowan B J and Koch J A 1993 *Appl. Phys. B* **57**, 309-311
[24] Nilsen J et al. 1993 *Ultrashort Wavelength Lasers II,* SPIE **2012**, ed. Suckewer S, 131-137
[25] Fill E E et al. to be published
[26] Lunney J G 1986 *Appl. Phys. Lett.* **48**, 891-893
[27] Kato Y et al. 1993 *Ultrashort Wavelength Lasers II,* SPIE **2012**, ed. Suckewer S, 12-21
[28] London R, Matthews D and Suckewer S eds. 1992 *Applications of X-Ray Lasers,* LLNL Proc.
[29] Schmahl G and Rudolph D, eds. 1984 *X-Ray Microscopy,* (Berlin: Springer)
[30] Da Silva L B et al. 1992 *X-Ray Lasers 1992,* Bristol: IOP, ed. Fill E E, 387-392
[31] Cauble R et al., these Proceedings
[32] DiCicco D et al. 1992 *X-Ray Lasers 1992,* Bristol: IOP, ed. Fill E E, 185-190
[33] Basu S et al. 1993 *Appl. Phys. B* **57**, 303-307
[34] Zhang J et al. 1994 *4th Intern. Coll. on X-Ray Lasers,* Williamsburg, May 16-20, 1994
[35] Burnett N H and Corkum P B 1989 *J. Opt. Soc. Am B* **6** 1195-1199
[36] Amendt P, Eder D C and Wilks S C 1991 *Phys. Rev. Lett.* **56** 2589-2592
[37] Nagata Y et al. 1993 *Phys. Rev. Lett.* **71** 3774-3777
[38] Borgström S et al. 1994 *4th Intern. Coll. on X-Ray Lasers,* Williamsburg, May 16-20, 1994
[39] Lemoff B E et al. 1994 submitted to *Phys. Rev. Lett.*
[40] Rocca J et al. 1994 submitted to *Phys. Rev. Lett.*

Inst. Phys. Conf. Ser. No 140: Section 6
Paper presented at Laser Interact. with Matter Conf., Oxford, 19–23 September 1994

© *1995 US Government*

Diagnosing High Density, Fast-evolving Plasmas using X-ray Lasers

R Cauble, L B Da Silva, T W Barbee Jr., P Celliers, S Libby, J. C. Moreno, S. Mrowka, D Ress, J E Trebes, A S Wan and F Weber

Lawrence Livermore National Laboratory, University of California, Livermore, CA 94550 USA

Abstract. As x-ray laser (XRL) research has matured, it has become possible to reliably utilize XRLs for applications in the laboratory. Laser coherence, high brightness and short pulse duration all make the XRL a unique tool for the diagnosis of laboratory plasmas. The high brightness of XRLs makes them well-suited for imaging and for interferometry when used in conjunction with multilayer mirrors and beamsplitters. We have utilized a soft x-ray laser in such an imaging system to examine laser-produced plasmas using radiography, moiré deflectometry, and interferometry. Radiography experiments yield 100-200 ps snapshots of laser driven foils at a resolution of 1-2 μm. Moiré deflectometry with an XRL has been used to probe plasmas at higher density than by optical means. Interferograms, which allow direct measurement of electron density in laser plasmas, have been obtained with this system.

1. Introduction

Production of laboratory soft x-ray lasers (XRL) is reliable enough that the high brightness and coherence of the XRL can now be confidently exploited. High brightness makes the XRL well-suited for imaging other bright sources such as laser plasmas, where the XRL can actually dominate much of the plasma self-emission. High brightness also makes it possible to use a number of components in an x-ray optical system, such as mirrors and beamsplitters. The XRL's coherence properties make it a viable source for deflectometry and interferometry. When these features are combined with short wavelength (the critical density > 10^{24} cm^{-3} at 155 Å), the XRL is seen to be an ideal plasma diagnostic. We have utilized a neon-like yttrium XRL at 155 Å (80 eV) [1] in an imaging system to perform radiography, deflectometry, and interferometry of fast-evolving laser-produced plasmas. The XRL duration of ~ 200 ps limits particle motion to ~ 10-20 μm during exposure. Fig. 1 shows the setup for interferometry. For radiography, the beamsplitters and two center mirrors are removed. For moiré deflectometry, the radiography setup is used with Ronchi gratings positioned near the CCD. [2]

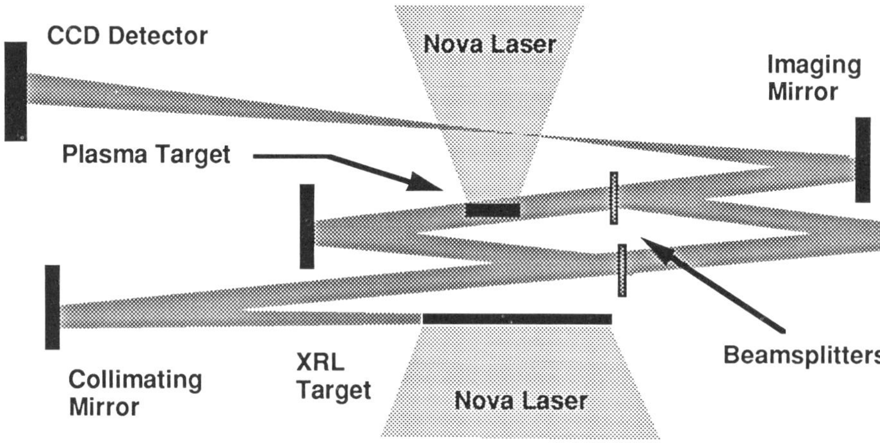

Figure 1. Experimental setup, showing the four necessary multilayer mirrors, for Mach-Zehnder interferometry of a plasma target using an x-ray laser on the Nova two-beam facility.

2. High Resolution Radiographic Images and Interferograms

Figure 2 shows an image of a laser accelerated foil backlit by the yttrium XRL halfway through a high energy 1 ns drive. The foil, which was 10 μm of CH on 3 μm Al, was irradiated on the CH side. Spatial resolution is better than 2 μm. Although the Nova drive beam was smoothed with steering wedges and a random phase plate, small-scale, 5-6 μm, structure is evident on the rear (Al) side. Larger-scale structure can also be observed. Some of the structure may be an artifact due to refraction of the XRL beam along extreme density gradients in the foil plasma, so that the actual extent of the plasma cannot be quantified from the image. However, the XRL beam is spatially quite smooth so the nonuniformities observed in the image are produced by density variations within the plasma.

Moiré deflectograms indicating peak electron densities exceeding 5 x 10^{20} cm^{-3} have been made of 3-mm-long laser produced blowoff plasma. [3] Since this density determination depends on a measurement of a line integral of the density *gradient,* it is more straightforward to perform interferometry, where the line density is measured directly in terms of the number of fringe shifts, N. The electron density n_e depends on N and the plasma length, L, as n_e = $1.4\text{x}10^{20}\, N\, /\, (L[\text{mm}]\ \lambda[155\ \text{Å}])\ \text{cm}^{-3}$. Because of recent improvements in the fabrication of beamsplitters, this is now possible. The silicon nitride beamsplitters we have used are 1 cm^2 in area with a thickness of 1000 Å onto which is put 8 layers of Mo/Si. The beamsplitter reflectivity and transmission are 20-25% at 155 Å. The beamsplitters are very fragile; they often do not survive the shock of the Nova laser irradiating the XRL target some distance away; those that do survive the shot rarely endure subsequent chamber pumpdown.

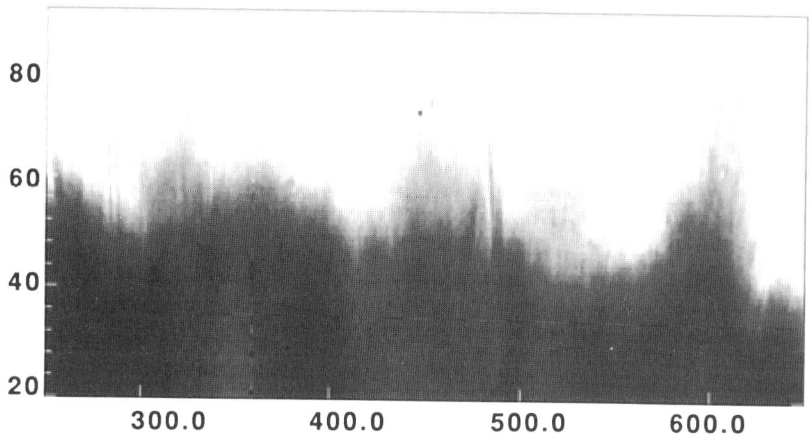

Figure 2. X-ray laser backlit image of a foil (10 μm CH on 3 μm Al) accelerated by a 1 ns Nova pulse taken at 0.5 ns. The picture shows the central region of the foil plasma, which is being driven upward in the picture. The scale is in microns; the 700 μm diameter foil was originally located at zero on the vertical scale. The Nova beam was smoothed, but there are the spatial nonuniformities in the plasma.

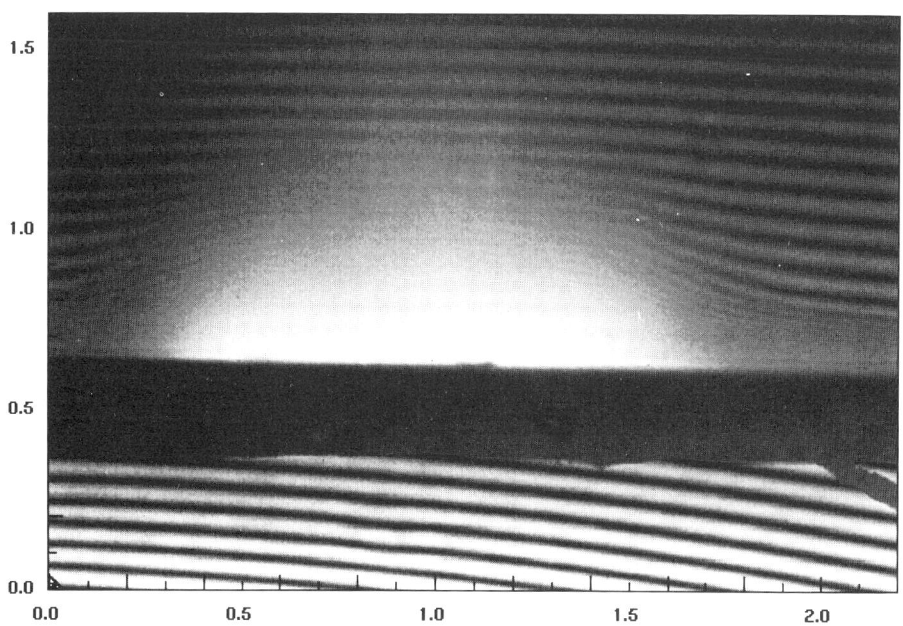

Figure 3. X-ray laser interferogram of a laser heated mylar disk viewed side-on. The laser is incident from above in the picture. The scale is in millimeters.

Figure 3 is the first x-ray laser interferogram of a laser-heated target, which was a 200 μm thick mylar slab. The mylar is irradiated from above in the picture at $1.5x10^{14}$ W/cm^2 over a 1 mm FWHM spot. The image view is side-on. The exposure is time-integrated with the XRL arriving near the end of the 1 ns heating pulse. Fringe visibility outside the region of intense self-emission is excellent. [Unfortunately, there is a strong oxygen line within the system bandpass so the emission region near the mylar, which contains oxygen, is bright and extensive. Subsequent plastic targets did not contain oxygen.] Regions with more than ten fringe shifts are visible 50 μm off the surface outside the central spot. It is possible to infer densities of 5 x 10^{20} cm^{-3} 300 μm off the surface above the spot, in good agreement with simulations. Between 300 μm and the brighter outer regions of self-emission, there is evidence of motion blurring of fringes due to the finite duration of the XRL.

Several changes will be made in order to improve the quality of plasma XRL interferometry. Gating the image will reduce self-emission by an order of magnitude. XRL signal can be enhanced relative to self-emission by narrowing the optical system bandpass. This can be accomplished by using different multilayer mirrors. In order to limit motion blurring and improve spatial resolution, the XRL pulselength needs to be shortened. Pulselengths as short as 45 ps have been demonstrated using a traveling wave pump. [4]

3. Conclusions

X-ray lasers possess unique features which make them ideal tools for studying fast-evolving plasmas. Among these characteristics are short wavelength, small bandwidth, high brightness, low divergence, short pulse duration, good coherence, and, because the wavelength is soft relative to high energy backlighters, the ability to use near-normal-incidence optics. High spatial resolution radiography using an XRL, which has revealed unexpected structure in driven plasmas, makes use of some of these properties. We can take advantage of most all or most of these properties in moiré deflectometry and interferometry. X-ray laser interferometry is expected to become a reliable and highly useful quantitative density diagnostic for laser-produced plasmas.

Acknowledgments

This work was performed under the auspices of the U. S. Department of Energy by Lawrence Livermore National Laboratory under contract number W-7405-ENG-48.

References

[1] Da Silva L, MacGowan B, Mrowka S, Koch J, London R, Matthews D, and Underwood J, 1993 Optics Lett. **18**, 1174

[2] Da Silva L, Cauble R, Frieders G, Koch J, MacGowan B, Matthews D, Mrowka S, Ress D, Trebes J, and Weiland T, 1993 in *Ultrashort Wavelength Lasers II*, SPIE Proceedings Vol. **2102**, 158

[3] Ress D, Da Silva L, London R, Trebes J, Mrowka S, Procassini R, Barbee Jr. T, and Lehr D, 1994 Science **265**, 514

[4] Da Silva L, London R, MacGowan B, Mrowka S, Matthews D, and Craxton R, 1994 to appear in Optics Lett. 15 September issue

Inst. Phys. Conf. Ser. No 140: Section 6
Paper presented at Laser Interact. with Matter Conf., Oxford, 19–23 September 1994
© 1995 IOP Publishing Ltd

Collisional-excitation X-ray lasers by multiple-pulse pumping

Y Kato, H Daido, R Kodama, K Murai, G Yuan, S Ninomiya, Y Oshikane, M Takagi, T Kanabe, C H Nam[1)], I W Choi[1)], and T Kawachi[2)]

Institute of Laser Engineering, Osaka University, Suita, Osaka 565 Japan
1) Korean Advanced Institute for Science and Technology, Taejon, Korea
2) Faculty of Engineering, Kyoto University, Kyoto 606 Japan

Abstract. By multiple, short pulse irradiation of solid targets, the plasma profiles can be controlled more flexibly and the plasma can be heated with less laser energy in comparison to single pulse irradiation. Soft x-ray amplification in Ne-like and Ni-like ions has been studied by irradiation of curved slab targets with multiple (2-4), short duration (100 ps) laser pulses of 1μm wavelength and 150 J/pulse energy. Intense soft x-ray laser pulse with narrow divergence has been generated for the J=0-1 line of Ne-like Ge for each pumping pulse except for the first pulse, suggesting a possibility for multiple-pass amplification with repetitive pumping. With the Ni-like ions of lanthanides Nd-Dy, amplification at 7.9-6.4 nm was observed with the gain-length products of 7.5. The efficiency obtained with the multiple-pulse pumping is approximately an order of magnitude higher compared to longer-duration, single-pulse pumping.

1. Introduction

Collisional excitation (CE) x-ray lasers have been quite successful for saturated amplification [1,2], high brightness [2], and broad wavelength coverage [3]. The CE x-ray lasers have been pumped so far with a single, high energy laser pulse of typically 1 ns duration. Improvement in the efficiency of the CE x-ray lasers is important in order to use them as laboratory instruments. With the Ne-like Ge laser, we have demonstrated [4] increase in the output power and reduction of the beam divergence using a curved slab target [5] where x-ray refraction in a gradient-density plasma is compensated. Improvement in the efficiency in terms of pumping energy may be feasible by pumping with shorter duration, multiple laser pulses, where a plasma generated by the first pulse is heated by the second pulse [6]. In this paper we report soft x-ray amplification in Ne-like Ge and Ni-like lanthanide elements using curved slab targets pumped by 100-ps duration, multiple laser pulses of 1-μm wavelength.

2. Curved slab targets

Since the optimum electron density of the gain region of the CE x-ray lasers is fairy high (of the order of 10^{20} cm^{-3}), propagation of the soft x-ray laser beam in the plasma is affected significantly by the electron density gradient. Refraction of the x-ray laser beam in the plasma limits an effective gain length and increases the beam divergence, thus reducing the spatial coherence of the x-ray laser beam. By bending a slab target to a curved profile, a waveguide to an x-ray laser beam is formed along the horizontal direction [5]. Analyses on Gaussian beam propagation through the waveguide shows that stable x-ray propagation and amplification is obtained resulting in narrow divergence in the horizontal direction and large divergence in the vertical direction [7].

A Ge stripe coated on a glass plate was irradiated with a 1-μm wavelength laser pulse of 1-ns duration at an irradiance of 1.7×10^{13} W/cm^2. For both of the J=0-1, 19.6 nm and the J=2-1, 23.6 nm laser lines, the x-ray laser intensity increased and the beam divergence decreased with increase in the bending angle of the target. An x-ray laser beam of 1 mrad divergence has been generated by double pass amplification in a curved target. The temporal pulse shapes of the lasing lines obtained with a flat target and a curved target (Fig.1) shows that the J=0-1 line maximizes before the peak of the pumping laser pulse (t=0) whereas the J=2-1 line maximizes at t=0.3 ns and the intensity enhancement due to curvature is more pronounced for the 0-1 line. These results suggests that, in previous experiments with plane targets, amplification of the 0-1 line over long gain lengths has been prevented, since excitation to the J=0 level is mostly electron-collisional monopole excitation in the high density plasma where x-ray refraction effect is significant.

2. Multiple-pulse pumping of Ne-like Ge laser

The early peaking of the J=0-1 line with respect to the 1-ns pumping pulse suggests a possibility for reduction of the pumping duration and thus the pumping energy, if the pumping intensity is kept at a similar value. Formation of a plasma before pumping is necessary in this case, since the density scale length at the gain region is too short with the single, short pulse pumping. We have irradiated a curved Ge slab target with two laser pulses of 100 ps duration separated by 300 ps. With the double pulse pumping, the J=0-1 line intensity exceeds

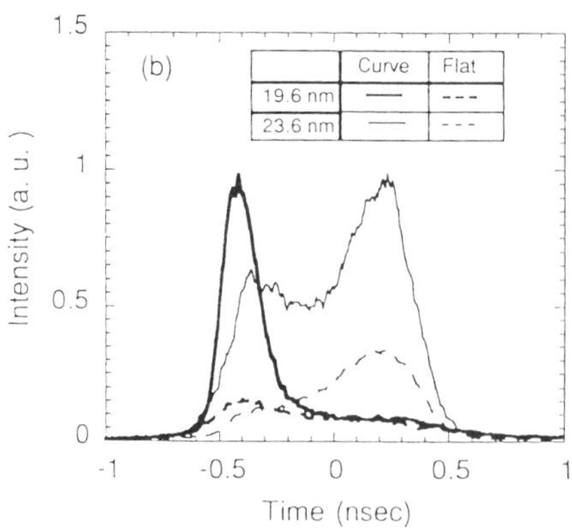

Fig.1 Pulse shapes of the 0-1 and 2-1 laser emissions generated with a flat and curved slab targets pumped by a 1-ns laser pulse peaked at t=0 ns.

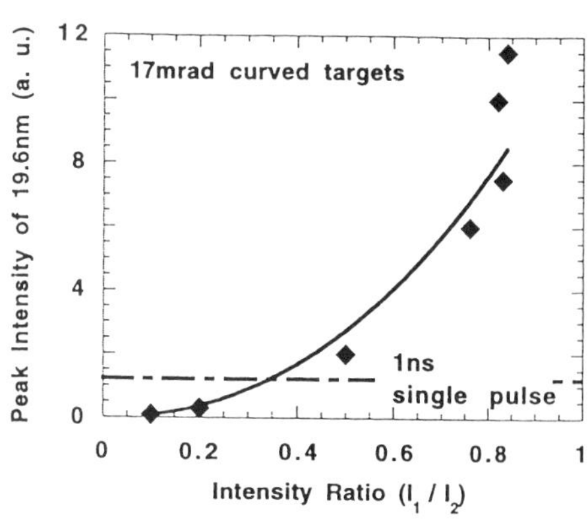

Fig.2 Peak intensity of the J=0-1 line generated with 1ns single pulse and 100ps double pulse irradiation.

completely over other lasing lines including the 23.2 and 23.6 nm laser lines which are the strongest with 1-ns, single pulse pumping. The lasing is observed only at the second pumping pulse showing that the preplasma formation is necessary for pumping with a short pulse. The energy of the second laser pulse was kept constant at 150 J over 6 cm, corresponding to an intensity on target of 2.6 x 10^{13} w/cm^2, whereas the energy of the first laser pulse was varied to investigate the effect of the preplasma formation. Note that the actual target length was up to 3 cm, above which the x-ray laser intensity did not increase due to the 100 ps pulse width. Figure 2 shows the peak intensity of the 19.6 nm 0-1 line plotted against the intensity ratio of the first to the second pumping laser pulses. This data shows that long-scale length plasma formation is very effective in enhancing the 0-1 line intensity and its intensity becomes approximately 10 times higher than the intensity obtained with the 1-ns, 1-kJ single pulse pumping (dashed line).

Multiple-pass amplification of the x-ray laser will become feasible if gain exists in the plasma for each pumping laser pulse. We have studied amplification of the J=0-1 line generated by pumping with 4 laser pulses separated by 400 ps for the first 3 pulses and 2 ns between the 3-rd and the 4-th pulse. The 2-ns separation corresponds to 60-cm beam propagation distance which is long enough to spatial filter only the coherent portion of the x-ray laser beam for amplification under the 4-th pumping pulse. A streak data of the angular intensity distribution shows that ASE is obtained even for the long pumping separation of 2 ns. The angular shift for the last x-ray laser pulse is larger by approximately 5 mrad than that for the first 2 pulses. These results indicate a possibility for the triple-pass amplification to generate a coherent x-ray laser beam which could be extracted out of the cavity formed by the two mirrors due to the larger refraction angle in the third amplification.

4. Multiple-pulse pumping of Ni-like lanthanide lasers

Figure 3 shows a simplified energy level diagram for the Ni-like ions. In previous works x-ray amplification at the two J=0-1 transitions have been observed. Based on collisional-radiative calculation, it is expected that the plasma parameters required for pumping of the Ni-like lanthanide elements are similar to those for the Ne-like Ge.

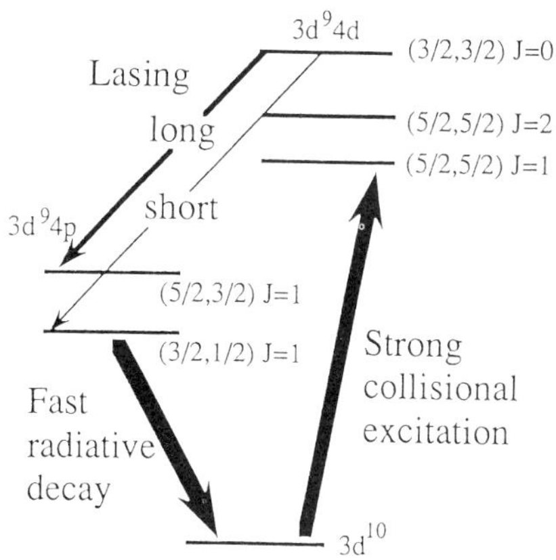

Fig.3 Schematic energy level diagram of the Ni-like laser.

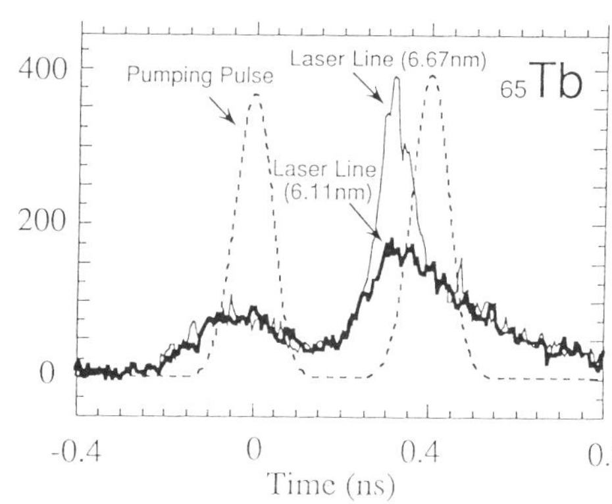

Fig.4 Temporal profiles of the pumping laser pulse, and the long- and short-wavelength components of of the lasing transitions.

We have tested soft x-ray amplification in Nd (Z=60), Sm (Z=62), Gd (Z=64), Tb (Z=65) and Dy (Z=66) by pumping with 2-3 laser pulses of 100 ps duration separated by 400 ps and 1-μm wavelength at an irradiance of 7.9 x 10^{13} W/cm^2. Careful handling of the targets were taken to prevent oxidization of the lanthanide coatings. Amplification in the longer wavelength component of the 0-1 transition was observed for all these elements, covering the wavelengths of 7.95-6.43 nm, with the gain coefficient of 2.5-3.1 cm^{-1} over the gain length of 2.5 cm, corresponding to the maximum gain-length product of 7.8 for Nd. Amplification at the shorter wavelength component of the 0-1 transition was observed, although weak, for Gd and Tb. This is consistent with the wavefunction calculation which shows that the oscillator strength for the shorter wavelength transition increases with Z and eventually becomes larger than that for the longer wavelength transition.

Figure 4 shows the pulse shapes of the pumping pulse and the long- and short-wavelength components of the x-ray laser pulses for Tb. The temporal profile of the pumping pulse was inferred from the pulse shape of the continuum (bremsstrahlung) emission observed simultaneously with the x-ray laser pulse. We find that the x-ray laser pulse is amplified in the rising part of the pumping laser pulse. Although exact cause for this experimental result is not yet clear at present, this may imply that further reduction of the width and energy of the pumping laser pulse may be feasible for observation of the similar gain value.

5. Summary

Experimental results on soft x-ray amplification in Ne-like Ge and Ni-like lanthanide ions using curved slab targets pumped by multiple, short laser pulses have been presented. It has been shown that under these pumping conditions the intensities of the lasing lines especially those arising from the J=0-1 transitions are enhanced significantly in comparison to the single, longer pulse pumping. The relative efficiency gl/E, where gl is the gain-length product and E is the pumping laser energy, is an order of magnitude higher than the efficiencies obtained with single, long pulse pumping.

References

[1] Carillon A, et al., 1992 Phys. Rev. Lett. **68,** 2917
[2] Da Silva L B, et al., 1993 Opt. Lett. **18,** 1174
[3] MacGowan B J, et al., 1990 Phys. Rev. Lett. **65,** 420
[4] Kodama R, et al., to be published.
[5] Lunny J G, 1986 Appl. Phys. Lett. **46,** 891
[6] Da Silva, et al., 1993 Proc. SPIE **2012,** 158
[7] Kato Y, et al., 4th Int. Colloq. X-Ray Lasers, May 16-20, 1994, Williamsburg VA.
[8] Dade H, et al., 4th Int. Colloq. X-Ray Lasers, May 16-20, 1994, Williamsburg VA.

Inst. Phys. Conf. Ser. No 140: Section 6
Paper presented at Laser Interact. with Matter Conf., Oxford, 19–23 September 1994
© 1995 IOP Publishing Ltd

Near field imaging of the C VI high gain recombination x-ray laser driven by a 20 J, 2 ps laser beam

J Zhang[1], M H Key[1,2], P A Norreys[2], G J Tallents[3], D Neely[2], A Behjat[3], C Danson[2], A Demir[3], L Dwivedi[3], M Holden[3], P B Holden[4], C L S Lewis[5], A G MacPhee[5], G J Pert[4], S A Ramsden[4], S J Rose[2,6], Y F Shao[7], O Thomas[2], F Walsh[2], Y L You[8]

[1]Department of Atomic and Laser Physics, University of Oxford, Oxford, OX1 3PU, UK
[2]Central Laser Facility, Rutherford Appleton Laboratory, Chilton, OX11 0QX, UK
[3]Department of Physics, University of Essex, Colchester, CO4 3SQ, UK
[4]Department of Physics, University of York, York, YO1 5DD, UK
[5]Department of Pure and Applied Physics, Queen's University, Belfast, BT7 1NN, UK
[6]Department of Physics & Space Science, University of Birmingham, Birmingham, B15 2TT, UK
[7]Scientific visitor from Institute of Applied Physics and Computational Mathematics, Beijing 100088, China
[8]Scientific visitor from Institute of Nuclear Physics and Chemistry, Chengdu 610003, China

Abstract. Recent experimental results of imaging of the high gain C VI recombination x-ray laser are described. Measurements of the near field distribution of the x-ray laser were made in the exit plane using a spherical multilayer mirror imaging system. The output power density was estimated to be 1.4×10^5 W/cm^2.

1. Introduction

Efficient production of high gain x-ray lasers at shorter wavelengths is a major goal in the development of x-ray lasers. Impressive progress has recently been made in different regimes [1-5] towards this goal.

Adiabatically cooled recombination lasers have long been promising in this respect but have suffered from a reduction of apparent gain coefficient with increase in plasma length. One way to tackle this problem is to produce high gain output in a smaller length of plasma by increasing the gain coefficient. Plasmas driven by picosecond laser pulses have a sufficient small scale length that heating occurs close to solid density, thus enhancing the volumetric expansion and adiabatic cooling of the plasma [6]. Based on this idea, an exceptional high gain operation has been demonstrated at 18.2 nm in a C VI recombination laser driven by a 2 ps, 20 J Nd-glass laser operating by chirped pulse amplification at DRAL [5]. In order to obtain information on the uniformity, size and distribution of the high gain region in the plasma, an imaging experiment has been conducted recently. We report here the measurements of the near field imaging of the C VI x-ray laser in the exit plane using a spherical multilayer mirror system.

2. Experimental set-up

A schematic diagram of the experimental set-up is shown in figure 1. A 2 psec, 20 TW CPA laser beam at 1.053 μm was generated using the Nd:glass laser VULCAN which was equipped with a chirped pulse amplification (CPA) system [7]. The final pulse recompression grating was located inside the target chamber under vacuum. The recompressed beam of diameter 140x88 mm was incident onto a partially reflecting turning mirror which transmitted 5% of the recompressed pulse for measurements of the laser pulse length, prepulse (contrast ratio), spectral distribution and divergence. An f/4 off-axis parabolic mirror reflected the beam to a spot focus which was imaged by an f/4 off-axis spherical mirror to produce a line focus 7 mm long and 20 μm wide.

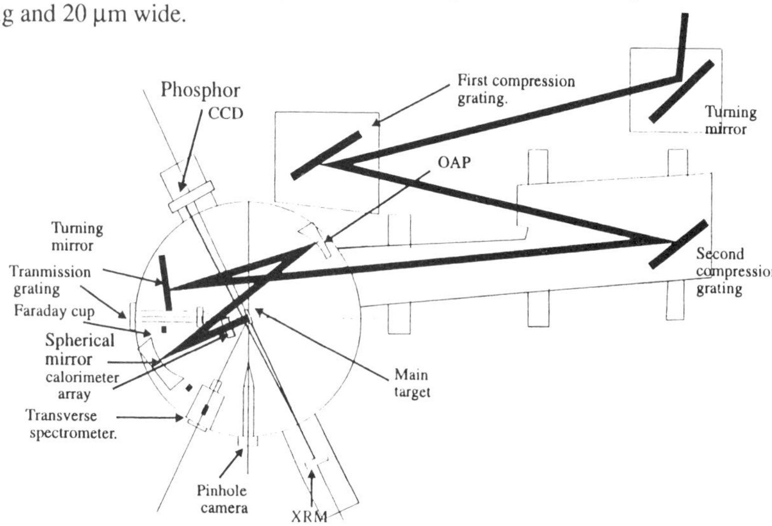

Figure 1. Schematic layout of the experimental set-up

It is important to minimise pre-pulse and post-pulse to avoid pre-forming a plasma and post-heating of the plasma respectively. Pre-forming a plasma may reduce the laser gain and both pre- and post-pulses can be expected to increase the plasma x-ray emission which is not associated with x-ray lasing. For this experiment, a large background contrast ratio of between 10^6 to 10^7 was measured [7]. The pulse length was monitored during the experiment by a single shot third order auto-correlator and was typically in the range of 2 ± 0.6 ps. The incident spatially averaged irradiance in the line focus was (6 ± 1)x10^{15} W/cm^2.

The targets used in the experiment were 7 μm diameter carbon fibres supported at one end [5]. The fibre targets were positioned with better than ± 2 μm spatial accuracy and ± 1 mrad angular accuracy using a modified split-field microscope system [5]. The irradiation of the fibre targets was accomplished by placing the fibres in the 20 μm wide line focus using a CW alignment laser. The free end of the fibre target was always placed well within the irradiated line focus length so as to avoid a cold end of the line plasma, which could absorb the x-ray radiation along the fibre length.

A 250 mm focal length concave x-ray multilayer mirror (central wavelength at 18.2 nm, bandwidth of 2.0 nm) situated at a distance 320 mm from the free end of the fibre target imaging at a magnification of 3.8x the output end of the lasing plasma onto an XUV sensitive phosphor that was coupled to a computer controlled CCD detector. The CCD detector was replaced for some shots by Kodak 101-04 film to calibrate the absolute output energy of the x-ray laser beam. A 1.5 μm thick Al filter provided rejection of optical emission.

Off-axis diagnostics included a flat-field grating spectrometer with a 2400 line/mm grating at 45 degrees off the fibre axis used to record C V and C VI resonance spectra in the range of 2.0 nm - 10.0 nm to diagnose the characteristics of plasma which could effect the soft x-ray lasing performance during adiabatic expansion [5,8]. The length and uniformity of the plasma were monitored with an x-ray pinhole camera recording an image of emission at wavelengths below 1 nm. A group of Faraday cups/calorimeters gave an estimate of the absorbed energy in the plasma and of the angular variation of ion emission.

3. Results

An image of emission around 18.2 nm from the output end of the lasing plasma is shown in figure 2. The image is dominated by Balmer α line emission rather than continuum as the line emission measured in the experiments is an order of magnitude stronger than the continuum emission when integrated over an 2.0 nm multilayer mirror bandwidth [5,9]. The peak intensity region has full widths at half maximum of 20 μm and 15 μm in the horizontal and vertical directions respectively. The spatial location of the emission peak was estimated to be displaced by 100 ± 50 μm towards the laser from the surface of the fibre by comparison with images of continuum emission from a Cu slab target. Modelling [10] predicts that the output intensity should peak at 35 μm from the target surface with a spatial range of 10 to 15 μm (full width at half maximum). This is broadly consistent with the observations within the ± 50 μm error in the referencing of the spatial location. The effects of beam refraction on the position of the beam in the near field at the electron densities (~10^{19} cm^{-3}) where the peak gain exists [6] due to estimated maximum refractive angles of ≤ 1 mrad are negligible.

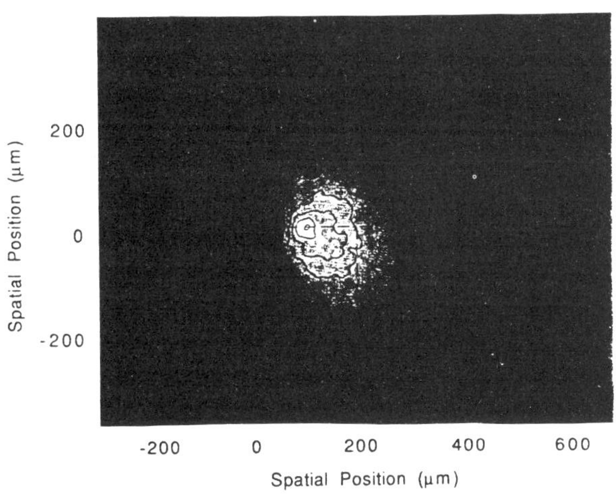

Figure 2. Image showing the intensity distribution of the Balmer α lasing beam at the exit plane of the recombining plasma. The positions are relative to the initial fibre axis. The contours representing 3, 5, 9, 14 , 28 and 32 relative Balmer α flux with the most intense central region only ~ 20 μm.

Further evidence that the observed image was that of ASE from the laser was obtained from the agreement of calculated and observed ASE output intensity for a gl product of 6. The calculated intensity was $\left(\frac{2h\upsilon^3}{c^2}\right)\frac{\exp(gl)}{\sqrt{gl}}\Delta\nu$. The bandwidth $\Delta\nu$ of the ASE was estimated assuming Doppler broadening with a temperature of 10 eV. The experimental intensity was obtained from the CCD image. The sensitivity of the CCD was calibrated to factor of a few accuracy by comparing its response with Kodak 101-04 film.

The output energy per steradian from the 5 mm long plasma was estimated from the exposure level on the recording film and the solid angle and component reflectivities of the spectrometer to be 1.4 J/sterad. A similar estimate of the energy collected by the larger (1cm) aperture of the mirror gave a figure of 2.2 J/sterad. The output aperture can be estimated from the image in figure 2 to be 2.3×10^{-6} cm^2. The duration of the emission was not measured but modelling predicts it to be about 50 ps. The power density of the source was then determined from the data to be 1.4×10^5 W/cm^2. It is of interest to compare this with the saturation intensity of the laser predicted by the modelling. The lifetime of the n = 3 level was assumed on basis of modelling to be dominated by Lyman β decay, and the line width of the transition was taken to be due to Doppler broadening at 10 eV temperature. The calculated saturation intensity is then calculated to be 2×10^9 W/cm^2. The onset of saturation for this laser is predicted using the modelling to occur when the plasma length is about 8 mm. This onset of saturation for a small increase in length reflects the fact that the peak gain is higher than the apparent gain so that with increasing length there should be an increase of apparent gain. A saturated laser of this type would generate a peak power of about 5 kW with a pulse length of about 50 ps and energy of about 200 nJ.

4. Conclusion

In conclusion, we have measured the near field distribution of the C VI recombination x-ray laser in the exit plane. The output power density from a 5 mm long plasma was estimated from experimental data to be 1.4×10^5 W/cm^2. A saturated and potentially useful x-ray laser beam at 18.2 nm can be produced with a small increase in plasma length and about 30 J driver energy. The rather low plasma density at the time of laser action implies little refractive distortion relative to collisional lasers so there would be better scope for achieving a diffraction limited beam.

References

[1] Nilsen J, MacGowan B J, *et al* 1993 *Phys. Rev.* A**48** 4682.
[2] Nagata Y, Midorikawa K, *et al* 1993 *Phys. Rev. Lett.* **71** 3774.
[3] Rocca J J, Shlyaptsev V, *et al* 1994 *Phys. Rev. Lett.* **73** 2192.
[4] Lemoff B E, Yin G Y, *et al* 1994 *Phys. Rev. Lett.* (to be published).
[5] Zhang J, Key M H, *et al* 1994 *Phys. Rev. Lett.* **73** no.23.
[6] Zhang J and Key M H, 1994 *Appl. Phys.* B**58** 13.
[7] Danson C D, Barzanti L J, *et al* 1993 *Opt. Commun.* **103** 392.
[8] Zhang J, Key M H, *et al* 1993 *Opt. Commun.* **95** 51.
[9] The mirror was manufactured by Changchun Institute of Optics and Fine Mechanics, Academia Sinica, Changchun, People's Republic of China
[10] Borovskiy A V, Holden P B, *et al* 1992 *J. Phys.* B **25** 4991.

Inst. Phys. Conf. Ser. No 140: Section 6
Paper presented at Laser Interact. with Matter Conf., Oxford, 19–23 September 1994

© 1995 IOP Publishing Ltd

Influence of dielectronic recombination on gain of X-ray lasers with Li-like ions

V. Stâncàlie (*), A. Sureau, A. Klisnick, C. Möller, H. Guennou, Y. Bérété.

(*) Institute of Atomic Physics, Laser Department, P. O. Box MG-6, Bucharest-Magurele, 76900 Romania.
Laboratoire de Spectroscopie Atomique et Ionique, URA du CNRS, Université Paris-Sud, Bât. 350, F- 91405 Orsay Cedex, France.

Abstract. A detailed analysis of dielectronic recombination of Li-like Al^{10+} ions in laser-produced plasmas is presented. Efficiencies of different recombining ways according to the electron temperature are compared. Numerical simulations taking into account the spatial and temporal dependences of the plasma density and temperature are used to calculate the intensity gain coefficients of the X-ray lasing lines associated to the transitions 3d-5f (λ = 10.57 nm), 3d-4f (λ = 15.46 nm) and 4f-5g (λ = 33.47 nm). These gain coefficients are about 10% smaller with dielectronic recombination than without.

1. Introduction

The analysis and the interpretation of the X-UV spectra emitted by laser-produced plasmas require the utilization of numerical models to evaluate the populations of the ionic excited levels and thence the intensity of the spectral lines emerging from the plasma. Local populations result from a balance between excitation and decay processes, the rate coefficients of which depend on plasma quantities such as density and temperature which are not constant during the plasma lifetime. Excitation and decay processes are the collision-induced transitions, ionizations, recombinations, the radiative transitions and recombinations, the dielectronic recombination. For Li-like ions in recombining plasmas, the last process is mainly due to the capture of free electrons by ions in $1s^22s$ and $1s^22p$ configurations producing autoionizing Be-like levels which emit a photon to give bound levels of $1s^22snl$ and $1s^22pnl$ configurations. Owing to the electron temperature during the recombination phase ($\leq$ 200 eV), the most probable ways involve the autoionizing levels $1s^22pnl\ ^{1,3}L'_{J'}$ and $1s^23pnl\ ^{1,3}L'_{J'}$ (with initial ion $1s^22s$) and $1s^23dnl\ ^{1,3}L'_{J'}$ (with initial ion $1s^22p$). They can be characterized respectively by the core transitions 2s-2p, 2s-3p, and 2p-3d.

In this work, a detailed study of the dielectronic recombination for Li-like Al^{10+} ions in recombining laser-produced plasmas is performed. Semi-empirical formulas (Burgess 1964 and Sobelman *et al* 1981) are compared. By means of numerical simulations above mentionned (time- and space-dependent density and temperature being taken into account), the influence of dielectronic recombination on gain coefficient of the radiation of spectral lines corresponding to 3d-5f (λ = 10.57 nm), 3d-4f (λ = 15.46 nm), and 4f-5g (λ = 33.47 nm) transitions are then examined.

2. Calculation method

For each way characterized by a core transition i-j, the population density of the initial level decays according to the relation $dN_i/dt = -N_i N_e K_d(i\text{-}j)$ where N_e is the free-electron density and $K_d(i\text{-}j)$ the total rate coefficient (sum over all possible intermediate and final states) of the considered way. During the time interval dt the total dielectronic recombination of the considered ionic species is then $\sum_{(i,j)} N_i N_e K_d(i\text{-}j)$. Complete detailed calculation of a K_d coefficient requires a lot of intermediate data (capture cross sections, rates, and transition probabilities involving a great number of states) and is therefore not easily manageable in a simulation calculating, for each time step, the populations of many levels in a number of plasma cells. However, convenient semi-empirical formulas can be used. In particular, for a Maxwellian free-electron velocity distribution, a general formula given by Burgess (1964) is

$$K_d(\alpha_0\text{-}\alpha) = 10^{-13}\, B_d\, \beta^{3/2} \exp(-\beta\chi_d) \quad (\mathrm{cm^3 s^{-1}}) \tag{1}$$

where $\beta = (z+1)^2 Ry/T$, z being the ionic charge, T the electron temperature,

$$B_d = 480\, f(\alpha_0\alpha) \left(\frac{z\chi}{z^2 + 13.4}\right)^{1/2} \left[1 + 0.105\,(z+1)\chi + 0.015\,(z+1)^2\chi^2\right]^{-1},$$

$$\chi_d = \chi\left[1 + 0.015\frac{z^3}{(z+1)^2}\right]^{-1}, \qquad \text{and} \qquad \chi = \frac{E\alpha - E\alpha_0}{(z+1)^2 Ry},$$

α_0 designates the initial core level, α the intermediate core level, $f(\alpha_0\alpha)$ is the oscillator strength of the corresponding transition and E_j is the energy of level j.

An expression similar to (1) can be deduced from a simplified model of electron capture described as a collisional excitation of the ionic core by an electron with kinetic energy below the excitation threshold. The missing energy is provided by the binding energy won by the capture of the incident electron on a bound orbital. The factor B_d is then given by an explicit sum over the quantum numbers *n* and *l* of terms containing cross sections which can be calculated numerically. This has been done by Sobelman *et al.*(1981) for H-like to Ne-like ions. Having fitted their results with a numerical formula, they have obtained

$$K_d(\alpha_0\alpha) = 10^{-13}\, Q_d(\alpha_0\alpha)\, \frac{A_d}{2l_0 + 1}\, \beta^{3/2} \exp(-\beta\chi_d) \quad (\mathrm{cm^3 s^{-1}}) \tag{2}$$

with $Q_d(n_0 l_0^m, n_0 l_0^{m-1}\, nl) = m$

$$Q_d(n_0 l_0^N\, nl^m,\, n_0 l_0^{N-1}\, nl^{m+1}) = N\left[1 - \frac{m}{2(2l+1)}\right]$$

the numerical values of A_d and χ_d being tabulated.

Formulas (1) and (2) give generally final results in close agreement. However, in our numerical simulations, we consider the population densities of the energy levels, for which the Burgess formula is well adapted, when the formula of Sobelman *et al.* is rather suited to the total configuration scheme. In addition, the A_d and χ_d coefficients of Eq. (2) are not given for core excitation 2p-3d in Li-like ions. We have therefore chosen to use Eq. (1) in calculations whose results are presented below.

3. Results and discussion

Since the $1s^2 2s$, $1s^2 2p_{1/2}$, and $1s^2 2p_{3/2}$ levels are in mutual Boltzmann's population ratio (Makhrov *et al.* 1994), it is possible to consider a global rate coefficient K_d of dielectronic recombination associated with the ground-level population N_1:

$\Sigma_{(i,j)} N_i N_e K_d(i\text{-}j) = N_1 N_e K_d$ with $K_d = \Sigma_{(i,j)} K_d(i\text{-}j) \frac{g_i}{g_1} \exp[-(\frac{E_i - E_1}{kT})]$. The values of K_d (cm^3s^{-1}) as a function of electron temperature (kT in eV) are shown in Fig. 1. The first maximum (near 10 eV) is principally due to the core transitions 2s-2p while the other two recombination ways (core transitions 2s-3p and 2p-3d) are dominant above 50 eV and maximum at 150 eV.

Numerical simulations with the current LSAI modelling described by Makhrov *et al.* (1994) have been performed to evaluate the influence of dielectronic recombination on the population densities of excited levels in a laser-produced plasma. In particular the spatial and

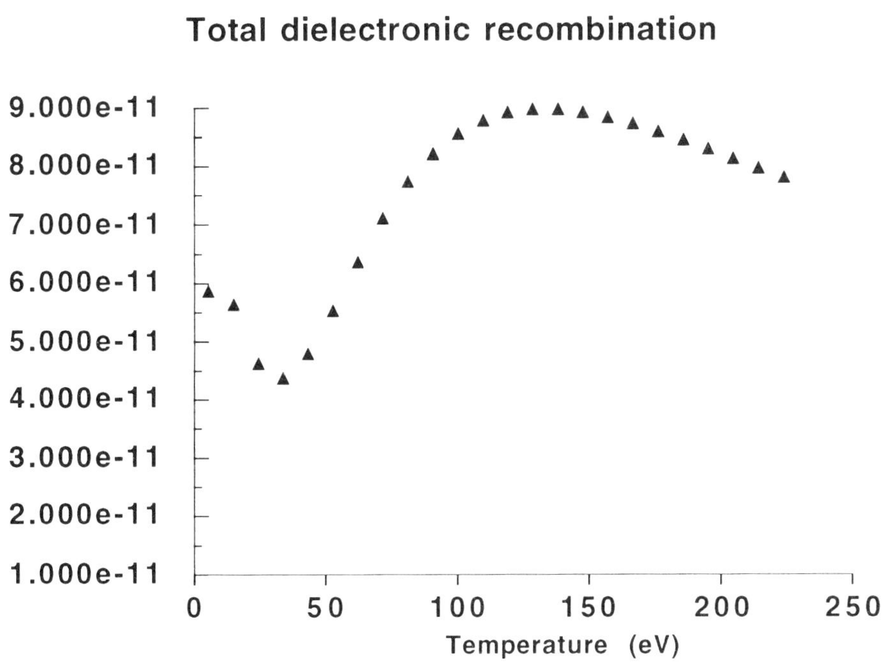

Figure 1. Dielectronic-recombination rate coefficient (cm^3s^{-1}) of Al^{10+} ion as a function of temperature (eV) according to the formula of Burgess (1964).

temporal evolutions of the gain coefficients (intensity gain with propagation length) for the 3d-4f, 3d-5f, and 4f-5g spectral lines of the Li-like Al^{10+} ion have been determined with and without dielectronic recombination respectively. The results are given in figures 2 to X. Figures 2 to X display results obtained from a one-dimensional simulation of a plasma generated by irradiating cylindrically an aluminium fiber of 7 μm diameter by a laser pulse of following characteristics: intensity 10^{13} w/cm^2, wavelength 0.53 μm, Gaussian temporal shape of 120 ps duration (full width at half maximum) with maximum at 70 ps. The total recombination, including radiative and collisional (three-body) processes, is faster with than without dielectronic recombination. At every time the Li-like ions are less abundant in all plasma regions, while the density of He-like ions is practically inchanged. Concerning the gain coefficients relative to Li-like transitions, the density lowering induces logically a lowering of inversion densities, and thence of the gains. However, it doesn't modify significantly the temporal profiles and appears only as a lowering of amplitude of about 8 % for each transition.

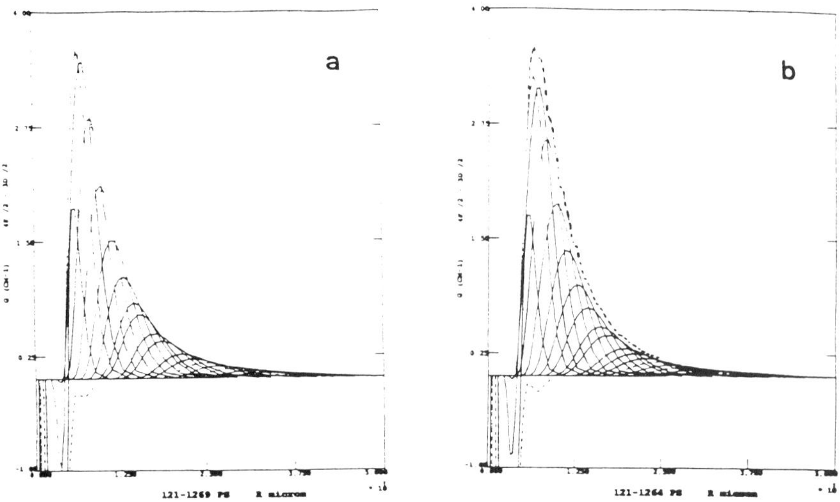

Figure 2. Spatial distribution of gain (per centimeter) on $3d_{5/2}$-$4f_{7/2}$ transition of Al^{10+} at different times (separated by 50 ps), as a function of the distance (µm) to the target.
a: without dielectronic recombination, b: with (the dotted line is the temporal envelope of curves a).

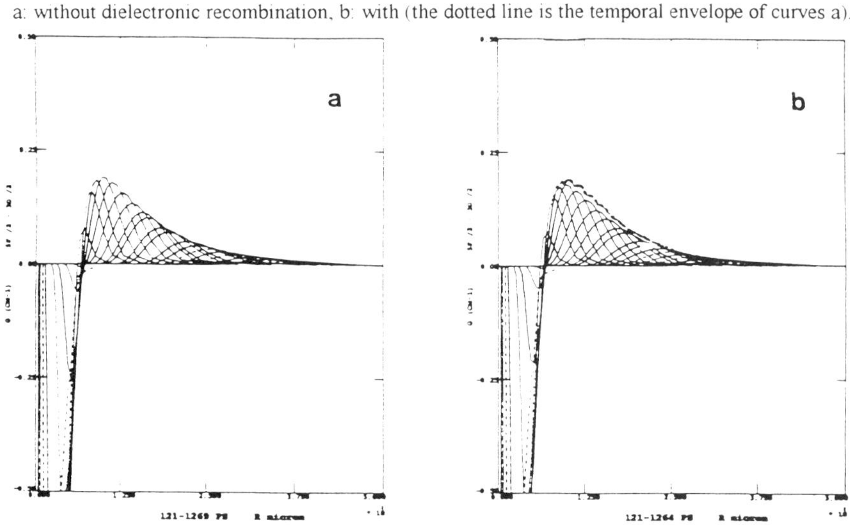

Figure 3. Spatial distribution of gain (per centimeter) on $3d_{5/2}$-$5f_{7/2}$ transition of Al^{10+} at different times (separated by 50 ps), as a function of the distance (µm) to the target.
a: without dielectronic recombination, b: with (the dotted line is the temporal envelope of curves a).

References

Burgess A 1964 *Astrophys. J.* **139** 776

Makhrov V, Roerich V, Starostin A, Stepanov A, Klisnick A, Sureau A and Guennou H 1994 *J. Phys. B: At. Mol. Opt. Phys.* **27** 1899

Sobelman I I, Vainshtein L A, Yukov E A 1981 *Excitation of atoms and broadening of spectral lines*, Springer-Verlag

Inst. Phys. Conf. Ser. No 140: Section 6
Paper presented at Laser Interact. with Matter Conf., Oxford, 19–23 September 1994
© *1995 IOP Publishing Ltd*

Photon reabsorption in recombining plasmas

Djamel Benredjem*, Alain Sureau, Philippe Zeïtoun, Annie Klisnick and Clary Möller

Laboratoire de Spectroscopie Atomique et Ionique, Université Paris-Sud, Centre d'Orsay, Bât. 350, 91405 Orsay Cedex (France)

*on leave from Université d'Annaba, Institut de Physique, BP 12, 23000 Annaba (Algeria)

Abstract. Photon reabsorption in recombining plasmas has been examined for a thin fibre irradiated by a laser with $\lambda = 1.06$ µm and pulse duration= 120 ps. Gain corresponding to lasing lines of Li-like sulfur in the X-ray range is calculated in a self-consistent manner by using the hydrodynamical results of 1D Lagrangian simulation as inputs. The reabsorbed photons are assumed to travel a maximum distance which is given by the inverse of the absorption coefficient. Reabsorption is found to reduce gain by a much smaller amount than is predicted by most calculations, giving better agreement with experiments.

1. Introduction

In laser-produced plasmas, photons due to radiative transitions from the lower level of a X-ray lasing transition to a lesser-energy level of the same ion may repopulate the former and destroy the population inversion, leading to an attenuation of the gain. The goal of our work is to examine this effect, which is called photon reabsorption or radiation trapping, for the case of a fibre irradiated by a laser, in the recombination scheme. Our model which generalizes the work of Brunner *et al.* [1] is self-consistent. It is applied to the lasing lines of Li-like sulfur at 95.5 and 65.3 *A*. Other calculations in the recombination scheme [2] as well as in the collisional excitation scheme [3] show an important reduction of the gain which is not confirmed by experiments. In our model reabsorbed photons originate from the resonance transition between the lower lasing-level l and the ground level g.

2. Model

A resonance transition is characterized by large Einstein coefficients. In most cases the level g is the most crowded and absorption overcomes induced emission, resulting in a positive absorption coefficient:

$$K_{gl}(\nu|\boldsymbol{r},t) = \frac{h\nu\Phi(\nu)}{c}\Big[N(Li,g|\boldsymbol{r},t)B_{gl} - N(Li,l|\boldsymbol{r},t)B_{lg}\Big], \tag{1}$$

where $\Phi(\nu)$ is the normalized line profile which will be of the Voigt form and the N's the population densities of Li-like ions. If we assume the excited levels (indexed by i) of Li-like ions to be in a quasi-steady-state (QSS) [4] we can show that the population densities are solutions of the following set of equations [5]:

$$N(Li,i|\boldsymbol{r},t) = \frac{R_i(\boldsymbol{r},t)}{\Gamma_i(\boldsymbol{r},t)} + c\frac{\delta_{il}}{\Gamma_i(\boldsymbol{r},t)}\iint d\nu\, d\Omega\; K_{gl}(\nu|\boldsymbol{r},t)n(\nu|\boldsymbol{r},\boldsymbol{u},t). \tag{2}$$

The integral above involves absorption and induced emission and $n(\nu|\boldsymbol{r},\boldsymbol{u})$ designates the spectral density of photons of wavelength λ_{lg} propagating in the direction of a unit vector $\boldsymbol{u}$. ν is the frequency and Ω is a solid angle associated with $\boldsymbol{u}$. Finally, Γ_i and R_i, which describe the radiative and collisional processes, are given by:

$$\Gamma_i(\boldsymbol{r},t) = \sum_{j<i}\left[A_{ij} + N_e(\boldsymbol{r},t)Y_{ij}(\boldsymbol{r},t)\right] + N_e(\boldsymbol{r},t)\sum_{k>i} X_{ik}(\boldsymbol{r},t) + N_e(\boldsymbol{r},t)S_c(Li,i;He,g|\boldsymbol{r},t) \quad (3)$$

$$R_i(\boldsymbol{r},t) = N_e(\boldsymbol{r},t)\sum_{j<i} N(Li,j|\boldsymbol{r},t)X_{ji}(\boldsymbol{r},t) + \sum_{k>i} N(Li,k|\boldsymbol{r},t)\left[A_{ki} + N_e(\boldsymbol{r},t)Y_{ki}\right] + \quad (4)$$
$$N(He,g|\boldsymbol{r},t)N_e(\boldsymbol{r},t)\left[A_r(He,g;Li,i|\boldsymbol{r},t) + N_e(\boldsymbol{r})A_c(He,g;Li,i|\boldsymbol{r},t)\right],$$

where the different collision rates are thoroughly discussed by Klisnick *et al.* [4].

The reabsorption correction to the population density of the lower lasing-level is given by:

$$H(Li,l|\boldsymbol{r},t) = \frac{c}{\Gamma_l(\boldsymbol{r},t)}\iint d\nu\, d\Omega\, K_{gl}(\nu|\boldsymbol{r},t)n(\nu|\boldsymbol{r},\boldsymbol{u},t). \quad (5)$$

At this stage some mathematical manipulation is needed [5]. In particular, we replace the integration over the line profile by the product of the peak value at the central frequency $\nu_0 + \Delta$ and the linewidth, *i. e.* $\int d\nu\Phi(\nu)f(\nu) \to \Phi(\nu_0+\Delta)f(\nu_0+\Delta)\delta\nu_{FWHM}$, where Δ is the frequency shift of the line [6] caused by electron collisions with the lasing ion. We examine photon reabsorption in cylindrical symmetry, which is the most appropriate for the case of a thin fibre target. It is reasonable to assume that H, K_{gl} and the population densities depend only on the radial variable r, *i. e.* on the distance from the plasma axis. We obtain

$$H(Li,l|r,t) = 3K_{gl}(\nu_0|r,t)\int_r^R dr'\,K_{gl}(\nu_0|r',t)/r'\int_0^{r'} dr''r''\left[H(Li,l|r'',t)(1-\Gamma_l(r'',t)/A_{lg}) + \overline{N}(Li,l|r'',t)\right], \quad (6)$$

in which the boundary $R = r + \alpha / K_{gl}(\nu_0|r,t)$, where α is a constant >1. $\overline{N}(Li,l)$ is given by:

$$\overline{N}(Li,l;\boldsymbol{r},t) = \int d\boldsymbol{r}'\,N^{(0)}(Li,l;\boldsymbol{r}',t)G(\boldsymbol{r},\boldsymbol{r}',t), \quad (7)$$

where

$$G(\boldsymbol{r},\boldsymbol{r}',t) = \frac{K_{gl}(\nu_0|\boldsymbol{r},t)}{4\pi|\boldsymbol{r}-\boldsymbol{r}'|^2}\exp\left[-\int_0^{|\boldsymbol{r}-\boldsymbol{r}'|} ds'\,K_{gl}(\nu_0|\boldsymbol{r}-s'\frac{\boldsymbol{r}-\boldsymbol{r}'}{|\boldsymbol{r}-\boldsymbol{r}'|},t)\right]. \quad (8)$$

In Eq. 7 $N^{(0)}$ designates the reabsorption-off population density of the lower lasing-level.

The variation of population of level l due to reabsorption yields a corresponding variation for all levels of the lasing ion, with regard to spontaneous emission, collisional excitation and de-excitation. Equations 2-4 give a first-order correction for levels of energy $E > E(l)$ and $E < E(l)$, respectively indexed by k and j,

$$\delta^{(1)}H_+(Li,k|r,t) = N_e H(Li,l|r,t)X_{lk} / \Gamma_k \quad (9a)$$

$$\delta^{(1)}H_-(Li,j|r,t) = N_e H(Li,l|r,t)(A_{lj} + N_e Y_{lj}) / \Gamma_j \quad (9b)$$

and the correction $H(l)$ becomes $H'(l)$

$$H'(l) = H(l) - \sum_{k>l}\delta^{(1)}H_+ - \sum_{j<l}\delta^{(1)}H_- \quad (9c)$$

The variations $\delta^{(1)}H_+$ and $\delta^{(1)}H_-$ modify the population of the level l, via the same processes as above, and we obtain a second-order correction for that level:

$$\delta^{(2)}H_+(Li,l|r,t) = N_e H(Li,l|r,t)\sum_{k>l} X_{lk}(A_{kl} + N_e Y_{kl}) / \Gamma_k\Gamma_l \quad (10a)$$

$$\delta^{(2)}H\_(Li,l|r,t) = H(Li,l|r,t)\sum_{j<l}(A_{lj} + N_e Y_{lj})(1 - N_e X_{jl} / \Gamma_l) / \Gamma_j \quad (10b)$$

and so on. $H(l)$ has to be replaced by the following $H'(l)$

$$H'(l) = H(l) - \sum_{k>l}\delta^{(1)}H_+ - \sum_{j<l}\delta^{(1)}H\_ + \delta^{(2)}H_+ + \delta^{(2)}H\_ + \ldots \quad (10c)$$

We can show that the inclusion of higher orders leads to a convergent series and $H(l)$ has finally to be replaced by $H'(l)$:

$$H'(Li,l|r,t) = H(Li,l|r,t) / (1 + df(Li,l|r,t)), \quad (11)$$

where df, which are called depopulation factors, are given by:

$$df(Li,l|r,t) = \sum_{k>l}\frac{N_e X_{lk}}{\Gamma_k}\left(1 - \frac{A_{kl} + N_e Y_{kl}}{\Gamma_l}\right) + \sum_{j<l}\frac{A_{lj} + N_e Y_{lj}}{\Gamma_j}\left(1 - \frac{N_e X_{jl}}{\Gamma_j}\right). \quad (12)$$

In our simulations the terms involving Γ^{-2} are much less than those involving Γ^{-1}, yielding positive depopulation factors, which lead to an attenuation of the reabsorption-feeding of the lower lasing-level.

The correction $H'(l)$ is obtained in a self-consistent manner. In a first step we calculate K_{gl}, where the population densities are deduced from a simulation with the code FILM (reabsorption-off). An initial guess for $H'(l)$ and the calculated $\overline{N}(Li,l)$ are introduced in the right-hand side member of Eq. 6 and a new correction is obtained. Population densities of levels l and g, where the latter is assumed to be insensitive to reabsorption, in regard to its relative importance, are used in Eq. 1 to obtain a new K_{gl} and so on. The variation of population $H'(l)$ obtained after convergence induces a corresponding variation $H(Li,u|\mathbf{r},t)$ for the upper lasing-level, due to spontaneous emission, collisional excitation and de-excitation. By considering Eqs. 2-4 we can calculate $H(u)$. Finally, we determine the gain correction $\Delta G(u,l|r,t)$ of the lasing line for which we assume a Voigt profile.

3. Results and discussion

In Fig. 1 we show the gain coefficient, as a function of the distance from plasma axis, for the lasing transition $3d_{5/2} - 4f_{7/2}$ of Li-like sulfur. The target consists of a thin fibre of diameter 7 μm irradiated by a laser of wavelength $\lambda = 1.06\ \mu m$, pulse duration= 120 ps and intensity $= 2.31\ 10^{13}\ W/cm^2$. Besides a peak-gain reduction of 20 % (27 % when the depopulation factors are neglected) we notice a slight shift of the maximum. The three curves converge in the outer region of the plasma where the coefficient of absorption and the electron density are small. Figure 2 represents the same for the lasing transition $3d_{5/2} - 5f_{7/2}$ of Li-like sulfur. The line 3-5 is less sensitive to reabsorption (10 % of relative difference at most). This behaviour was already noticed for the same line of aluminium [5]. Finally, the shift is negligible.

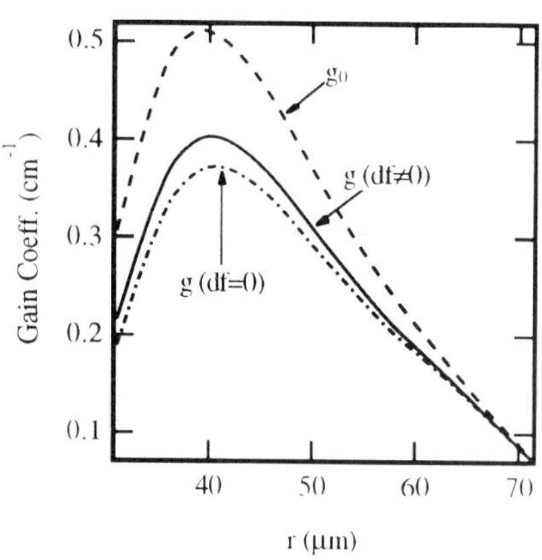

1

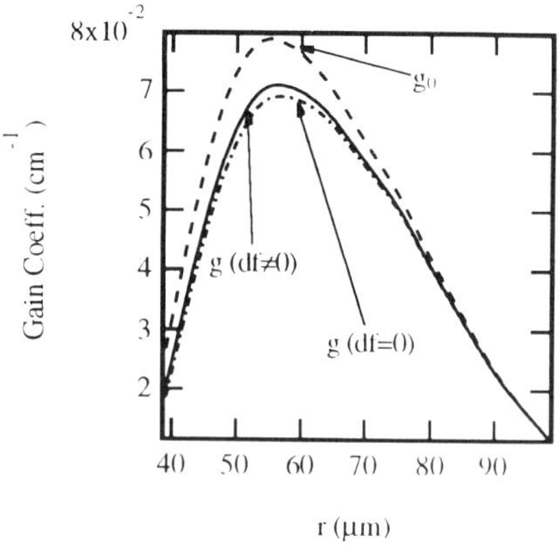

2

Figure 1. Gain coefficients vs distance from plasma axis for the lasing transition $3d_{5/2} - 4f_{7/2}$ (95.5 Å) of Li-like sulfur, at time of peak gain (240 ps after the peak of the laser pulse). g_0 and g are the gain coefficients without and with reabsorption, respectively. df≠0 and df=0 show the effect of the depopulation factors. Laser and target characteristics are given in the text.

Figure 2. Same as Fig. 1 for the lasing transition $3d_{5/2} - 5f_{7/2}$(65.3 Å) of the same ion, at time of peak gain (290 ps after the peak of the laser pulse).

In summary, photon reabsorption has been examined by a self-consistent method in the recombination scheme. Gain has been calculated for two lasing lines of Li-like sulfur in the case of a thin fibre. The main result is that the gain is reduced by a much smaller amount than predicted by most of the calculations, in agreement with experimental results. Moreover, the shift of the maximum gain reported by Makhrov*et al* [2] and us [5] for Li-like Al is confirmed in the case of sulfur.

4. References

[1] Brunner W, John R W, Paul H and Steudel H 1988 *Laser and Particle Beams* **6** 723

[2] Makhrov V, Roerich V, Starostin A, Stepanov A, Klisnick A, Sureau A and Guennou H 1994 *J. Phys. B.* **27** 1899

[3] Eder D C, Scott H A, Maxon S and London R A 1992 *Appl. Opt.* **31** 4962

[4] A Klisnick, A Sureau, H Guennou , C Möller and J Virmont 1990 *Appl. Phys. B* **50** 153

[5] Benredjem D E, Sureau A, Möller C and Guennou H 1994*International Colloquium on X Ray Lasers* (Williamsburg, VA)

[6] Sobel'man I I, Vainshtein L A and Yukov E A 1981 *Excitation of Atoms and Broadening of Spectral Lines* (Berlin: Springer Verlag)

Inst. Phys. Conf. Ser. No 140: Section 7
Paper presented at Laser Interact. with Matter Conf., Oxford, 19–23 September 1994
© *1995 IOP Publishing Ltd*

Harmonic generation and dynamics in high intensity fields: from Over-Barrier-Ionization to Stabilization

Pablo Moreno, Luis Roso, Luis Plaja and Victor Malyshev†

Departamento de Física Aplicada, Universidad de Salamanca.
37008 Salamanca, Spain. Phone (34) 23 294678; Fax (34) 23 294584

†Also at Vavilov State Optical Institute, 199034, St. Petersbourg, Russia

Abstract. Comparing quantum simulations with a naïve classical model we are able to identify several situations where the harmonic generation cutoff is increased or decreased.

1. Introduction

One of the most peculiar features of the harmonic spectra resulting from the interaction of an intense laser with a diluted gas, is the presence of a plateau at high harmonic frequencies followed by an abrupt cutoff. In the tunneling limit, the electron wave function overlaps with continuum states and therefore part of the wave function is not bound. It has been shown recently [1, 2] both quantum and classically that this unbounded part of the wave function absorbs energy from the field during part of the cycle, and is able to release it, in the form of harmonic radiation as rescattering with the nucleus. We try to extend the study of the harmonic generation to a wider range of parameters of the incident field, looking for the highest photon energies in the spectra. We compare now the spectra obtained from the numerical solution of the Schrödinger equation and the maximum energies predicted by the classical model [2] for a free electron immersed in a classical electric field.

2. Low frequency regime

The regimes of multiphoton and tunneling ionization are quite well known. The maximum kinetic energy of an electron ionized in such ways and coming back to the nucleus is the "magic" quantity $3.17U_p$, where $U_p = E_0^2/4\omega^2$ is the ponderomotive energy. As it is well known the classical motion may be split into two parts, the drift motion and the

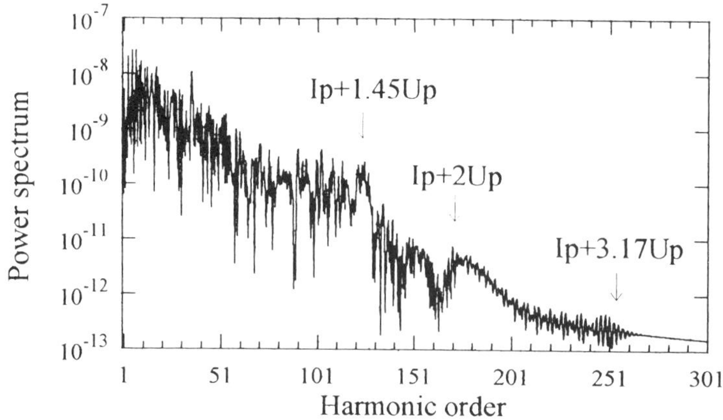

Figure 1. Harmonic spectrum corresponding to a case of barrier suppression, $E_0 = 0.26a.u.$ and $\omega = 0.06a.u.$

ponderomotive motion referring to constant and time dependent velocity respectively. The classical dynamical equations are (in atomic units):

$$\begin{aligned} \ddot{z} &= -E_0 sin(\omega t + \phi) && (1) \\ \dot{z} &= \dot{z}_0 + E_0/\omega(cos(\omega t + \phi) - cos(\phi)) && (2) \\ z &= z_0 + \dot{z}_0 t + E_0/\omega^2(sin(\omega t + \phi) - \omega t cos(\phi) - sin(\phi)) && (3) \end{aligned}$$

The phase ϕ refers to our lack of knowledge about the exact moment in which the electron is released from the nucleus. The previous numerical simulations [3] and the experiments [1] have given a cutoff energy which extends until the order corresponding to an energy of $I_p + 3.17U_p$ where I_p is the ionization energy. However, as we go beyond a given value of the intensity we find in the spectra an unexpected feature: the cutoff energy becomes lower and lower and several plateaus appear (see figure 1) . The reason is that the effective potential (i.e. atom plus field) is distorted in such a way that no bound state is present and the population is ejected suddenly into the continuum, therefore no further tunneling process is possible.

The classical picture predicts qualitatively this reduction of the maximum kinetic energies of the electron coming back to the nucleus. The electron that rescatter with the nucleus with the maximum kinetic energy is precisely this one which has tunneled to the continuum when the field is close to its maximum. For large fields, this electron has already been released before the field reaches this value, called saturation field, and consequently its maximum kinetic energy coming back to the nucleus is smaller. In the classical equations this means a restriction in the initial phase to satisfy the condition, $E_0 sin\phi < E_s$, where E_s is the saturation field corresponding to the complete potential barrier suppression (i.e. no minimum is present).

3. High frequency regime

When we work with fields of photon energies higher than the ionization energy, we have the well known phenomenon of photoelectric effect. This is not an interesting regime for

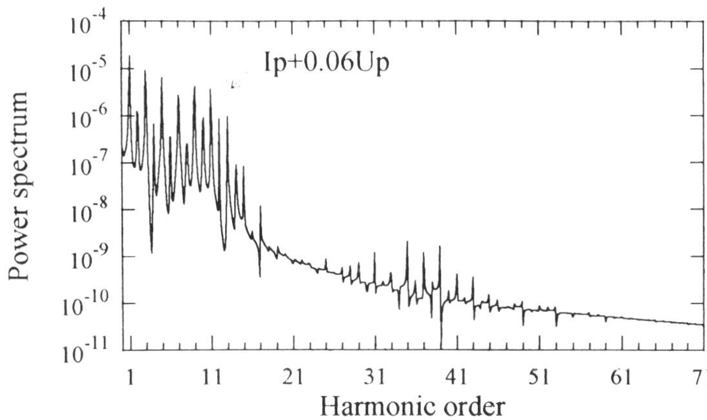

Figure 2. Harmonic spectrum corresponding to a case of stabilization, $E_0 = 75a.u.$ and $\omega = 2a.u.$

us because it predicts only the generation of the fundamental frequency. But when we go to very strong fields we find the phenomena of stabilization. The most remarkable effect in this regime is the inhibition of the ionization process. If we look at the harmonic generation in this regime, the numerical simulation gives only the lowest harmonic orders (see figure 2). Translating to U_p units we get photon energies of the order of one tenth (however, in eV this energy is much bigger than any photon energy obtained in the range of low frequencies). We consider from the classical point of view that stabilization corresponds to an extremely fast case of Over-Barrier-Ionization where the phases have been reduced so much that the coming back electron is able to give back to the nucleus a small quantity of the energy it can absorb from the field.

4. Is possible to get higher photon energies?

We may conclude from the previous sections that we cannot obtain indefinitely large photon energies only by increasing the intensity of the field. The maximum energy we expect is this given by the magic law $I_p + 3.17U_p$. In the case we increase the intensity we find that the maximum photon energy decreases instead of being increased. However we propose a system to allow for higher photon energies. This is based on the classical picture again. We know from the classical equations that the maximum kinetic energy of an electron ionized by tunnel effect is $8U_p$ far from the nucleus at some fixed distances given by the expression $z_r = (2n+1)\pi\alpha$ with n=0,1,2,..., where $\alpha = E_0/\omega^2$ is the classical excursion of an electron in a classical field. Therefore, if we add another ion located at these distances we expect to find from the solution of the Schrödinger equation, harmonic orders until an energy of $I_p + 8U_p$. The result of our numerical calculations confirm this fact (figure 3).

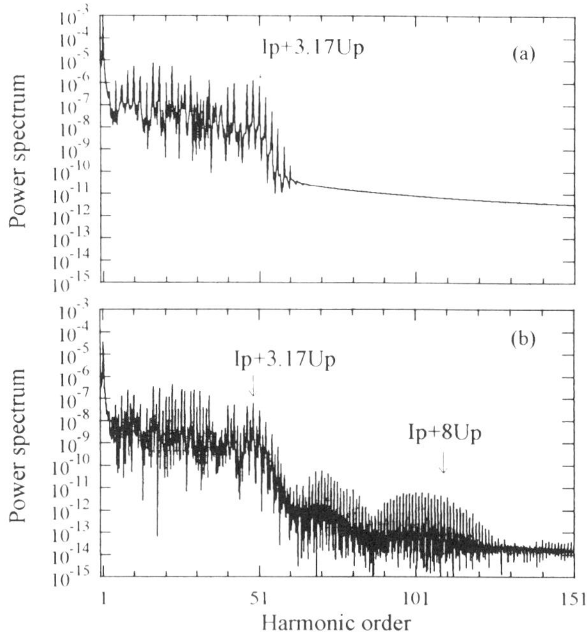

Figure 3. Harmonic spectra corresponding to the case of (a) one atom; (b) two atoms for a case of tunneling ionization.

Acknowledgments

We all acknowledge the support from the Spanish Dirección de Investigación Científica y Tecnológica (under grant PB93-0632). P.M. is grateful for the support from the Universidad de Salamanca, and V.M. also acknowledges support from Spanish Ministerio de Educación y Ciencia (SAB93-0157).

References

[1] A. L'Huillier, M. Lewenstein, P. Salières, Ph. Balcou, M. Yu. Ivanov, J. Larsson and C. G. Wahlström, Phys. Rev. A 48, R3433 (1993).

[2] P. B. Corkum, N. H. Burnett and F. Brunel, Phys. Rev. Lett. 62, 1259 (1989).

[3] J. L. Krause, K. J. Schafer and K. C. Kulander, Phys. Rev. A 45, 4998 (1992).

Paper presented at Laser Interact. with Matter Conf., Oxford, 19–23 September 1994
© 1995 IOP Publishing Ltd

X-ray spectra of optical-field ionized plasmas

S Borgström[1], E Fill[2], J Larsson[1], T Starczewski[1], J Steingruber[2], S Svanberg[1], and C-G Wahlström[1]

[1]Department of Physics, Lund Institute of Technology, S-221 00 Lund, Sweden
[2]Max-Planck-Institut für Quantenoptik, D-85748 Garching, Germany

Abstract. Soft X-ray spectra of He, N, C, O, S and Ar ions are reported. The emission was generated by focusing pulses of a high-power Ti:sapphire laser into a gas jet. The observed spectra are consistent with the predictions of the classical barrier suppression model. Indications of gain on lines between excited states are observed. Time-resolved observation revealed peculiar emission on He^{+} Lyman-α lines.

1. Introduction

The classical barrier suppression model [1] predicts that intensities of a few 10^{16} W/cm^2 should be sufficient to strip all the electrons from the outer shell of a number of low-Z elements (see table 1). By three-body recombination ions of lower ionization stages are generated, the X-ray spectra of which can be recorded. The conditions in such optical-field ionized plasmas are predicted to be favourable to inversion in the XUV and soft X-ray regions. [2,3]

2. Experiments

Linearly polarized laser pulses (λ = 794 nm, pulse duration 150 fs, pulse energy 150 mJ [4]) were focused into a pulsed gas jet with a density of about 10^{18} cm^{-3}. By using a lens of 50 cm focal length an intensity of 5 x 10^{16} W/cm^2 was achieved in a 50 μm diameter spot. Time-integrated soft X-ray spectra transverse and longitudinal to the laser beam were taken by means of a 1 m grazing incidence spectrometer. In separate experiments the time-dependent emission on isolated lines was recorded with an X-ray streak camera coupled to a flat-field spectrometer.

A time-integrated longitudinal spectrum of helium (fig. 1) shows strong Lyman-α emission and relatively weak emission on Lyman-β, suggesting gain on the Lyman-α line. However, since transverse and longitudinal spectra looked very similar, the existence of gain on that line is unlikely.

Table 1: Threshold intensities according to the barrier suppression model.

Ion	Electron configuration	I_{th} (W/cm^2)
He^{2+}	bare nuclei	8.8×10^{15}
C^{4+}	He-like	4.3×10^{15}
N^{5+}	He-like	1.5×10^{16}
O^{6+}	He-like	4.0×10^{16}
F^{7+}	He-like	9.6×10^{16}
S^{6+}	Ne-like	6.7×10^{15}
Ar^{8+}	Ne-like	2.6×10^{16}

In contrast to helium, the spectra of the other gases exhibited relatively strong emission from higher energy states. A transverse spectrum of nitrogen (fig. 2) shows that the emission on the 2p-4d line is almost as intense as on the 2p-3d resonance line. Similar observations were made for carbon, argon and sulphur.

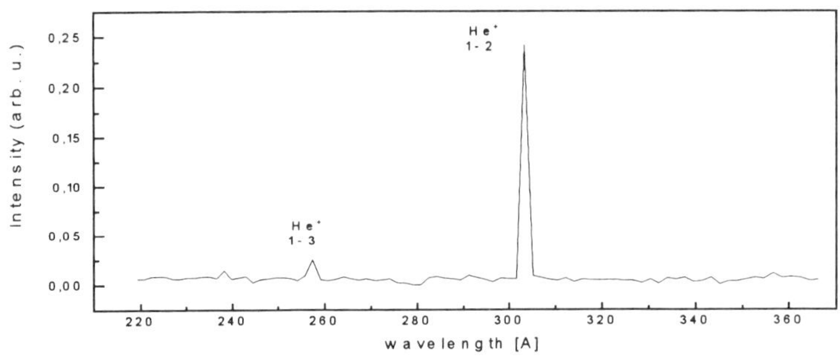

Figure 1: Longitudinal spectrum of helium showing Lyman-α and Lyman-β lines.

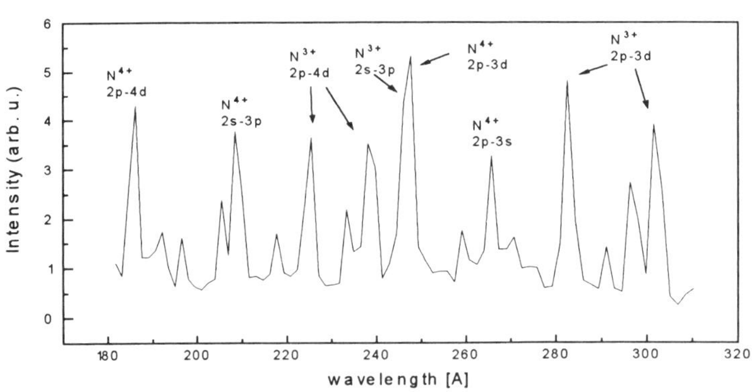

Figure 2: Transverse spectrum of nitrogen in the spectral region of resonance transitions, showing lithium-like and beryllium-like lines.

The possibility of gain was investigated on various lines by comparing spectra taken longitudinally and transversally to the laser beam. No clear indication of gain on any of the lines leading to a ground state was observed. However, for lines between excited states strong indications of gain were seen. In particular, on the 3d-5f transition in lithium-like nitrogen (λ = 512 Å) the emission in the longitudinal direction was considerably stronger than in the transverse direction. A gain measurement was carried out by generating a gas jet with an elongated cross-section, yielding a gain coefficient of about 30 cm^{-1} [5].

3. Time-resolved measurements

In separate experiments the time history of the emission on several lines was monitored by means of a streak camera coupled to a Jobin-Yvon flat-field spectrometer. A trigger for the streak camera was derived from the laser pulse, using an ultrafast diode. The low intensity of the lines required integration of the signal over typically about 3000 shots. The time resolution (which is mainly determined by the streak camera jitter) was measured by recording the emission on the 25th harmonic in argon. A resolution of about 100 ps was derived from these data.

The emission of lithium-like nitrogen lines was found to exhibit the expected fast rise after the laser pulse and subsequent slow decay. The decay time corresponded well to the time of depletion of helium-like nitrogen ions due to recombination, as calculated by simulations.

The emission of He^+ Lyman-α exhibited quite peculiar behaviour (fig. 3): After an initial spike of about 200 ps duration the emission again slowly rises. A second peak (not in the time window of the streaks) occurs after approximately 3 ns.

4. Discussion and conclusion

In general, the results of the time-integrated measurements confirm the validity of the barrier-suppression model. Single and double electron spectra are observed for helium, nitrogen, carbon, sulphur and argon, as expected. For oxygen the observation of lithium-like lines was marginal, whereas lithium-like fluorine lines were not observed at all, consistently with the fact that the maximum intensity reached in our experiment is lower than the threshold intensity required to generate the helium-like parent ion (table 1).

The dissimilarity between the relative intensity of the first and second resonance lines in helium and the other gases is partially accounted for by computer simulations. Assuming that the initial electron temperature is determined by ATI heating [6] one can explain the helium spectrum as an excitation spectrum, whereas the nitrogen spectrum is due to recombination. The low electron density in the case of helium (2×10^{18} cm^{-3}) leads to a low rate of excitation of the high energy states, whereas the opposite is true in the case of three-body recombination.

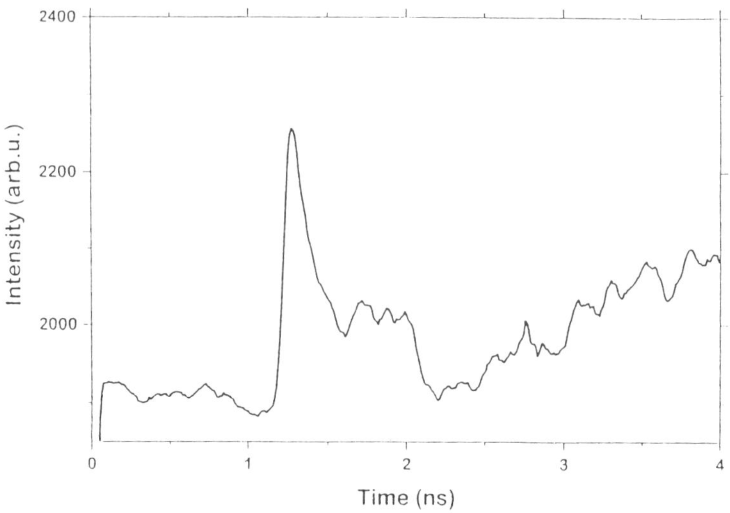

Figure 3: Temporal evolution of the He^+ Lyman–α emission, showing an initial spike and a second rise.

The temporal evolution of the helium Lyman-α emission is not yet fully understood. A possible explanation is that the radiative decay time of the n = 2 level of hydrogenic helium (133 ps) is not too far from the observed decay time of the first spike. In this model, after optical field ionization, part of the helium is left in the n = 2 level, from which it decays. The subsequent rise in the emission is then caused by electronic excitation from the hydrogenic ground state.

Acknowledgments: We thank A. Persson for operating the laser. We are grateful to A. L'Huillier for the loan of the spectrometer. This work was supported by the Swedish National Science Research Council and was performed within the framework of the European X-Ray Laser Network.

References

[1] Augst S, Strickland D, Meyerhofer D D, Chin S L and Eberly J H 1989 *Phys. Rev. Lett.* **63**, 2212–2215

[2] Burnett N H and Corkum P B 1989 *J. Opt. Soc. Am.* **B6**. 1195–1199

[3] Amendt P, Eder D C and Wilks S C 1991 *Phys. Rev. Lett.* **66**, 2589–2592

[4] Svanberg S, Larsson J, Persson A and Wahlström C-G 1994 *Physica Scripta* **49**, 187–197

[5] Borgström S et al., *Proc. Intern. Coll. X-ray Lasers,* Williamsburg, May 1994

[6] Penetrante B M and Bardsley J N 1991 *Phys. Rev. A* **43**, 3100–3113

Inst. Phys. Conf. Ser. No 140: Section 7
Paper presented at Laser Interact. with Matter Conf., Oxford, 19–23 September 1994
© 1995 IOP Publishing Ltd

Fokker-Planck simulations of short pulse laser solid experiments

R P J Town†, A R Bell† and S J Rose‡§

† Blackett Laboratory, Imperial College of Science, Technology and Medicine, London, SW7 2BZ, UK

‡ Rutherford Appleton Laboratory, Chilton, Didcot, Oxfordshire, OX11 0QX, UK

§ Department of Physics and Space Science, University of Birmingham, Birmingham, B15 2TT, UK

Abstract. This article describes simulations of short-pulse-laser-solid experiments using a one-dimensional Fokker-Planck code coupled self-consistently to a simplified ionisation model.

1. Introduction

The availability of short-pulse high-intensity lasers has opened up an exciting new regime of plasma physics [1, 2]. These ultra intense lasers produce plasmas at solid density with temperatures of the order of 1keV. At such temperatures the electron mean free path of the heat carrying electrons is sub-micron and is of the same order as the temperature and density scalelengths. This would lead to a strongly non-Maxwellian distribution of electrons in velocity space. Thus the heat flow would be poorly approximated by the Spitzer-Härm heat flow [3]. Furthermore during the interaction an initially solid density cold, un-ionised plasma is heated to a hot, ionised plasma. Ionisation is, thus, an important mechanism in the time development of the plasma.

In a previous paper [4] we reported simulations using a fully ionised Fokker-Planck (FP) code. These simulations showed that the non-local nature of the heat flow was an important consideration in correctly modelling short-pulse laser-solid interactions. We found that limiting the Spitzer heat flux to 0.1 of the free streaming value (as is commonly done in long pulse experiments) led to too severe a reduction in the heat flow. A better fit was found with a flux limit of 0.3, at least in the body of the target. We also investigated the ionisation dynamics by means of a non-LTE ionisation dynamics model [5]. Our simulations showed that for the carbon targets considered full ionisation in the heated region took the rise time of the 350 fsec, $10^{15} W/cm^2$ pulse. Further into the target the carbon never reached the fully ionised state. We identified the dominant ionisation

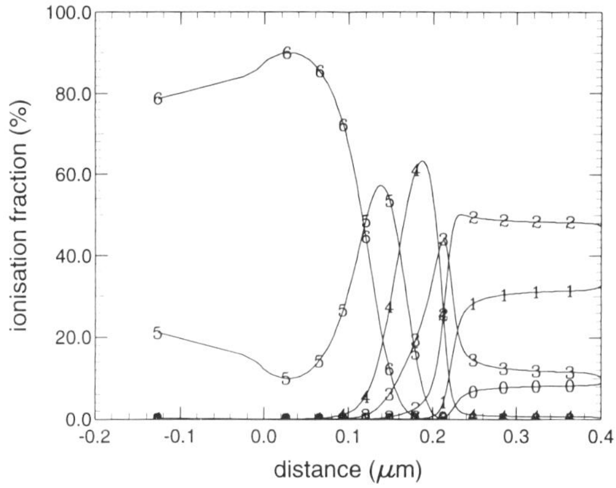

Figure 1. The spatial distribution of ionisation states in the carbon target at the end of the simulation. The number refers to the ionisation state ranging from 0 (un-ionised carbon) to 6 (fully ionised carbon).

mechanism to be collisional ionisation and the dominant recombination process to be three body recombination. However, the ionisation model was used as a post-processor.

Recently we have included a simplified ionisation model self-consistently with our FP code [6]. We have incorporated the collisional ionisation cross-section data of Bell *et al* [7] and Lennon *et al* [8] and the three body recombination cross-sections obtained by assuming LTE in the ionisation dynamics equation and the evolution of the distribution function, which have been written down in general form by Marchand and Matte [9]. In this article we will report on simulations undertaken using this code on the interaction of short-pulse lasers with solid targets.

2. Simulation Results

We shall consider the interaction of a 300 fsec KrF laser pulse incident on a $0.4\mu m$ thick carbon target. The target was initially at 20eV. The initial ionisation state was chosen such that the ionisation and recombination processes balanced. The laser energy was taken up by the low velocity collisional electrons. The average absorbed intensity was $7.5 \times 10^{14} W/cm^2$.

Figure 1 shows the spatial distribution of ionisation states at the end of the simulation (600 fsec). The laser was incident from the left hand side. The left hand side of the target was initially located at $0.0\mu m$. There has been considerable hydrodynamic motion during the laser interaction and the target has expanded by $0.15\mu m$. This is remarkable considering the short time scales, but suggests that hydrodynamics cannot be

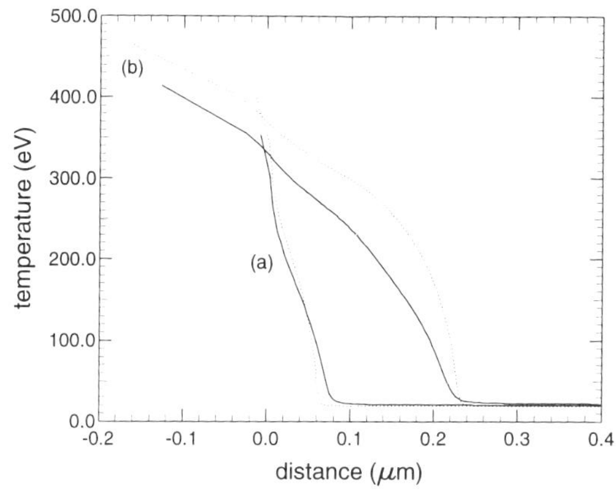

Figure 2. The temperature profiles at (a) 180 fsec and (b) the end of the simulation for the variable ionisation simulation (solid line) and a fixed fully ionised simulation (dotted line).

ignored even on short pulses. In the region where the energy was absorbed the majority of the ions are in the fully ionised state. Slightly further into the target the +5 state dominates. In the region where the heat front has penetrated the majority of the carbon ions are in the +4 state. There is a narrow region (at approximately $0.22\mu m$) where the +3 state is dominant. This corresponds approximately to the position of the heat front. The region on the extreme right (from $0.22\mu m$ and larger) the heat front has not arrived and the initial distribution of ionisation states (dominated by the +2 state) still exists.

Figure 2 shows the temperature profile after 180 fsec and at the end of the simulation (600 fsec). The dotted lines on figure 2 refers to a fully ionised simulation whose absorbed total energy was the same as the variable ionisation simulation. Notice that at the end of the simulation the temperature in the variable ionisation case is below that of the fixed ionisation case. Although the absorbed energy was the same in both simulations the variable ionisation case has to use energy in order to ionise the plasma, thus for a given absorbed energy this simulation will reach a lower temperature. In general the variable ionisation simulation's temperature has a marginally less convex profile in the body of the target.

The effect of variable ionisation on the heat flow (and temperature profile) can be thought of as two competing effects. Firstly, the lower average ionisation state can lead to enhanced heat flow, since the electron mean free path is inversely proportional to the ionisation state. Secondly, energy must be expended in ionising the plasma, this will reduce the temperature and, consequently reduce the heat flow. The evidence for a small low temperature foot is greater in the variable ionisation case than in the fixed ionisation simulation. The foot is due to the long mean free path electrons from the heated region penetrating the unheated target. The variable ionisation simulation's unheated target is at a lower ionisation state compared to the fully ionised simulation, thus the long mean free path electron can penetrate further into the target. At early times the main heat front arrives earlier than the fully ionised simulation. This is again due to the lower average ionisation state in the unheated plasma. However, at later times the energy expended in ionisation takes its toll. The temperature is lower and the heat front penetration slows. In fact at the end of the simulation the fully ionised heat front arrives before the variable ionisation heat front.

3. Conclusions

In this article we have presented results from our one-dimensional FP code which self consistently includes the dominant ionisation and recombination processes. We have shown that the effect of ionisation alters the temperature profile and, in a time dependent way alters the penetration of the heat front into the target.

Acknowledgments

This work was supported by the United Kingdom's Engineering and Physical Sciences Research Council and the Ministry of Defence.

References

[1] M Chaker *et al* 1991 *Phys. Fluids* **B3** 167

[2] U Teubner *et al* 1993 *Phys. Rev. Lett.* **70** 794

[3] L Spitzer Jr & R Härm 1953 *Phys. Rev.* **89** 977

[4] R P J Town, A R Bell & S J Rose 1994 *Phys. Rev. E* **50** 1413

[5] A Djaoui & S J Rose 1992 *J. Phys. B: At. Mol. Opt. Phys.* **25** 2745

[6] R P J Town, A R Bell & S J Rose 1994 *submitted to Phys. Rev. Lett.*

[7] K. L. Bell, H. B. Gilbody, J. G. Hughes, A. E. Kingston & F. J. Smith, Phys. Chem. Ref. Data **12**, 891 (1983).

[8] M. A. Lennon, K. L. Bell, A. E. Gilbody, J. G. Hughes, A. E. Kingston, M. J. Murray & F. J. Smith, Phys. Chem. Ref. Data **17**, 1285 (1988).

[9] R. Marchand & J. P. Matte, J. Comput. Phys. **97**, 352 (1991).

Inst. Phys. Conf. Ser. No 140: Section 7
Paper presented at Laser Interact. with Matter Conf., Oxford, 19–23 September 1994
© *1995 IOP Publishing Ltd*

ELECTRON TEMPERATURE MEASUREMENTS OF OPTICALLY IONISED PLASMAS.

S.G.Preston(1), W.J.Blyth(1), A.E.Dangor(2), A.Djaoui(3), M.H.Key(3,1), A.Modena(2), Z.Najmudin(2), A.A.Offenberger(4), J.S.Wark(1).

(1)Department of Physics, Clarendon Laboratory, University of Oxford, Oxford, OX1 3PU, U.K.;
(2)Department of Physics, Blackett Laboratory, Imperial College of Science, Technology and Medicine, London SW7 2BZ, U.K.;
(3)Rutherford Appleton Laboratory, Chilton, Didcot, Oxon OX11 0QX, U.K.;
(4)Department of Electrical Engineering, University of Alberta, Edmonton, Alberta, Canada T6G 2G7.

Abstract. We report on optical ionisation experiments using the 350fs and 12ps beams of the Sprite KrF laser at the Rutherford Appleton Laboratory producing ultraviolet laser radiation at focused intensities up to 10^{18}W/cm^2. The experiments are concerned with determining the electron temperature of optically ionised gases and are motivated by the possibility of developing XUV recombination lasers based on this technique . Thomson scattering and stimulated Raman scattering, from gas densities varying from 10^{17} to 10^{20} atoms/cm^{-3}, have been made and compared with modelling calculations of heating. A particular objective is to identify the relative importance of inverse bremsstrahlung absorption and SRS in determining the temperature of the electrons. Results for 350fs and 12ps pulses are compared and the importance of using sub-ps, short wavelength lasers to minimise electron temperature is confirmed.

Theoretical studies have shown the potential of recombination in optical field ionised (OFI) plasmas for XUV laser action in transitions to the ground state in hydrogen-or lithium-like ions[1, 2]. As the possible gain that can be produced in such a system is so greatly dependent on the temperature, it is important to study the actual electron temperatures produced by such laser-plasma interactions and to compare these measurements with theoretical calculations.

We have performed experiments[3, 4] using the Sprite KrF laser at the Rutherford Appleton Laboratory. The laser has been used in both Raman amplified mode[5] (4J in 12ps at 268nm focused to 3×10^{17}W/cm^2) and chirped pulse amplification mode[6] (250mJ in 350fs at 248nm focused to 10^{18}W/cm^2). The two gases investigated were helium (to produce a uniform ionisation profile) and neon (as a higher Z material suitable for a recombination X-ray laser scheme). Low density targets (10^{16}-10^{18} atoms/cm^3) were provided by using a static gas fill inside the target chamber, while higher densities (10^{18}-10^{20} atoms/cm^3) were produced using a solenoid-valve gas jet with a 1mm diameter cylindrical nozzle.

The electron density and temperature diagnostic at low densities was primarily Thomson scattering at 90° and at 180° (150° for the 350fs experiment). At higher densities, the unstimulated Thomson scattering spectrum gradually changed to a stimulated Raman scattered spectrum. The form of the Thomson spectrum can be compared to a computational model of the scattering expected from a Maxwellian distribution of electron temperatures[7]. This gives us an electron density, an electron temperature, and the ratio of the electron temperature to the ion temperature. Typical Thomson scattered spectra for static fill Helium from the 12ps experiment are shown along with theoretical fits in figure 1. The α parameter in the figure is given by $\alpha = 1/k\lambda_D$, where k is the plasma wavevector produced by scattering of the laser light, and λ_D is the Debye length. The densities implied by the data fit are not those we expect if we simply assume an electron density given by twice the density of the neutral gas. This is because the focused laser beam produces radial ponderomotive force that pushes the electrons and ions away from the centre (high intensity region) of the focus. The Medusa[8] simulation shown in figure 2 gives a qualitative representation of this effect in both helium and neon. Medusa is a Lagrangian code and does not allow the net separation of electrons from the ions. This requirement for neutrality stops us seeing the full effect of the rapid expulsion of the electrons. Nevertheless, the effect is still very strong for a 12ps pulse.

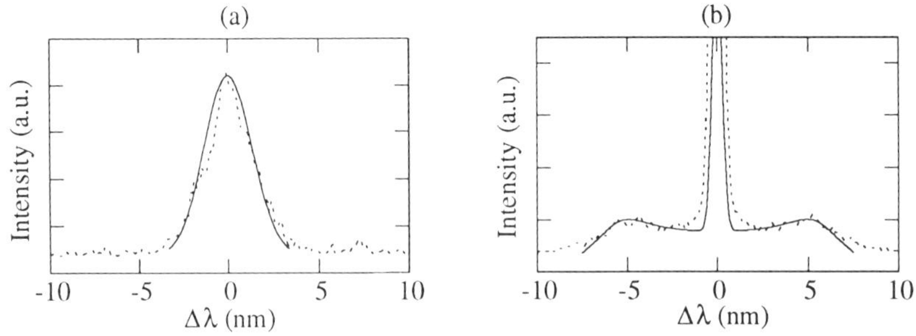

Figure 1. Theoretical fitting of Thomson scattering data from static fill helium in the 12ps experiment. The two plots shown are: (a) 2 torr, α=0.2, Te=5eV; and (b) 100 torr He, α=1.2 Te=25eV. Experimental data is shown by a dotted line, the theoretical profile by a solid line.

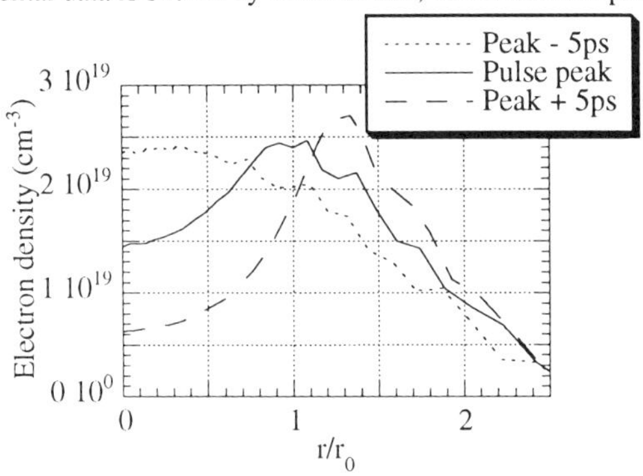

Figure 2. Medusa simulation of the ponderomotive expulsion of electrons (and ions) from the focus of a Gaussian laser pulse with a temporal FWHM of 12ps, a spatial FWHM ($2r_0$) of 6.5μm, and a peak intensity of 3×10^{17}W/cm^2. The figure shows the electron density profile 5ps before the peak of the pulse, at the peak, and 5ps after the peak.

At 350fs the ponderomotive expulsion does not have time to significantly reduce the plasma density at the centre of the focus. However, we are now in a regime where the electrons have not had time to thermalise during the period of the pulse. The electron-electron collision time for an electron density of 2.6×10^{19}cm^{-3} (100 torr of Ne^{8+}) is of the order of 1ps, and the electron-ion collision time is over 500ps. When the plasma is created by OFI, each ionisation stage produces a (Gaussian) above threshold ionisation (ATI) electron energy distribution. So, before thermalisation has occurred, the electron distribution will be the superposition of several (two for helium, eight for neon) ATI peaks. At this point our theoretical model of the Thomson scattered spectra breaks down. It can still be used, however, to give an approximate measure of the plasma parameters. Figure 3 shows typical Thomson scattering spectra from the 350fs experiment. Note the enhanced red-shifted 'electron feature' for 100 torr Helium. This could be due to the early appearance of SRS with the higher intensity (10^{18}W/cm^2) of the 350fs beam compared to the 12ps beam. The blue shifted feature is not amplified and is still solely due to unstimulated Thomson scattering.

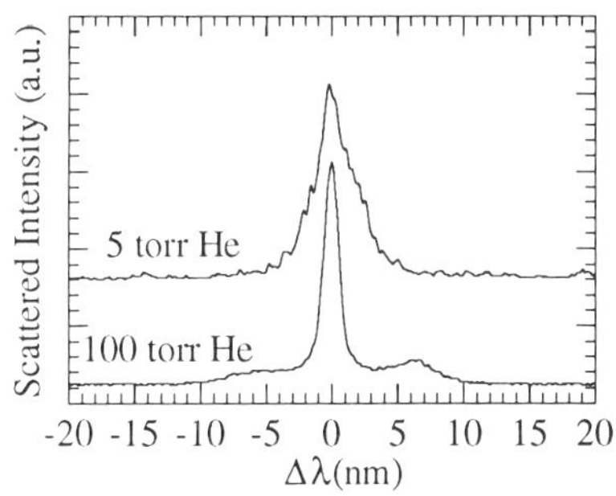

Figure 3. Thomson scattering from 5 torr and 100 torr helium in the 350fs experiment. The relative intensities and baselines of the two spectra are arbitrary.

As we move to higher densities in the static fill, we see the electron temperature gradually increase from an ATI temperature to one dominated by inverse bremsstrahlung (IB) heating. Figure 4 shows this increase in temperature for both 12ps and 350fs pulses. The 350fs temperatures are higher than expected because of the SRS heating mentioned earlier. It can clearly be seen, however, that reducing the pulse length from 12ps to 350fs has significantly reduced the electron temperatures.

At higher densities (produced by the gas jet), IB heating continues to increase, and SRS also becomes significant. Figure 5(a) shows the increase in SRS reflectivity with gas jet backing pressure in the 12ps experiment (the gas density at the laser focus is linearly dependent on the jet backing pressure). Note the rapid increase and then saturation of the reflectivity. This saturation is probably caused by the onset of stimulated Brillouin scattering which disrupts the electron plasma waves. Figure 5(b) shows the energy dependence of the SRS reflectivity of neon at electron densities of about 5×10^{20}cm^{-3} in the 350fs experiment. Taking a maximum reflectivity of about 2% and a wavelength shift of the order of 50nm, we calculate the mean energy per electron dumped into the laser focus (of approximately 10μm diameter by 100μm length) to be 1.2keV. However, the strong dependence of reflectivity on energy means that, if we reduce the laser intensity, by a factor of 3, to the threshold intensity for producing helium-like neon we obtain SRS heating of only 0.01eV per electron. Hence, SRS heating should not cause a problem for lithium-like neon recombination schemes. However, if we wish to move to shorter wavelengths by using higher Z materials, the intensity required scales approximately as Z^6. This implies that SRS heating will restrict Lithium-like OFI recombination schemes to atomic numbers less than 12, and thus to wavelengths longer than about 60Å.

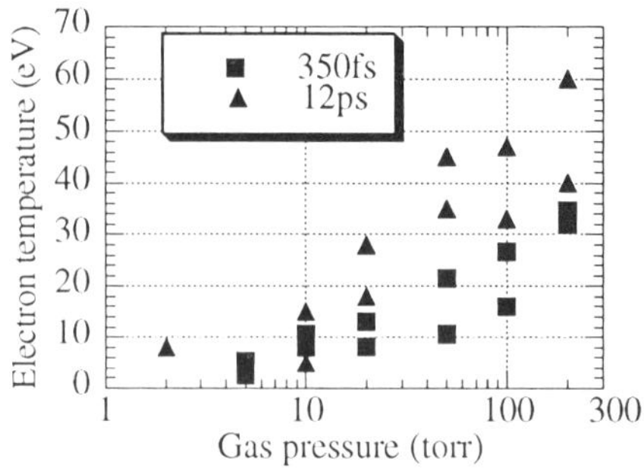

Figure 4. Electron temperatures measured for helium static gas fills in both the 12ps and 350fs experiments. Upper and lower bounds on each temperatures are shown.

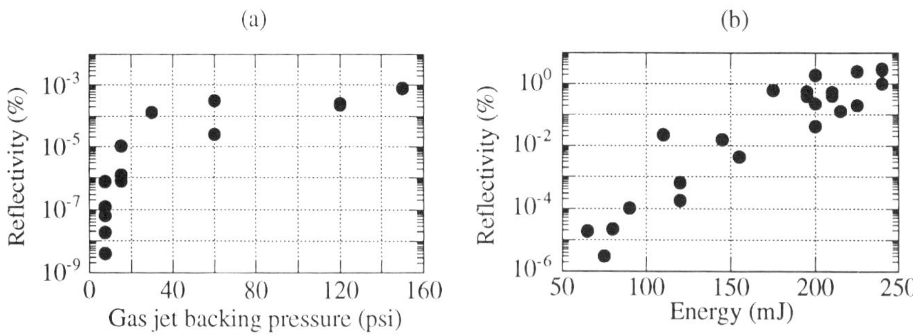

Figure 5. The variation of SRS reflectivities in neon with (a) gas jet backing pressure in the 12ps experiment, and (b) with energy on target at an electron density of $5\times10^{20}cm^{-3}$. Note the saturation of reflectivity in (a) and the strong energy dependence in (b).

References.

1. Burnett N.H., Enright G.D. IEEE J. Quant. El. 1990;26(10):1797-1808.
2. Eder D.C., Amendt P., Wilks S.C. Phys. Rev. A 1992;45(9):6761-6772.
3. Offenberger A.A., Blyth W., Dangor A.E., *et al.* Phys. Rev. Lett. 1993;71(24):3983-3986.
4. Blyth W.J., Preston S.G., Offenberger A.A., *et al.* Submitted to Phys. Rev. Lett. 1994.
5. Shaw M.J., Bialolenker G., Hirst G.J., *et al.* Optics Lett. 1993;18(16):1320.
6. Ross I.N., Bialolenker G., Evans J., *et al.* In: Baldis H, ed. Proceedings of SPIE - The International Society for Optical Engineering, SPIE Proceedings Vol. 1860 - Short-Pulse High-Intensity Lasers and Applications II. (Los Angeles, California, 1993) 21.
7. Wark J.S., Offenberger A.A., Blyth W.J., *et al.* In: Richardson MC, ed. Proceedings of SPIE - The International Society for Optical Engineering, SPIE Proceedings Vol. 2015 - Applications of laser plasma radiation. (San Diego, California, 1993) 270-276.
8. Christiansen J.P., Ashby D.E.T.F., Roberts K.V. Comput. Phys. Comm. 1974;7:271.

Inst. Phys. Conf. Ser. No 140: Section 8
Paper presented at Laser Interact. with Matter Conf., Oxford, 19–23 September 1994
© *1995 IOP Publishing Ltd*

Strongly Driven Ion Acoustic Waves in a Laser Produced Plasma

H.A. Baldis, C. Labaune, N. Renard, E. Schifano, S.D. Baton, A. Michard
Laboratoire pour l'Utilisation des Lasers Intenses, Ecole Polytechnique, France

W. Seka, R. Bahr
Laboratory for Laser Energetics, University of Rochester, USA

B.S. Bauer, K. Baker
Plasma Physics Research Institute, University of California, USA

K. Estabrook
Lawrence Livermore National Laboratory, USA

Abstract. This paper presents an experimental study of ion acoustic waves with wavenumbers corresponding to stimulated Brillouin scattering. Time resolved Thomson scattering in frequency and wavenumber space, has permitted to observe the dispersion relation of the waves as a function of the laser intensity. Apart from observing ion acoustic waves associated with stimulated Brillouin scattering, a strong second component is observed at laser intensities above 10^{13} Wcm^{-2}.

The study of parametric instabilities driven by a high power laser has become particularly important in recent years as a result of the vigorous research effort in laser driven inertial confinement fusion (ICF). The instabilities of interest are stimulated Brillouin scattering (SBS), stimulated Raman scattering (SRS) and filamentation [1][2][3]. An important component is the study of ion acoustic waves (IAW) and electron plasma waves (EPW) associated with the scattering instabilities, in an effort to understand the key issues of saturation and coupling between instabilities. The SBS instability can scatter a substantial fraction of the incident laser light affecting drive efficiency and drive symmetry in ICF targets. The saturation mechanism, the limiting factor in controlling the reflectivity, is likely to be associated with the nonlinear behavior of the waves when they are strongly driven.

We present in this paper observations of IAW associated with SBS, which show a transition in behavior as a function of laser intensity. These results are part of a study of IAW and EPW at the LULI laser facility, using Thomson scattering as the key diagnostic. Thomson scattering[4] is a powerful technique to study plasma waves, either in the linear or non-linear regime[5][6], permitting the direct measurements of frequency, wavenumber and location of waves. The results presented here had been obtained using the six beam laser facility of LULI, an interaction beam with a wavelength of 1.05 μm, and performed on well characterized preformed plasmas[7].

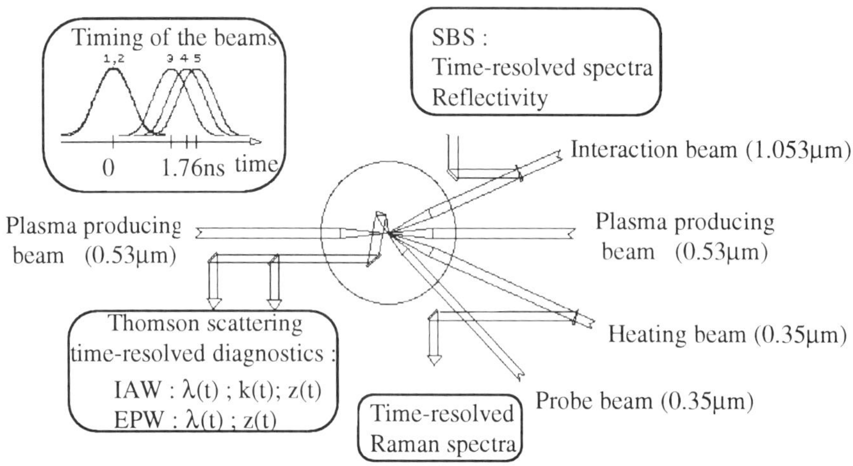

Figure 1. Experimental setup, showing beam configuration and array of diagnostics. The relative timing between the different beams is shown in the insert.

The laser beam configuration is shown in Fig. 1. The timing between the beams was as follows : plasma producing beams arrived at t=0, plasma heating beam at t=1.2 ns, Thomson scattering probe beam at t=1.48 ns, and interaction beam at t=1.72 ns. The pulse duration for all beams was 600 ps. The geometry for Thomson scattering was chosen to observe the IAW and EPW associated with backscattered SBS and SRS. Random phase plates (RPP) were used on the beams producing and heating the plasma to obtain plasmas as uniform and reproducible as possible. The targets were CH disks, with a diameter of 400 μm and a thickness of 1.5 μm. The choice of target and time delay between the different beams assured that all the target material has been ablated at the time of interaction, and that the plasma has expanded to the desired maximum electron density.

Particular attention was paid to the characterization of the plasma. Backscattered SRS light from the plasma heating beam provided information on the temporal evolution of the plasma density, from the long wavelength cut-off of the SRS spectra which corresponds to the maximum electron density along the plasma profile. The inferred electron densities agreed well with 2D Lasnex simulations, and these were used to calculate the evolution of the electron density profile during the interaction time. The electron density at the peak of the plasma profile evolved from 25% n_{crit} to 8% n_{crit} from the beginning to the end the interaction beam (see Fig. 2), where n_{crit} is the critical electron density for $1\omega_0$ light. Figure 3 shows the temporal evolution of the electron density and the density profile along the axis of the interaction beam calculated from Lasnex.

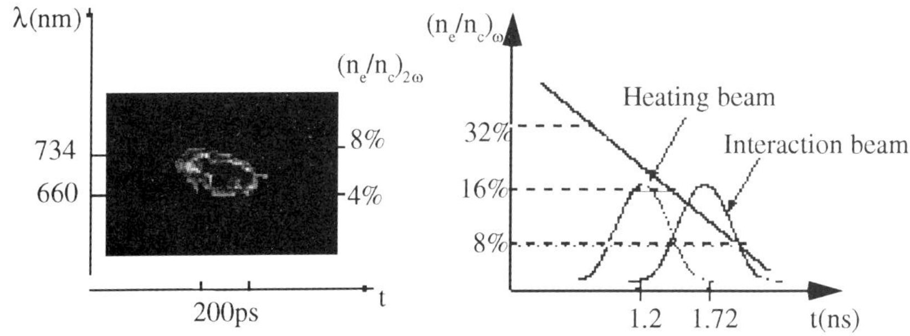

Figure 2. Temporal evolution of the SRS light from the heater beam, used as electron density diagnostic. The figure on the right shows the electron density at the time of the interaction.

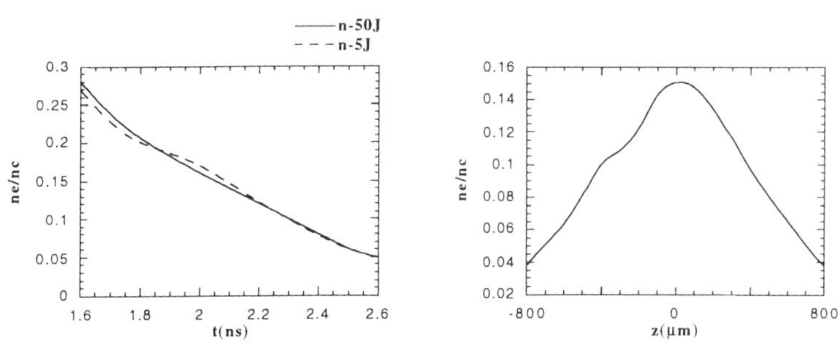

Figure 3. Temporal evolution of the maximum electron density, and electron density profile from 2D hydrodynamic simulation at t = 2 ns as predicted by Lasnex.

The electron temperature (T_e) of the plasma was obtained from thermal Thomson scattering in the absence of the interaction beam. The shift $\delta\lambda$ of each ion peak in the Thomson spectra corresponds to one ion acoustic frequency, thus permitting to obtain a value for T_e:

$$\delta\omega = k_{ia}\, c_s = 2|k_{probe}|\sin\left(\frac{\theta_{scat}}{2}\right) c_s \quad \text{or} \quad \delta\lambda_{(\text{Å})} = 5.3 \sin\left(\frac{\theta_{scat}}{2}\right) T^{1/2}_{e(keV)},$$

where k_{probe} is the wavenumber of the Thomson scattering probe, θ_{scat} the scattering angle, c_s the sound speed and k_{ia} the wavenumber of the IAW. The measured T_e was between 0.4 and 0.5 keV in agreement with the observed Landau cut-off of the backscatter SRS light from the plasma heater beam as well as with the Lasnex simulations.

Thomson scattered light was collected with a parabolic mirror which covered a range of IAW wavenumbers from $1.2k_o$ to $3.8k_o$, where k_o is the wavenumber of the interaction beam, in order to observe scattered waves from IAW as well as from EPW. The scattered light was analyzed by an array of three separate detection systems to obtain simultaneously time resolved spectra of IAW and EPW in a number of different configurations. Two separate spectrometer and streak camera arrangements permitted to obtained either frequency spectra of the waves collected over a range of wavenumbers, or wavenumber spectra for a specific frequency, thus providing complimentary information on the dispersion of the waves. A third detection system, consisting of a combination of streak camera and color filters, was used to obtain time resolved images of the location of the waves along the axis of the interaction beam. Other diagnostics included time resolved backscattered SBS light, and backscattered SRS from the heater beam to provide information on the electron density as mentioned earlier. In the present paper we will limit the discussion to the observation of IAW, leaving for a future publication a complete description of the study.

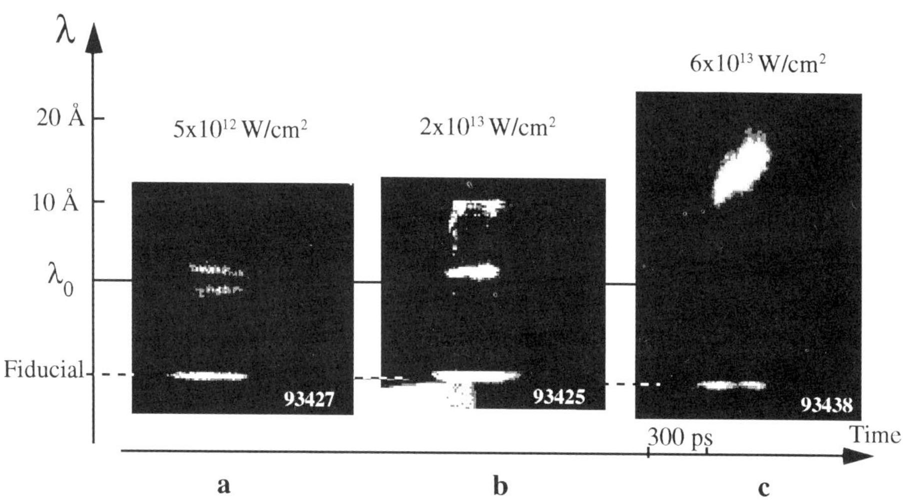

Figure 4. Time resolved Thomson scattered spectra of ion acoustic waves for three laser intensities. Different optical attenuation have been use in each case. The vertical axis is the shift from the probe wavelength λ_o.

The temporal evolution of the spectrum of IAW for three different intensities of the interaction beam is shown in Fig. 4, obtained with Thomson scattering and integrated over all wavenumbers collected by the parabolic mirror. The scattering region inside the plasma was of the order of 50-100 μm along the axis of the interaction beam. At low intensities ($I \leq 8 \times 10^{12}$ Wcm^{-2}) the spectrum shows two ion peaks (Fig. 4a), which are the characteristic ion satellites observed from scattering from thermal fluctuations, and similar to the spectrum observed in the absence of interaction beam. Each satellite corresponds to ion waves traveling in opposite directions, with magnitude Δk as set by the scattering geometry, and frequency shift corresponding to one ion acoustic frequency ω_{ia} from the probe frequency $3\omega_o$ (λ_{probe} = 0;35 μm). As the interaction beam intensity was increased, the satellite that corresponds to waves moving along the direction of the interaction beam is enhanced by the presence of IAW associated with SBS. Only one satellites is observed in Fig. 4b since optical attenuation was required to keep the signal within the dynamic range of the streak camera. Also shown in Fig. 4b is a second component in the spectrum, with a much larger spectral shift, which strongly depends on the pump intensity and which becomes dominant at intensities $I = 3 \times 10^{13}$ Wcm^{-2}. The Thomson scattering peak intensity in Fig. 4c corresponds to an increase in the scattered signal of 10^5 from the one shown in Fig. 4a, for a change of interaction beam intensity of the order of 10. The frequency shift also increased with laser intensity, with a shift of more than 10 times the ion acoustic frequency at an intensity of $I = 10^{14}$ Wcm^{-2} while the component shifted only by one ion acoustic frequency may still be present.

At low intensities the IAW have frequency shifts and wavenumbers corresponding to waves associated with the SBS decay process and they satisfy the linear dispersion relation $\omega_{ia}=k_{ia}\, c_s$, where $k_{ia} \approx 2\, k_o$. At higher intensities, the second spectral component has a frequency shift much larger than ω_{ia} but only a moderate shift in wavenumber. These waves do not satisfy the linear ion acoustic dispersion relation. The frequency shifts of this second component are observed to vary with the laser intensity, either by varying the interaction beam energy from shot to shot or during its temporal evolution . However, this behavior can not be explained in terms of strong coupling of IAW [8] which requires intensities of $\geq 10^{14}$ Wcm^{-2}. Even if the IAW will be strongly coupled for some part of the beam (due to hot spots or speckles), this effect can account at most for a factor of 2 in the frequency shift. Regarding the simultaneous presence of the two spectral components, one can speculate on the possibility that the spectral component with the large shift may be associated with strong levels of fluctuations driven by the interaction beam, but not related to SBS.

The observed IAW spectra are broad both in frequency and wavenumber, with a width comparable to the shift itself, presenting no indication of modulation or structure.

These results are different from results observed in earlier experiments using a CO_2 laser beam as interaction beam [9], where well defined harmonics were observed in frequency and wavenumber, with both components satisfying the dispersion relation. The difference between these two experimental results can be explained in terms of the level of fluctuations $(\delta n/n)_H$ required for the generation of harmonics [10] (λ_D = Debye length):

$$\left(\frac{\delta n}{n}\right)_H = (24)^{1/2}\,(k_{ia}\lambda_D)^2 = 8.86\text{x}10^{-2}\,(T_{e(keV)})^{1/2}\,\left(\frac{n_c}{n_e} - 1\right)^{1/2}$$

In the case of CO_2, the electron temperature was low, thus requiring that the fluctuation level δn/n associated with the IAW be larger than 3.5%, smaller than the measured value of 10%. In the case of the present experiment, on the other hand, the required fluctuation level has to be larger than 20%, a value much higher than that estimated (a few percent), due to the higher electron temperature.

In conclusion, these experiments have provided a detailed characterization of IAW in a laser plasma, in terms of frequency, wavenumber and location. In addition to observing IAW that corresponds to waves produced by SBS, a strong second component has been observed at high laser intensities. The frequencies and wavenumbers of this component do not satisfy the dispersion relation for IAW and is not clear that they are related to SBS.

References

[1] Kruer W L, 1988, *Physics of Laser Plasma Interactions*, (Addison-Wesley, New York).

[2] Baldis H A et al, 1991, *Physics of Laser Plasmas,* Handbook of Plasma Physics, Vol.3, Chap.9 (Elsevier Science Publiers).

[3] Labaune C et al, 1982, *Phys. Fluids* **B2** ,166.

[4] Shefield J ,1975,*Plasma scattering of electromagnetic radiation* (Academic Press Inc., New York, N.Y.)

[5] Baldis H A et al , 1986, *Can. J. Phys.* **64**, 961.

[6] Villeneuve D M et al , 1991, *J. Opt. Soc. Am.* **B8**, 895.

[7] Labaune C et al, 1993, Sections A4, A5 and A5, *Rapport Scientifique du Laboratoire pour l'Utilisation des Lasers Intenses.*

[8] Forslund D W et al, 1975, *Phys. Fluids* **18**, 1002.

[9] Walsh C J and Baldis H A, 1982, *Phys. Rev. Lett.* **48**, 1483.

[10] Denis P, 1993, chap. VII, p.597, *L'Interaction laser-matière* in La fusion thermonucléaire inertielle par laser, Vol.1, Ed. R. Dautray and J.P. Watteau (Eyrolles, Paris).

Inst. Phys. Conf. Ser. No 140: Section 8
Paper presented at Laser Interact. with Matter Conf., Oxford, 19–23 September 1994

© *1995 IOP Publishing Ltd*

Laser-Plasma Instability in Hohlraums

Juan C. Fernández†, J. A. Cobble†, W. W. Hsing†, B. H. Wilde†, B. H. Failor§, R. G. Hockaday†, P. L. Gobby†, D. A. Montgomery‡, H. N. Kornblum‡ and J. L. Miller‡

† Los Alamos National Laboratory, Los Alamos, NM, USA

‡ Lawrence Livermore National Laboratory, Livermore, CA, USA

§ Physics International, San Leandro, CA, USA

Abstract. A gas-filled hohlraum designed so as to approach plasma conditions expected in future ignition hohlraums has been fielded at the Nova laser. Radiation hydrodynamics modeling of these Nova hohlraums predicts reasonably well the measured plasma parameters. The measured reflectivity of a probe beam by Stimulated Brillouin scattering[Forslund *et al.*, PRL 30(1973)739] is modest. Some observed dependencies of reflectivity on laser and plasma parameters are understood theoretically, while others are not.

1. Introduction

Controlling laser-plasma instability is important for the success of indirect-drive laser fusion in the National Ignition Facility (NIF). Because of the relatively long duration of NIF laser pulses, ignition hohlraums are expected to fill with large (size $\sim$ few mm) underdense plasmas (electron density $n_e \sim 5 - 15\%$ of n_{cr}), with electron temperatures in the KeV range. ($n_{cr} = 9 \times 10^{21}\text{cm}^{-3}$ is the critical density above which the 351 nm light to be used in the NIF cannot propagate.) In these plasmas, Stimulated Brillouin scattering (SBS)[1] could be important, particularly back-scatter, which is expected to have the highest gain.[2] In order to understand SBS in these large plasmas, it is important to study simpler, open targets such as gas bags (see MacGowan *et al.*, these proceedings) or foams. However, in order to extrapolate to NIF conditions, it is also important to study the more complex hohlraum plasmas. The wide range of densities, temperatures and gradient lengths in such a system, particularly with a wall of high-atomic-number material, could lead to various instabilities which could either seed each other or compete with each other. This paper presents results from a Los Alamos program to study hohlraum SBS using the Nova laser at Lawrence Livermore.

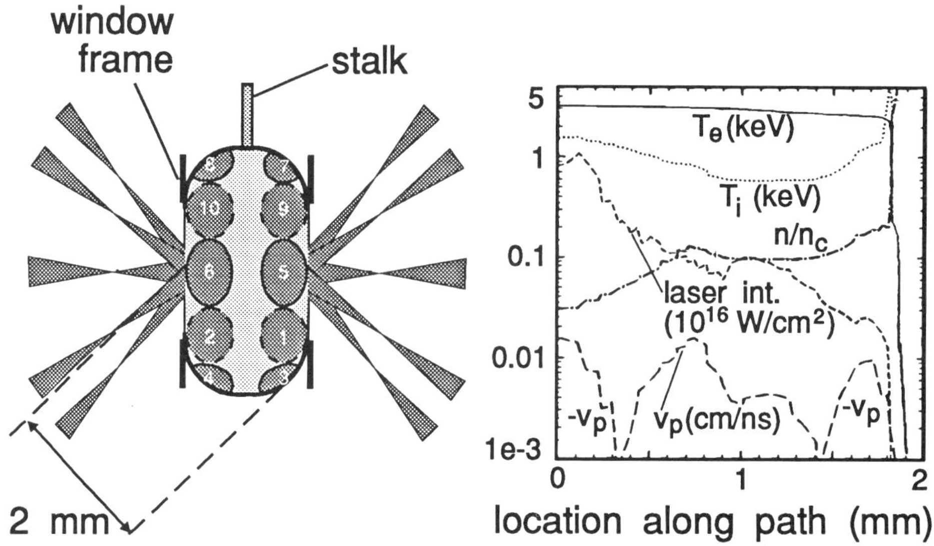

Figure 1. Left: The toroidal gas-filled hohlraum is illustrated along with the footprints of the ten Nova beams incident on the inner wall. Footprints on the far wall are drawn with a dashed outline. Right: Calculated parameters along a beam path are plotted for time 1 ns.

2. Hohlraum design

A hohlraum for the Nova laser was designed at Los Alamos (see Fig. 1, left) so as to approximate plasma conditions expected in NIF hohlraums. Despite the much lower energy of Nova when compared to NIF, various key plasma parameters in these gas-filled gold Nova hohlraums theoretically approach simultaneously NIF-like values. These design parameters include relatively high electron temperature ($T_e \approx 3$ keV), radiation temperature ($T_r \approx 200$ eV) and electron density ($n_e \approx 0.1 n_{cr}$), but ion temperature $T_i \approx T_e/2$. These conditions are maintained for about 0.5 ns with density and velocity gradient scale lengths down from NIF scales by only a factor of ≈ 2. Fig. 1 (right) shows radiation hydrodynamic modeling predictions for various plasma parameters along the beam path when the hohlraum is filled with 1 atm of neopentane gas (C_5H_{12}). In the experiment, nine 351 nm f / 4.3 Nova beams are turned on at time 0 to a power level of 2 TW each. Power linearly ramps up towards 3 TW at time 1.4 ns, when the beams are turned off. The beams diverge for a distance of 3.4 mm from best focus before hitting the hohlraum wall. The pulse shape on the probe beam (# 7) has been varied, but is typically a 1 ns square pulse which turns on at time 0.4 ns.

The evolution of T_r has been measured with the Dante diagnostic.[3] Peak T_r measurements in four hohlraums filled with neopentane and illuminated using the same laser pointing yield 202 ± 5 eV. T_e has been diagnosed radially

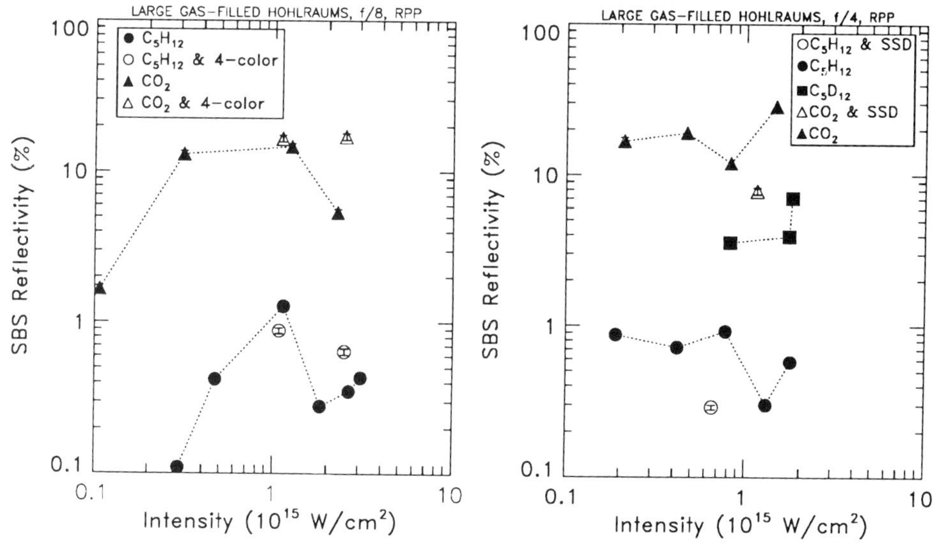

Figure 2. SBS reflectivity of a probe beam versus beam intensity from toroidal hohlraums filled with various gases. On the left plot, the probe-beam focusing lens had an f / 8 optic, while on the right it had f / 4.3.

at an axial position near the hohlraum midplane by Cr/Ti isoelectronic X-ray spectroscopy[4] (see Gobby *et al.*, these proceedings) using time-resolving Bragg-diffraction spectrometers.[5] Around the hohlraum center, measurements yield $T_e = 3 \pm 0.5$ keV. The biggest discrepancy between modeling and measurement has been in electron density. Whereas it is predicted that $n_e \approx 0.10 - 0.15 n_{cr}$ where any of the beams cross the midplane, Stimulated Raman sidescatter and $(3/2)\omega$ emission (from two-plasmon decay) indicate the presence of $n_c/4$ there.

3. Reflectivity measurements

The SBS reflectivity of the probe beam is shown in Fig. 2. The 351 nm probe beam generally had a random-phase plate to spatially smooth the beam.[6] In some f / 4.3 experiments, 3 Angs. of bandwidth was also added to the probe beam for temporal smoothing (SSD).[7] The beam was normally focused at the hohlraum midplane, but sometimes near the laser-entrance hole. The probe beam is divided into four quadrants. For some f / 8 experiments, the laser wavelength of successive quadrants has been displaced by 2 Angs. (8 Angs. separation between the bluest and the reddest quadrant). This four-color configuration is a good approximation to the NIF, where four beamlets form an effective f / 8 cone.

Fig. 2 shows a dramatic effect of the ion species on SBS behavior, presumably due to differences in the Landau damping of the daughter ion wave. Based on the T_i from modeling, it is calculated that in neopentane the H ions are very effective in Landau damping the ion waves, whose acoustic speed is dominated by the C ions. As the ratio of ions masses in the mix gets closer to unity for C_5D_{12} and for CO_2, the damping rate decreases.[8] The saturated level (at higher intensity) of reflectivity is strongly anticorrelated with the damping level, as expected theoretically. It is noteworthy that present NIF designs use a He/H mixture, which projects to an even higher level of Landau-damping than neopentane.

The qualitative shape of the reflectivity curves versus intensity for the f / 8 probe beam is similar to that predicted theoretically.[9] There is an onset critical intensity I_c where SBS turns on sharply and saturates. And as observed, with an increasing normalized damping rate, the calculated I_c increases because it is necessary to drive SBS harder to get significant gain. The reasons why, for f / 8, the reflectivity decreases at the highest intensities could include significant scatter outside the beam cone, which we do not measure.

The f / 4.3 data are more puzzling. As the f-number decreases, I_c should theoretically increase, as long as the velocity-gradient scale-length is much longer than the length of hot-spots ($l_s \approx 8f^2\lambda$, where λ is the laser wavelength[10]). So with l_s about four times smaller with f / 4.3 than with f / 8, I_c should be higher for f / 4.3. Yet, I_c for f / 4.3 has been bounded by a very low intensity.

In conclusion, the SBS reflectivity from neopentane-filled toroidal hohlraums (the most similar to NIF hohlraums) is low, even for high intensities. The observed reflectivity dependence on ion species and the SBS-onset intensity observed for f / 8 are qualitatively in agreement with theoretical expectations. But the lower value for the SBS-onset intensity for f / 4.3 is not understood.

The authors wish to thank the Nova operations team and the Los Alamos technicians supporting Nova for their excellent help. Mike Salazar (LANL) and Gary Stone (LLNL) did copious and dedicated work supporting the target gas-fill operations. This work is supported by the US Department of Energy.

References

[1] FORSLUND, D. W., KINDEL, J. M. ., and LINDMAN, E. L., Phys. Rev. Lett. **30** (1973) 739.

[2] KRUER, W. L., *Physics of Laser Plasma Interactions*, Addison-Wesley, N.Y., 1988.

[3] KORNBLUM, H. N., KAUFFMAN, R. L., and SMITH, J. A., Rev. Sci. Instr. **57** (1986) 2179.

[4] MARJORIBANKS, R. S., Rev. Sci. Instr. **66** (1995).

[5] FAILOR, B. H., HOCKADAY, R., and HSING, W. W., Rev. Sci. Instr. **66** (1995).

[6] DIXIT, S. M., THOMAS, I. M., WOODS, B. W., MORGAN, A. J., HENESIAN, M. A., WEGNER, P. J., and POWELL, H. T., Appl. Optics **32** (1993) 2543.

[7] SKUPSKY, S., SHORT, R. W., KESSLER, T., CRAXTON, R. S., LETZRING, S., and SOURES, J. M., J. Appl. Phys. **66** (1989) 3456.

[8] VU, H. X., WALLACE, J. M., and BEZZERIDES, B., Phys. Plasmas (1994), subm.

[9] ROSE, H. A. and DUBOIS, D. F., Phys. Rev. Lett. **72** (1994) 2883.

[10] BERGER, R. L., LASINSKI, B. F., KAISER, T. B., WILLIAMS, E. A., LANGDON, A. B., and COHEN, B. I., Phys. Fluids B **5** (1993) 2243.

Inst. Phys. Conf. Ser. No 140: Section 8
Paper presented at Laser Interact. with Matter Conf., Oxford, 19–23 September 1994

© 1995 IOP Publishing Ltd

The influence on stimulated Raman scattering of steep intensity gradients in short pulse laser drivers

H.C. Barr, T.J.M. Boyd, F.I. Gordon, and S.J. Berwick

Physics Department, University of Essex, Colchester, England CO4 3SQ

Abstract. Stimulated Raman scattering driven by sub-picosecond laser pulses is analyzed, in particular the effects of steep intensity gradients at the leading and trailing edges (intensity 'boundaries') on the nature of the instability and its emission signatures. A finite level of mode conversion due to intensity gradients is sufficient to close a feedback loop within the pulse and sustain absolute growth in the pulse frame. Relativistic effects limit the instability growth. In the case of forwards scattering, relativity causes the Raman emission to be confined within the pulse.

1. Introduction

Stimulated Raman scattering (SRS) from a plasma is a parametric instability in which an electromagnetic wave (laser driver) (ω_0, k_0) decays into a scattered electromagnetic wave (ω_s, k_s) and an electron plasma wave (ω, k). It has long been a process of concern in ICF schemes. Short pulse lasers have given rise to new regimes of SRS, relevant to recombination X-ray lasers and plasma accelerator schemes. The shortness and the high intensity of the pulse give rise to novel physics as seen in recent experiments [1,2].

A model of SRS driven by short laser pulses has been developed which is time-asymptotic in the pulse frame and caters for arbitrary pulse shapes and density profiles (each static in the pulse frame). The SRS daughter waves for both back and forward scattering emerge from the trailing edge of the pulse; in the pulse frame their group velocities have the same direction and hence the instability is always convective according to the standard parametric instability analysis [3]. SRS in (static) inhomogeneous density profiles has been extensively analysed. However, the short pulse duration, the high intensity, the presence of intensity inhomogeneities (i.e. the pulse shape) as well as relativistic effects also give rise to material differences in the Raman scattering process. The high intensities mean that the Raman backscattering (SRBS) is in a strongly coupled regime where the linear dispersion characterstics of the daughter waves no longer hold. This regime may also be approached for Raman forward scattering (SRFS). Inhomogeneities in density or intensity give rise to mode conversion within the pulse. This is sufficient to complete a feedback loop which can sustain absolute growth in the pulse frame. Then it is not the subpicosecond pulse duration that is important but the long plasma lengths. Relativity places a limit on the driver strength as well as giving an effective density reduction and hence a quenching of the instability in strongly relativistic regimes. For SRFS, the emission as seen in the pulse frame is created near its own critical density so that any effective density reduction due to the relativistic effects will confine it within the pulse.

The elements are explained most simply for the case of a homogeneous preformed plasma through which propagates a rectangular laser pulse in which the only inhomogeneities are confined to the pulse boundaries. Regions inside and outside can then be analysed in terms of their respective dispersion relations. The code makes no such assumptions and can be used for arbitrary pulse and density profiles. Henceforth we assume $1\mu m$ laser light and an intensity of $10^{18}Wcm^{-2}$ unless otherwise stated.

2. Results

The relativistic Raman equations are Lorentz transformed to the frame of the pulse group velocity $u = k_0c^2/\omega_0$. The laser driver is then near dipole-like at the slow frequency $\omega_0' = \omega_p = \omega_0/\gamma_u$ where $\gamma_u = (1 - u^2/c^2)^{-1/2}$ and $k_0' = 0$. Excluding the relativistic corrections for now the pulse frame dispersion relation for waves inside the pulse is

$$\left((\omega' + k'u)^2 - \frac{\omega_p^2}{\gamma_u^2}\right)(\omega_s'^2 - \omega_p^2 - k'^2c^2) = (k' + u\omega'c^{-2})^2\omega_p^2v_0^2$$

and the same with $v_0 = 0$ for the waves outside the pulse. ω_p is the electron plasma frequency, $2v_0$ the electron quiver velocity, primes denote pulse frame variables and $\omega_0' = \omega_s' + \omega'$. Electron temperature is omitted from this expression although catered for in the code where the full kinetic form of the electron susceptibilty is used . This is an essential inclusion at low densities where SRBS transmutes into stimulated Compton scattering (SCS). For SRFS both Stokes and anti-Stokes components must be included.

From the dispersion relation we obtain the growth rates ν' assuming k' real for convective SRS (i.e. no boundary effects). The growth rates as measured in the laboratory frame are $\nu = \gamma_u\nu'$. These are not the same as those obtained from a laboratory frame analysis. For densities ranging from $0.1n_c$ to $10^{-5}n_c$ (n_c is the laser critical density), SRBS pulse frame growth rates are reduced by a factor between 0.75 and 0.55 relative to a laboratory frame analysis. For SRFS, the growth rates appear much enhanced in the pulse frame analysis: by a factor of 2 at $0.1n_c$ to nearly 300 at $10^{-5}n_c$. Alternatively, the growth rate measured with respect to laboratory time but recorded by a detector moving with the pulse is $\nu_g = \nu - uk_i = \nu/\gamma_u^2 = \nu'/\gamma_u$. This is the pulse frame growth rate ν' with the time dilation removed. Continuity of phase with the detected waves outside the pulse produces detected frequencies ω_d distinct from ω_s.

For the parameters of recent short pulse experiments [2], the SRBS growth rate can easily be of the order of or greater than the plasma frequency. At the highest intensities, this may also be the case for SRFS. In the strong coupling regime ($|\omega' + k'u| >> \omega_p/\gamma_u$), the SRBS growth rate is easily shown to be $\nu' = \sqrt{3}(v_0^2\omega_p^2\omega_0/c^2)^{1/3}/2\gamma_u$. This expression is valid at higher densities where the shift of the detected frequency (for maximum growth) from the laser frequency is less than ω_p. The detected range of frequencies is significantly broadened relative to that expected from the scattered frequencies measured within the pulse. A significant blue wing is predicted and observed, reducing towards lower density. Figure 1 displays the growth rate ν versus the detected frequencies ω_d

(solid curves) and also the scattered frequencies ω_s (dashed curves) as generated within the pulse. At lower densities where Landau damping becomes significant, the unstable detected frequency range is narrow and solely red-shifted; the peak shift is equal to $2k_0v_t$, greater than ω_p, corresponding to SCS. v_t is the electron thermal velocity. SRFS shows a large red shift in its Stokes line (and a corresponding blue shift in its anti-Stokes line). However relativity may prevent emission from escaping the pulse (see below).

Since both decay waves propagate in the same direction both SRBS and SRFS are convective in this pulse frame analysis. The net amplification is $\exp(G)$ where $G = \nu' L'/(v'_{gs}v'_{gp})^{1/2}$ where $L' = c\gamma_u\tau_L$ is the pulse proper length, τ_L is the pulse duration and v'_{gs} (v'_{gp}) is the scattered (plasma) wave group velocity. For SRBS the latter are both near c while for SRFS the scattered light has a group velocity near to the pulse group velocity allowing a longer interaction time although at a slower growth rate than for SRBS. Using the numerically obtained values of ν', SRBS requires $15, 40, 250 fs$ of pulse duration per e-folding at densities of $10^{-3}n_c$, $10^{-4}n_c$ and $10^{-5}n_c$ respectively. The corresponding figures for SRFS are 30, 200 and $560 fs$ of pulse. Figure 2 displays the growth rate ν_g for SRBS versus detected frequency. The largest growth rates are for convective SRBS as predicted by the above dispersion relation [1]. A density of 0.01 of the critical density and a temperature of 50eV are assumed. The two lower curves are obtained from the code which incorporates the physical effects of the pulse boundaries and represent the absolutely growing eigenfrequencies. This growth can exponentiate for as long as the pulse is sustained or until saturation occurs. The spectrum of eigenmodes is the same as for the convective regime when well above the absolute threshold. Recent experiments [1,2] have exceeded this absolute threshold.

The frequency of the electromagnetic wave which propagates from trailing to leading edge of the pulse is upshifted in the lab frame to a frequency $\approx 4\gamma_u^2\omega_0 = 4\omega_0^3/\omega_p^2$ for the SRBS instability. In very underdense plasma this is a large photon "acceleration". Relativistic effects become significant when $v_0 \approx c$. The modifications may be simply estimated in that only a relativistic mass correction need be applied to the plasma frequency and to the quiver velocity. The mass correction is $\gamma_a \simeq (1 + \alpha e^2 A_0^2/m^2c^4)^{1/2}$ where A_0 is the vector potential of the laser pulse and α is a number near unity,

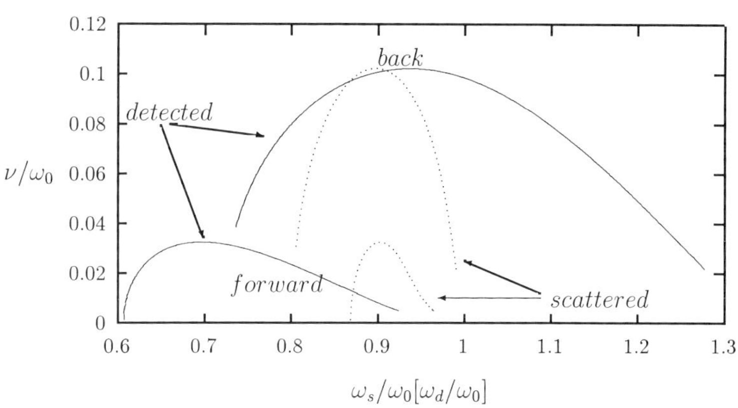

Figure 1. Growth rates at a density $n = 0.01n_c$ for SRBS, SRFS

depending on details of the polarization. Relativity gives an effective density reduction by a factor γ_a. The quiver velocity is self-limiting: $2v_0/c = a_0/\gamma_a < 1$ where $a_0 = eA_0/mc^2$. Figure 3 shows how relativity quenches SRBS. This is also the case for SRFS but since, in the pulse frame, the scattered light is created near its critical density the relativistic reduction in density means that the emission sees overdense plasma outside of the pulse and is thus confined within the pulse. There is then total feedback of emission. The detailed spectral features for SRFS are more complex than SRBS.

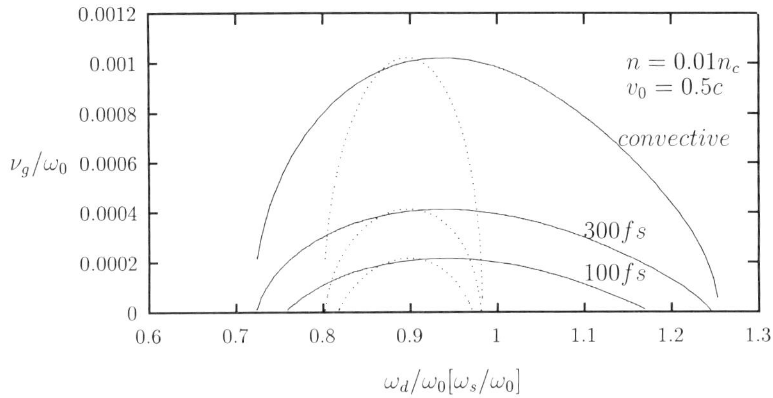

Figure 2. Convective and absolute growth rates for SRBS

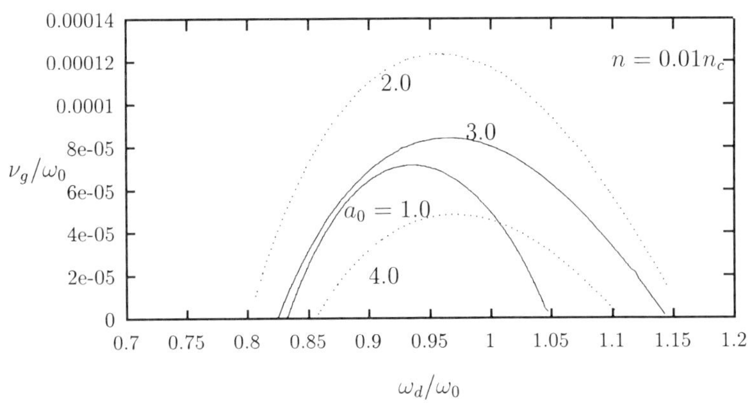

Figure 3. Relativisitic quenching of absolute SRBS growth

References

Tripathi V K and Liu C S 1991 *Phys. Fluids B* **3**, 468

Najmudin Z, Dangor A E, Malka V, Marques J-R, Modena A, Norreys P, and Taday P 1994 *Rutherford Appleton Laboratory Annual Report* RAL-94-042,23

Darrow C B, Coverdale C, Perry M D, Mori W B, Clayton C, March K and Joshi C 1992 *Phys. Rev. Lett.* **69**, 442

Inst. Phys. Conf. Ser. No 140: Section 8
Paper presented at Laser Interact. with Matter Conf., Oxford, 19–23 September 1994

© 1995 IOP Publishing Ltd

Laser – produced plasma clouds for laboratory simulation of explosive – type cosmophysical phenomena.

Antonov V.M., Afonin Yu.V., Zakharov Yu.P., Maksimov V.V., Orishich A.M., Ponomarenko A.G., Posukh V.G. Snytnikov V. N.

Institute of Laser Physics,630090,Novosibirsk,Russia

Abstract. This paper is devoted to laser production of spherical plasma clouds of large masses ($10 - 100\mu g$) and their using in laboratory simulations of interactions between exploding plasmas and magnetized media in space. The results of experiments conducted at CO_2 - Laser Simulation Facility $KI - 1$ are presented. Plasma clouds of an axially-symmetrical or near-spherical forms and energy up to 500 J has been obtained by two-sided irradiation of cylindrical and spherical targets. Laser pulses had energy up to 1 kJ and duration range of $0.1 \div 1\mu s$. Also a laser pre-evaporation of target has been used. Experimental data concerning dynamics, instabilities and diamagnetism of spherical plasma clouds in uniform and dipole magnetic fields of 1-2 kG are presented as well as a collisionless generation of large amplitude disturbances in background plasma at high Alfven-Mach numbers. Data obtained have been used for analysis of natural explosive-type phenomena in space plasma like Supernova bursts, Barium releases of AMPTE type or huge plasma releases of "Starfish" type in Earth magnetosphere.

1. Introduction

The idea of laser-produced plasma application for simulation of explosive-type astrophysical phenomena was suggested by Dawson [1] as long ago as 1964 (simultaneously with the suggestions of its application for ICF) and soon the first experiments [2] of this kind were carried out. But many years of investigations have passed (both in plasma physics, space physics and physics of the laser plasma itself) so that to go over from a general establishment of unique similarity of laser plasma cloud (LPC) with some astrophysical objects to more substantialed comparison of their characteristics [3], as well as to design the laboratory experiments [4] for simulation by LPC the collisionless deceleration of superalfvenically expanding Supernova remnants (SNR). With the help of such type of experiment it's convenient to analyze the main problems (and methods of their solving) while using a LPC-simulation method [5] for investigation of interaction processes between space plasma clouds and magnetized background or magnetic field.

The main difference between clouds of laboratory and space plasmas lies in the scales (sizes of the interaction region R_*, clouds'mass M and so on). So, for example, at comparable usual velocities of LPC expansion ($v_0 \approx 10^8 cm/s$) and SNR (v_0 up to 10^9 cm/s) even for weak interstellar fields ($B_0 \approx 1\mu G$) and characteristic laboratory ones ($B_0 \approx 1$ kG) there may be a basic difference in such important similarity criterion as the degree of ion magnetization $\varepsilon_* = \frac{R_H}{R_*}$ (with directed ion Larmor radius R_H). For SNR with $\varepsilon_* \ll 1$ and their observable deceleration by background may be formally described by a MHD - model (although $\lambda \gg R_*$), but for usual LPC conditions $\varepsilon_* > 1$ (at $R_* \leq 1$ cm defined by size of photo ionization region of the neutral background).

At high Alfven - Mach numbers $M_A = v_0\sqrt{4\pi\rho_*}/B_0 \gg 1$, the collisionless deceleration of SNR by background plasma with density ρ_*, can be caused by inductive electric field E_φ the action of which must accelerate the background ions on the scale $R_* = (\frac{3M}{4\pi\rho_*})^{1/3}$, for SNR'total mass M. An expanding diamagnetic cavity created by spherical cloud should effectively generate this electric field through $rot\bar{E}\varphi = -\frac{1}{c}\cdot\frac{\partial\bar{B}}{\partial t}$. Therefore, for laboratory investigations of such magnetic laminar mechanism - MLM [4] it is necessary to create spherical LPC with parameters ($M, v_0, m/z$ of ions), which allow to achieve a condition of MLM - interaction $\varepsilon_* = \frac{R_H}{R_*} < 1$. A little can be achieved by decreasing R_H through B_0 - increasing or v_0 - decreasing (because decreasing of M_A) and beside it there is a some threshold on v_0, connected with plasma generation processes. So, do solve this problem a way of creating of spherical clouds with large masses $M \geq 10\mu g$ and moderate velocities $v_0 \approx 10^7 cm/s$ has been choosed [4,5].

2. Plasma clouds'production for simulation experiments at $KI-1$ facility.

The multi-purpose simulation facility $KI-1$ [5] is designed to investigate a wide range of non-stationary processes in space plasma and it includes a large-scale (Ø1.2 m×5 m), high-vacuum (10^{-7} torr) interaction chamber and system of CO_2-laser and CO_2-amplifier with output energy $Q_r \approx 1$ kJ. The chamber is supplied by sources of an axial magnetic field (B_0 up to 1 kG) and of impulse dipole field as well(with moment 10^7 G$\cdot cm^3$). There is also a quasi-stationary source of background plasma to fill the volume $\sim 1\ m^3$ of chamber by highly-ionized density up to $10^{14}\ cm^{-3}$. The electrical and optical schemes of laser system allows to irradiate one or several targets by 4 beams both in simultaneous mode and in regime with controlled delay up to several μs between them. All this in combination with developed method [5] of controlling the form and duration of the CO_2-amplifier pulse ($\tau_r \approx 0.05 \div 3\mu s$) allows us to vary in a wide range the expansion geometry, energy $\mathcal{E}_0$ and the density profile $n(r)$ of LPC.

To solve the problem of production of spherical LPC with large mass a number of method [5-7] of two-sided irradiation of thin filament-target (Ø0.3 mm) or spherical target (Ø$3 \div 4$ mm, tying by metallic wire Ø0.1 mm), both made of Nylon $(C_6H_{11}OH)_n$ have been developed. We have established [6,7] that in the mode of single pulse irradiation ($\tau_r \approx 0.1 \div 0.5\mu s$) of filament under condition for beam size $\O_r \geq 2\O_t$ an axial-symmetrical cloud is generated with deviation $\delta \leq 15\%$ for n and v_0 (in the plane $\perp$ to filament). Expansion of such LPC in the plane of filament is characterized by angular distribution of ion flow $j_r = env_0 \sim \sin^\alpha\theta$ (with $\alpha = 2 \div 3$ typical for plane target), whereas their parameters at average flux $q \sim 10^{10}\ W/cm^2$ consist : $\mathcal{E}_0 \approx 50\ J$, M=10 μg and $v_0 \approx 1.5 \cdot 10^7$ cm/s (for the main H^+, C^{+3}, C^{+4}-ion species with total number $N_i \approx 10^{18}$).

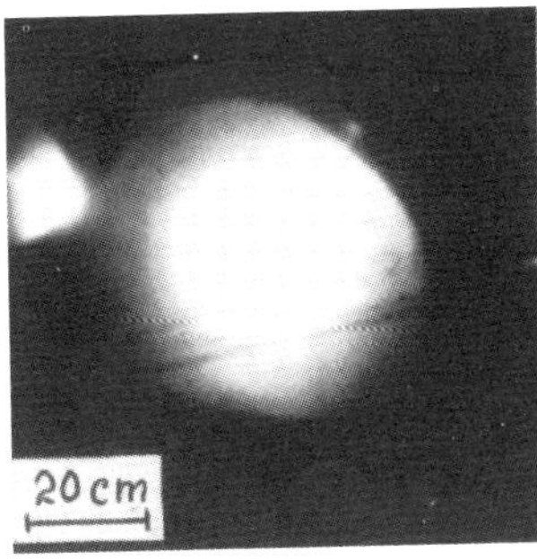

Fig. 1

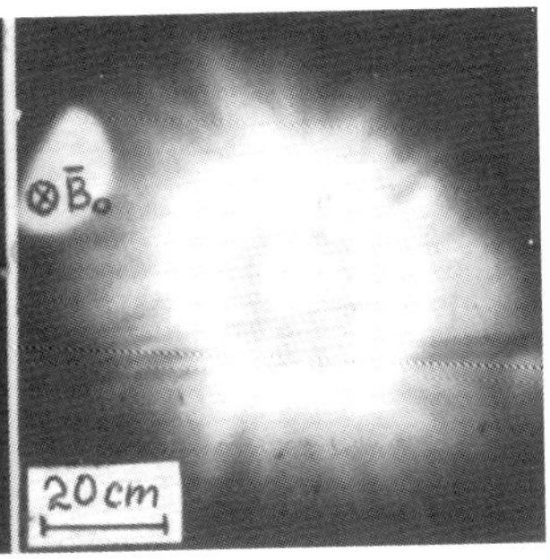

Fig. 2

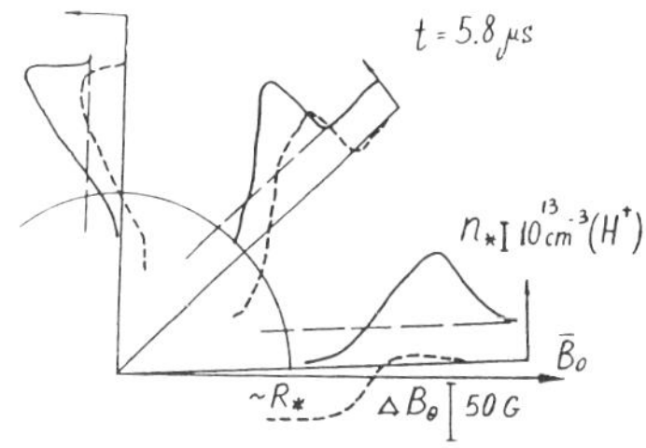

Fig. 3

Note, that additional application of the method [8] of preliminary target CO_2 evaporation for preparation of gas cloud with $n_i \sim n_{cr(10^{19}\ cm^{-3})}$ allowed us to generate the LPC with the same record number $N_i \sim 10^{19}$ (and $\mathcal{E}_0 \approx 500J$) as at NRL [8], but having an axial symmetry of filament target type.

Finally we pay attention to our results [7], concerning rather effective method of quasispherical LPC-production by means of two-sided and single-pulse irradiation of massive spherical target with $\O_t \leq (0.5 \div 0.7)\ \O_r$. Under conditions $\tau_r v_0 \geq \O_t$ (corresponding also to similar "Janus" [9] experiments) we have obtained for the first time the LPC with suitable degree of symmetry $\delta \leq 20 \div 30\%$ in all planes.

3. Study of collisionless interaction between plasma clouds and magnetized media for simulation of some cosmophysical phenomena.

The MLM-interaction of cloud, expanding with $M_A \gg 1$ into background plasma was investigated for the first time under condition $\varepsilon_* < 1$ in the experiments at $KI-1$ [5,15]. The apparent deceleration of LPC and formation of a strong disturbance like a spherical shock wave ($\Delta n \sim n_* \gg n$, see Fig.3) has been observed in contrast to results of similar experiments on high-power laser facilities"Helios" [13], "Pharos III" [12] or "Vulcan" [14]. It allows to speculate [15] about initial, pre-"Sedov" phase deceleration of historical SNRI by means comparing the deceleration low $R(t)$ of LPC with corresponding astronomical data (in dimensionless units $S_* = R/R_*$ and $\tau_* = tv_0/R_*$). Diamagnetic interaction of plasma clouds with vacuum magnetic field $\vec{B}_0$ at $M_A \ll 1$ largely depends on geometry of plasma expansion in the directions $\perp \vec{B}_0$. So, a number of experiments with symmetrical LPC has been carried out during last years various groups [8-11]. Recently such kind of experiments have been conducted at $KI-1$ with quasi-spherical plasma clouds for the first time. At has been showed that in this case with $\varepsilon_B = R_H/R_* \leq 1$ cavity is almost spherical (Fig.4a) and reaches its maximum value $R_B = (3\mathcal{E}_0/B_0^2)^{1/3}$. The results obtained (Fig.2) on dynamics of cloud deceleration, cavity size (R_B) and its "life-time" ($\nu_{eff} \approx 0.3\omega ce$) as well as the scale as of developed flutes with mode $K \geq 8\pi/\varepsilon_B$ are in good agreement [5,15] with AMPTE Barium release data obtained in geomagnetic tail (also at $M_A \ll 1$ and $\varepsilon_B \approx 1$).

Finally, it should be noted that this processes of diamagnetic interaction of clouds with dipole magnetic field can play important role [15] in dynamics of huge plasma releases (of "Starfish" kind) in Earth magnetosphere. Experiments [15] conducted at

$KI-1$ with dipole magnetic field reveal new effects of this interaction - cavity deformation (Fig.4b) and its moving as a whole away from dipole with velocity $\sim v_0$.

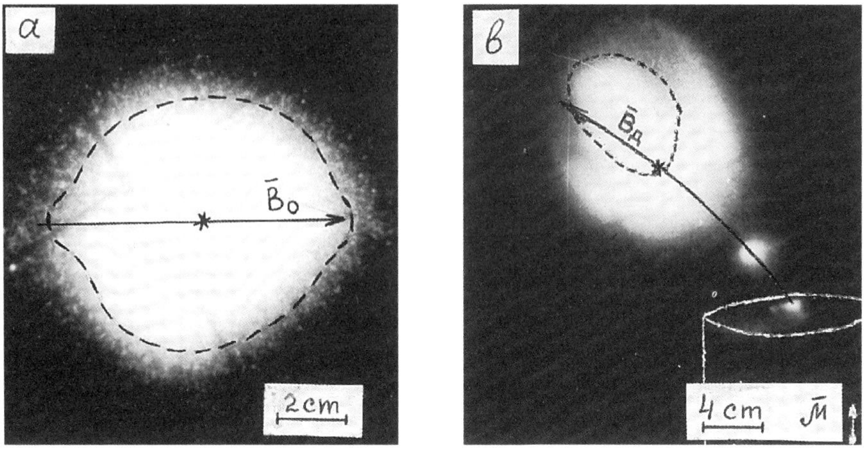

Fig.4

References

[1] Dawson J M 1964 *Phys. Fl.* **7** 981-7

[2] Tsuchimori N Yamanaka T Yamanaka Ch 1968 *Jpn. J. Appl. Phys.* **7** 84.

[3] Rose S J 1991 *Laser and Particle Beams* **9** 869-79

[4] Zakharov Yu P Ponomarenko A G 1980 *Laser Interaction With Matter* (Novosibirsk: Inst. Pure and Appl. Mech.) p. 4-39-In Russian.

[5] Zakharov Yu P Orishich A M Ponomarenko A G 1988 *Laser-Produced Plasma and Simulation of Non-Stationary Space Processes* (Novosibirsk: Inst. Pure and Appl. Mech.) - In Russian.

[6] Antonov V M Zakharov Yu P Maksimov V V et. al 1985 *Thermal Phys. of High-Temperature.* **23** 649-52 - In Russian.

[7] Avdyeva A A Zakharov Yu P Maksimov V V et al 1990 *J. Appl. Mech. and Techn. Phys.* **30** 892-5.

[8] Pechacek R E Greig J R Raleigh N et. al 1981 *Rev. Sci. Instr.* **52** 371-6

[9] Dimonte G Wiley L G 1991 *Phys. Rev. Lett* **67** 1755-8

[10] Zakharov Yu P Orishich A M Ponomarenko A G et al 1986 *Sov.J. Plasma Phys.* **12** 674-82

[11] Ripin B H Huba J D Mc Lean E A et al 1993 *Phys. Fl.* **B5** 3491-506

[12] Ripin B H Manka C K Peyser T A et al 1990 *Laser and Particle Beams* **8** 183-90

[13] Borovsky J E Pongratz M B Russell-Dupre R A Tan T H 1984 *Astrophys J.* **280** 802-8

[14] Bell A R Choi P Dangor A E 1988 *Phys. Rev. A. Gen Phys.* **38** 1363-9

[15] Zakharov Yu P Orishich A M Ponomarenko A G et al 1986 *Proceed Int. Workshop on Plasma Astrophys, Sukhumi (ESA Report SP-251)* 37-40

[16] Zakharov Yu P Orishich A M Ponomarenko A G et al 1992 *Proc. Int. Conf. on Plasma Phys., Innsbruck Pt. III (Europhys. Conf. Abs. Ser. 16C)* 1689-92

Fig.1 LPC at $t = 2.4\mu s$, sphere target, $B_0 = 0$ ($\tau_r \approx 0.5\mu s$)

Fig.2 LPC at $t = 5.7\mu s$, filament target, $B_0 = 400\ G$ preimpuls,$P_0 = 10^{-3}$ torr.

Fig.3 Backgrounde plasma desturbances from LPC with $M_A = 5$.

Fig.4 Expansion of LPC with $\mathcal{E}_0 = 10$ J in the vacuum uniform (a) and dipole (b) magnetic fields; dashed line is boundary of diamagnetic cavity (on 50%-level), sphere target suited at the asterisk and irradiated by pulse with $\tau_r = 0.1\mu s$ from the top and bottom in both cases ($t_a \approx 0.3\mu s, t_b \approx 0.8\mu s$).

Inst. Phys. Conf. Ser. No 140: Section 8
Paper presented at Laser Interact. with Matter Conf., Oxford, 19–23 September 1994
© *1995 IOP Publishing Ltd*

MAGNETIC FIELDS IN A PLASMA INDUCED BY CIRCULARLY POLARIZED LASER LIGHT

S. Eliezer, E. Kolka, Y. Horowitz and Y. Paiss

Plasma Physics, Soreq Nuclear Research Center, Yavne 70600, Israel

Abstract. A scheme of Magnetically Insulated Confinement Fusion (MICF) induced by circularly polarized laser light (CPLL) is suggested. The magnetic field profile during the laser pulse and the classic absorption of the laser light are calculated. First measurements of the magnetic field induced by the CPLL using a ferrite ring are reported.

1. Introduction

Generation of Megagauss magnetic field by CPLL was suggested[1]

$$B(\text{Gauss}) = 2\cdot 10^4 \cdot [I_L/10^{14}\ \text{W}\cdot\text{cm}^{-2}]\cdot[\lambda/\mu\text{m}]^3\cdot[n/10^{21}\text{cm}^{-3}] \qquad (1)$$

where B is the magnetic field, λ is the laser wavelength, I_L is the laser intensity and n is the plasma density. For example, taking λ=1.06μm, as for a neodymium laser, and $n=10^{21}\text{cm}^{-3}$, $I_L = 10^{16}$ W/cm^2, gives B = 2 MG. These megagauss magnetic fields can be used for magnetic insulation.

The new concept presented in this paper relies upon the hybrid use of inertial and magnetic confinements and megagauss field generation by CPLL. The schematic structure of the suggested configuration is as follows: a DT plasma is created inside a cylindrical or a spherical heavy conductor (or superconductor) shell with a hole. The plasma is irradiated by an intense circularly polarized laser beam. The CPLL creates a toroidal current in the plasma[2,3] which in turn induces an opposite current in the wall (during the plasma current creation). The currents induce poloidal magnetic fields inside and outside the plasma (see fig. 1) in addition to the toroidal magnetic field formed by the grad n x grad T mechanism.

The plasma is heated resonantly by the CPLL from a temperature of ~ 1 keV to ~5 keV during a time of few nanoseconds. The CPLL heats the plasma also by a resonant absorption mechanism between the electrons and the circularly polarized photons. It was calculated[4] that the efficiency of absorption of the CPLL is given by

$$\eta_R = \frac{1}{2}\left(\frac{\omega_{pe}}{\omega}\right)^2 \qquad (2)$$

independent of the laser irradiance. Therefore, even for large values of I_L (e.g. 10^{16} W/cm^2), the absorption efficiency at the critical density ($\omega_{pe} = \omega$) is 50%. The high absorption efficiency of the CPLL makes it possible to heat the plasma to high temperatures such as 5 keV necessary for fusion.

Linearizing the equations of motion for a gaussian CPLL that propogates in the z direction and assuming a gaussian density profile with the same width as the laser one gets for the second order pertrubed toroidal current

$$j_\theta(r) = -1/2 \exp(-1.5\, r^2/L^2)(e^3E^2/m^2\omega^3)(n_o \cdot r/L^2) \qquad (3)$$

where E is the magnitude of the electric field, n_0 is the maximum plasma density and L is the laser (gaussian) width. As a consequence a current j_θ^W is induced into the wall (see fig. 1).

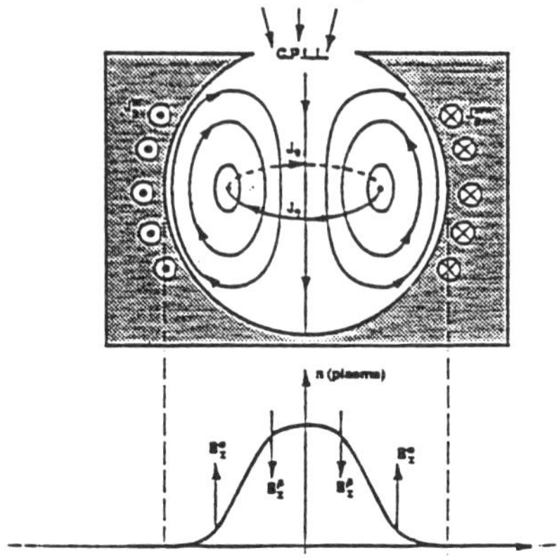

Fig. 1: A description of the suggested scheme. CPLL is the circularly polarized laser light, J_θ is the plasma current and J_θ^W is the induced current into the vessel wall B_z^P and B_z^O are the poloidal magnetic fields for the plasma density n.

The wall current is produced in such a way that the penetration of the magnetic field into the wall is avoided. This statement is accurate for a superconducting shell, however for a conducting vessel the magnetic field diffuses into the wall. In fig. 2 $B_z(r,z=o)$ is calculated (the solid line) for the following parameters:laser wavelength 0.35 µm, irradiance 10^{18} W/cm^2 and a gaussian width L=0.5 mm, peak plasma density 10^{22} cm^{-3}, the vessel inner radius and half height are 1 and 3 mm accordingly. The magnetic field changes direction at $r_o = 0.55$ mm. The magnetic fluxes through $0 \le r \le r_{o1}$ (Φ_{out}) and $r_{o1} \le r \le b$ (Φ_{in}) are in opposite directions and have equal magnitudes (see fig. 2).

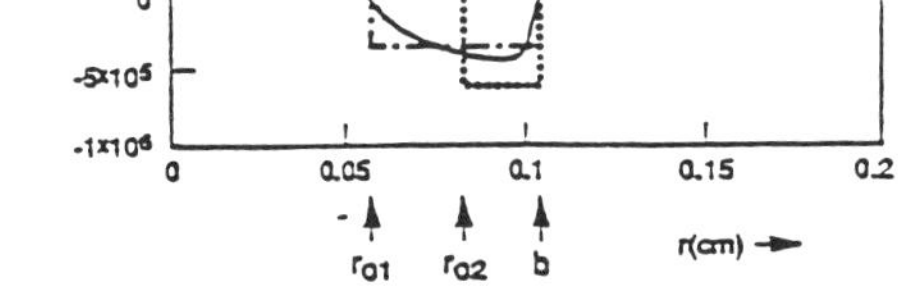

Fig. 2. The poloidal magnetic field for z = o (the center of a cylindrical vessel) as calcualted numerically is described by the solid line. The line-dot and the dot curves are two steps approximations for different times.

The plasma expands towards the wall due to its thermal pressure. Assuming magnetic flux conservation the magnetic field which confines the plasma increases. A schematic description is given in fig. 2: the calculated magnetic field profile (solid curve) is approximated by a constant field in each region (the line-dot curve) with the same fluxes Φ_{in} and Φ_{out}. During the plasma expansion one can see the increase in the magnetic field in the region $r_{01} \leq r < b$ (the dotted curve in fig. 2). This magnetic field (in the region $r_{02} \leq r < b$) is responsible for (a) a better plasma confinement (b) thermal insulation and (c) smoothening of the plasma pressure transferred to the wall.

2. The confinrment time

Three time scales should be considered to estimate the plasma confinement. First is the diffusion time (τ_B) of the magnetic field. From Maxwell equations the diffusion time of B to a distance l_B

$$\tau_B \approx \frac{4\pi\sigma l_B^2}{c^2} \tag{4}$$

where the electrical conductivity σ is given by Spitzer's formula. Taking $l_B \sim 0.5$ mm, of the order of the plasma described above, and $\eta_{max} \approx 2.10^{-18}$s one gets

$$\tau_B(\mu s) \geq 40 \left(\frac{l_B}{mm}\right)^2 \quad . \tag{5}$$

The second time scale to be considered is the heat diffusion time into the wall. The classical heat conduction of the electrons across the magnetic lines is given by

$$\tau_H(ns) \approx 400 \left(\frac{l_H}{mm}\right)^2 \left(\frac{T}{keV}\right)^{1/2} \left(\frac{B}{MG}\right)^2 \left(\frac{10^{21}\ cm^{-3}}{n}\right) \quad . \tag{6}$$

where ρ_e is the Larmor radius and τ_e is the typical electron collision time. For a 5 keV plasma, a density $n = 5 \cdot 10^{21}$ cm^{-3} and a confinement magnetic field B = 2 MG one gets $\tau_H \approx 180$ ns.

The third time scale is the hydrodynamic time. The plasma pressure creates a shock wave into the vessel. For $n = 5 \cdot 10^{21}$ cm^{-3} and T = 5 keV one gets a pressure of 160 Mbars. The shock wave induced by this pressure is moving inside the vessel with a speed U_s and flow velocity U_p. Using the Hugoniot equation of state for a copper vessel and the Hugoniot relation one gets $U_p = 3 \cdot 37 \times 10^6$ cm/s . Therefore, the hydrodynamic time is b/U_p where b is the vessel inner radius. For the above example one gets $\tau_h \approx 15$ ns. However, for a better approximation one has to take into account the plasma formation at the walls and its expansion towards the hot expanding plasma.The hydrodynamic confinement time of the plasma is changed to

$$\tau_h \approx \frac{1}{(1-\eta)} \left(\frac{b}{U_p}\right) \tag{7}$$

where η is the ratio between the wall plasma flow due to ablation and the hot plasma flow outwards. For the example under consideration one gets $\tau_h \approx 32$ ns.

In summary, there are 3 time scales describing the plasma confinement in this scheme: the magnetic diffusion time - τ_B, the heat diffusion time - τ_H and the inertial confinement time - τ_h . For the parameters under consideration we get $\tau_B > \tau_H > \tau_h \sim 30$ ns.

3. Experimental evidence of CPLL induced magnetic field

The experimental setup is described schematically in fig 3(a). Through a hole in a ferrite a CPLL is irradiated with intensities in the domain of $5*10^9$ to $5*10^{13}$ W/cm^2. The laser pulse duration (FWHM) is 7ns and the spot diameter is 200μm. A CPLL induced toroidal current is created in the plasma which creates an axial magnetic field associated with an appropriate magnetic flux. This flux penetrates into the ferrire and its time derivative induces a voltage measured by a coil connected to a GHz osciloscope. The ressponse time of the system and the callibration was done with a fast pulse generator.

The characteristic of CPLL induced voltage is the change in polarity when a left handed CPLL is changed to a right handed CPLL. This result is demonstrated in fig. 3(b) and 3(c) where the laser light had opposite polarization.The figures describe an average of about 10,000 laser shots with an irradiance of $5*10^9$ W/cm^2. The same phenomena was also observed for the higher laser irradiance, however this effect was not always observed.A possible explanation may be the fact that the grad n x grad T magnetic field might have a component in the axial direction due to plasma ripple instabilities. Another observed phenomena at 10^{13} W/cm^2 is the destruction of the ferrite and a drastic reduction of the magnetic flux penetrating into the ferrite. In this experiments the induced magnetic fields are larger than predicted by eq. 1 by a factor of about 10. Equation 1 seems to be an under estimation because it assumes a smooth density and laser spatial profiles. Large density or laser profile gradients (due to hot spots and filamentation), like in our experiment, are expected to increase the axial magnetic field.

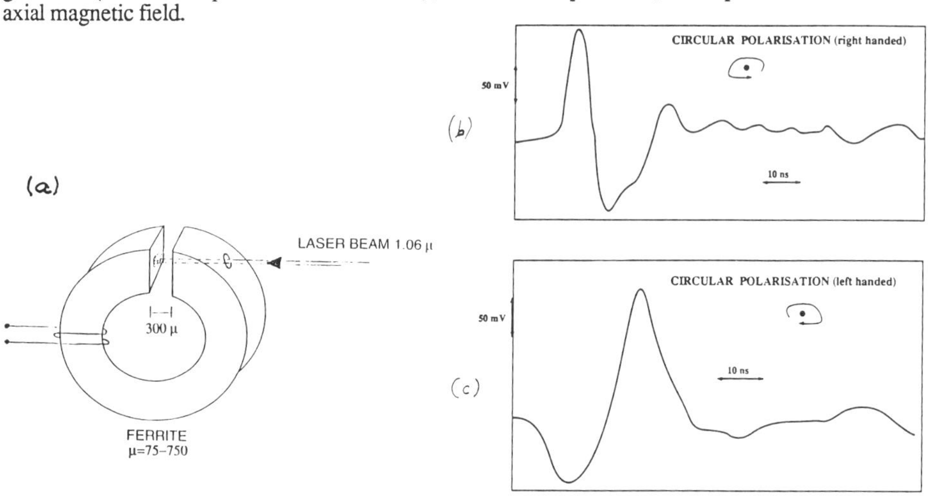

Fig. 3. (a) The experimental setup for CPLL induced magnetic field measurements.
(b) The time profile of the magnetic induced voltage from right handed CPLL.
(c) The time profile of the magnetic induced voltage from left handed CPLL.

REFERENCES

1. S. Eliezer, Y. Paiss and H. Strauss, Phys. Lett. A 164 (1992) 416.
2. A. Hasegawa, K. Nishihara, H. Daido, M. Fujita, R. Ishizaki, F. Miki, K. Mima, M. Murakami, S. Nakai, K. Terai and C. YAmanaka, Nucl. Fusion 28 (1988) 369.
3. H. Daido, F. Miki, M. Fuhita, K. Sawai, H. Fujita, Y. Kitagawa, S. Nakai and C. Yamanaka, Phys. Rev. Lett. 56 (1986) 846.
4. E. Kolka, S. Eliezer and Y. Paiss, Phys. Lett. A 180 (1993) 132.

Inst. Phys. Conf. Ser. No 140: Section 8
Paper presented at Laser Interact. with Matter Conf., Oxford, 19–23 September 1994

© *1995 IOP Publishing Ltd*

Evidence for "moving focus" in laser interaction with long scalelength plasmas

A. Giulietti, V. Biancalana, C. Danson[1], D. Giulietti[2], L. A. Gizzi, E. Schifano[3], and O. Willi[4].

Istituto di Fisica Atomica e Molecolare via del Giardino, 7 - 56127 Pisa, Italy.

[1] Central Laser Facility, Rutherford Appleton Laboratory, Chilton, Didcot, UK.
[2] Dipartimento di Fisica, Università di Pisa.
[3] Laboratoire pour l'Utilization des Lasers Intenses, Ecole Polytechnique, Palaiseau, France.
[4] The Blackett Lab., Imperial College of Science, Technology and Medicine, London,UK.

Abstract. Laser interaction with a preformed expanding plasma has been experimentally studied. The plasma had an $n_c/10$ bulk surrounded by a large low-density region. Time-resolved spectra of the light backscattered during interaction show that, after a short phase of laser coupling with the whole plasma, the laser ligth decouples from the main plasma bulk. In fact, the interaction region moves outside at supersonic velocity due to dynamical self-focusing of the laser beam. Beam smoothing with RPP does not modify appreciaby the effect. This observation is relevant to the general problem of laser energy deposition in a plasma.

1. Introduction

Efficient laser energy deposition into a dense plasma surrounded by a low density corona is a goal for many applications, including ICF. In this latter case uniformity of energy deposition is also required. Among the problems to be faced in order to obtain both uniformity and efficiency, two instabilities play a crucial role: filamentation and stimlated Brillouin scattering (SBS). The usual picture simply relates filamentation to the non-linear growth of non uniformities in laser irradiation and SBS to a poor laser energy deposition due to possible high levels of backscattering. In fact, most of the theorical and experimental works considered the two instabilities as separated or simply co-existing phenomena.

Recently, the interplay between filamentation and SBS was considered theoretically [1] and some interesting numerical simulations showed that the reciprocal influence of the two instabilities is strong in particular conditions [2,3]. A few experiments have been also addressed to clarify this aspect. It has been proved that the filamentation threshold acts very often also as an *effective* threshold for SBS, and consequently the beam smoothing techniques, if properly used, can reduce the backscattering level via the suppression of filamentation [4]. An important piece of information to better understand the connection between the two instabilities is the location of the backscattering active region in the plasma. Very recent mesurements attempted to obtain such information using a Thomson scattering technique together with the usual SBS backward spectroscopy [5].

Usually a low level of backscatterig is considered as a signature of an efficient laser-plasma coupling. This is not always true. In this paper we present SBS data obtained in condition of strong filamentation, mostly in the sense of whole beam self focusing (WBSF), with a low level of backscattered energy. SBS spectra show however a substantial decoupling of the laser beam from the main absorbing bulk of the plasma. This observation supports the idea that in presence of a deep low density corona, dynamical self-focusing can make laser energy deposition very inefficient. In this case the backscattering level fails to be a good *indicator* of the laser energy deposition and a great amount of the laser light is expected to be scattered and diffracted sideward [2].

2. The experiment

The experiement essentially consisted in the study of the interaction of a powerful laser beam with a preformed plasma and was performed at the Central Laser Facility (Rutherford Appleton Laboratory,UK). Four 600-ps, 1.053-μm beams of the Vulcan laser were used to preform the plasma. These *heating* beams were superimposed in two opposed pairs, providing an irradiance of 6.0 10^{13} W/cm^2 on each side of the target. Target consisted of 0.5-μm-thick 400-μm-diam aluminium dots coated onto a 0.1-μm plastic stripe.

The evolution of the plasma was carefully characterised in terms of electron density and temperature [6]. The electron temperature was obtained from time-resolved x-ray spectra by line-ratio measurements, accounting for opacity effects. Two-dimensional electron density maps were reconstructed from Nomarski interferograms taken at different delays with respect to the heating pulse. They showed that the expansion was axially symmetric and reproducible shot by shot. The plasma was found to be substantially free from small-scale density inhomogeneities.

The 600-ps, 1.053-μm interaction beam was delayed by 2.5-ns with respect to the heating pulse and focused with an f/15 optics into the preformed plasma along the main symmetry axis of the exploded target, at an irradiance up to to 5.0 10^{14} W/cm^2. The focal spot diameter was typically 100-μm, i.e., smaller than the plasma transverse scalalength. The electron temperature of the preformed Al plasma, 2.5-ns after the heating pulse was measured to be 400 ± 50 eV. The electron density profile along the interaction axis, as inferred from interferometric data [6] at 3.0-ns and 4.3-ns after the heating pulse, is shown by the plots of Fig.1. The maximum density is close to $n_c/10$ and a plasma bulk extends along the interaction axis up to a distance comparable with the target diameter. A region of density lower than $n_c/100$ surrouds the denser plasma for a few millimetres. The corresponding profiles at 2.5-ns are expected to be very close to the ones shown in Fig. 1 for t = 3-ns.

The effects, on the interaction, of beam manipulation or smoothing were also studied. The unsmoothed beam spot nonuniformities were limited to an amplitude of about 20% of the mean irradiance. The spot patterm was modualted in some mesurements using specially designed phase plates [7]. Other series of measurements were performed with the interaction beam smoothed using Randon Phase Plates (RPP), Induced Spatial Incoherence (ISI), or Smoothing by Spectral Dispersion (SSD) devices respectively.

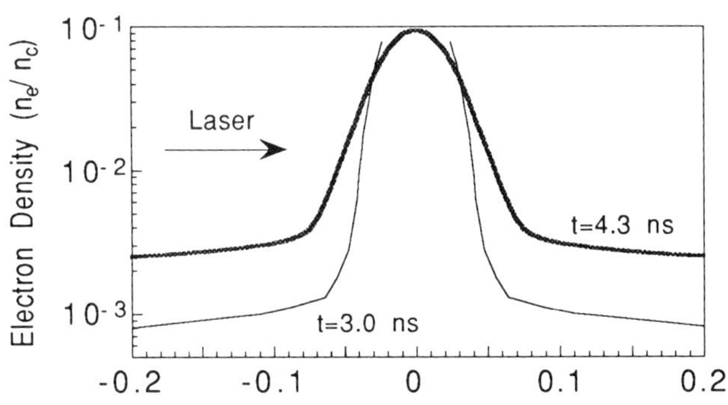

Fig.1 - Electron density profile v/s distance, 3.0-ns and 4.3-ns after the heating pulse. Density is in unit of the critical density.

3. Backscattering measurements

Time resolved spectra of backscattered light were obtained with a 1-m spectrometer and an optical streak-camara. Some reference spectra were taken by directly irradiating the cold target with only the interaction beam. One of these spectra is shown in Fig. 2 (a). They are some 20 Å wide and their mean wavelength moves from the blue to the red side with respect to the laser wavelength. They show similar features to other SBS spectra obtained by direct irradiation of solid targets [8]. Also the rate of energy backscattered in the solid angle of the focusing optics (a few percent of the laser energy) is close to values obtained in similar conditions.

On the contrary, the spectra obtained by interaction with the preformed plasma show novel features, highly reproducible in a large number of shots. A typical spectrum obtained from irradiation of the preformed Al plasma is shown in Fig.2 (b). Two temporally separated phases can be observed in these spectra. First, a short flash of backscattered light is observed with a broad spectrum, more than 30 Å wide. The temporal resolution of our detection system was limited by the spectral dispersion to 120-ps, and this is the upper limit we can give to the duration of the first phase of the backscattering process. In the second phase, the backscattering ligth is rather narrow at a given time (a few Ångstroms). The spectrum moves from the red side of the previous broadband flash to the blue side at an almost constant rate. The level of energy backscattered in the same solid angle was very low in this case, oscillating shot by shot from 7 10^{-5} to 2 10^{-4} of the laser energy.

A limited number of measurements on backscattering were also performed with the interaction beam smoothed by RPP, ISI or SSD methods respectively. The comparison between spectra obtained with smoothed or unsmoothed interaction beam at the same irradiance, show a relevant feature: as the interaction beam was narrow-band (<10^{-2}Å), time resolved spectra were substantially similar to the one shown in Fig. 2 (b), no matter if unsmoothed, RPP or ISI smoothed beam was used. The amount of backscattering was still very low, 3 10^{-4} in the case of RPP smoothing, 1 10^{-4} in the case of ISI. With the broadband beam (13 Å), the wide laser spectrum did not allow observation of spectral details as those shown in Fig. 2 (b). The backscattered energy was not measured in this case.

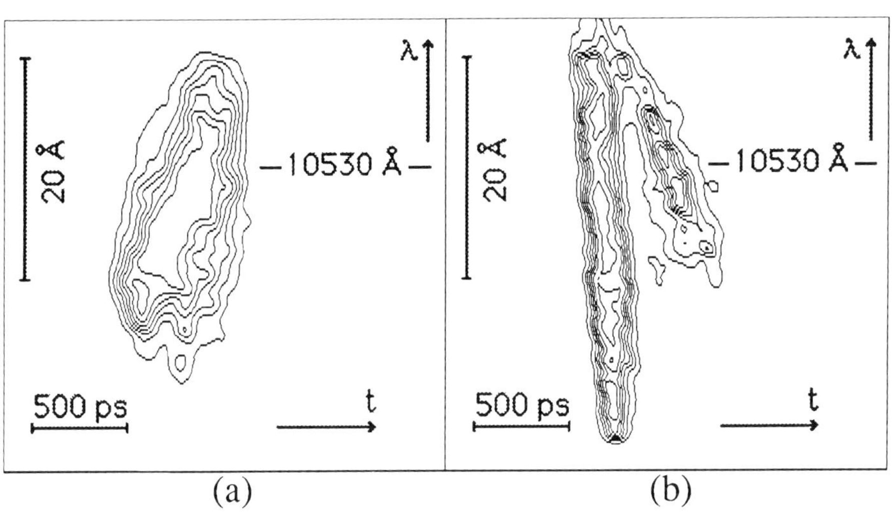

Fig.2 - Time-resolved spectra of light backscattered during interaction.
(a) 500 nm Al dot target; no preformed plasma; irradiance of the interaction beam: 4.0 10^{14} W/cm^2.
(a) 500 nm Al dot target; irradiance of the heating beams: 5.0 10^{13} W/cm^2; interaction beam: 4.0 10^{14} W/cm^2; delay between heting and interaction pulses: 2.5ns.

4. Discussion

Due to the Doppler effect of the expanding plasma, the relative red shift of the SBS backscattered light is expected to be (2/c)(S+F) where S is the ion-sound velocity and F the flow velocity, which is positive in the direction of propagation of the laser beam. The absolute value of F is expected to increase about linearly with the distance from the maximum density position (see Fig.1). Moving into the expanding plasma in the direction of laser light propagation one find the supersonic region, where $F < - S$, giving blue shifted backscattered light; the first sonic layer, $F = - S$, no shift; the subsonic region $0 > F > - S$, red shift; the maximum density layer, $F = 0$, relative shift (2/c)(S); the region beyond the maximum density, $F > 0$, with increasing red shift.

In terms of Doppler modified SBS spectra, and considering a temperature of 400-eV, the first short flash of backscattered light which is visible in Fig. 2 (b) should originate in a plasma region between the layers $F \approx 0$ and $F \approx - 3$ S. In the second phase the plasma region involved at a given instant is much narrower and moves regularly from the $F \approx 0$ to the $F \approx -1.5$ S. An estimation of the flow velocity profile at 2.5-ns after the heating pulse, shows that the first flash of backscattering should involve a region from the maximum density to a distance of the order of 1.5-mm in the direction of the focusing lens. The pure Doppler picture could be inadequate to describe this first phase of the scattering process, which is rather intense and some non linear effects on SBS cannot be excluded. On the contrary we believe that the second phase can be described in terms of linear SBS. We are in presence of a *moving focus* produced by dynamical self-focusing from the maximum density to a distance of the order of 0.8 mm. This *focus* actually is the interaction volume were the intensity is above threshold for SBS, and moves at a rate of about $4\ 10^8$ cm/s into the plasma. The direction of this motion was correctly predicted in the "supersonic limit" approximation by the model of Lontano *et al* [9]. Simulations based on a laser beam focused on the surface of an omogeneous plasma slab lead to a motion in the opposite direction [3,10], which is not realistic to describe the interaction with a corona of decreasing density.

Time resolved spectra show that there is a very short time of effective coupling of the laser light with the plasma up to the maximum density region where an important level of absorption is expected. Suddenly, as a consequence of the local heating of the plasma, the beam suffers a process of self focusing with a growth-time of the order of 100-ps. This process progressively decouples the laser beam from the plasma. This is the reason of the very low level of backscattered energy measured in this condition. Actually the backscattered power is much higher in the first short flash and drops during the moving focus phase. This is due to the low density of the plasma regions involved at this stage and to the reduced volume of the moving focus. On the other hand it is possible that an important fraction of the laser energy is side-scattered in our experimental condition, as expected from 2-D simulation in the case that the interaction length is comparable with the beam size [2], which is the case with strong self focusing.

In conclusion the SBS data presented here seem to be very interesting both in terms of the study of SBS instability in flowing plasmas and in terms of the general problem of the laser coupling with coronal plasmas.

This work was supported by CNR, Italy, SERC, UK, and ECC in the framework of the Uman Capital and Mobility Programme.

References

[1] Frycz P. *et al.* 1992, *Comm. Plasma Phys. Contr. Fusion* **15** 19-38
[2] Amin M. R. *et al.* 1993, *Phys. Rev. Lett.* **71** 81-84
[3] Amin M. R. *et al.* 1993, *Phys. Fluids B* **5** 3748-3764
[4] Willi O. *et al.* 1990, *Phys. Fluids B* **5** 1318-1324
[5] Labaune C. *et al.* 1994, *Rapport Scientifique 1993 du LULI* p. 33-39
[6] Gizzi L. A. *et al.* 1994, *Phys. Rev. E* **49** 5628-5643
[7] Biancalana V. and Chessa P. 1994, *Appl. Optics* **33** 3465-3477
[8] Young P. E. *et al.* 1993, *Phys. Fluids B* **3** 1245-1249
[9] Lontano M. *et al.* 1989, *Phys. Fluids B* **1** 901-907
[10] Andreev N. E. *et al.* 1993, *Phys. Fluids B* **5** 1986-1999

Inst. Phys. Conf. Ser. No 140: Section 9
Paper presented at Laser Interact. with Matter Conf., Oxford, 19–23 September 1994

© 1995 IOP Publishing Ltd

Influence of fill factor on the uniformity of line foci produced with a cylindrical-lens array

J.E. Balmer, A. Glinz, and A. Präg

Institute of Applied Physics, University of Berne, CH-3012 Berne, Switzerland

Abstract: The uniformity of line-focused laser beams has been investigated. A cylindrical-lens array (CLA) in combination with a spherical lens was used to produce a line focus with a uniformity of the irradiance along the line of better than ± 3%. Maximum uniformity of the intensity along the line focus can be achieved if the incident laser beam overfills the aperture of the CLA. The uniformity of the X-ray emission from from copper and aluminum plasmas produced with the CLA is found to be of order ± 15%.

1. Introduction

Long, narrow plasma columns are a basic requirement for X-ray laser studies. It was shown that uniform plasma conditions are crucial for the achievement of maximum gain, particularly in recombination-pumped schemes [1]. Plasma columns can be produced, for example, by focusing a high-power laser beam into a line. The uniformity of the plasma column depends on the uniformity of the irradiance along the line focus because plasma production depends on the irradiance. However, a high uniformity of the irradiance along the line focus does not necessarily result in plasma conditions of the same uniformity. Plasma instabilities such as self-focusing can degrade the initial degree of uniformity [2]. Line focusing of a laser beam is usually realized by utilizing a pair of crossed cylindrical lenses, a spherical lens in combination with a cylindrical lens or a spherical lens together with an off-axis spherical mirror. A short coming of all these methods is that the intensity along the line focus is a continuous mapping of the intensity distribution of the laser beam. In this way a circular beam of uniform intensity cannot been brought to a line focus of uniform intensity. One possibility to solve this problem is to use a cylindrical-lens array (CLA) [3]. As shown in figure 1, it consists of an array of

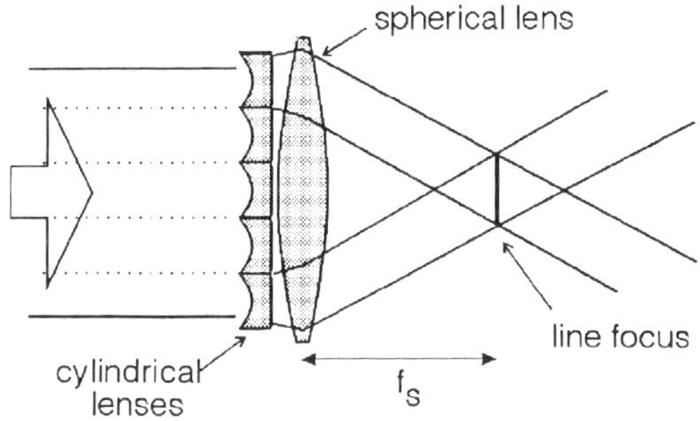

Fig. 1. *Schematic of the focusing configuration of the cylindrical-lens array (CLA)*

cylindrical lenses in combination with a spherical lens. The CLA splits the incident pump laser beam into as many parts as there are cylindrical lens elements in the array. Each part of the split beam is then separately focused to a line and all these individual line foci overlap at the same location, forming a common line focus. Due to the superposition of the individual beamlets the intensity variations of the incident beam are averaged out, leading to a smooth line focus. The goal of the work presented here is to demonstrate that maximum uniformity of the intensity along the line focus can be achieved if the incident laser beam overfills the clear aperture of the CLA.

2. Experiments

The experiments were originally performed with the 42-mm diameter beam of a Nd:glass laser system (λ = 1054 nm). The beam was line focused by means of a CLA having a focal length $f_c = -80$ mm, in combination with a 50-mm diameter achromatic lens doublet of focal length $f_s = 100$ mm. The CLA consisted of five cylindrical lenses, each of them 9 mm wide, yielding a total clear aperture of 45 mm. According to geometrical optics, the length L of the resulting line focus is given by

$$L = d\,(f_s/f_c)$$

where d is the width of a cylindrical lens element. With the parameters given above, our line focus is calculated to be 11 mm long. The dimensions and the uniformity of the line focus were measured by focusing the attenuated laser beam directly onto the chip of a cooled, slow-scan CCD camera having a dynamic range of better than 10^3. The spatial resolution of this measurement is given by the 25-μm center-to-center spacing of the camera pixels.

The Nd:glass laser facility used in these experiments was recently upgraded to nominally 120 J/1 ns by the addition of a 90-mm diameter rod amplifier. The focusing optics for the corresponding 84-mm diameter beam was designed to produce a 13-mm long line focus. This was achieved by combining a 500-mm focal length, plano-convex lens with a CLA made up of seven 11-mm wide cylindrical lenses having a focal length of $f_c = -400$ mm. The total width of the CLA of 77 mm was thus less than the diameter of the laser beam. It is shown below that this overfilling of the clear aperture of the CLA results in the maximum possible irradiance uniformity along the line focus. The energy loss introduced by clipping part of the beam edge was calculated to be less than ~10%. The Fresnel diffraction introduced by overfilling the aperture of the CLA does not present a problem since it just adds to the diffraction features from each element.

3. Experimental results

The width of the line focus measured with the CCD camera was found to be less than 50 μm (FWHM) for the 42-mm diameter beam, and ~60 μm for the 84-mm beam. Figure 2 shows plots of the intensity along the line focus of the 42-mm beam as recorded for a 100-psec pulse at 10^{13} W/cm^2. These intensity plots along the line focus have been averaged over a width of four pixels, i.e., 100 μm. A characteristic feature of the intensity distribution is the strong modulation observed at both ends. This may be shown in a straightforward manner to be due to Fresnel diffraction at each of the cylindrical lenses. Curve a in figure 2 shows the uniform-

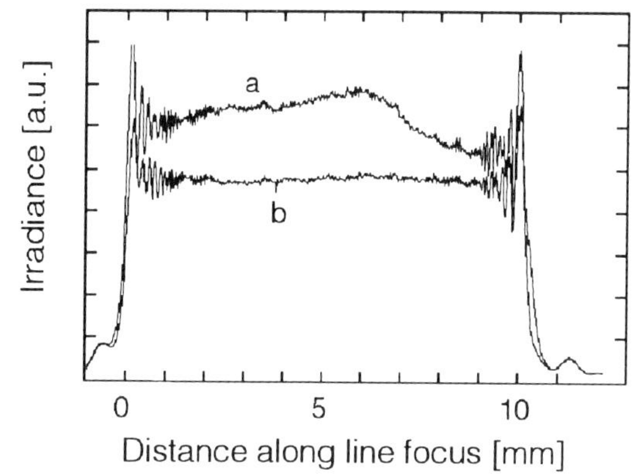

Fig. 2. *Scans of the irradiance along the line focus, (a) when the incident laser beam (42-mm diameter) is smaller than the cylindrical-lens array and, (b) after the two outermost cylindrical lenses of the array had been masked.*

ity between the modulations at the ends to be about ± 8%. It corresponds to the case where the incident laser beam with its diameter of 42 mm did not completely fill the 45-mm wide cylindrical-lens array. Curve b in figure 2 shows the intensity distribution along the line focus after the two outermost cylindrical lenses of the array had been masked so that only the part of the laser beam falling onto the three cylindrical lenses in the center of the array was focused. With this masking about 25% of the incident energy was lost, but the uniformity between the modulated ends was improved to better than ± 3%. It can be readily shown that the increase in modulation results from the incompletely illuminated edge elements of the CLA. The consequence is that, in order to obtain the maximum possible uniformity along the line focus, the incident laser beam should slightly overfill the aperture of the CLA. This rule was obeyed for the design of the CLA for the 84-mm diameter beam.

When assembling and testing the 77-mm aperture CLA for the 84-mm diameter beam, it was found that the line foci of the individual cylindrical lens elements did not exactly overlap because of fabrication tolerances. By selecting the seven elements having their line foci closest to each other, the best possible superposition of the individual line foci was achieved. Figure 3 shows scans of the irradiance along the line focus of the 84-mm beam as recorded with a 700-ps pulse at $\sim 10^{12}$ W/cm^2. Again, the curves have been averaged over a width of four pixels, i.e., 100 μm. Curve a in figure 3 shows the uniformity of the irradiance from a single CLA element to be approximately ± 5%, while the uniformity of the full beam (curve b) is approximately ± 10%. At best superposition, the full-beam irradiance shows a substantial fine-scale non-uniformity which is caused by interference of the radiation from different lens elements. The slight sloping of the curve towards the right is due to some asymmetry in the intensity distribution of the incident beam caused by improper alignment. The origin of the modulation seen in the single-element case is not clear at present.

Copper and aluminum slab targets were irradiated with 100-ps and 700-ps laser pulses, line focused to approximately $3 \cdot 10^{12}$ W/cm^2 with the 42-mm beam. The plasma radiation was imaged with a double-slit camera onto a P20 phosphor screen and recorded with a CCD camera. To restrict the investigated plasma radiation to the keV range, a 14-μm thick Be filter was optionally inserted. The magnification of the double-slit camera was 1x and 20x in the di-

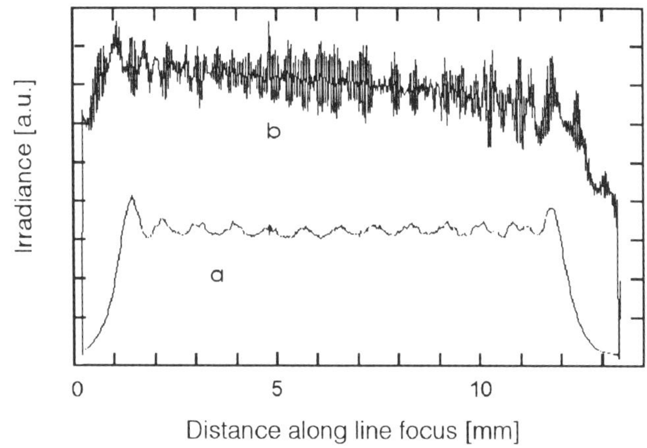

Fig. 3. *Scans of the irradiance along the line focus of the 84-mm beam, (a) for one single element of the CLA and, (b) after selecting and assembling seven cylindrical lenses for optimum superposition.*

rections parallel and perpendicular to the plasma column, respectively. The uniformity of the recorded X-ray emission was found to be highest (of order ± 10%) in the case of Cu targets irradiated with 700-ps pulses and lowest (of order ± 20%) for Al targets irradiated with 100-ps pulses [4].

4. Conclusions

We have shown that, with the aid of a cylindrical-lens array, line foci can be produced having an irradiance non-uniformity as low as ± 3%. Maximum uniformity requires all the elements of the CLA to be fully illuminated. This is most easily achieved if the laser beam slightly overfills the aperture of the CLA. The uniformity of X-ray emission from copper and aluminum plasmas created by such a line focus was found to be of order ± 15%.

5. References

[1] Dumont H, Bourgade J L, Bruneau J, Desenne D, Dulieu A, Louis-Jacquet M, Berthet L, Decoster A, Jacquemot S, Carillon A, Jaeglé P, Jamelot G, Raucourt J P, and Gauthé B 1993 *Opt. Commun.* **96** 87–93

[2] Herbst M J, Stamper J A, Whitlock R R, Lehmberg R H, and Ripin B H 1981 *Phys. Rev. Lett.* **46** 328-331

[3] Chen W, Wang S, Chen B, Xu A, and Mao C 1992 *Chinese Physics* **12** 403-407

[4] Glinz A, Präg A, and Balmer J *Opt. Commun.*, to be published

Modelling X-ray radiation in laser-plasma

S.A.Bel'kov, G.V.Dolgoleva

Russian Federal Nuclear Centre - All Russia Scientific Research Institute of the Experimental Physics (VNIIEF), Arzamas-16, Nizhniy Novgorod Region, Russia.

Abstract. Numerical modelling of laser plasma emission processes in a plane and spherical geometry has been performed for experimental conditions of laser facilities Iskra-4 and Iskra-5. In calculations the following processes have been taken into consideration: a two-temperature gasdynamics, laser radiation absorption, nonequilibrium and nonstationary ionization kinetics, nonequilibrium radiation transport. The results of the simulation are presented.

As a part of ICF problem, there have been vigorous discussions in the recent years on the ways to initiate fusion reactions in indirect-drive targets (so called X-ray targets), i.e. when laser or ion beam energy is to be initially converted by suitable devices to soft X-ray [1] which finally interacts and implodes the DT-filled microballoons. This technology, on the one hand, can substantially improve the microsphere implosion symmetry, and on the other, referring to Laser Fusion problem, would make it possible to escape from issues concerning nonlinear interactions of ultrahigh-power coherent fluxes and works in the region when classical linear physic is applicable.

Several successful experiments with targets of this kind have been performed at different facilities around the world, such as Nova (USA) [2], Gekko-XII (Japan) [3], Foebus (France) [4], Iskra-5 (Russia) [5] where were shown that this trend is promising, and moreover seems even more preferable now in terms of the feasibility of ignition facility projects despite high the laser energy requirements.

In view of the above situation, increasingly more vital is the ability to adequately describe and numerically simulate physical phenomena occurring in these targets, from laser beam absorption by the converter to DT-filled microsphere acceleration and implosion, and, eventually, fuel ignition and burn. This is particularly important in cases when a target must be designed and optimise to have as high performance as possible with minimum energy input.

This paper briefly summarises 1D radiation gasdynamics code package SNDP [6], developed by the authors, and also illustrates the calculations for plane and spherical geometry's, as made using this code, in comparison with the experimental data on laser to X-ray conversion in different target experiments on Iskra-4 and Iskra-5 laser facilities [7,8].

The authors have developed SNDP package providing simulations of the following physical phenomena: 1D, two-temperature gasdynamics in plane, cylindrical and spherical geometry's allowing for viscosity forces; electrons and ions heat transfer (including nonlocal heat transfer approximation) with limited thermal flux; inverse bremsstrahlung and resonance absorption of laser radiation, phenome-

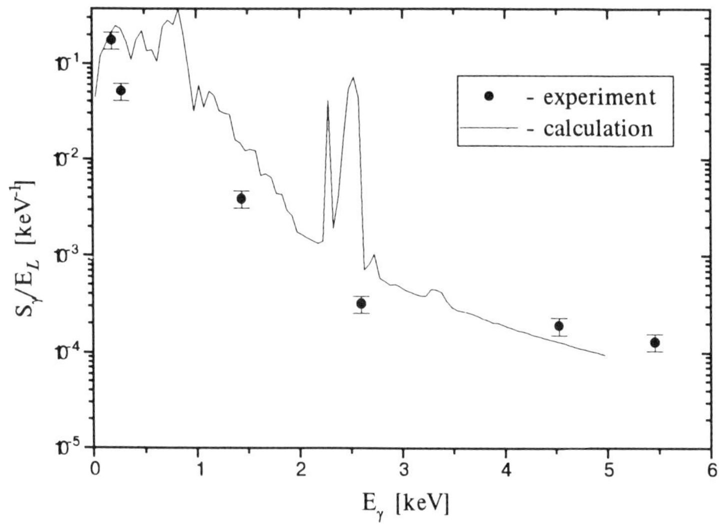

Fig. 1

nologically accounting for density profiles getting steeper due to ponderomotive force of the electromagnetic field; generation and deceleration of "hot" electrons phenomenologically including the partly loss of deposited energy by acceleration of the "fast" ions; non-equilibrium and nonstationary kinetics of excited state population numbers and multicomponent and multicharged plasma atoms ionisation in the average-ion radiation-collision approximation; nonequilibrium, nonstationary and spectral X-rays transport in a quasidiffusion approximation including free-free, bound-free and bound-bound transitions of electrons interacting with radiation field.

These codes were used to analyse and simulate iodine laser beam conversion experiments (wave-length λ=1.315 μm at pulse fundamental frequency) for Iskra-4 facility using plane Au converter irradiated at 1ω and 2ω harmonics [9].

Fig. 1 compares the SNDP-calculated spectrum to be emitted by this converter as it is irradiated by a laser pulse having width (FWHM) $\tau_{0.5}$=0.25 ns and peak intensity $I_L=10^{14}$ W/cm^2, with the experimental one observed by multichannel spectrometer in the experiment with laser pulse having nearly the same parameters as in the calculations. The limitation factor for electronic heat conduction involved in the calculations was f_{lim}=0.03. With

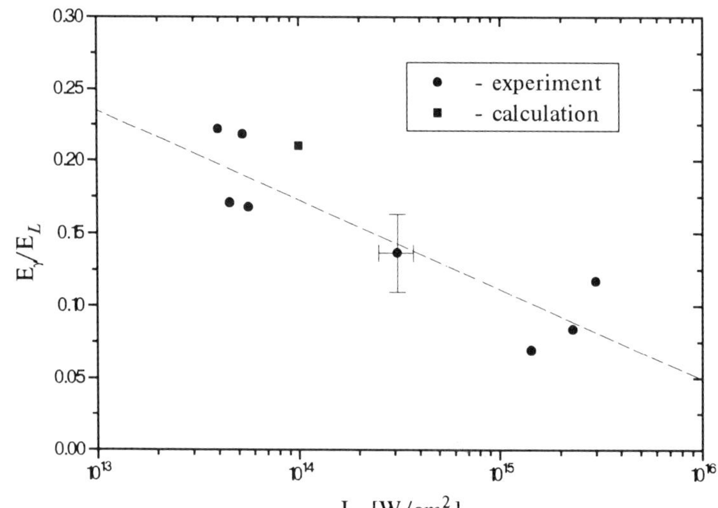

Fig. 2

these conditions, the calculated conversion ration was $k_\gamma=E_\gamma/E_L=20\%$, this being in a good agreement with the experimental data (Fig. 2).

Fig. 3 shows calculated and experimental radiation spectra from the converter representing a spherical copper cavity with its inside covered with Au, that was irradiated by a laser pulse from 12-beam Iskra-5 system. In the experiment, laser pulse had the following parameters: wave-length $\lambda=1.315$ μm, FWHM $\tau_{0.5}=0.7$ ns, and energy input inside the target $E_L=9$ kJ. Laser radiation input was performed through corresponding holes with area percentage <13%. The same input holes were used to observe the X-ray spectrum emitted from the cavity [10].

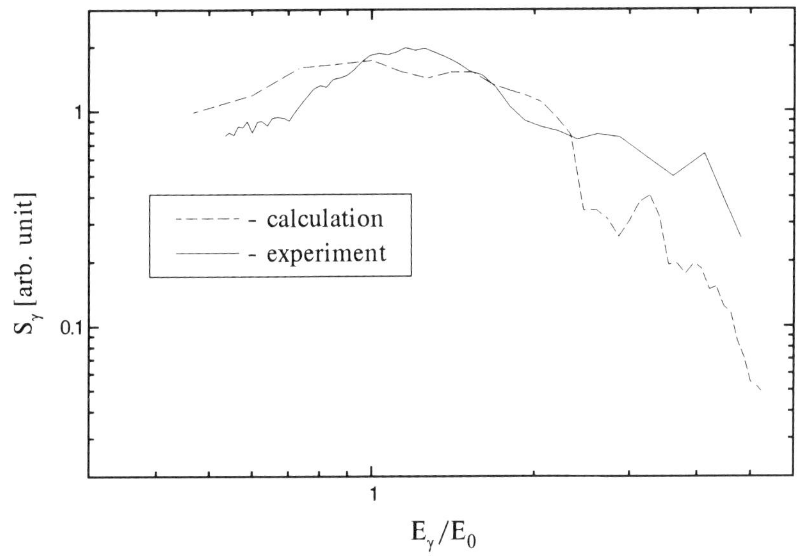

Fig. 3

Generally, what should be concluded is reasonable agreement between numerical simulations and experimental data obtained using Iskra lasers for different target types and over a wide range of experimental parameters.

References

[1] M.Murakami, J.Meyer-ter-Vehn. Nuclear Fusion, 31, 1315 (1991)

[2] J.D.Kilkenny and et al. *Plasma Phys. and Controlled Nucl. Fusion Res.*, **3**, 29 (1989)

[3] S.Nakai and et al. *Nuclear Fusion*, **30**, 1779 (1990)

[4] M.DeCroisette. *Zababahinskie nauchnye chtenija.* Thesis of the reports. Cheljabinsk - 1992.

[5] F.M.Abzaev and et al. *Pis'ma v ZETF*, **58**, 28 (1993)

[6] S.A.Bel'kov and et al. *Voprosy atomnoj nauki i tehniki.* Ser.: Matematicheskoe modelirovanie fizicheskih processov, vyp.1, 76 (1990)

[7] S.B.Kormer. *Izv. AN SSSR*, Ser. fizicheskaya, **44**, 2002 (1980)

[8] V.I.Annenkov and et al. *Kvantovaja electronika*, **18**, 536 (1991)

[9] S.A.Bel'kov and et al. *Czechoslovak Journal of Physics*, **42**, 969 (1992)

[10] S.A.Bel'kov and et al. *Kvantovaja electronika*, **21**, 271 (1994)

Paper presented at Laser Interact. with Matter Conf., Oxford, 19–23 September 1994
© 1995 IOP Publishing Ltd

High-power laser beam smoothing using binary-phase Fresnel zone plate arrays

T H Bett[1], C N Danson[2], P Jinks[1], D A Pepler[2], S Rivers[2], I N Ross[2] and R M Stevenson[1]

[1] *AWE plc, Aldermaston, Berkshire, RG7 4PR, UK*
[2] *Central Laser Facility, Rutherford Appleton Laboratory, Chilton, Didcot, Oxon. OX11 0QX*

Abstract. We present the design and optimisation of novel binary phase Fresnel zone plate arrays, for high power laser focal spot smoothing applications. We have demonstrated that these devices can be accurately simulated using a commercial propagation code.

1. Principle of operation

The binary-phase Fresnel zone plate array (PZP) is a diffractive optic, but can be understood using geometric optics, as demonstrated in figure 1.

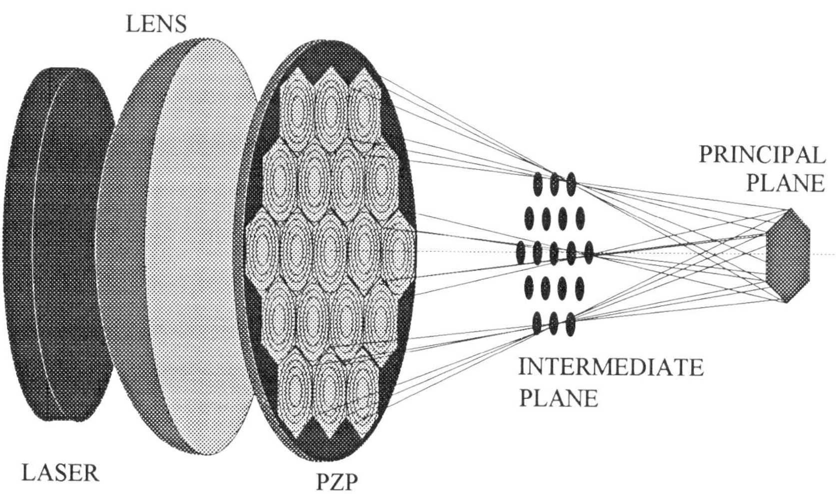

Figure 1 Schematic of the PZP operation

Each zone plate in the array acts as a lens and effectively subdivides the input beam into an array of beamlets. These are then brought to an intermediate focal plane using a principle focusing lens. Each of the focal spots are allowed to defocus where they overlap in the focal plane of the principal focusing lens to produce a top-hat envelope. Smoothing is achieved from the overlap of the many beamlets averaging the intensity distribution.

The focal spot is formed primarily by the ± 1st order diffraction, with 82% of the diffracted light contained within the 1st order area. The focal spot size is given by $W_S = W_Z F_P / F_Z$, where: W_Z is the size of each cell; F_P and F_Z the focal lengths of the principal lens; and Fresnel lens respectively.

In practice [1] it was found that a 0th order component produces a central spike in this target plane which can be minimised by using a target plane just beyond focus.

2. Simulation v experiment

The Phase Zone Plate (PZP) arrays were simulated using the propagation code GLAD™. This is a commercially available physical optics analysis code designed to analyse beam trains and laser devices including the effects of diffraction, apertures, lenses and aberrations. Analysis of the PZP array considered the diffraction from an array of binary level Fresnel lenses with superimposed phase delays of 0 or π radians onto a uniform input beam. This phase plate was then combined with a principal focusing lens to simulate the conditions used for the measurements on manufactured plates. In order to get sufficiently high fidelity reproduction of the zones in the computations the number of beamlets that can be modelled is restricted by 2D array sizes. In the example described this limit is found to be 5 zones accross a beam diameter and hence approximately 15 - 20 beamlets.

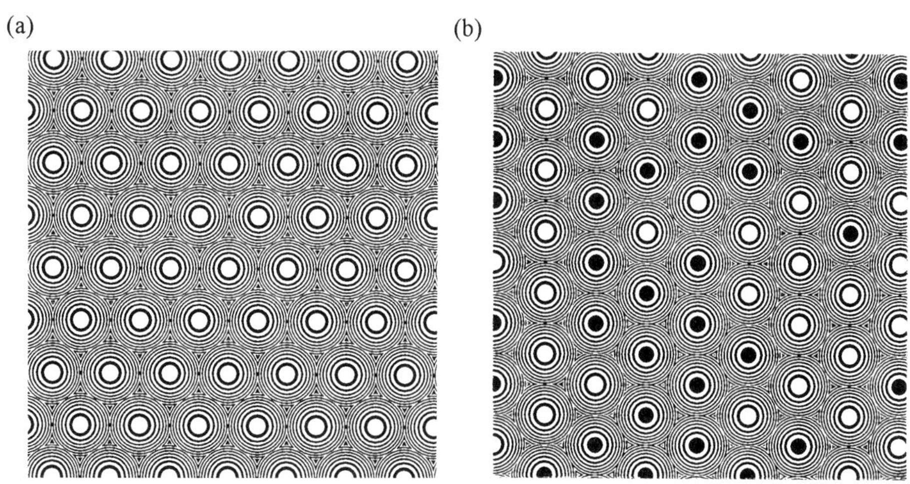

Figure 2 Sections of PZP mask
a) With fixed phase elements b) With random phase elements

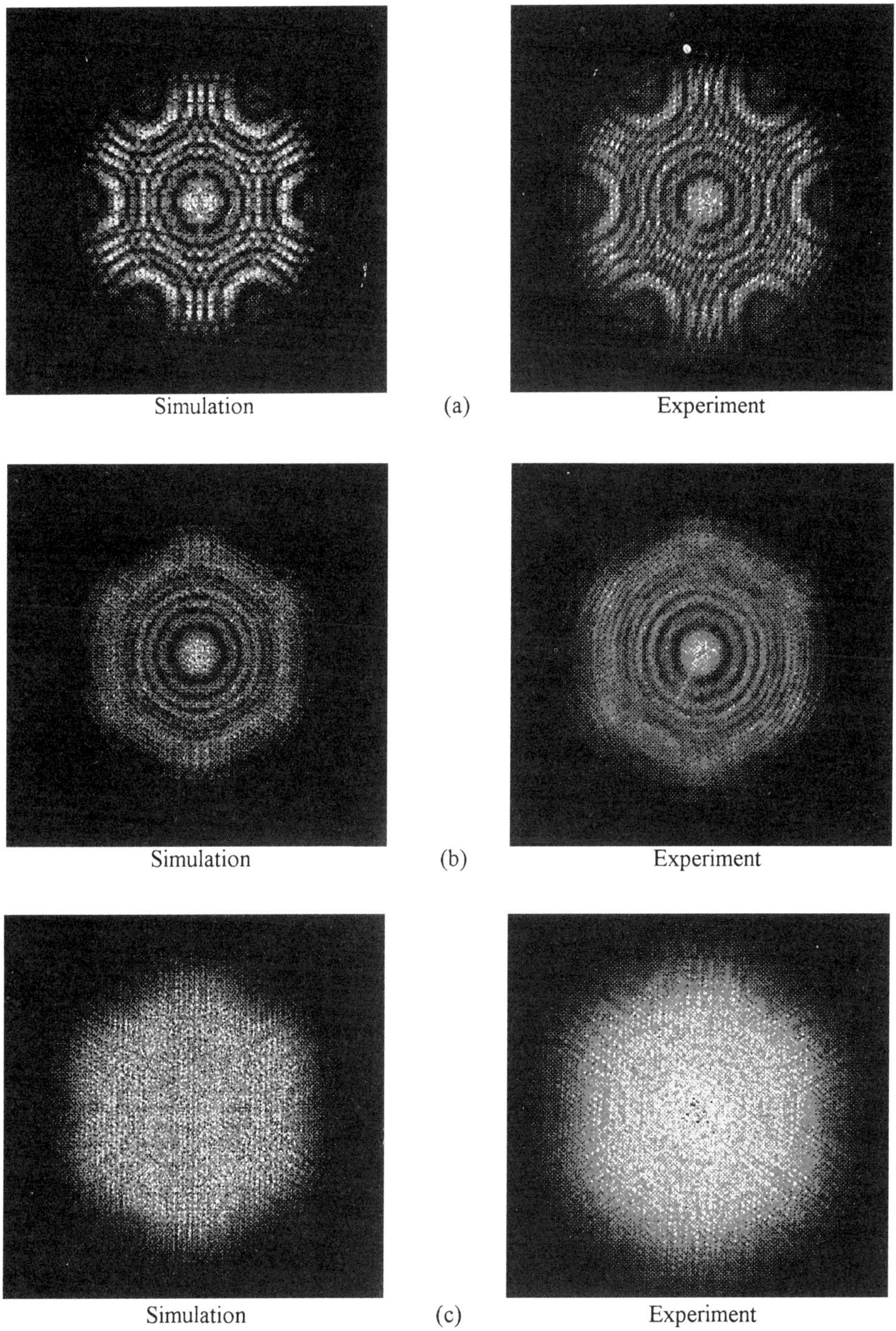

Figure 3 Simulated and experimental PZP focal spot distributions in the principal plane a) with constant phase b) with random phase and c) in the optimum plane with random phase

To compare the simulations with test plates an experimetal arrangement was set up using a large aperture (70 mm) collimated green HeNe focused by a 1.35 m focal length lens. The focal profile was analysed using a cyrogenically cooled high dynamic range CCD coupled to an image capturing system.

Two cases were analysed:

> The first considered an array whose phase was fixed such that every zone plate within the array was identical.
>
> The second case randomly set the phase of each zone plate to either 0 or π radians, such that equal numbers of zones were of each phase.

The masks used in the experiment to compare these two cases are shown in figures 2 (a) and 2 (b) respectively. The black and white areas would correspond to a π and 0 radian phase shift respectively.

The GLAD™ images are compared with experimental results in figure 3, measured in both the principal and optimal planes. In the fixed phase case strong modulation is observed in the principal plane, but this is partially 'washed out' if the target plane is positioned beyond focus in the optimum plane. In the case where random phases are used the modulation seen in the principal plane is less strong and again a much smoother focal profile is seen in the optimum plane. These images demonstrate the excellent agreement between the simualtions and experimental results. This will allow further theoretical optimisation of the PZPs with reasonable confidence in the results.

3. Discussion

We have shown that Phase Zone Plate arrays produce uniform intensity focal envelopes, with good efficiency (~82%). They are simple to use and can be interchanged on a shot by shot basis to generate a wide range of focal spots, for many laser applications.

Manufacturing methods are similar to those used in the production of Random Phase Plates (RPP) [2], where photoresist techniques make them inexpensive for large aperture lasers (<400mm).

References

[1] Stevenson et al 1994 *Optics Letters* Vol. 19, No. 636-365

[2] Pepler DA et al 1993 *Proc SPIE Conf. on Laser Coherence Control*, SPIE Vol. 1870, 76-87

TM Applied Optics Research, Tuscon, AZ, USA.

Inst. Phys. Conf. Ser. No 140: Section 9
Paper presented at Laser Interact. with Matter Conf., Oxford, 19–23 September 1994
© *1995 IOP Publishing Ltd*

Electromagnetic solver for short pulse laser-matter interaction

L. Bonnet, A. Decoster and S. Jacquemot

Commissariat à l'Energie Atomique
Centre d'Etudes de Limeil-Valenton
94195 Villeneuve Saint-Georges Cedex, France

Abstract. The resolution of Maxwell's equations describing the propagation of an electromagnetic wave normally incident onto a given medium has been coupled to the hydrodynamical code CHIVAS in order to simulate short pulse laser-matter interaction. A novel finite difference scheme, based on the Poynting vector and particularly suitable for the calculation of laser deposition, has been developed and can be easily extended to obliquely incident light. The sensitivity of the hydrodynamical history of the target to the modelling of laser absorption is pointed out, and is illustrated by calculations of H-like recombination gain in Aluminium at 38.8 Å.

1. Introduction

One of the crucial points in laser-matter interaction simulations is the calculation of the laser energy deposition. Many processes theoretically couple the electromagnetic (noted EM) wave and the target, but usually the hydrodynamical codes have only considered inverse bremsstrahlung and resonant absorption, in the geometrical optics WKB approximation. This is no longer valid for *short* laser pulses because of the strong density gradients in the little expanded plasma. Maxwell's equations for the propagation of the laser EM wave must then be considered and coupled to the hydrodynamics equations describing the evolution of the target [1]. As neither an EM energy density nor a conservation law can be found in our plasmas, the laser deposition by Joule effect is traditionally obtained from the product of the electric field $\vec{E}$ by the current density vector $\vec{J}$. However, for a monochromatic wave, a "pseudo-conservation" relation, based on the energy flux vector $\vec{S}$, can be obtained.

After a brief presentation of the general equations, we describe the numerical method specially developed for the resolution of the second order Helmholtz wave equation for normally incident laser, as well as the various expressions for the medium complex permittivity ϵ' [2]. X-Ray Lasers calculations illustrate the sensitivity of gain coefficients at 38.8 Å in H-like aluminium to conductivity and laser absorption models.

2. Propagation equations

$\vec{H}$ is the magnetic field. It is assumed that the characteristic time of the hydrodynamics evolution is much larger than the laser period and the characteristic propagation time.

For a monochromatic EM wave of frequency ω incident on a medium where the spatial dispersion of the permittivity can be neglected[1], the solutions of Maxwell's equations can be sought in the form $\vec{\mathbf{E}} = \mathcal{R}e\left(\vec{\mathrm{E}}(\vec{\mathrm{r}})\,\mathrm{e}^{\imath\omega t}\right)$ and $\vec{\mathbf{H}} = \mathcal{R}e\left(\vec{\mathrm{H}}(\vec{\mathrm{r}})\,\mathrm{e}^{\imath\omega t}\right)$ where $\vec{\mathrm{E}}(\vec{\mathrm{r}})$ and $\vec{\mathrm{H}}(\vec{\mathrm{r}})$ are the complex vectors solutions of

$$\mathrm{rot}\,\vec{\mathrm{H}}(\vec{\mathbf{r}}) = \frac{i\omega}{c}\epsilon'(\omega,\vec{\mathbf{r}})\vec{\mathrm{E}}(\vec{\mathbf{r}})\ ,\ \mathrm{rot}\,\vec{\mathrm{E}}(\vec{\mathbf{r}}) = -\frac{i\omega}{c}\vec{\mathrm{H}}(\vec{\mathbf{r}})\ ,$$
$$\mathrm{div}\left(\epsilon'(\omega,\vec{\mathbf{r}})\vec{\mathrm{E}}(\vec{\mathbf{r}})\right) = 0\ ,\ \mathrm{div}\left(\vec{\mathrm{H}}(\vec{\mathbf{r}})\right) = 0\ . \qquad \text{(Maxwell)}$$

ϵ' being related to the electric permittivity ϵ and conductivity σ by the relation $\epsilon'(\omega,\vec{\mathbf{r}}) = \epsilon(\omega,\vec{\mathbf{r}}) - \imath\frac{4\pi\sigma(\omega,\vec{\mathbf{r}})}{\omega}$. Combining this set of equations leads to the following well-known equations describing the propagation of the fields

$$\Delta\vec{\mathrm{E}}(\vec{\mathbf{r}}) - \mathbf{grad}\left(\mathrm{div}\,\vec{\mathrm{E}}(\vec{\mathbf{r}})\right) + \frac{\omega^2}{c^2}\epsilon'(\omega,\vec{\mathbf{r}})\,\vec{\mathrm{E}}(\vec{\mathbf{r}}) = 0$$
$$\Delta\vec{\mathrm{H}}(\vec{\mathbf{r}}) - \frac{1}{\epsilon'}\mathbf{grad}\,\epsilon'\times\mathbf{rot}\,\vec{\mathrm{H}}(\vec{\mathbf{r}}) + \frac{\omega^2}{c^2}\epsilon'\,\vec{\mathrm{H}}(\vec{\mathbf{r}}) = 0\ . \qquad \text{(propagation)}$$

However, for normally incident lasers, the divergence of $\vec{\mathrm{E}}(\vec{\mathrm{r}})$ is equal to zero and we get the *Helmholtz wave equation*

$$\frac{\mathrm{d}^2\mathrm{E}(\mathrm{z})}{\mathrm{dz}^2} + \frac{\omega^2}{c^2}\epsilon'(\omega,\mathrm{z})\,\mathrm{E}(\mathrm{z}) = 0$$

3. Expressions for the deposited power

Consider, on one hand, a continuous distribution of charges in a given volume V, characterized by its current density vector $\vec{\mathbf{J}}$. The temporal mean of the integral $\int_V \vec{\mathbf{J}}.\vec{\mathrm{E}}$ over a laser period represents the conversion of the wave energy into thermal or mechanical energy. The evaluation of $\vec{\mathbf{E}}$ and $\vec{\mathbf{J}}$ at each time and space allows the calculation of the deposited power $\mathrm{P_{las}}(\vec{\mathbf{r}}) = \left\langle\vec{\mathbf{J}}\cdot\vec{\mathbf{E}}\right\rangle = \frac{1}{2}\sigma\left|\vec{\mathrm{E}}(\vec{\mathbf{r}})\right|^2$. In order to get the total laser energy deposition in the plasma, a summation over the grid is required $\mathrm{P_{las}^{tot}} = \sum_{\mathrm{mesh}}\mathrm{P_{las}}(\vec{\mathbf{r}})$ which can unfortunately lead to an unaccurate laser energy balance.

However, on the other hand, from Maxwell's equations in a material medium, the following relation can be written $\frac{1}{4\pi}\left[\vec{\mathbf{E}}\cdot\frac{\partial\vec{\mathbf{D}}}{\partial t} + \vec{\mathbf{H}}\cdot\frac{\partial\vec{\mathbf{B}}}{\partial t}\right] + \mathrm{div}\,\vec{\mathbf{S}} + \vec{\mathbf{J}}\cdot\vec{\mathbf{E}} = 0$, where $\vec{\mathbf{S}} = \frac{c}{4\pi}\left(\vec{\mathbf{E}}\times\vec{\mathbf{H}}\right)$ is the energy flux vector also called *Poynting vector*. Taking the temporal mean of the above relation leads to $\boxed{\mathrm{div}\left\langle\vec{\mathrm{S}}\right\rangle + \left\langle\vec{\mathrm{J}}\cdot\vec{\mathrm{E}}\right\rangle = 0}$. The second term of this equation is just $\mathrm{P_{las}}(\vec{\mathbf{r}})$. It means that, from a practical point of view, the deposited energy in a given cell is just given by the *difference between its two borders* of the Poynting vector temporal mean. $\mathrm{P_{las}^{tot}}$ is then immediately obtained from the difference of $\left\langle\vec{\mathbf{S}}\right\rangle$ between the two edges of the plasma and no summation is required: the equality between the incident power on one side and the som of reflected+deposited powers on the other side is automatically achieved. Moreover, as these properties have been established directly from the equations of propagation *without any assumption on the angle of incidence or on the polarization* of the laser, they are also *valid for obliquely incident lasers, either S or P polarized*, provided they are monochromatic.

[1] This is possible if $\mathrm{k\,d} \ll 1$ where k is the wave vector and d is the caracteristic correlation length depending upon the medium.

4. Numerical schemes

Two numerical schemes have been implemented. The first one is derived from atomic physics methods but has been adapted to our case of a given and moving hydrodynamical grid: it's a *variable mesh Numerov-like finite difference scheme.* It is not detailed here because it is not well adapted to the Poynting method of determining laser energy deposition. The following notations are used: E_n is the value of $E(z)$ at the center of cell n and $f_n = \frac{\omega^2}{c^2}\epsilon'(z)$; $h_{n+1/2}$ is the distance between centers of cells n and n+1.

The *Poynting finite difference Helmholtz wave equation* is (precision $O(h^3)$)

$$h_{n+1/2}E_{n-1} - (h_{n+1/2} + h_{n-1/2})\frac{1 - f_n\frac{h_{n+1/2}+h_{n-1/2}}{4}}{1 + f_n\frac{h_{n+1/2}+h_{n-1/2}}{4}}E_n + h_{n-1/2}E_{n+1} = 0 \qquad (*)$$

The Poynting vector temporal mean is then located exactly at the interface between two cells $\bar{S}_{n+1/2} = \frac{c^2}{8\pi\omega}\frac{E^*_{n+1}E_n - E_{n+1}E^*_n}{2ih_{n+1/2}}$. The absorbed power is defined at the center of the cell. And the finite difference equation for it is

$$\frac{\bar{S}_{n+1/2} - \bar{S}_{n-1/2}}{\frac{h_{n+1/2}+h_{n-1/2}}{2}} + \frac{\omega}{8\pi}\frac{Im(\epsilon'_n)|E_n|^2}{\left|1 + \epsilon'_n\frac{\omega^2}{c^2}\frac{h_{n+1/2}\times h_{n-1/2}}{4}\right|^2} = 0$$

It's also easy to demonstrate that this numerical scheme provides the *exact reflexion coefficient* at the interface between two media of different density.

At a given time, the hydrodynamical grid results from the evolution of the target and is therefore not suitable for an accurate resolution of the wave equation. An *EM grid* must then be defined. In each hydro. cell n, the number of EM sub-zones is determined by two conditions: a minimum number of points per laser wavelength (21 in practise, for integration accuracy) and an *odd integer* number of sub-cells: the borders of the hydro. cell are then borders of EM sub-cells and hydrodynamical parameters in sub-cells are easy to determine from those of cell n.

The two *initial conditions* required for the resolution of the *second-order* wave equation are defined in the critical part of the plasma where the electric field amplitude is known to be exponentially increasing. The solution is then normalized at the amplitude of the incident laser in vacuum.

5. Permittivity model

The permittivity model should account for every state of the material during the interaction, from solid to plasma at any temperature and density. A standard description $\epsilon'(\omega) = 1 - i\frac{4\pi}{\omega}\frac{\sigma(0)}{1-i\omega/\nu_{eff}}$ has been used with two possible expressions for the zero frequency conductivity $\sigma(0)$ and effective frequency ν_{eff} involved:

1. the kinetics (labelled plasma) point of view based on the Boltzmann equation for the electronic distribution function, the equilibrium one being Maxwellian; ν_{eff} is then $\nu_{eff} = \nu_{ei} = \frac{4}{3}\sqrt{2\pi}\frac{e^4}{\sqrt{m_e}}\frac{Z^{*2}N_i\ln(\Lambda)}{T_e^{3/2}}$ and $\sigma(0) = \frac{Z^*N_ie^2}{m_e\nu_{eff}}$ where T_e is the electronic temperature, N_i the ion density, Z^* the mean ionization and $\ln(\Lambda)$ the Coulomb logarithm for inverse bremsstrahlung [3]
2. the complete model of YT Lee and RM More (noted Lee-More) $\sigma(0) = \frac{Z^*N_ie^2}{m_e\nu_{eff}}A^\alpha(\mu/T_e)$ where $A^\alpha(\mu/T_e)$ is function of the chemical potential μ and ν_{eff} depends upon the state of the material (see [4] for details).

6. Applications to X-Ray Lasers

The use of short pulse, high intensity lasers should help, by creating an initially totally ionized high density plasma, increasing the efficiency of H-like recombination schemes. Figure 1 shows the space and time evolution of Balmer α gain coefficient at 38.8 Å obtained by irradiating a 4 μm thick massive Al target with a 0.3 ps/10^{17} W/cm^2 gaussian clean pulse and assuming "Lee-More" conductivity model for the resolution of the Helmholtz wave equation (nominal case); *Voigt profiles* are assumed for the lasing line and *reabsorption* is taking into account. Gain coefficients appear to be a real signature of the laser absorption modelling: neither the localization, nor the duration are similar from WKB to "Helmholtz+Lee-More" modellings. So is the value of maximum gain: from 0.09/cm for WKB to 130/cm for "Helmholtz+Lee-More" up to 210/cm for "Helmholtz+plasma". However, this parameter is also sensitive to the ionization model used in the hydro. code: for example, using the pseudo-coronal model CORAT instead of the non LTE screened hydrogenic ion one NOHEL leads to a maximum gain of 230/cm instead of 130/cm. What is interesting is that the overall aspect of space and time gain evolution does not vary from NOHEL to CORAT.

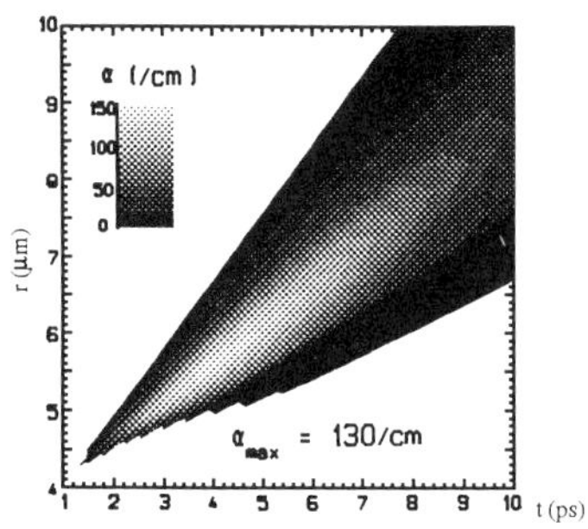

Figure 1: Time and space evolution of nominal Balmer α local gain coefficient.

A preliminary result evaluating the effect of space integration on gain, the propagation of the XRL signal along the plasma being treated in a retarded mode and a 1 ps resolution assumed, has shown that, despite the relatively narrow gain region, the damage resulting from integration is limited to a factor 4 on the maximum value: the nominal integrated gain peaks then at 30/cm and has a FWHM of about 8 ps.

7. Conclusions

Several improvements are still in progress either in normal or oblique incidences: the resolution of the finite difference Helmholtz equation by tri-diagonal methods will allow us to treat different lasers configurations; the extension of Poynting modelling to S or P polarized light (in progress with J.-C. Gauthier's group (LULI)) will help getting more reliable hydrodynamical histories because a lot of experimental results on the absorption of Al targets irradiated with obliquely incident lasers are available. Even if many uncertainties still remain on the behaviour of solid targets at those laser intensities, the preliminary results exhibiting an integrated gain of 30/cm at 38.8 Å are encouraging.

References

[1] Alley WE 1992 *LLNL Annual Report* UCRL-LR-105820-92 160
[2] Jackson JD 1975 *Classical Electrodynamics* (NY: John Wiley & Sons)
[3] Skupsky S 1987 *Phys. Rev. A* **36** 5701–5712
[4] Lee YT and More RM 1984 *Phys. Fluids* **27** 1273-1286

Inst. Phys. Conf. Ser. No 140: Section 9
Paper presented at Laser Interact. with Matter Conf., Oxford, 19–23 September 1994

© *1995 IOP Publishing Ltd*

Effects of random phase plates on aberrated laser beams

T. Ceccotti, S. Bastiani, A. Giulietti, D. Giulietti† and C. Danson*

Istituto di Fisica Atomica e Molecolare, Via del Giardino 7, 56127 Pisa, Italia
† Dipartimento di Fisica, Università di Pisa, Piazza Torricelli 2
* Central Laser Facility, Rutherford Appleton Laboratory, Chilton, Didcot, UK

Abstract. Far-field images produced using random phase plates on a pulsed laser beam have been recorded by a CCD and then analysed. Spot sizes agree with diffraction theory while mean sizes of the pattern structures has been found larger than diffraction limit. Probability density distribution of intensity in the focal plane has not, in most cases, the expected exponential behaviour. Images obtained with reduced beam diameter show a definite central symmetry. Preliminary measurements of the optical breakdown threshold intensity in air show that threshold can be determined with much higher accuracy with RPP smoothed beams.

1. Introduction

Laser beams are usually affected by considerable aberrations arising mainly from the amplifier chain (optical inhomogeneities, thermal nonuniformities). Complex phase modulations in the beam produce focal spots whose size is much larger than the diffraction limit and with large intensity modulations on a scale comparable to the spot size itself. These can easily cause the growth of non linear plasma instabilities (SBS, SRS, filamentation, etc.). In order to smooth the intensity distribution in the laser spot, random phase plates[1] (RPP) are the simplest technique invented so far. RPP are plane bidimensional array in which half of the cells introduces a dephasing of π radians for the operative wavelength, in a way to produce on focal plane a speckle pattern with so high spatial frequency to prevent or reduce the instability growth. The far-field intensity distribution of a beam with diameter D, focused with a focal length f, can be roughly represented as a set of randomly distributed beamlets, each having a cross section of $2\lambda f/D$. The whole focal region has a cross section of $2\lambda f/d$, where d is the cell size. A detailed description of the far-field pattern can be obtained both analytically or numerically[2]. We studied the far-field of a 1.064 mm, nanosecond pulsed laser beam, produced by a Nd-YAG oscillator (operating) and three amplifiers[3]. The laser beam was modulated by different RPP's and focused directly on a CCD array. The CCD video signal was synchronised with both the freeze gate of a grabber and the laser trigger.

2. Spot and mean speckle sizes

Using the diffraction theory developed by Abbe[4], we can write the focal plane intensity related to the electric field produced by a RPP as

$$I \propto \left(\frac{\sin(\pi dx/\lambda f)}{\pi dx/\lambda f}\right)^2 \left(\frac{\sin(\pi dy/\lambda f)}{\pi dy/\lambda f}\right)^2 \left|\sum_{m,n}^{N-1} e^{-i2\pi\left[(x_0-nd)\left(\frac{x}{\lambda f}\right)+(y_0-md)\left(\frac{y}{\lambda f}\right)\right]} e^{i\varphi_{m,n}}\right|^2 \qquad (1)$$

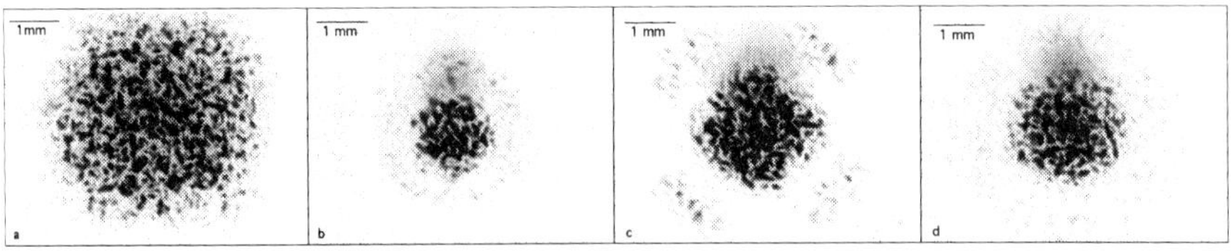

Fig.1. Spots produced by RPP with (a) 1mm square cell, (b) 3 mm circular cell, (c) 2 mm square cell, (d) 2 mm circular cell.

where φ is the random dephasing $(0,\pi)$ and N is the cell number for each side of the RPP. The envelope of the focal spot corresponds to the far-field diffraction figure produced by a single cell: for a square RPP, the envelope is the sinc^2 function. The images we have collected (some of wich are reported in fig. 1) have substantially confirmed this theoretical prediction. Particularly interesting for applications is the knowledge of speckle mean size. A rather large number of speckles were analysed, and their size was measured from the contour plot obtained at different fraction of the maximum intensity of each speckle. A large spread in speckle size distribution was found. In the graphs in fig. 2, we report the mean size for several intensity fractions and their best fit obtained with a $\cos^2$ function according to equation (1). As expected, the four series of data fit essentially to the same curve because the speckle size does not depend on RPP features. The mean speckle size (zero to zero) is about 280 μm while the diffraction limit was about 180 μm. The same kind of measurements was made on a set of images obtained reducing input beam diameter with a variable aperture. Experimental values show an inverse dependence from the beam diameter and the ratio between experimental and theoretical values is about the same for each point.

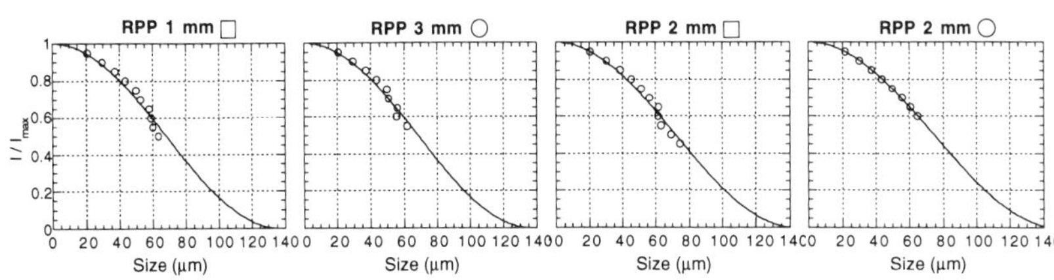

Fig. 2. Mean speckles size at several intensity fractions with different RPP's.

3. Statistical distribution of the intensity

Electric field amplitude produced in the far-field by a RPP can be seen as the sum of beamlets coming from each single cell. Beamlet phases are independent from each other with the same probability to have 0 or π value. For such an ensemble of phases, the following relation holds for the electric field ($E=E_r+iE_i$):

$$\langle E\rangle = 0 = \langle EE\rangle \Rightarrow \langle E_r^2\rangle = \langle E_i^2\rangle$$

Central limit theorem allows to consider the real and imaginary part of electric field as Gaussian variables. So, it is possible to derive the following expression for intensity probability density

$$p(I) = \frac{1}{I}\exp\left(-\frac{I}{\langle I\rangle}\right) \quad I\geq 0 \tag{2}$$

In fig. 3 we report four "intensity histograms" related to the images of fig. 1. Only the distribution produced by 1 mm square RPP resembles the exponential behaviour expected from eq. (2). The three other curves are very similar to each other but quite different to the exponential; also the contrast parameter ($\sigma_I/\langle I\rangle$) is far from the theoretical value.

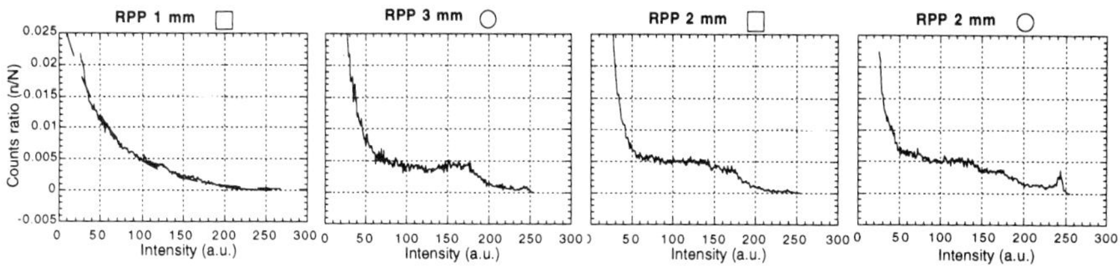

Fig. 3. Histograms showing pattern percentage (expressed in pixel count ratio) vs. intensity (expressed in gray levels) related to images of fig. 1.

We attribute this anomalous behaviour to the presence of aberrations in the laser beam. Because the far-field pattern is related to the input field through Fourier transform, it is possible that some "parasite" spots add, with variable weight, to the principal one. Their presence can be just revealed by the intensity distributions. In fact, if there are a number M of distinct electric field distributions that produce the speckle pattern in place of a single one, the expression (2) changes to[5]

$$p_M(I) = \frac{x^{M-1}M^M}{(M-1)!\langle I\rangle}\exp(-Mx)$$

where $x = I/\langle I\rangle$. Fig. 4 shows an example of how such a model can work.

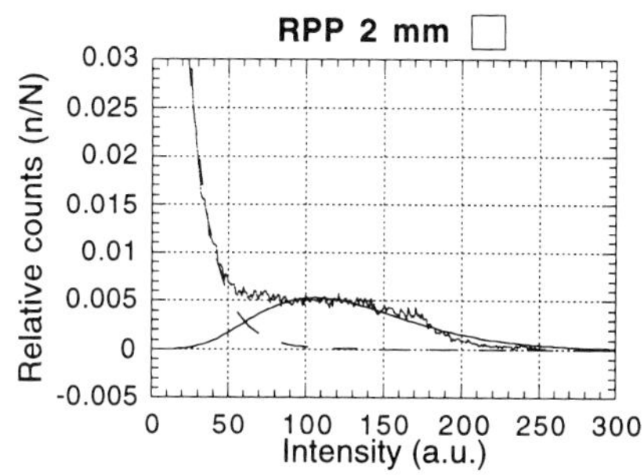

Fig. 4. The third histogram of fig. 3 can be represented as superposition of the distribution due to the principal pattern (dashed line) and that due to "parasite" patterns (solid line) (M=4).

4. Pattern symmetry

Fig. 5a,b and c show a group of images obtained progressively reducing the laser beam diameter with a variable aperture. We can see how the patterns acquire increasingly a central symmetry. This can be explained looking at expression (1) that, under the hypothesis of perfect coherence, is central symmetric. In fact the aperture increases the coherence in the beam portion wich is left out. When reducing the beam diameter through a telescopic system that symmetry is lost as shown in fig.5d.

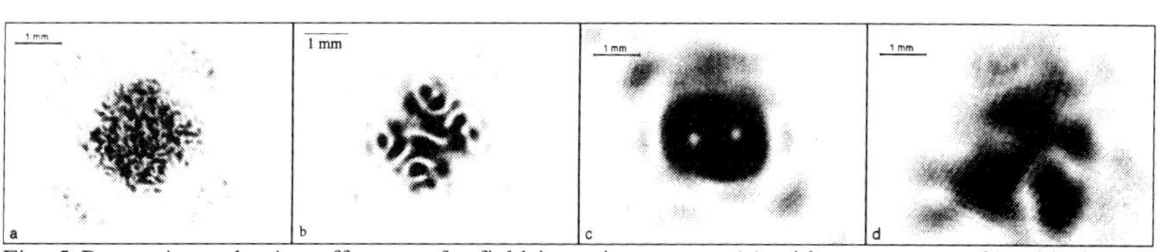

Fig. 5 Beam size reduction effects on far-field intensity pattern. (a) without aperture; (b) with the 1.2 cm diameter aperture; (c) with the 0.4 cm diameter aperture; (d) with the telescope

5. "Wrong" dephasing

Another interesting effect we have detected concerns the action on the pattern of a RPP cell dephasing different from π. It's easy to see how in this case the pattern on focal plane is a superposition of the usual random speckles pattern and the aberrated spot of the untreated laser beam[6]. The action of a "wrong" RPP introducing a dephasing equal to δ can be visualized as a superposition of two virtual RPP, one with the right dephasing and the other with no dephasing at all. The vectorial sum of the phasors of the two virtual RPP must be equal to the phasors of the real one. So the coherent central spike takes a fraction of total intensity corresponding to $\cos^2(\delta/2)$. This effect is dramatically evident with a 0-π/2 dephasing RPP, as shown in fig. 6.

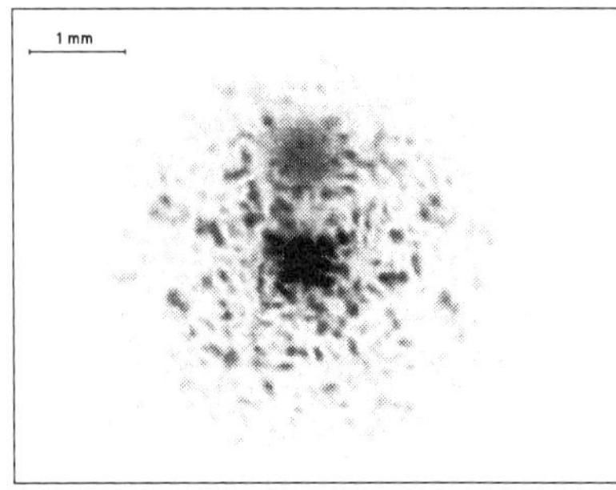

Fig. 6. Far-field pattern obtained with the 0-π/2 dephasing 1 mm square RPP

6. Optical breakdown

The use of RPP makes possible to have reproducible and well fixed patterns in the focal plane. This is very useful in determining threshold intensity of optical breakdown in gases. In fact, difficulties in associating a value to spot size or mean intensity inside the spot itself, create a large uncertainty related to threshold measurements results. As a preparatory measurements we found the threshold intensity of air at S.T.P., modulating the beam with RPP. The beam was also apertured to change the sizes of speckles. There is a net difference in the uncertainty with or without RPP: it was reduced up to 50%[3]. We also found that the use of RPP reduces the shot by shot fluctuations of the nominal threshold as shown by the example of fig. 7.

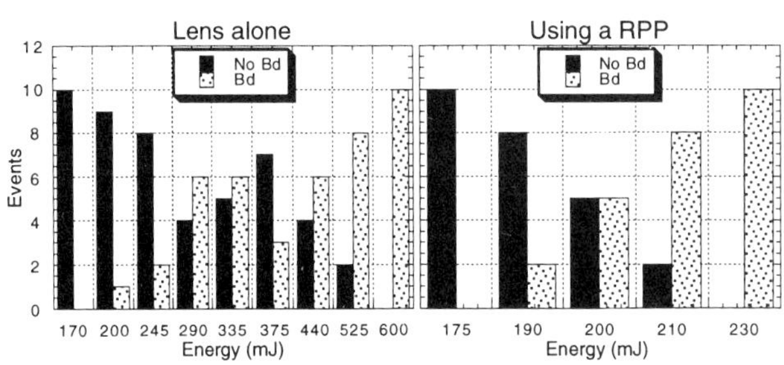

Fig.7. Energy threshold with and without RPP

References

[1] Kato Y et al 1984 *Phys. Rev. Letters* **53** 1057-1060

[2] Biancalana V and Chessa P 1994 *Appl. Opt.* **33**

[3] Ceccotti T 1994 *Studio delle proprietà di un fascio laser trattato con lamine a fase casuale* Degree Thesis University of Pisa

[4] Born M and Wolf E 1970 *Principles of Optics* chapt. 8 Pergamon Oxford

[5] Goodman J W 1984 *Laser speckle and related phenomena* Springer-Verlag New York

[6] Burckhardt C B 1970 *Appl. Opt.* **9** 695-700

Inst. Phys. Conf. Ser. No 140: Section 9
Paper presented at Laser Interact. with Matter Conf., Oxford, 19–23 September 1994
© 1995 IOP Publishing Ltd

TWO PHOTON INTERACTION: SO(3,2) SYMMETRY AND EXACT SOLUTIONS

J. M. Cerveró and J.D. Lejarreta
Departamento de Física Teórica. Facultad de Ciencias
Universidad de Salamanca. E-37008 Salamanca. SPAIN

1. Introduction

In the last few years a great deal of new techniques for solving exactly time dependent hamiltonians has been intensively developped. Since the pioneering papers of Lewis [1] and Lewis and Riesenfeld ([2]. See also [3]) many others similar ideas have been put forward specially in connection with boson hamiltonians exhibiting such phenomena as Squeezing [4], Berry phases [5] and the like. The main idea lying behind these techniques is related to the existence of an Invariant Operator which allows us to construct exact solutions of the correspondent time dependent hamiltonians ([6]-[9]). This so called Ermakov Invariant is a manifestation itself of the underlying dynamical symmetry [10] which can also be exploited by directly constructing unitary operators acting on initial states and leading to general dynamical states that represent the exact evolution of the physical system. This approach has been followed by the authors in a previous series of papers ([11]-[12]) for hamiltonians invariant under the SO(2,1) dynamical group of symmetry. In quantum optical applications this leads to one mode boson hamiltonians with exact and calculable squeezing and non trivial topological phases. However, for a more detailed treatment of phenomenological squeezing [13] we need to deal with a wider class of hamiltonians including two mode interaction (For a review see [14] and [15]). There exist some other similar papers in the literature dealing in one way or another with these aspects of the exact solutions of time dependent hamiltonians. We address the interested reader to references [16]- [25]. An expanded version of the results appearing in the present short communication is now being prepared and it will be published elsewhere [26].
Our purpose in this work is to generalize the above results to the two mode boson hamiltonian. Due to the well-known SO(3,2) dynamical symmetry that this two boson system exhibits, we are able to clasify the interesting subgroups under wich a possible solution is both feasible and possible. Also we pay some attention with partial succes to the general case for which no solution is yet known and we advance some possible strategy to obtain such an exact solution.

2. Two mode hamiltonian and the SO(3,2) symmetry

Let us first consider the most general form of the two mode boson hamiltonian. This hamiltonian reads:

$$H(t) = 2s_1(a_1^+a_1 + \frac{1}{2}) + 2s_2(a_2^+a_2 + \frac{1}{2}) + g_1(a_1^+)^2 + g_1^*a_1^2$$
$$+g_2(a_2^+)^2 + g_2^*a_2^2 + 2ha_1^+a_2^+ + 2h^*a_1a_2 + 2la_1^+a_2 + 2l^*a_1a_2^+ \tag{1}$$

where g_1, g_2, s_1, s_2, h and l are arbitrary complex functions of time. Now we construct the dynamical SO(3,2) symmetry as a function of the creation and anihilation operators of the two mode hamiltonian (1). Setting

$$J_{12} = \frac{1}{2}[a_1^+ a_1 - a_2^+ a_2] \qquad J_{13} = \frac{1}{2}[a_1^+ a_2 + a_2^+ a_1]$$
$$J_{14} = \frac{1}{4}i[(a_1^+)^2 - a_1^2 + (a_2^+)^2 - a_2^2] \qquad J_{15} = \frac{1}{4}[(a_1^+)^2 + a_1^2 + (a_2^+)^2 + a_2^2]$$
$$J_{23} = \frac{1}{2}i[a_1^+ a_2 - a_2^+ a_1] \qquad J_{24} = -\frac{1}{4}[(a_1^+)^2 + a_1^2 - (a_2^+)^2 - a_2^2]$$
$$J_{25} = \frac{1}{4}i[(a_1^+)^2 - a_1^2 - (a_2^+)^2 + a_2^2] \qquad J_{34} = -\frac{1}{2}[a_1^+ a_2^+ + a_1 a_2]$$
$$J_{35} = \frac{1}{2}i[a_1^+ a_2^+ - a_1 a_2] \qquad J_{45} = \frac{1}{2}[a_1^+ a_1 + a_2^+ a_2 + 1]$$

It is a trivial task to verify that the J-generators fulfill the Lie algebra of the SO(3,2) group $[J_{ab}, J_{cd}] = i(g_{ad}J_{bc} + g_{bc}J_{ad} - g_{ac}J_{bd} - g_{bd}J_{ac})$ where $g_{ij} = 0$ if $i \neq j$, $g_{ii} = 1$ if $i = 1, 2, 3$ and $g_{ii} = -1$ if $i = 4, 5$. Let us now construct the following linear combinations

$$C_0 = J_{12} = \frac{1}{2}[a_1^+ a_1 - a_2^+ a_2]$$
$$C_+ = J_{13} - iJ_{23} = a_1^+ a_2$$
$$C_- = J_{13} + iJ_{23} = a_2^+ a_1$$
$$D_0 = J_{45} = \frac{1}{2}[a_1^+ a_1 + a_2^+ a_2 + 1]$$
$$D_+ = -J_{34} - iJ_{35} = a_1^+ a_2^+$$
$$D_- = -J_{34} + iJ_{35} = a_1 a_2$$

$$E_0 = \frac{1}{2}(J_{12} + J_{45}) = \frac{1}{2}[a_1^+ a_1 + \frac{1}{2}]$$
$$E_+ = \frac{1}{2}(J_{15} - J_{24}) - \frac{i}{2}(J_{14} + J_{25}) = \frac{1}{2}(a_1^+)^2$$
$$E_- = \frac{1}{2}(J_{15} - J_{24}) + \frac{i}{2}(J_{14} + J_{25}) = \frac{1}{2}a_1^2$$
$$F_0 = \frac{1}{2}(J_{45} - J_{12}) = \frac{1}{2}[a_2 a_2^+ + \frac{1}{2}]$$
$$F_+ = \frac{1}{2}(J_{15} + J_{24}) - \frac{i}{2}(J_{14} - J_{25}) = \frac{1}{2}(a_2^+)^2$$
$$F_- = \frac{1}{2}(J_{15} + J_{24}) + \frac{i}{2}(J_{14} - J_{25}) = \frac{1}{2}a_2^2$$

Using the commutation relations it is easy to verify that the following Lie-subalgebra structure holds for the SO(3,2) Lie group

GENERATORS	**SUBALGEBRA**
$C = \{C_0, C_+, C_-\}$	$SO(3)$
$D = \{D_0, D_+, D_-\}$	$SO(2,1)$
$E = \{E_0, E_+, E_-\}$	$SO(2,1)$
$F = \{F_0, F_+, F_-\}$	$SO(2,1)$
$E \oplus F$	$SO(2,1) \otimes SO(2,1)$
$C_0 \oplus D$	$SO(2) \otimes SO(2,1)$
$D_0 \oplus F$	$SO(2) \otimes SO(3)$

3. The Group Elements and the Time Evolution Operators

As it is well known one can express any element of $SO(3,2)$ through the exponential mapping as $exp\{\sum \alpha_{ij} J_{ij}\}$, $\{i,j = 1 \text{ to } 5\}$, and we shall take henceforth the α_{ij} to the arbitrary functions of time. Since we are looking for a suitable factorization of the unitary operator U as a function of the group elements of the form above described, one can show that a quite convenient parametrization turns out to be

$$U = S_1 S_2 S_{12} T R_1 R_2$$

with the following definitions

$$S_1(\beta_1, a_1) = exp\{\frac{1}{2}[\beta_1 (a_1^+)^2 - \beta_1^* a_1^2]\} \qquad S_2(\beta_2, a_2) = exp\{\frac{1}{2}[\beta_2 (a_2^+)^2 - \beta_2^* a_2^2]\}$$

$$S_{12}(\beta, a_1, a_2) = exp\{\beta a_1^+ a_2^+ - \beta^* a_1 a_2\}$$

$$T(\tau, a_1, a_2) = exp\{\tau a_1^+ a_2 - \tau^* a_1 a_2^+\} \qquad R_i(\Theta_i, a_i) = exp\{\frac{i}{2}\Theta_i [a_i^+ a_i + \frac{1}{2}]\}$$

again β_i, β and τ are arbitrary functions of time while $\Theta_i \{i = 1 \text{ or } 2\}$ must be real in order to fulfill the unitary condition. If the following notation is used

$$\begin{array}{lll} \beta_i = r_i e^{i\phi_i} & \beta = r e^{i\phi} & \tau = q e^{i\Phi} \\ \eta_i = -\tanh(r_i) e^{i\phi_i} & \eta = -\tanh(r) e^{i\phi} & \mu = \tan(q) e^{i\Phi} \\ \gamma_i = \ln(1 - \eta_i \eta_i^*) & \gamma = \ln(1 - \eta\eta^*) & \rho = \ln(1 + \mu\mu^*) \end{array}$$

the operators take the form

$$S_i(\eta_i, \gamma_i, a_i) = exp\{\frac{\eta_i}{2}(a_i^+)^2\} exp\{\frac{\gamma_i}{2}(a_i^+ a_i + \frac{1}{2})\} exp\{-\frac{\eta_i^*}{2}(a_i)^2\} \qquad i = 1,2$$

$$S_{12}(\eta, \gamma, a_1, a_2) = exp\{\eta a_1^+ a_2^+\} exp\{\frac{\gamma}{2}(a_1^+ a_1 + a_2^+ a_2 + 1)\} exp\{-\eta^* a_1 a_2\}$$

$$T(\mu, \rho, a_1, a_2) = exp\{\mu a_1^+ a_2\} exp\{\frac{\rho}{2}(a_1^+ a_1 - a_2^+ a_2)\} exp\{-\mu^* a_1 a_2^+\}$$

We have been able to solve exactly the following cases

CASE I)

$$g_1 = g_2 = h = 0 \qquad SO(3) \otimes SO(2) - symmetry$$

CASE II)

$$g_1 = g_2 = l = 0 \qquad SO(2,1) \otimes SO(2) - symmetry$$

CASE III)

$$h = l = 0 \qquad SO(2,1) \otimes SO(2,1) - symmetry$$

CASE IV)

$$g_1 = g_2 = s_1 = s_2 = 0 \qquad No\ definite\ symmetry$$

CASE V)

$$g_1 = g,\ s_1 = s_2 = s,\ g_2 = ge^{-2i\arg(l)};\quad \arg(l) = constant \quad No\ definite\ symmetry$$

CASE VI): *General case with a constraint among the functions*

More details can indeed be found in [26]

This work has been partially supported by DGICYT under contract PB92-0302

REFERENCES

1. H. R. Lewis, *J.Math. Phys.* **9**, 1976-1982 (1968)
2. H. R. Lewis and W. B. Riesenfeld, *J.Math.Phys.* **10**, 1458-1473 (1969)
3. E. Pinney, *Proc. Am. Math. Soc.* **1**, 681 (1950)
4. H. P. Yuen, *Phys. Rev* **13A**, 2226-2243 (1976)
5. M. Berry, *Proc. R. Soc.* **A392**, 45-57 (1984)
6. J. G. Hartley and J. R. Ray, *Phys. Rev.* **D25**, 382-386 (1982)
7. J. R. Ray, *Phys. Rev* **D25**, 3417-3419 (1982)
8. C. C. Gerry and S. Silverman, *J.Math.Phys.* **23**, 1995-2003 (1982)
9. I. A. Pedrosa, *Phys. Rev* **D36**, 1279-1280 (1987)
10. B. G. Wybourne, *Classical Groups for Physicist* **J. Wiley and sons**, (1973)
11. J. M. Cerveró and J. D. Lejarreta, *J. Phys.A: Math. Gen.* **22**, L663-L666 (1989)
12. J. M. Cerveró and J. D. Lejarreta, *Quantum Optics* **2**, 333-349 (1990)
13. F. J. Bermejo and P. García Fernández, *Phys. Rev* **A39**, 1869-1875 (1989)
14. C. M. Caves and B. L. Schumaker, *Phys. Rev* **A31**, 3068-3092 (1985)
15. B. L. Schumaker, *Phys. Rep.* **135**, 317-408 (1986)
16. J. M. Cerveró and J. Villarroel, *J. Phys.A: Math. Gen.* **17**, 2963-2971 (1984)
17. A. Vourdas and R. M. Weiner, *Phys. Rev* **D38**, 2209-2217 (1988)
18. Jing-Bo Xu et al, *Europhys. Lett* **15**, 119-123 (1991)
19. Xiao-Chun Gao et al, *Europhys. Lett* **17**, 485-490 (1992)
20. Xiao-Chun Gao et al, *Europhys. Lett* **19**, 69-72 (1992)
21. S. Baskoutas, A. Jannussis and R. Mignani, *Phys. Lett* **A164**, 17-22 (1992)
22. T. Lisowski, *J. Phys.A: Math. Gen.* **25**, L1295-L1298 (1992)
23. C. F. Lo, *Europhys. Lett* **24**, 319-323 (1993)
24. A. Dewisne and S. Bouquet, *J. Math. Phys.* **34**, 997-1006 (1993)
25. Fu-li Li et al, *J. Phys.A: Math. Gen.* **27**, 985-992 (1994)
26. J. M. Cerveró and J. D. Lejarreta, *Preprint* **DFTUS 37/94**, (1994)

Inst. Phys. Conf. Ser. No 140: Section 9
Paper presented at Laser Interact. with Matter Conf., Oxford, 19–23 September 1994
© 1995 IOP Publishing Ltd

Time-of-flight ion energy analyser with magnetic charge separation.

V K Chevokin , A V Kilpio, V A Podvyaznikov

General Physics Institute of Russian Academy of Sciences, Vavilov str.38, Moscow, 117942, Russia

Abstract. In [1] a Thomson parabola analyser with ion mass and charge separation by the parallel electric and magnetic fields was suggested and described. We have suggested and realised in experiment the new scheme of time-of-flight ion energy analyser (IEA)with magnetic separation of charges.

The IEA is constructed on the basis of the small-size X-ray streak tube (XRST) with demountable photocathodes [2].Our XRST has an inverted-type focusing system with electron-optical magnification equal to unity that does not depend on photocathode potential. At the input the XRST has a special flange connecting it with a vacuum chamber. In the centre of flange there is a hole for installation of demountable holders, containing a substrate which is covered by photocathode. As photocathode in XRST MCP was used. XRST was connected with image intensifier and image from the screen was recorded on the film. The general view of XRST is shown in fig. 1.

Since under bombardment in MCB emission current predominate slow electrons there is a principle possibility to detect the ion fluxes with picosecond temporal resolution [3]. Our system allows to record per single laser shot the temporal behaviour of plasma ion fluxes separately with mass and charge resolution of the order 1/5 and energy resolution better than 0.1%.

The principle of operation of the ion energy-analyser with the magnetic separation of charges is shown in fig. 2.

Consider now its principle characteristics. From equation of the ion motion (Z is the charge, v is the velocity) in a transverse magnetic field B we have a curvature radius of the R trajectory:

$$R = \frac{M \cdot v}{Z \cdot e \cdot B} \tag{1}$$

Knowing a longitudinal size of magnet l_B ,deviation shoulder l_0, we can determine the deviation y in the conditions of small deviation angle ($\alpha \leq 0.1$):

$$y = l_b \cdot l_0 \cdot \frac{Z \cdot e \cdot B}{M} \left(v_0^2 + \frac{2 \cdot Z \cdot e \cdot u}{M} \right)^{-1/2} \tag{2}$$

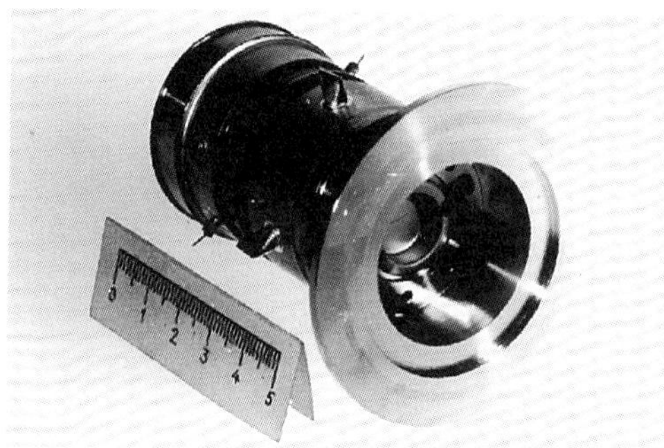

Fig. 1 Small size X-ray streak tube

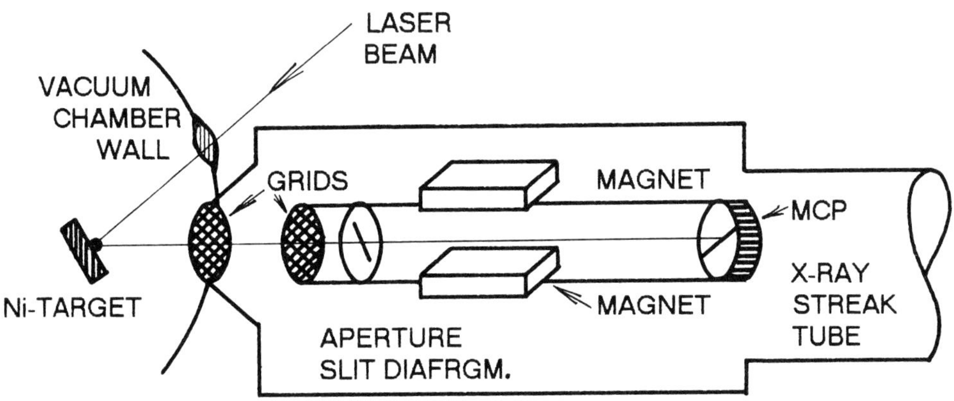

Fig. 2 The principle of the ion energy-analyser

where v_0 is the ion velocity at the entrance into the system, u is the accelerating voltage. If kinetic energy of the input electron $Mv_0^2/2$ is small as compared with the energy of the acquired in energetic field $Z \cdot e \cdot u$, expression (2) may be simplified:

$$y \approx l_B \cdot l_0 \cdot B \cdot \sqrt{\frac{Z}{2M \cdot u}} \tag{3}$$

or in the form which is more suitable for calculations:

$$y \approx 0.022 \cdot l_B \cdot l_0 \cdot B \sqrt{\frac{Z}{A \cdot M}} \tag{3`}$$

Here A is the atomic weight of ion, y, l_B and l_0 are expressed in millimetre, B in kGs, u - in kV. In our set up l_B = 15-20 mm, l_0 = 55-50 mm, so that for ion registration from Z = 1 to Z = 4 with XRST slit height $h \approx$ 8 mm the magnetic field voltage should not exceed:

$$B_{Z=1\div4} \le 0.36\sqrt{A \cdot U} \tag{4}$$

For a fixed value of magnetic field (constant spectrograph magnitude), a variation of the accelerating of dispersion when a mass of the ion registered is modified.

As magnetic field decreases the analyser dispersion is linearly decreased which leads to a loss of charge resolution. In the lack of focusing of the ion beam (aperture slit width is 0.3 mm) the apparatus function of spectrograph (one charge monoenergetic ion beam width) cannot be less than 1 mm.

The main reason of beam broadening with high density of the ion current ($j \ge 1$ mA/cm^2) is the Coulomb melting of the bulk charge. Let us estimate the value of broadening:

$$\frac{\Delta d}{d} = 1.2 \cdot 10^6 \cdot e^2 \left(\frac{A}{Z}\right)^{1/2} \cdot j \cdot \left(\frac{E_i}{Z}\right)^{-3/2} \tag{5}$$

By substituting l = 7 cm - flight path, j = $3 \cdot 10^{-4}$ A/cm^2 - current density of the ion component of nickel plasma with Z = 1 and energy E_i = 3000 eV we derive Δd ≈ d, which indicates that the present mechanism of the ion beam broadening in essential.

The ion beam width perpendicular to XRST slit restricts spectrograph resolution with respect to mass and ion charge. By differentiating formula for linear dispersion, and assuming the ion beam width to be Δd = 0.5÷1 mm, we find mass and charge resolutions:

$$\frac{\Delta Z}{Z} = \frac{\Delta M}{M} = \frac{2 \cdot \Delta d}{y(M, Z, u, B)} \tag{6}$$

In our system where the dispersion angles are small the beam deviations in magnetic field are $y(M,Z,u,B) \approx h \approx$ 10mm, which makes it possible to obtain the following resolution $z/\Delta z = M/\Delta M$ = 5÷10. Thus, charge-analysing system with magnetic deviation in the form which was used in our experiment, enables one to detect the ion components of laser plasma with charges $Z \le 5$ and ion mass difference exceeding 10-20% separately.

Coulomb "melting" of ion packet in longitudinal direction limits system resolution in flight time and hence in kinetic of ions consisting analyser:

$$\frac{\Delta t}{t} = \frac{\Delta v}{v} = \frac{\Delta E}{2E} \tag{7}$$

The time "melting" of ion packet with a longitudinal size Δl is the following:

$$\Delta t = \frac{\Delta l}{v} = t_A \cdot \frac{\Delta l}{l_0 + l_B} \tag{8}$$

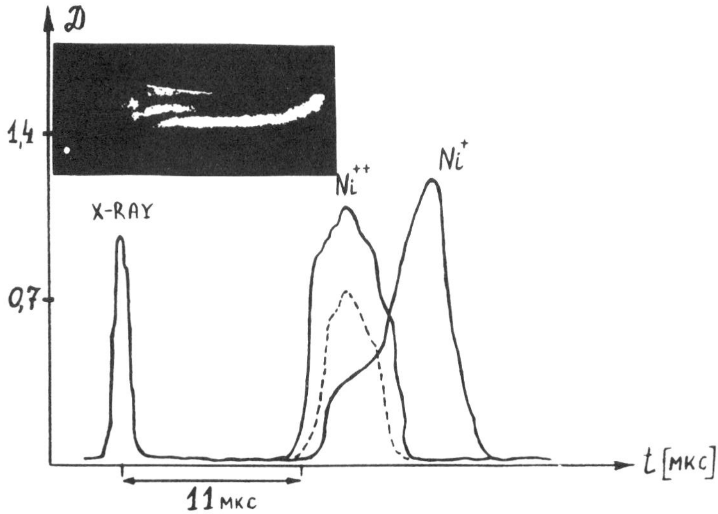

Fig. 3 Time behaviour of charge components of ion flew

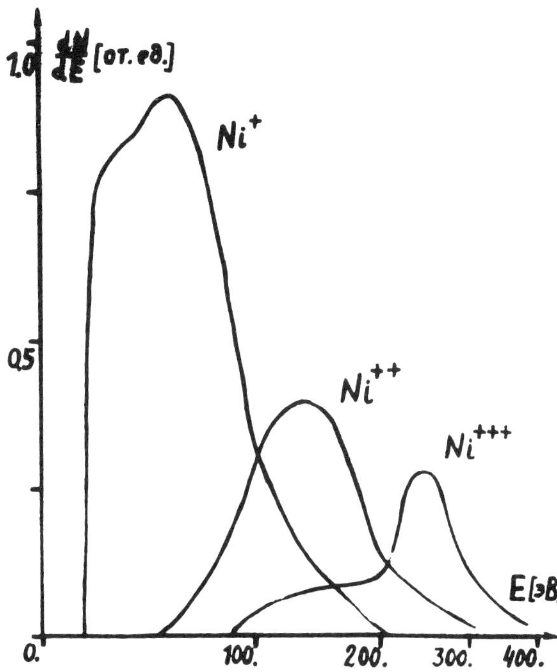

Fig. 4 Energy spectrum of Ni ions

Here t_A is the flight time of ions in analyser which does not exceed usually 1μs. Assuming that Δl =0.5÷1 mm we find the required time resolution of the XRST: Δt < 10 ns which is easily reached.

During experiments caring out with a time-of-flight analyser we could observe simultaneously two signals: static image of the slit irradiated by UV and X-ray plasma radiation, at the moment of plasma production, and streak of slit image bombarded by MCP ions.

IEA was successfully used in experiments on investigation of charge composition and energy spectrum of the ions of Ni-target at the power density on the surface up to 10^{12} W/cm^2 for the following laser pulse parameters: λ= 1.06 μm, E = 10 mJ, τ = 0.1 ns.

The time behaviour of charge components of ion flux is show in fig.3 and energy spectrum of ions is shown in fig.4.

Acknowledgements

We would like to acknowledge support from the Russian Basic Research Foundation RBRF (Pr. 94-02-03876-a) and International Science Foundation (MM 1000).

References

[1] Olsen J N et al. Journal of Applied Physics, may 1973, vol. 44, 5, 2275-2283

[2] Dashevsky B E et al. Advances in Electronics and Electron Physics. vol. 74, Academic Press,1988, p.233

[3] Kanaya K, Ono S, Japanese Journal Applied Physics, 1974 vol.13, **6**, 944-949.

Inst. Phys. Conf. Ser. No 140: Section 9
Paper presented at Laser Interact. with Matter Conf., Oxford, 19–23 September 1994
© *1995 IOP Publishing Ltd*

Hydrodynamic study of ponderomotive force effect on plasma flow at oblique incidence of high intensity laser radiation

N N Demchenko and V B Rozanov

Lebedev Physics Institute, Leninsky prospect, 53, 117924 Moscow, Russia

Abstract. Numerical simulations of the laser plasma flow under strong ponderomotuve action of the laser radiation have been performed by using the hydrodynamic code RAPID(SP). It has been demonstrated that the friction between the beatings of ponderomotive potential and plasma leads to the deceleration of underdense plasma. Then the sound point appears in this region, and strong scattering of the incident radiation takes place. We have shown that the competition between ponderomotive potentials of transverse and longitudinal fields may lead to the reflectivity nonregular time dependence.

1. The physical model of RAPID(SP) code

The code RAPID(SP) (SP means "short pulse") enables one to simulate the interaction process on the base of hydrodynamic equations with account for the ponderomotive force, limited ion viscosity and the energy transport by ions, thermal and superthermal electrons, and on the base of Maxwell equations for oblique incidence of s and p-polarized waves (generalized model of [1]). To calculate collisionless shock waves artifical viscosities Ω_e and Ω_i have been used for the electrons and ions. The relation $\Omega_i/\Omega_e = 73.4(A/Z)^{1/2}x^{3/2}e^{-x}$ has been derived using the Landau damping decrements for electrons and ions (here $x = ZT_e/2T_i$; Z and A, the charge and atomic weight of ions).

Maxwell equations have been used on the assumption that condition $L_p < \lambda c/c_s$ should be met (here L_p is the plasma extent; λ, the wave length; c and c_s, the light and sound velocities). This condition implies that the time of light propagation in plasma is less than the period of sound wave. As this takes place, the time derivative of slowly varying amplitudes can be neglected.

Oscillation energy of electrons is taken into account in plasma permittivity $\epsilon = \epsilon_1 + i\,\epsilon_2$. Real part $\epsilon_1 = 1 - \omega_p^2/\omega^2$, where $\omega_p^2 = 4\pi e^2 n_e/m_0\gamma_0$ and $\gamma_0 = [1+0.5(e|E|^2/\omega m_0 c)^2]^{1/2}$ is a relativistic factor. Imaginary part $\epsilon_2 = \omega_p^2\nu/\omega^3$, $\nu = \nu_{ei} + \nu_h$. Electron-ion collision frequency $\nu_{ei} = (\omega/2\pi)\int_0^{2\pi/\omega} \nu_{ei}(v(t))dt$ is averaged over the field oscillation period. The effective frequency ν_h does not equal zero behind the turning point. It describes the generation of superthermal electrons at the resonance point. For the nonrelativistic case we have:

$$\frac{\nu_h}{\omega} = \left[\frac{e\alpha_0|H_c|}{\pi m_0\omega^2 L_c}\right]^{1/2} - \frac{\nu_{ei}}{\omega} \qquad (1)$$

where ω is the laser frequency; $\alpha_0 = sin\theta_0$, θ_0, the angle of incidence; H_c and L_c, the magnetic field and characteristic inhomogeneity length at the critical point; e and m_0, the electron charge and mass. The relativistic expression used in the code has the form:

$$\frac{\nu_h}{\omega} + \frac{\nu_{ei}}{\omega} = \frac{e\alpha_0|H_c|}{\omega m_0 c}\left(\left(1 + \frac{\pi e\alpha_0|H_c|L_c}{2m_0c^2}\right)^2 - 1\right)^{-1/2} \tag{2}$$

Formulas (1)-(2) are not suitable for a very steep density profile when $L_c \to 0$. The case $L_c = 0$ had been studied by Brunel [2]. We have used a limit $\nu_h/\omega = 1/4$ deduced from Brunel's capacitor model. The energy flux transported by superthermal electrons is:

$$q_h = \frac{\nu_h}{\nu} q_{abs}^{res} \tag{3}$$

where $q_{abs}^{res} = (1/8)\alpha_0^2|H_c|^2\omega L_c$ is the energy flux absorbed due to resonance mechanism [3]. We have used one energy group approximation for superthermal electrons transport. The characteristic energy of electrons generated in the resonance region equals to the oscillation energy in the longitudinal field E_{zc} [2,4]: $\epsilon_h = (\gamma_h - 1)m_0c^2$, where $\gamma_h = [1 + (p_h/m_0c)^2]^{1/2}$, $p_h = e|E_{zc}|/\omega$. The energy of the superthermal electrons is mainly deposited in the dense part of the target. The ionization losses of energy have been taken into account for the superthermal electrons. The ponderomotive force has a potential for arbitrary angle of incidence and polarization:

$$p_r = \frac{1}{16\pi}\left((1-\alpha_0^2)|E_x|^2 + |E_y|^2 + (1-2\epsilon_1)|E_z|^2 + |H_x|^2 + |H_y|^2\right) \tag{4}$$

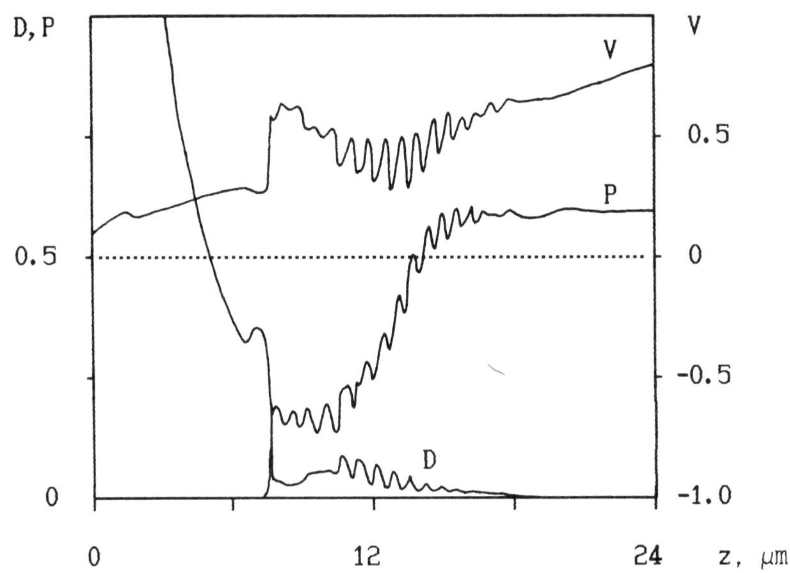

Fig. 1. The profiles of density (D), velocity (V) and ponderomotive potential (P) at t=28 ps. Maximum axes values: D - $3.0 \cdot 10^{-2} g/cm^3$; V - $1.5 \cdot 10^8 cm/s$; P - $8.7 \cdot 10^{12} erg/cm^3$.

2. Calculations of the laser pulse interaction with the solid target

The statement of problem: the plane aluminum target has been irradiated by p-polarized light of Nd-laser. The angle of incidence, $\theta_0 = 20°$, the pulse power growing exponentially in time from $10^{12}\ W/cm^2$ at $t = 0$ to $5{\cdot}10^{16}\ W/cm^2$ at $t = 32ps$.

Fig.1 shows the profiles for the moment when strong scattering takes place (see ponderomotive potential). This effect is due to friction between the plasma and the radiation with an oscillating amplitude [5]. Initially the oscillations of the amplitude are caused by reflection at the turning point. The sound point appears due to the plasma deceleration in the subturning region, and the scattering is increased. The light attenuation at the turning point results in the amplitude oscillations decrease. The friction force is reduced, and the sound point disappears. Then the process is repeated. The resonant absorption is an additional mechanism of the reflectivety oscillations. It is well known that the sound point (in the radiation frame [5]) is placed at the maximum of the ponderomotive potential. Since longitudinal fild at the critical point is diminished with the density gradient increase

$$|E_{zc}|^2 = \pi m_0 \omega^2 \alpha_0 L_c |H_c|/e \quad \text{for} \quad L_c > L^* = 16e\alpha_0|H_c|/\pi m\omega^2 \tag{5}$$

$$|E_{zc}|^2 = 16\alpha_0^2|H_c|^2 \quad \text{for} \quad L_c < L^* \tag{6}$$

there may occur the sound point transition from the longitudinal maximum at the critical point to the transverse one in the subturning region. The local stationary flow is disturbed in this region. The interaction of the scattering and resonant absorption oscillations causes nonregular time dependence of the reflected radiation. This interaction is enhanced with the radiation flux density increase.

References

[1] Demchenko N N et al 1990 Fizika Plazmy **16** 812-817
[2] Brunel F 1987 Phys. Rev. Lett. **59** 52-55
[3] Ginzburg V L 1967 Propagation of electromagnetic waves in a plasma (Moscow: Nauka)
[4] Gibbon P and Bell A R 1992 Phys. Rev. Lett. **68** 1535-1538
Wilks S C et al 1992 Phys. Rev. Lett. **69** 1383-1386
[5] Demchenko N N and Rozanov V B 1993 JETP **76** 997-1001

Inst. Phys. Conf. Ser. No 140: Section 9
Paper presented at Laser Interact. with Matter Conf., Oxford, 19–23 September 1994
© *1995 IOP Publishing Ltd*

Investigation of laser-produced plasma in external magnetic field

V M Dyakin* A Ya Faenov* A Kasperczuk S Ya Khakhalin* L Karpiński J Makowski M Paduch P Parys T Pisarczyk I Yu Skobelev* K Tomaszewski J Wołowski J.Wyzgał

Institute of Plasma Physics and Laser Microfusion 00-908 Warsaw 49, P.O. Box 49, Poland

*National Scientific and Research Institute for Physical-technical and Radio-technical Measurements Mendeleevo, Moscow region, 141570, Russia

Abstract. Hitherto results of the formation of an elongated plasma column by combining laser-produced plasma with an external strong magnetic field were published earlier . The paper presents results of measurements obtained with the use of a modernized magnetic field generator which makes it possible to measure magnetic field distribution in each discharge. The plasma was generated by means of a Nd-glass laser which was focused on solid targets located on the axis of a single coil providing magnetic field of up to 50T in the laser plasma region. The X-ray spectroheliograph technique, based on large aperture spherical mica crystal ($10, 30mm$ size with $R = 100mm$), placed according with FSSR-2D scheme was used for obtaining monochromatic images of plasma in spectral range of $9.2\AA$. An interesting influence of the magnetic field on the plasma dynamics and shape was observed. Strong interaction of the expanding plasma with the magnetic field (plasma density increases and formation of a characteristic semicircle shape occurs) was observed for the times longer than $10ns$. It is shown that it is possible to create enough dense and long plasma jet of multicharged MgXI ions in the case of interaction of expanding laser-produced plasma with a strong axial magnetic field. The electron density of plasma in such jet at the distance of $5mm$ from the target was at least 10 times higher in comparison with the free expanding plasma and was about $2 \cdot 10^{19} cm^{-3}$.

1. Introduction

The decision of some problems, beeing now of considerable interest, requires the creation of disintegrating multicharged ions plasma bunches with different configurations. For example, in the short wavelength laser research this configuration has to be sufficiently homogeneous, elongated, and thin column of a recombinating plasma. For multicharged ions injector, a dense plasma jet of multicharged ions with high concentration in a phase space is required. In our opinion, it is possible to use a laser-produced plasma for most of such applications. On the one hand it is sufficiently easy to obtain any initial ionic distribution for a laser plasma. On the other hand one of the known way can be used to provide the control of the plasma decay parameters: colliding a laser-produced plasma stream with obstacles [1, 2, 3] or plasma expansion in a magnetic field [4-12]. The aim of the present work is to show experimentally the possibility of creating a very dense and sufficiently elongated jet of a plasma containing multicharged MgXI ions by means

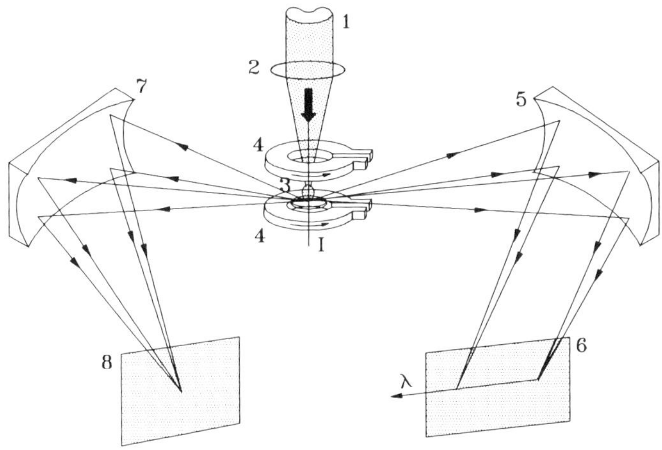

Fig.1 Schematic diagram of the experimental setup.

of expansion of the laser plasma bunches in an axial magnetic field of $B \sim 20T$. This work continues the investigations [5-12]. It shows that electron density in the jet is $\sim 2 \cdot 10^{19} cm^{-3}$ at distances of $\sim 5mm$ from the target, that is more than ten times higher than the electron density for free plasma expansion.

2. Experimental setup

Schematic diagram of the experimental setup is shown in Fig.1. Nd: glass laser beam 1 ($\lambda = 1.064\mu m$, pulse: $6J/1ns$) of $40mm$ in diameter is focused through the coil turn 4 by the lens 2 onto the magnesium target surface 3 . The focusing spot diameter is $100\mu m$. The target is placed into the single turn coil which is powered by the low inductive capacitor bank. The axial component B_z amplitude of the magnetic field pulse generated by this coil is $\sim 20T$. The heating laser and the magnetic field generator are synchronized so that the laser pulse comes onto the target surface at the moment of reaching the maximum magnetic field. The magnetic field pulse duration is about $5\mu s$ so that the field can be treated as uniform on the time scale of observed phenomena ($\sim 100ns$). The coil has 8mm wide vision slit perpendicular to its axis for the observation of the region of laser interaction with the target. In contradistinction to previous experiments, the expansion of the laser plasma into the magnetic field was observed using a three-frame interferometry [13]. For the registration and the analysis of interferograms, CCD cameras and a multichannel image acquisition system are used. The way of computer analysis of interferograms is presented in [14]. Electron density profiles of the plasma expanding into the magnetic field were determined for the time from $1ns$ to $50ns$. The parameters of the formed plasma object were determined from the soft X-ray emission. Two high-resolution large-aperture spectrographs with mica crystals which spherical surfaces have the radius curvature $R = 100mm$ were used in our measurements. The first of them was used as FSSR-1D 5 [15] and it recorded X-ray spectra on RAR2490 film or DEF-2 film 6 in the spectral range near $9.2\AA$ with one-dimensional spatial resolution along the laser plasma expansion axis. The second spectrograph was used as FSSR-2D 7 [15] and it formed two-dimensional monochromatic X-ray images on DEF-2 film 8.

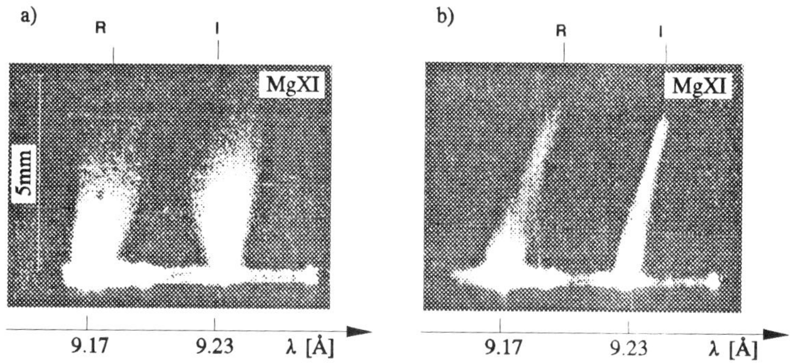

Fig.2 Spectroheliogram of laser-produced plasma in the resonance and intercombination lines of He-like Mg ion: a) free expansion, b) expansion in magnetic field.

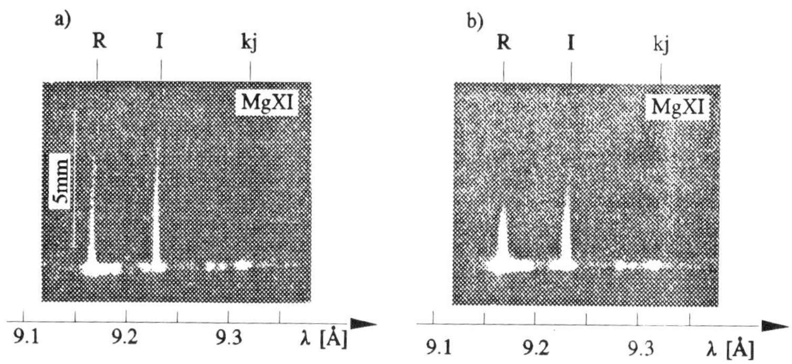

Fig.3 Spectra in the vicinity of resonance and intercombination lines of Mg: a) free expansion, b) expansion in magnetic field.

3. Results

The examples of spectroheliograms and spectrograms obtained are shown in Fig.2 and Fig.3, respectively. The magnetic field influence on the laser-produced plasma dynamics can be qualitatively seen from the spectroheliograms of the plasma radiation emitted in resonance and intercombination lines of He-like Mg ion. These spectroheliograms are presented in Fig.2 for a laser-produced plasma jet in the absence (a) and in the presence (b) of a magnetic field. Note that the results for these two cases are greatly different: without the field the laser-produced plasma expands into the cone with the angle of $\sim 33^o$, but in the presence of magnetic field at the distance at $1mm$ from the target the plasma jet is formed on the axis of the coil and propagates to the surface of upper half of the coil (i.e. up to the distance $5mm$ from the target) without noticeable decrease in density.

The spectrograms and spectroheliograms obtained enabled us to estimate some plasma parameters. For central luminous region near the target surface, we have the well resolved dielectronic satellite lines in the vicinity of the resonance and intercombination lines of the helium-like ion. With the aid of traditional methods (e.g., see [16]) we

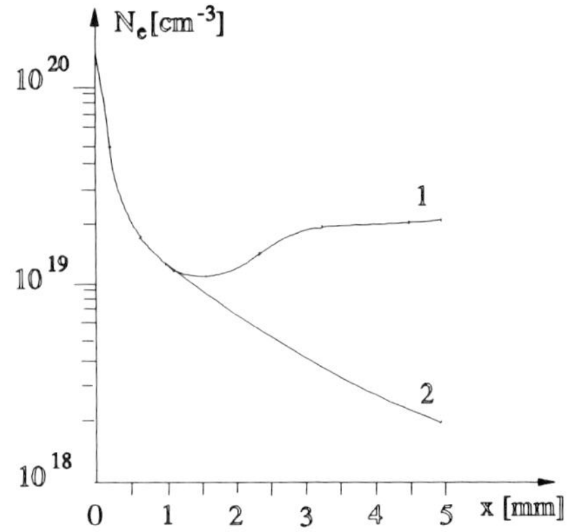

Fig.4 The dependence of electron density on the distance from the target: 1 - expansion in magnetic field, 2 - free expansion.

obtained $T_e \sim 300eV$, $N_e \sim 10^{20}cm^{-3}$, the values being practically independent on the presence of magnetic field. The temperature dependence of the resonance to intercombination lines intensity ratio is very weak, so we could find the electron density along the laser jet. The resulting curves for the two different situations (with and without the field) is shown in Fig.4. Unfortunately, for the distances longer than $\sim 1mm$ from the target, corresponding to plasma density less then $\sim 10^{19}cm^{-3}$, the resonance to intercombination intensity ratio does not depend on the plasma parameters. Extrapolation of the curves to the region with $x > 1mm$ was made with taking into consideration the space form of the plasma cloud (see spectroheliogram in Fig.2. As it can be seen in Fig.4, in case of the magnetic field present, starting with $x \sim 1mm$ (plot 1), the electron density stops decreasing. In the range of $(2-3)mm$, we can observe that the electron density increases a few times. Then, for all the luminous plasma up to $x \sim 5mm$, we have the homogeneous enough plasma jet with ion density higher (4 times higher for $x \sim 3mm$ and 10 times higher for $x \sim 5mm$) than corresponding density values for free expanding plasma without field.

4. Conclusions

Thus, it was shown in present report that it is possible to create dense and longitudinal plasma jet of MgXI multicharged ions in the case of interaction of expanding laser-produced plasma with a strong (about 20 T) axial magnetic field. Such a quazihomogeneous jet may be used not only in short-wavelength active media designe (see for example [5, 6]), but also for solving the transport problem in ions laser injector.

References

[1] Bhagavatula V A and Yaakobi B 1978 *Optics.Comm.* **24** 331

[2] Boiko V A et al 1982 *Preprint FIAN* **237**

[3] Boiko V A et al 1983 *Preprint FIAN* **181**

[4] Suckewer S et al 1980 *J.Appl.Phys* **51** 1922

[5] Bryunetkin B A et al 1991 *Kvantovaya Elektronika* **18** 1003

[6] Pisarczyk T et al 1992 *Laser and Particle Beams* **10** 767

[7] Pisarczyk T et al 1992 *Inst.Phys.Conf.Ser. No.125 Section 7. Paper presented at Int. Colloquium on X-ray Laser, Schliersee, Germany*

[8] Faryński A et al 1991 21st *ECLIM Book of Proceedings, Warsaw*

[9] Bryunetkin B A et al 1991 *ZhTF* **17** 82

[10] Begimkulov U Sh et al 1991 *ZhTF* **17** 63

[11] Bryunetkin B A et al 1992 *Kvantovaya Elektronika* **19** 246

[12] Begimkulov U Sh et al 1992 *Laser and Particle Beam* **10** 723

[13] Kasperczuk A et al *Proc.SPIE (in press)*

[14] Pisarczyk T et al 1994 *Laser and Particle Beams* **12**

[15] Bryunetkin B A et al 1992 *Laser and Particle Beam* **10** 849

[16] Boiko V A et al 1985 *J.Soviet Laser Research* **6** 81

Inst. Phys. Conf. Ser. No 140: Section 9
Paper presented at Laser Interact. with Matter Conf., Oxford, 19–23 September 1994
© 1995 IOP Publishing Ltd

X-ray spectroscopic investigation of recombination in laser-produced aluminium plasma expanding in external magnetic field

V M Dyakin* A Ya Faenov* L Karpiński S Ya Khakhalin* P Parys T A Pikuz* T Pisarczyk I Yu Skobelev* J Wołowski E Woryna

Institute of Plasma Physics and Laser Microfusion 00-908 Warsaw 49, P.O. Box 49, Poland

*National Scientific and Research Institute for Physical-technical and Radio-technical Measurements Mendeleevo, Moscow region, 141570, Russia

Abstract. A high-luminosity spectrograph with a spherical mica crystal was used to obtain the high spectrally and spatially resolved images of aluminium laser-produced plasma with and without external magnetic field. Particular attention was paid to the investigation of high-order members of the Lyman spectral series which were presumed to be optically thin to provide the reliable model interpretation. The monochromatic images of plasma were obtained with the aid of the X-ray spectroheliograph technique. The high-order Lyman transitions of H-like and He-like aluminium ions give an information for the study of recombination processes in plasma. The analysis of the experimental spectra is based on the collisional-radiative kinetics calculations. The traditional X-ray diagnostics methods and ion measurements were used along with high spectral members investigation for determining the plasma parameters (electron temperature and density, plasma ionization state). It is shown in the present report that the space-resolved X-ray spectroscopy of high excited levels is a helpful diagnostic technique.

1. Introduction

The development of practical X-ray lasing schemes [1] and searching for the optimal design of multicharged ions injector [2, 3] motivated the recent attemps to create the elongated and homogeneous plasma configurations with proper ionization degree and high density in the phase space. Magnetic field is one of the means practically used to affect dynamic characteristics of expanding laser-produced plasma [4]. In the magnetic field, the plasma cloud often evolves into a plasma jet (for example, see Fig.2 in paper [5] these proceedings), which is easily seen in the photographs. The ability to control the physical parameters of such plasma configurations closely depends on the extent of the theoretical insight into ionization-recombination effects. A complete theoretical analysis of laser-plasma recombination in the magnetic field requires detailed calculations in which ionization-recombination dynamics is coupled with magnetohydrodynamics. Lacking such full-scaled complex model, we analyse the laser-plasma expansion for different target materials to compare its behavior in the presence and in the absence of magnetic field. The spectral studies of spatial and temporal plasma cloud parameters distribution are of great interest, making it possible to analyse ionization-recombination processes and optical characteristics of plasma by means of traditional X-ray diagnostic methods.

2. Setup

The aim of the present report is to give the preliminary comparison of aluminium laser-produced plasma expansion with and without the external magnetic field. For this purpose we concentrate on the most reliable X-ray spectroscopic methods and laser interferometry methods for the plasma diagnostics. Experimental spectroscopic results are compared with collisional-radiative kinetics calculation. For the sake of simplicity of such analysis, particular attention was paid to the investigation of high-order members of the spectral series which are presumed to be optically thin to provide the reliable registration and model interpretation in the wide range of plasma conditions: from the most dense and hot region on the target to the farthest part of plasma cloud. The experimental arrangement was quite analogous to the one described in report [5] of these proceedings: the laser beam ($\lambda = 1.064\mu m$,pulse energy $6J$, pulse length $1ns$, focus spot $100\mu m$) and the magnetic field ($15-20T$, duration $5\mu s$) were synchronised in the same way. One spherical mica cristal ($R = 100mm$) was used as FSSR-1 [6] and the other one as FSSR-2 [6], forming 2-dimentional monochromatic X-ray images of laser-produced plasma. The CCD camera and the multichannel image acquisition system were used for registration and analysis of interferograms.

3. Results

Figs.1-4 present the examples of recorded spectra and 2D monochromatic images. Estimates of electron temperature were made from lithium-like satellite and helium-like resonance lines intensities [7]. The temperature was about 350 eV near the target and did not depend on the presence of magnetic field. The density-sensitive intercombination to resonance line ratio was used for measuring the electron density [8, 9]. The density near the target surface was $\sim 2 \cdot 10^{20} cm^{-3}$ from those measurements. The same value of N_e follows from the merging of highest members of Lyman series lines and He-like lines $1s^2 - 1snp$ [10, 11]. The population density distribution over the excited levels of AlXII and AlXIII ions was in good correspondence with the theoretical calculations, performed in collisional radiative model [12-17]. The calculated population densities of excited upper levels for AlXII and AlXIII Lyman series lines as functions of electron density depend on N_e as follows: the coronal limit ($N_e \rightarrow 0$), the gradual and then more rapid growing with the increasing of N_e (capture-cascade mode) and finally the saturation phase (at $N_e \rightarrow \infty$). Except the coronal limit, all the other stages are well seen in the spectra, registered with the space resolution in the direction normal to the target surface. The intensity distribution among the resonance series lines of H-like and He-like aluminium ions is in enough good coincidence with the one predicted by simulations. The length of luminous laser-plasma flow (Fig.2) is much larger in the case of Fig.2b (field is turned on). At large distances from the target, the population inversion over excited levels is established in correspondence with calculations. It is clear from Fig.3 that with the magnetic field we could observe the emission lines that were very weak without the field at the same laser beam intensities. For example the line on the long-wavelength side of the $1s6p - 1s^2$ line in the presence of magnetic field is even more intense at the target than the He-like ions line itself. Another type of effect is illustrated in Fig.1. The laser-produced plasma, expanding in magnetic field (Fig. 1b), exhibits the structure of emission which differs both from the free-expansion case (Fig. 1a) and from the magnesium laser-plasma jet [5] in the presence of field. It is obvious

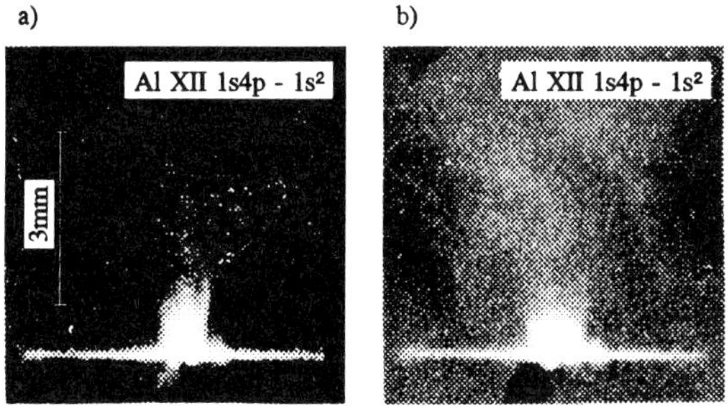

Fig.1 Spectroheliogram of aluminium laser-produced plasma in the $1s^2 - 1s4p$ emission radiation: a) without magnetic field, b) with magnetic field.

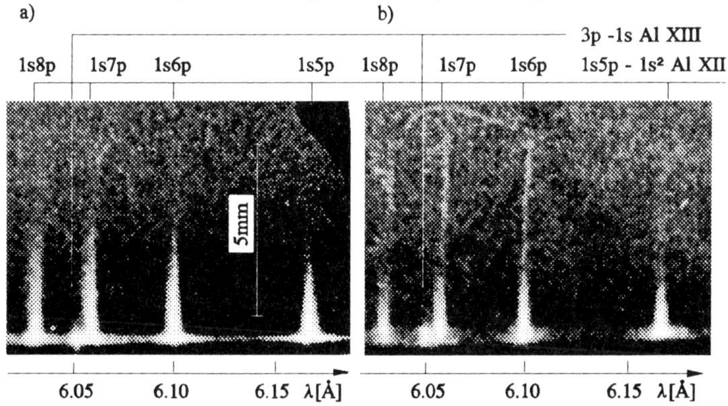

Fig.2 Aluminium emission line, spatially resolved in the direction normal to the target: a) without magnetic field, b) with magnetic field.

that gasdynamical behavior of laser-produced plasma is quite different in the two cases (with and without the field), which drastically affects the picture of spatially resolved spectra. It can be seen from Fig.4 that even adjacent wavelengths give 2D plasma images with some specific detailes, corresponding to the differences in excitation mechanisms of such lines in different space regions. Nevertheless, the very extreme difference of plasma forms in Fig. 1a and 1b is qualitative, giving rise to the very different spectra for the cases with and without the field.

4. Conclusions

We have observed spectroscopically the X-ray emission of predominantly recombining expanding laser-produced aluminium plasma with good spectral and spatial resolution. Line intensities and the space shape of plasma flow in the presence of magnetic field were found to be contrasted drastically with the case of free plasma expansion.

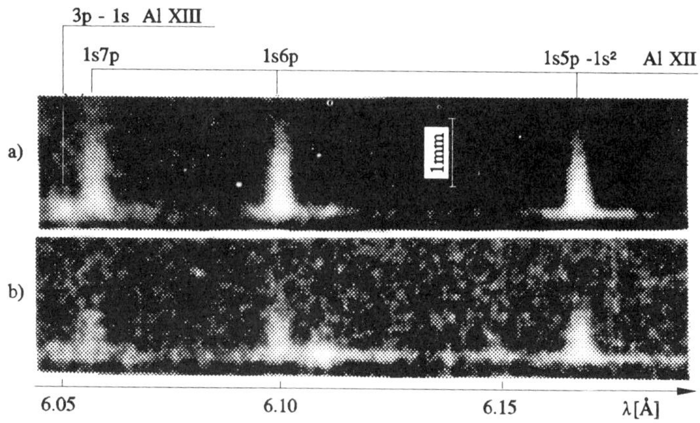

Fig.3 Aluminium spectra near the target surface: a) without magnetic field, b) with magnetic field.

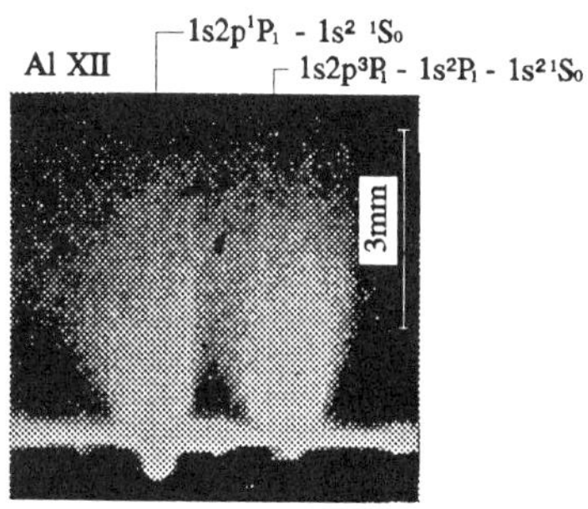

Fig.4 Spectroheliogram of aluminium plasma in the vicinity of the resonance and intercombination lines of He-like AlXII ion.

References

[1] Elton R C et al 1990 *X-ray Lasers, New York, Academic Press*

[2] Mróz W et al 1994 *Rev.Sci.Instr.* **65** 1272

[3] Haseroth H and Hora H 1992 *Ion Sources, Report CERN/PS 92-93 (HI)*

[4] Pisarczyk T et al 1994 *Physica Scripta* **50** 72

[5] Dyakin V M et al *at this conference proceedings*

[6] Bryunetkin B A 1992 *Laser and Particle Beams* **10** 849

[7] Gabriel A H and Jordan C 1969 *Nature* **221** 947

[8] Vainshtein L A Sobelman I I Yukov E A 1979 *Excitation of atoms and spectrl lines broadening Nauka Moscow*

[9] Boiko V A et al 1983 *Proceedings of VNIIFTRI* Moscow 8

[10] Grim G 1974 *Academic, New York*

[11] Inglis D R and Teller E 1939 *Astrophys. J.* **90** 439

[12] Skobelev I Yu et al 1986 *Proceedings of VNIIFTRI* Moscow 4

[13] Skobelev I Yu et al 1986 *Multicharged ions spectroscopy, Moscow* 164

[14] Volkov G S et al 1988 *Proceedings of VNIIFTRI* Moscow 43

[15] Skobelev I Yu et al 1988 *Proceedings of VNIIFTRI* Moscow 49

[16] Skobelev I Yu et al 1990 *Proceednigs of VNIIFTRI* Moscow 3

[17] Bayanov V I et al 1988 *Kvantovaya Elektronika* **9** 1842

Inst. Phys. Conf. Ser. No 140: Section 9
Paper presented at Laser Interact. with Matter Conf., Oxford, 19–23 September 1994

© 1995 IOP Publishing Ltd

EM66

Coupled rate equations in the modeling of non-LTE equations of state

S. Eliezer* and E. Mínguez

Institute of Nuclear Fusion, ETSII
Madrid Polytechnical University, Spain

* Visiting Professor from Soreq N.R.C., Yavne 70600 Israel

Abstract. Plasmas produced by intense laser pulses incident on solid targets often do not satisfy the conditions for local thermodynamic equilibrium, and so cannot be modeled by transport equations relying on equations of state. A proper description involves an excessively large number of coupled rate equations connecting many quantum states of numerous species having different degrees of ionization. Here we pursue a recent suggestion to model the plasma by a few dominant states perturbed by a stochastic driving force. The driving force is taken to be a Poisson impulse process, giving a Langevin equation which is equivalent to a Fokker-Planck equation for the probability density governing the distribution of electron density. An approximate solution to the Langevin equation permits calculation of the characteristic relaxation rate. An exact stationary solution to the Fokker-Planck equation is given as a function of the strength of the stochastic driving force. This stationary solution is used, along with a Laplace transform, to convert the Fokker-Planck equation to one of Schrödinger type. We consider using the classical Hamiltonian formalism to obtain the time-dependent solution.

1. Introduction.

The interaction of intense laser radiation with a solid target creates several spatial regions characterized by physical situations very far from the local thermodynamic equilibrium (LTE) [1].

Also with the advent of short laser pulses[2], a non-steady-state plasma is created. Moreover, for these short pulses (~ psec) even the electrons might be far from LTE so that an electron temperature is not always properly defined.

In a LTE regime the equations of state (EOS) are well defined[3] and one can obtain the pressure, energy and entropy, knowing the density and temperature of the plasma. The EOS relations are necessary in order to solve the plasma fluid equations. It is evident that in a plasma where LTE is not satisfied, such as a laser-produced corona or a picosecond laser-produced plasma, one cannot define a temperature and therefore EOS relations do not exist. In these situations the standard approach should require the solution of transport equations, which in most of the cases is extremely difficult and maybe prohitive for existing computers.

In this paper we follow the recently suggested idea[4-6] to describe a non-LTE plasma by few ion states in interaction with electrons sustained in a medium of "noise". The noise describes the extremely large number of states not considered explicitly as well as all the collision terms not taken into account.

2. The kinetic atomic rate equations.

The general kinetic rate equations are for each type of ion with a density N_i and with A_{ji} and B_{ij} being the total rate for the transition from level j to i and i to j, respectively, containing both collisional and radiative terms, are the followings:

$$\frac{dN_i}{dt}=\sum_j (A_{ji}N_j-B_{ij}N_i) \quad i=1,......,\nu, \tag{1}$$

We will follow a scheme with three species of ion coupled through ionization (I) and recombination rates (R), neglecting both excitation and deexcitation. N_{j-1}, N_j, and N_{j+1} are the number densities for these three ion species. It is also assumed that rates are due only to collisions with free electrons, avoiding the three-body recombination that works near the LTE region. Also, radiation terms are neglected. An algebraic combination of the kinetic equations[5] for the ionization $x=n_e/N_t$, with N_t being the total number density and n_e the electron density, is the following:

$$\frac{dx}{dt}=Ax^2+Bx \tag{2}$$

where we have assumed $\frac{d^2}{dt^2}(\ln x)\approx 0$. A and B are given by a combination of rate coefficients[6], and A<0 and B>0.

The potential ϕ describing the nonlinear damped anharmonic oscillator of equation (2) is given by

$$\phi(x)=-\frac{A}{3}x^3-\frac{B}{2}x^2 \tag{3}$$

that has a minimum at x=-B/A>0. The equation of motion in the electron density space is

$$\frac{dx}{dt}=-\frac{\partial\phi}{\partial x}. \tag{4}$$

and the equation of "motion" (2) is of Bernouille type and has the following solution

$$x(t)=[(\frac{1}{x_0}+\frac{A}{B})e^{-Bt}-\frac{A}{B}]^{-1} \tag{5}$$

where x_o is the inital condition (x(t=0)). For $t\to\infty$, or more practically $t\gg t_R$ where t_R is the relaxation time given by 1/B, one gets the steady state solution.

Let us consider a lithium-like iron plasma (Z=26, j=23) with an electron energy distribution about 600 eV and ρ= 0.002 $g.cm^{-3}$ (i.e. n_i= 2.1 x 10^{19} cm^{-3}). For E_{23}=-850 eV and E_{24}=-1699 eV, obtained with an hydrogenic atomic model[7], one gets,

$$\mathbf{A=-2.29x10^7s^{-1};\ B=5.22x10^8s^{-1}} \tag{6}$$

Using Eq. (5) the average ionization is obtained: x=22.8, and the relaxation time for the numerical example is $t_R=B^{-1}\approx 2x10^{-9}$ s. The relaxation time scales with electron plasma density n_e as Eq. (7). For example, for a critical plasma density of $9x10^{21}$ cm^{-3} (i.e. for 3ω laser frequency obtained from a Nd:YAG) one gets a value of 100 ps for the relaxation time.

$$t_R=(2ns)(\frac{2.1x10^{19}cm^{-3}}{n_e})\ x \tag{7}$$

3. A random noise.

It is suggested that the collission terms, radiative and all other interaction sources not taken explicitly into account in Eq. (2) be described by a "noise term". In particular Eq. (4) is changed into a Langevin equation

$$\frac{dx}{dt}=Ax^2+Bx+\sqrt{g}W \tag{8}$$

where the noise g W is a random "force" in density space g being the size of the random impulse. The random fluctuation W satisfies

$$< W(t) >=0;\ <W(t)W(t')>=\delta(t-t') \tag{9}$$

where <> means a statistical average over the direction of the "impulses" and the time. The equation (8) can be written via the potential ϕ, giving a Langevin equation, which is equivalent to the Fokker-Planck equation[8-10].

$$\frac{df}{dt}=\frac{\partial}{\partial x}(\frac{\partial\phi}{\partial x}f)+\frac{1}{2}g\frac{\partial^2 f}{\partial x^2} \tag{10}$$

where f(x,t) is the probability density for the variable x. This equation is assumed to be equivalent to the full set of rate equations. The diffusion term, second term on the r.h.s. of Eq.(10) in density space, describes the extremely large number of states not taken explicity into account. Experimentally we expect to measure an average ionization <x> given by

$$<x>(g,t)=\int f(x,t)x dx\ , \tag{11}$$

so that we have a one-parameter theory described by g.

By substituting into (10), the following solution

$$f(x,t)=e^{-\phi(x)/g}\psi_\lambda(x)e^{-\lambda t/g} \tag{12}$$

one obtains a Schrödinger equation for $\Psi_\lambda(x)$, where the Hamiltonian H is given by

$$H=-\frac{1}{2}g^2\frac{\partial^2}{\partial x^2}+\Phi \tag{13}$$

and the new potential Φ is related to the old potential by

$$\Phi=\frac{1}{2}(\frac{\partial\phi}{\partial x})^2-\frac{1}{2}g\frac{\partial^2\phi}{\partial x^2}=\frac{1}{2}(A^2x^4+2ABx^3+B^2x^2)+gAx+\frac{g}{2}B. \tag{14}$$

For g=0 the potential Φ has a minimum at x= -B/A, the same as that of ϕ. For a "small" noise, i.e. $g \ll g_0$, one can find a minimum near -B/A. Using perturbation theory one gets for the minimum of Φ the following value

$$\hat{x} = -\frac{B}{A}\left(1 + \frac{A^2}{B^3} g\right) \tag{15}$$

where x has the physical meaning of equilibrium ionization. Since B>0 and A<0 the "shift" due to "noise" in the equilibrium ionization is positive. So noise increases the equilibrium average ionization.

4. Summary.

In a "complex" plasma, e.g. a laser-induced corona from a high-Z target, the number of ion species is extremely large so that a very large ($\rightarrow\infty$) number of rate equations must be considered. This approach is very difficult to follow and may be prohibitive on present available computers. Therefore, we suggest describing a non-LTE plasma by a few ion states in interaction with electrons sustained in a medium of "noise". The noise simulates the states not considered explicitly as well as the collision terms not taken into account. This noise is assumed to modify the damped anharmonic oscillator equation into a Langevin equation for the electron density. This Langevin equation is equivalent to the Fokker-Planck equation. A transformation is made to obtain a Schrödinger equation so that the set of the original rate equations is described by a Schrödinger equation with a potential Φ. This potential includes the "noise" so that we have a one parameter (g) Hamiltonian which should describe the time development of the densities. An experimental confirmation of this analysis must be tested in a near term.

Acknowledgements

Author E.M. wish to acknowledge the Spanish Direccion General Investigación Científica y Técnica (DGICYT) for partial support of this work, under research contract No. PB 90-0317.

References

[1] S. Eliezer, A.D. Krumbein and D. Salzmann, J. Phys D **11**, 1963 (1978).
[2] H.M. Milchberg et. al. Phys. Rev. Lett. 61, 2364 (1988); A. Zigler et. al. Appl. Phys. Lett. **59**, 534 (1991)
[3] S. Eliezer, A.K. Ghatak and H. Hora, **An Introduction to Equations of State: Theroy and Applications** (Canbridge University Press, Cambridge, U.K.) 1986; S. Eliezer and R.A. Ricci eds., **High Pressure Equations of State. Theory and Applications** (North Holland, Amsterdam) 1991.
[4] S.Eliezer and E. Mínguez, Laser and Particle Beams **10**, 495 (1992).
[5] E. Mínguez S. Eliezer and R. Falquina, JQSRT **51**, 229 (1994).
[6] S. Eliezer, R. Falquina and E. Mínguez, Phys. Rev. E, **49**, 4735 (1994)
[7] G. Velarde et al, in Radiative Properties of Hot Dense Matter III, p. 433, B. Rozsnyair, C. Hooper, R. Cauble, R. Lee, J. Davis, eds. World Scientific 1987.
[8] R. Kubo et. al. J. Stat. Phys. **9**, 51 (1973).
[9] H. Haken, Rev. Modern Physics **47**, 67 (1975).
[10] N.G. Van Kampen, Adv. Chem. Phys. **34**, 245 (1976).

Inst. Phys. Conf. Ser. No 140: Section 9
Paper presented at Laser Interact. with Matter Conf., Oxford, 19–23 September 1994
© *1995 IOP Publishing Ltd*

Volume ignition targets driven by picosecond laser pulses

S. Eliezer*, H. Hora, J.M. Martínez-Val, J.J. Honrubia, M. Piera, G. Velarde**

Institute of Nuclear Fusion, ETSII
Madrid Polytechnical University, Spain

* Visiting Professor from Soreq N.R.C., Yavne 70600 Israel
** Visiting Professor from University of New South Wales, Sydney 2033, Australia

Abstract. Plasma foils of subcritical density illuminated by short and powerful laser pulses can undergo a huge internal explosion as a consequence of the strong outward acceleration suffered by the electrons at both edges of the plasma. The result of this type of non-linear interaction beween the laser electric field and the plasma electrons is the division of the plasma into two halves, each one flying outwards from the initial center plane. If the laser intensity is high enough and the refraction index inside the plasma show the appropriate shape, both halves will be acceleratecd to very high speeds. The acceleration phase takes a few picoseconds.

Such a plasma foil can be the pre-conditioned shell of an ICF target. Its internal electrostatic explosion will produce an outward flying semi-shell and an inwards flying one. The latter will act as the imploding fuel of that target. The central void of the shell closes after the acceleration phase. If the acceleration has been uniform across the 4π stereoradians, the fuel central collision will produce the untampered compression of the fuel due to inertial forces. Fusion conditions are achieved after a compression phase of some hundreds of picoseconds and a final fusion burst happens which lasts some tens of picoseconds.

Preliminary results show that this model of laser direct drive of ICF can yield outstanding target performances. The main requirements are the uniformity of the laser intensity in order to produce uniform spherical accelerations and the very high laser power, of the order of 10^{17} W/cm^2 during a time period of few picoseconds.

1. Introduction.

While the usual hydrodynamic or thermokinetic laser ablation reaches hydroefficiencies of 5% to 10% only it was noted that the rocket effect in heavy ion ablation can come up to 30% efficiency [1]. The last case is the reason why heavy ion driving with pulses of only 1.6 MJ produce gains of 120 MJ fusion energy [1]. There is also a possibility to have very high efficiencies with a laser driver by using the nonlinear forces. In this case the laser energy is converted directly into mechanical energy of plasma motion.

Since the early times of laser-plasma interaction studies, it was evident that nonlinear processes exist and that on top of the gas dynamic interaction, a nonlinear force interaction should exist at high laser intensities [2]. This forces can dominate the plasma dynamics. The nonlinear forces consist of ponderomotive and nonponderomotive terms if the complete theory for dispersive and dissipative plasmas is applied. The first proposal to use these forces for an inertial confinement fusion (ICF) efficient compression was put forward in 1979 [3]. This idea was revived by Tabak et al. [4] by using laser pulses of few picoseconds which can be converted nearly without thermalizing losses into directed motion of plasma [3] inducing a spark ignition scheme of fusion.

2. Volume Ignition.

The main idea of ICF is to achieve very high compression (up to several thousand times the liquid density) of a small pellet containing a mixture of hydrogen isotopes surrounded by a coating of pusher material. The fuel is strongly accelerated toward the center of the void during the implosion phase. When the void collapses, the kinetic energy of the fuel begins to be transformed into internal energy. This process lasts some tenths of a nanosecond (for 1 mg class DT targets) and can evolve through different patterns. At the end of this phase, a final compression state is reached, and the features of this state can vary signficantly from one case to another. For some cases a spark configuration is sought at the end of the implosion with a central region of the fuel (10% to 20% of it) at very high temperature. However, in other cases, the central hot spark does not appear. These cases can be called *volume ignition targets*.

In the central spark ignition model, temperatures of 5 keV are only required in the innermost part of the fuel, which can account for about 10% of it, the bulk of the fuel remaining at much lower temperatures but reaching much higher densities. The central spark has to be thick enough (above 0.3 g/cm^2) for the alpha-particles to deposit their energy inside the spark, so overcoming radiation losses. From this initial situation, the fuel evolves very rapidly: on the one hand, the central spark becomes hotter because the energy source terms are much stronger than the energy sink terms, so increasing the fusion reaction rate. On the other hand, the outer fuel also becomes hotter as it is affected by energy transfer from the spark due to electron, radiation and neutrons conduction. Hence, the bulk of the fuel also reaches high temperatures and high burnup fraction, which means a high gain.

Volume ignition can be defined as any of the possible configurations which differ from the spark ignition scheme. The "temperature-density" "phase-space" for volume ignition is thus more flexible in comparison with the "rigid" spark ignition model. It should be clear that high compressions ($\rho/\rho_0 > 1000$) are needed for volume ignition. The fuel mass has to be large enough and highly compressed so that the bremsstrahlung radiation is reabsorbed and part of the neutron energy heats the nuclear fuels.

Contrary to the spark ignition it was found in 1977 [5] that the whole volume can ignite like in a Diesel engine due to the selfheat of the nuclear reaction products and favored by some re-absorption of bremsstrahlung in the reacting sphere. Instead of the extraordinary profiles of the spark ignition, nearly adiabatic natural (monotonous) profiles are then used. These profiles had been easily produced in experiments with stagnation free compression [6]. The highest fusion gains for simple burn could be achieved [7] with volume ignition. Analytic estimations supported by simulations suggest that the volume ignition requirement for 0.3 g/cm^2<ρR<4.7 g/cm^2 is given by

$$\rho/\rho_o \geq C_{T_o}\left(\frac{M_f}{mg}\right)^{-1/2}\left(\frac{T}{keV}\right)^{3/4} \tag{1}$$

where $C_1=9.9\cdot 10^3$, $C_{1.5}=860$, and $C_2=270$ corresponds to a driver heating of $T_0=1$, 1.5 and 2 keV. For example, for two mg of fuel, if the driver compresses the target to about 40,000 times the liquid density then an initial heating of 1 keV will be sufficient to achieve volume ignition. For compressions of 3500 and 1000, an initial heating of 1.5 and 2 keV is repectively necessary.

For ρR>4.7 g/cm^2 the volume ignition requirement is modified because of the contribution of neutron heating

$$\rho/\rho_o \geq A_{T_o}\left(\frac{M_f}{mg}\right)^{-1/2}\left(\frac{T}{keV}\right)^{3/4} \tag{2}$$

where A_1= 3500, $A_{1.5}$=300 and A_2=95 for T_0= 1, 1.5 and 2 keV respectively. It is worth underlining that A_{T_0} is smaller than the corresponding C_{T_0} by a factor close to 3. For the above-mentioned temperatures of T_0, and ICF target with 2 mg of fuel must reach a compression of 14,000 (=ρ/ρ0), 1200 and 350 respectively in order to start a fusion burst.

3. Fast Pushing by Nonlinear Forces.

In plasmas of electron densities below the critical density, a ponderomotive non-linear force appears and becomes dominant over the thermalization interaction phenomena in ultrafast and powerful laser pulses. The force law can be approximated by

$$f_{NL}=\frac{E^2}{16\pi}\frac{\omega_p^2}{\omega^2}\frac{\partial}{\partial x}\left|\frac{1}{\tilde{n}}\right|=m_i n_i\frac{du}{dt} \tag{3}$$

where it is assumed that the role of collisional mechanism is negligible (v≃0). E is the laser electric field in vacuum, ω is the laser angular frequency, ω_p is the plasma frequency, m_i is the ion mass in the plasma with an ion density n_i, u is the plasma velocity. The index of refraction is approximated by

$$\tilde{n}^2=1-\frac{n_e}{n_c} \tag{4}$$

where n_e is the electron density and n_c is the criticial density.

For analytic calculations, a bi-Rayleigh density profile can model the plasma to obtain the following refraction index

$$\tilde{n}=\frac{1}{1+\alpha x} \tag{5}$$

where α is a parameter that characterizes the density gradient. The final speed is given to a good approximation from eqs. (3)-(5)

$$u\simeq\left(\frac{I_L}{2c}\right)\left(\frac{Z}{A}\right)\frac{\alpha}{m_p n_c}\tau_L \tag{6}$$

where Z is the degree of ionization, A the atomic mass number of the plasma, m_p the proton mass and τ_L is the laser pulse duration and

$$I_L=\frac{c}{8\pi}E^2 \tag{7}$$

In this type of very short interaction (τ_L<<100 ps) α must be properly chosen in order to obtain a high acceleration. For instance, if n_e/n_c ≃0.99, it follows that αx≃10. Thus, for x~1μm, α must be of the order of 10^5 cm^{-1}.

In order to increase the plasma density and to improve the uniformity requirements, an indirect illumination through X rays can be applied [8]. In this case, the laser pulse impinges on a very thin gold plate which reaches a very high temperature and radiates X-rays at more than 200 eV. The X-ray rear-side emission goes through the plastic thin foil and creates the plasma.

Should the ultrafast interaction occur as predicted by this theory, the plasma will be split into two parts, each one flying outwards at a speed of $\sim 10^7$ cm/s. The charge separation in both edges will lead to a coulombian explosion with a very uniform speed inside each half of the plasma.

Preliminary simulations with picosecond laser pulses suggest density-temperature profiles consistent with the volume ignition scheme. In the simulation, a solid DT shell was irradiated by a prepulse of 6 kJ (or 25 kJ using the direct-indirect drive scheme [8]) in order to produce the bi-Rayleigh plasma ready for the main neodymium glass laser pulse of 0.43 MJ energy and 2 ps duration for pushing the inner half of the shell to a velocity of 5.87×10^7 cm/s which after collapse and compression results in a volume ignition with an output of 38.7 MJ fusion energy.

In summary, picosecond laser pulse interacting with a subcritical plasma target is suggested for ultrahigh acceleration leading to fusion ignition. Due to the ponderomotive non linear forces a very high hydrodynamic efficency (~50%) is achieved. The main requirements to get high gains are a high laser power (about 1 MJ energy in 10 ps) with a good uniformity.

References.

[1] J.M. Martínez-Val, S. Eliezer, G. Velarde, H. Hora, J.J. Honrubia, M. Piera, P. Velarde, J.M. Perlado, E. Mínguez, and J.M. Aragonés, *Il Nuovo Cimento*, **106A**, 1873 (1993).

[2] H. Hora, *Phys. Fluids*, **12**, 182 (1969).

[3] H. Hora, R. Castillo, R.G. Clark, E.L. Kane, V.F. Lawrence, R.D.C. Miller, M.F. Nicholson-Florence, M.M. Novak, P.S. Ray, J.R. Shepanski and A.I. Tsivinky, *Conf. Plasma Physics and Nuclear Fusion*, Insbruck Austria 1978 (IAEA, Vienna 1979) Vol. III, p. 237.

[4] M. Tabak, J. Hammer, M. Glinsky, W.L. Kruer, S.C. Wilks, J. Woodworth, E.M. Campbell et al., *Physics of Plasma*, **1**, 1626 (1994).

[5] H. Hora, *Plasmas at High Temperature and Density* (Springer, Heidelberg 1991). S. Eliezer and H. Hora, *Nuclear Fusion by Inertial Confinement: A Comprehensive Treatise*; G. Velarde, J.M. Martínez-Val and Y. Ronen, Eds. (CRC, Boca Raton 1993), p. 43.

[6] C. Yamanaka, S. Nakai, T. Yamanaka et al., *Laser Interaction and Related Plasma Phenomena*, H. Hora and G.H. Miley, Eds. (Plenum, New York 1986) Vol. 7, p. 357.

[7] M.M. Basko, *Nuclear Fusion*, **30**, 2443 (1990); M.M. Basko, *Laser and Particle Beams*, **11**, 733 (1993).

[8] S. Eliezer, J.J. Honrubia and G. Velarde, Phys. Lett. **A166**, 249 (1992).

Inst. Phys. Conf. Ser. No 140: Section 9
Paper presented at Laser Interact. with Matter Conf., Oxford, 19–23 September 1994

© British Crown Copyright—1994 MOD. Published with the Permission of the Controller HMSO

Feasibility of Hugoniot EOS Measurements on HELEN.

A.M. Evans, N.J. Freeman, S.D. Rothman.

AWE Aldermaston, Reading, Berks., RG7 4PR, U.K.

Abstract. The AWE HELEN laser is being used to study the feasibility of obtaining high pressure Hugoniot data by the impedance match method. Indirect drive is used to generate pressures of up to 10MBar in the aluminium reference material.
Improvements in target fabrication and characterisation, streak camera calibration and data analysis, coupled with investigations into shock planarity and attenuation, lead us to conclude that an accuracy of ±2% in shock velocity is achievable.
Preliminary experiments at slightly less accuracy have been performed with copper and chlorinated plastic. Our results are compared with theory, and, for copper, with US and Russian gas-gun, explosive and nuclear driven experiments.

1. Introduction.

Equation of state (EOS) data are of fundamental importance to the modelling of hydrodynamic, astrophysical and geophysical phenomena and experiments. Pressures below a few MBar can be studied using gas-guns, e.g. [1],[2], and chemical explosives [3],[4] while underground nuclear tests have been used to explore pressures above 5-10MBar, e.g. [1],[2],[5],[6]. These methods offer an accuracy of about 1% in shock velocity. The few laser experiments done - from a few MBar, e.g. [7], to hundreds of Mbars [8] - have been of lower accuracy (~10% or worse), and are less useful for comparison with existing theories or with other experimental data. This work attempts to identify the problem of measuring precise shock velocities in laser driven experiments and to show that ~2% accuracy is achievable.

2. Experimental technique.

The impedance match method [9] requires simultaneous shock velocity measurements in a standard and unknown material. Shock velocity can be calculated from the time taken for a shock to transit a known step, where the shock breakout times at base and step are found by recording the surface luminosity with optical streak cameras. We chose aluminium as our standard material, and have conducted experiments on copper and chlorinated plastic (Parylene-C) as unknowns.

Targets (witness plates) were produced at AWE by vapour deposition. Typically these consisted of 600μm diameter discs with a 5-6μm Al ablator, 2-3μm Au preheat shield and a 20μm Al base with 2 separate steps of 10μm of Al and 8μm of Cu. The steps were made using suitable masks and had 50-150μm rectangular wells between them. Flat targets with a 26μm Al layer and wedges with either 5-40 or 5-75μm Al in place of the Al base and steps were also produced, the latter at LLNL and Rocky Flats. In some cases, Parylene-C was coated as a 15μm layer over targets with either a single Al step or both Al and Cu steps.

Step heights were measured using a Zeiss laser scanning microscope with a theoretical precision of ±0.05μm. However, relatively few measurements were made on each batch of targets and the typical errors were ±0.2μm, or ~2%. Surface finish was measured with a Zygo interferometer, capable of 5Å resolution, and was found to be typically 10-50nm. Density measurements on evaporated samples have been made to ~±5%, although X-ray diffraction measurements indicate that the lattice parameters of Al and Cu in a sample were within 0.05% of their values at standard density. This only means that the crystals are of solid density, but does not put any figure on porosity.

Indirect drive from a hohlraum heated by two 450J, 0.53μm nominal 1ns laser pulses was used. The hohlraum was cylindrical with the beams entering through fill-holes, offset up and down, in the curved walls, and the witness plates were mounted at the centre of one circular end-wall. Calculations indicated that the flux on the end-wall was uniform to better than 1% within a 167μm radius, and that the pressure generated in the witness plate was an approximately constant 9MBar from 1-1.5ns relative to the peak of the laser pulse.

Two optical streak cameras monitored the rear (outside) face of the witness plates with 100μm slits orientated horizontally and vertically (step edges were horizontal). These cameras had a wavelength range, defined by filters, of 580 to ~750nm, a spatial resolution at the target of 10μm with 12-15X magnification and a temporal resolution of 14ps at a streak rate of ~80ps/mm.

The cameras were calibrated by producing trains of short pulses with an étalon. Because the slit images produced by each pulse on the camera intensifier were curved due to electron time-of-flight effects, a full 2-d calibration was performed. For this, the peaks of the pulses as a function of position along the slit were tracked by computer using a method adapted from [10]. The final result was a map of relative times against position on the intensifier. The accuracy of this method was assessed by analysing a calibration shot using this time data. This should have showed pulses at constant time and separated by the étalon spacing. From 8 pulses the calibration accuracy was estimated to be ±0.7%.

Streak records were analysed by a method, developed from the calibration pulse tracking, which found the 50% of maximum shock luminosity points (taken to be shock breakout time). This resulted in 100-150 data points per streak. These data were then fitted by polynomials to correct for variations in shock planarity.

Shock planarity was confirmed by measuring shock breakout times from flat witness plates. A good result (#6075) shows a shock breakout time over the central 400μm of 887±4ps.

The time-dependence of the shock pressure was checked with wedged targets - breakout time should be proportional to position along the wedge slope for a constant shock. The breakout times of #5315 were best fitted by a line with an rms error of ±8ps over a ~1ns interval (or ~30μm thickness of Al). We have thus assumed constant shock velocity to 1% or better. This also indirectly confirms that the witness plates are not significantly preheated - as any expansion of the rear surface would have affected the breakout times.

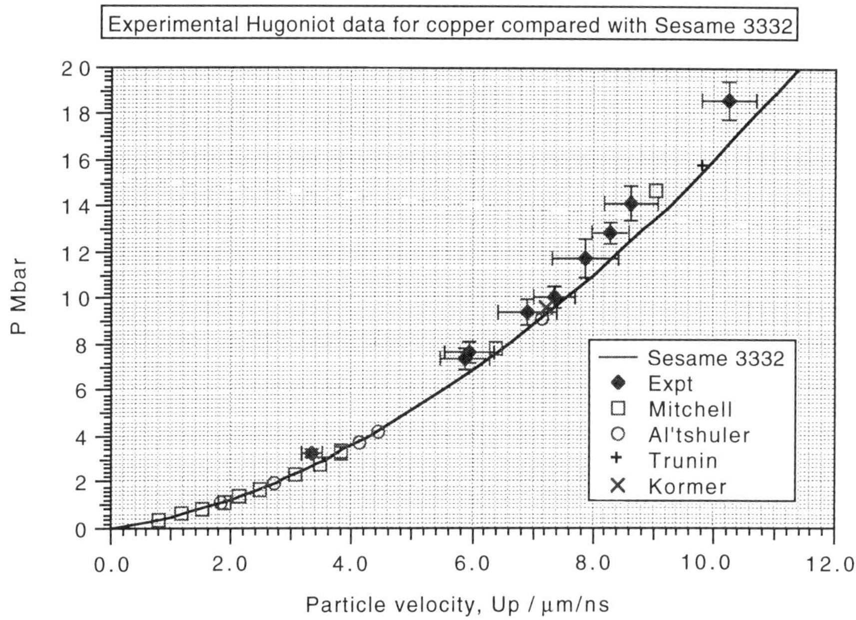

Figure 1. Experimental Cu Hugoniot points compared with SESAME EOS and data from references [1]-[6].

3. Results and analysis.

From the shock velocities measured for 9 shots on Cu targets the shock parameters in the Al were calculated assuming a solid density of 2.70gcm^{-3} and the SESAME 3715 EOS. Cu Hugoniot points were evaluated using a solid density of 8.852gcm^{-3}. The points are compared in fig. 1 with the SESAME 3332 EOS, and with data from Mitchell [1],[2] and Russian sources [3],[4],[5],[6]. It is apparent that the 2-3% errors in shock velocities have yielded 5-7% in pressure and particle velocity after the analysis process, but the scatter in our data is small by comparison. There could be a systematic error in our results and we are investigating this possibility. Preheat of the target, porosity of the Al and camera calibration errors have been considered, but do not move our data significantly toward the SESAME EOS and the previous experimental data.

Similar data for Parylene-C (density taken to be 1.29gcm^{-3}) is presented in fig. 2, where it is compared with an AWE CHARTD [11] calculation.

The errors in the shock velocity measurement can be broadly divided into four areas. Firstly, target metrology errors (1-2%). Secondly, errors in fitting to breakout times - which include noise in the streaks, witness plate non-uniformities and spatial variations in shock pressure - with ~1% rms deviations of data from fit. Camera calibration errors are 0.7%. Finally, shock attenuation is estimated from the wedged target data to have less than a 1% effect, and has not been included in the total error.

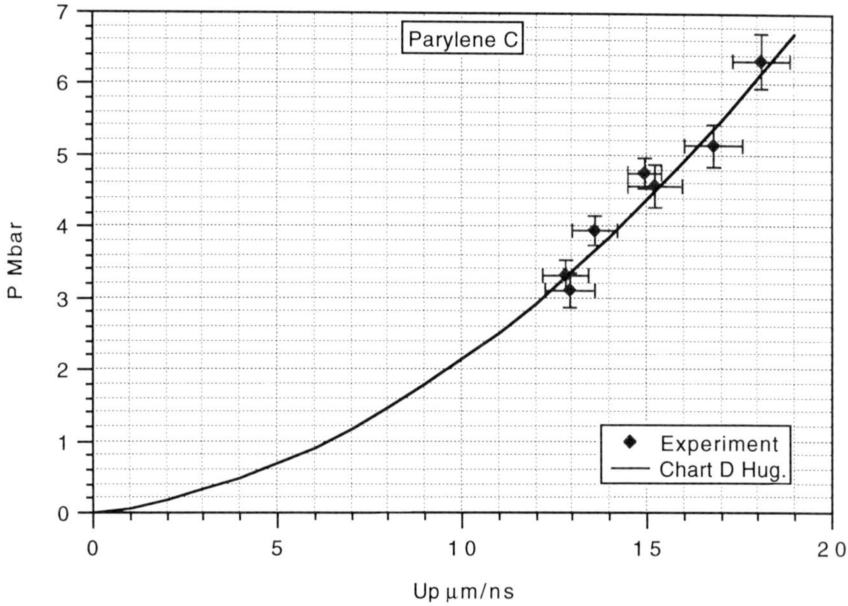

Figure 2. Parylene-C Hugoniot points and AWE Chart-D calculation.

4. Summary and conclusions.

We have demonstrated the ability to make EOS measurements in the pressure range 1-20MBar. Our errors are currently 2-4% in shock velocity, 5-7% in eventual pressures and particle velocities, but we have identified several sources of errors and methods of improvement and we believe we can improve our accuracy to 1-2% in velocity and better than 5% in Hugoniot pressures.

References.

[1] Mitchell A C, Nellis W J, Moriarty J A, Heinle R A, Holmes N C, Tipton R E and Repp G W 1991 *J. Appl. Phys.* **69** 2981-2986.

[2] Nellis W J, Moriarty J A, Mitchell A C, Ross M, Dandrea R G, Ashcroft N W, Holmes N C and Gathers G R 1988 *Phys. Rev. Lett.* **60** 1414-1417.

[3] Al'tshuler L V, Kormer S B, Bakanova A A and Trunin R F 1960 *Sov. Phys. JETP* **11** 573-597.

[4] Kormer S B, Funtikov A I, , Urlin V D and Kolesnikova A N 1962 *Sov. Phys. JETP* **15** 477-488.

[5] Al'tshuler L V, Bakanova A A and Trunin R F 1962 *Sov. Phys. JETP* **15** 65-74.

[6] Trunin R F, Podurets M A, Moiseev B N, Simakov G V and Popov L V 1969 *Sov. Phys. JETP* **29** 630-631.

[7] Ng A, Parfeniuk D and DaSilva L 1985 *Phys. Rev. Lett.* **54** 2604-2607.

[8] Romain J P, Cottet F, Hallouin M, Fabbro R, Faral B and Pépin H 1986 *Physica* **139 & 140B** 595-598.

[9] Zel'dovich, Ya. B and Raizer Yu. P 1967 *Physics of Shock Waves and High-Temperature Hydrodynamic Phenomena* (New York: Academic Press).

[10] Montgomery D S, Lerche R A, Laumann C W, Wiedwald J D and Drake R P 1992 *Rev. Sci. Inst.* **63** 4322-4326.

[11] Thompson S L 1970 SC-RR-70-28

Inst. Phys. Conf. Ser. No 140: Section 9
Paper presented at Laser Interact. with Matter Conf., Oxford, 19–23 September 1994

© *1995 IOP Publishing Ltd*

New screening coefficients for the hydrogenic ion model including *l*-splitting up to *4p*

G. Faussurier, C. Blancard and A. Decoster

Commissariat à l'Energie Atomique
Centre d'Etudes de Limeil-Valenton
94195 Villeneuve Saint-Georges Cedex, France

Abstract. New screening coefficients obtained from numerical fits are given for calculating energy levels of many-electrons atoms in the hydrogenic ion approximation including *l*-splitting with integer electronic populations. This model allows a fast calculation of various atomic parameters (one-electron energies, ionization potentials,...) for all ionization stages of elements with $1 \leq Z \leq 36$, with *nl* subshells up to 4*p* state.

1. Introduction

The evaluation of energy levels in many-electron atoms is needed in many applied physics domains as, for example, dense laser-plasmas interactions where a large number of ionic species can be found simultaneously due to temperature effects. The study of electronic properties of such systems depends on the calculation of numerous levels, ionization potentials, transition rates... For hydrodynamic laser-plasma simulations, the level of detail of the ionic structure does not exceed the configuration description. Hence, the atomic physics model must describe all energy configurations of various ionization stages of a given element. It must be also sufficiently fast to compute energy level populations in all space cells at each time-evolution of the plasma without particular hypothesis on the plasma equilibrium state.

The screened hydrogenic model is well-adapted to such calculations. Slater [1] was the first to use it to estimate binding energies of inner-shell electrons in atoms and ions. Later, Mayer [2] extended Slater's ideas to compute any one-electron energy in a given ion. A significant improvement has been proposed by More [3], with screening coefficients (SC) adjusted in order to reproduce accurate ionization potentials (IP) of a large number of ionic species. Compared with earlier models, the one-electron states deduced from More's coefficients are degenerate with respect to angular momentum and populations of a given shell n may vary continuously from 0 to $2n^2$. Recently, two new adaptations of More's model have been published. The first one, suggested by Perrot [4], includes *l*-splitting as a correction to the one-electron energies. The second one, proposed by Marchand et al. [5], is better suited to a detailed configuration accounting of an ion in which the energy levels of the various possible ionization stages are populated by an integer number of electrons. The need of fast atomic parameter calculations and the use of high speed computers have motivated us to revisit the screened hydrogenic model including the angular momentum contribution.

2. Description of the model

The screened hydrogenic model is based on the independent particle approximation where each bound electron is taken to move in an effective central potential representing the Coulomb interaction with the nucleus and the N-1 other electrons. The total energy of a given configuration is written as (Rydberg unit): $E_T = -\sum_{nl} p_{nl}\frac{(Z-Snl)^2}{n^2}$ where p_{nl} is the integer population of the nl subshell varying from 0 to $2(2l+1)$ and $Z - S_{nl}$ is the screened nuclear charge seen by one electron belonging to the nl subshell. In this model, the S_{nl} factor is written as: $S_{nl} = \sum_{n'l'} \sigma_{nl,n'l'}(p_{n'l'} - \delta_{nl,n'l'})$ where $\delta_{nl,n'l'}$ is the Krönecker symbol and $\sigma_{nl,n'l'}$ is a SC independent of the ion. This formulation ensures exact energy levels for H-like ions in the non-relativistic approximation.

The 630 IP of elements with $1 \leq Z \leq 36$ are deduced from [6]. Each IP has been calculated as the difference between total energies of the ground levels of consecutive ionization stages. The IP of neutral atoms have been initially excluded of the data base. A part of the excitation energies (EE, 1635 values) came from [7] where one-electron excitations have been retained only. Higher ionization stages deduced from this reference are limited to B-like ions. EE from He- to Be-like ions have been calculated, with the SUPERSTRUCTURE code [8] in the monoconfigurational approximation, for all one-electron excitations up to $4p$ from the ground configuration of each ions (1051 values by including inner-shell excitations). Each EE is calculated as the difference between average energies of two distinct configurations.

3. Results

The minimization procedure has been realized in several steps and for each step, two minimization methods have been used: a least-squares fit which gives the first estimation of SC and a Downhill Simplex Method [9] which stabilizes the constants set by minimizing the maximum relative error. A part of SC are deduced from [10]. A preliminary set of SC up to $4p$ is given in table 1.

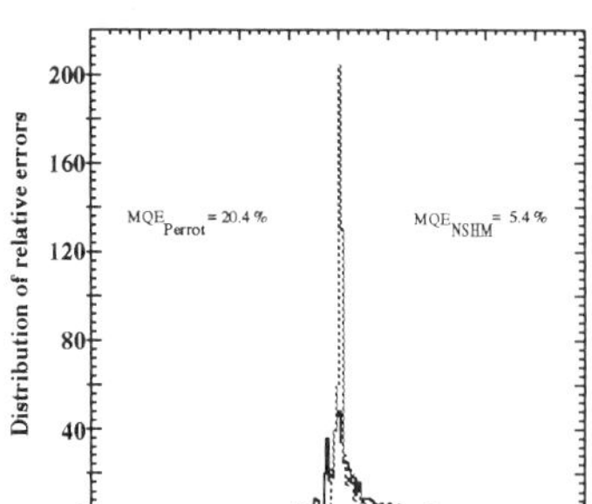

Figure 1: Relative error distributions of the ionization potentials of the 630 ions and the respective mean quadratic errors (MQE), without neutral atoms, for Perrot's model (full line) and the new screened hydrogenic model (dotted line).

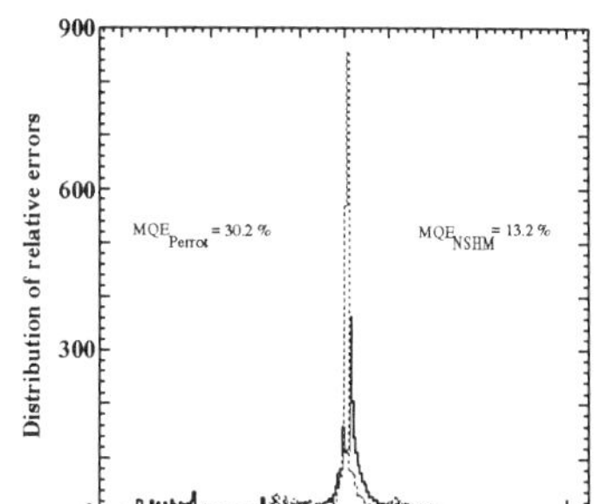

Figure 2: Relative error distributions of the monoelectronic excitation energies and the respective mean quadratic errors (MQE), without neutral atoms, for Perrot's model (full line) and the new screened hydrogenic model (dotted line).

Figure 1 presents relative error distributions of the ionization potentials calculated over IP of all ionic species with $1 \leq Z \leq 36$ and the respective mean quadratic errors (neutral atoms excluded) for Perrot's model and the new screened hydrogenic model (NSHM). The mean square error of the distribution deduced from the NSHM is better than the previous one by a factor 4. This difference is less if IP of neutral atoms are included : the mean square errors are then respectively equal to 13.4% and 26.3% for the new and Perrot's models.

Figure 2 shows relative error distributions of the monoelectronic excitation energies (neutral atoms discarded) and the respective mean quadratic errors for Perrot's model and the new screened hydrogenic model. The distribution obtained from the SC given in table 1 is characterized by a mean square error equals to 13.2% which is better than the one deduced from Perrot's model (30.2%). The NSHM presents a distribution with a main sharp structure centered around the origin. In each distribution, two other structures can be detected, one centered near 25% and the other one around -25%; they correspond to small EE (namely $\Delta n=0$) in few ionized atoms. Note that Perrot's distribution is far more noisy for values lower than -60%.

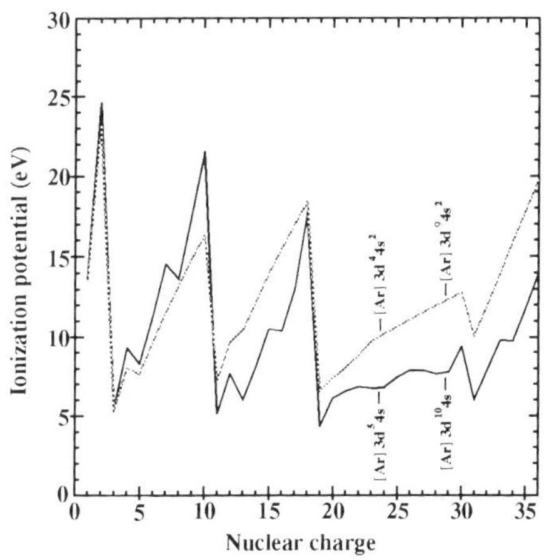

Figure 3: Comparison of the ionization potential of neutral atoms between experimental data (full line) and calculated values obtained from the new model (dotted line).

Using the new SC, the minimum energy configuration of neutral atoms have been found with respect to integer subshell populations, then IP of each configuration have been calculated and compared with experimental values [11]. Results are presented on figure 3. Ground configurations of neutral atoms are well reproduced even through the transition elements sequence, except for Cr ($Z=24$) and Cu ($Z=29$) where the model predicts respectively the fillings $3d^4 4s^2$ and $3d^9 4s^2$ instead of $3d^5 4s$ and $3d^{10} 4s$.

4. Conclusion

This screened hydrogenic model gives a more realistic description of *nl*-configuration energies compared to previous models. Both ionization potentials and monoelectronic excitations energies, even for $\Delta n=0$ transitions, are correctly predicted for any ion with $1 \leq Z \leq 36$. This set of screening coefficients have been enlarged to deal with higher excitation states and *nl*-configuration energies in heavier elements.

The authors would like to thank J.C. Gauthier for his helpful comments during various phases of this work.

$\sigma_{nl,n'l'}$	1s	2s	2p	3s	3p	3d	4s	4p
1s	0.311	0.020	0.001	0.000	0.000	0.000	0.000	0.000
2s	0.726	0.309	0.422	0.000	0.000	0.000	0.000	0.000
2p	0.932	0.206	0.343	0.039	0.000	0.000	0.000	0.000
3s	0.951	0.851	0.650	0.311	0.335	0.144	0.000	0.000
3p	0.980	0.852	0.851	0.270	0.349	0.367	0.000	0.000
3d	1.000	0.987	0.950	0.539	0.328	0.364	0.075	0.025
4s	1.000	0.987	0.976	0.806	0.796	0.595	0.308	0.391
4p	1.000	0.988	0.978	0.854	0.842	0.786	0.194	0.330

Table 1: New screening coefficients $\sigma_{nl,n'l'}$ where *nl* and *n'l'* are respectively the screened and screening states.

References

[1] J. C. Slater 1930 *Phys. Rev.* **36** 57

[2] H. Mayer 1947 *unpublished report* LA-647 Los Alamos Laboratory (Los Alamos, NM)

[3] R. M. More 1982 *J. Quant. Spectrosc. Radiat. Transfer* **27** 345

[4] F. Perrot 1989 *Phys. Scripta* **39** 332

[5] R. Marchand S. Caillé and Y. T. Lee 1990 *J. Quant. Spectrosc. Radiat. Transfer* **43** 149

[6] S. Fraga, J. Karwowski and K. M. S. Saxena 1979 *Handbook of Atomic Data* (Elsevier Scientific Publishing Company, Amsterdam)

[7] S. Fraga, K. M. S. Saxena and J. Karwowski 1980 *Atomic Energy Levels for Parametric Calculations* (Elsevier Scientific Publishing Company, Amsterdam)

[8] W. Eissner and H. Nussbaumer 1969 *J. Phys. B* **2** 1028

[9] W. H. Press, B. P. Flannery, S. A. Teukolsky and W. T. Vetterling 1992 *Numerical Recipes* (Cambridge: Cambridge University Press)

[10] C. Blancard and A. Decoster *unpublished*

[11] B. H. Brandsen and C. J. Joachain 1983 *Physics of Atoms and Molecules* (New York: Longman Scientific & Technical)

Inst. Phys. Conf. Ser. No 140: Section 9
Paper presented at Laser Interact. with Matter Conf., Oxford, 19–23 September 1994
© 1995 IOP Publishing Ltd

Experimental Measurements and Numerical Modelling of Plasmas Heated by Picosecond KrF Laser Pulses

R Fedosejevs, P A Naik[†], Y Y Tsui and C E Capjack

Department of Electrical Engineering, University of Alberta, Edmonton, Alberta, T6G 2G7, Canada

Abstract. Experiments using a 1 ps KrF laser pulse focused to an intensity of 10^{12} to 10^{14} W/cm^2 have been carried out. Angular dependence of the reflectivity from solid targets has been studied for both s and p polarized radiation, using an Ulbricht sphere and backscattering diagnostics. The results of the experiments are compared with results from 1-D hydrodynamic simulations including a wave solver for various angles of incidence and polarizations.

1. Introduction

Over the past several years a number of experiments have been carried out to investigate the absorption of picosecond and subpicosecond ultraviolet laser pulses in high density plasmas [1-2]. In order to understand the absorption and heating process in detail, numerical modelling of the plasma hydrodynamics using a full wavesolver is required [3]. The present study is aimed at carrying out a comparison of experimentally measured and calculated laser energy absorption for plasmas heated by picosecond KrF laser pulses.

2. Experimental setup

The laser system used in the experiments consists of a XeCl pumped dye laser giving short pulse output at 496nm. This is then amplified in a double pass KrF amplifier after frequency doubling to 248nm in a BBO crystal. The full output beam has an energy up to 30mJ with a pulse duration varying between 1-2 ps. For the present experiment a uniform portion of the beam with an energy of a few millijoules was selected using a 1cm diameter aperture.

The laser beam was focused using a 25cm doublet lens to a focal spot of 25μm diameter on solid targets of aluminum which were thermally evaporated onto glass slides. The energy of the beam was monitored using both calorimeters and cross calibrated photodiodes and the laser pulsewidth was monitored with a two photon absorption setup.

In the plasma absorption experiments, a 28cm diameter Ulbricht sphere coated on the inside with Kodak white reflectance paint was used to collect the scattered light. The sphere had ports for specular reflection at 0^0 to 180^0 at 15^0 intervals. To avoid nonlinear response, multiple calcium fluoride diffusers followed by neutral density filters were employed in the specular reflection detectors. UV filters were also used to cutoff any fluorescence.

Measurements of the frequency shift of the specularly reflected laser light were carried out by focussing the laser pulse and the specularly reflected light onto the entrance slit of a 93 cm Ebert type UV spectrograph with a resolution of about 0.5 Å.

[†]on leave from: Centre for Advanced Technology, Indore 452013 India

3. Simulations

A 1-D hydrodynamic code was used to simulate the experimental results for various angles of incidence and polarizations. The deposition of the laser radiation is calculated by solving the electromagnetic wave equations in a stratified medium using a finite element approach [2]. In our calculation, the electrical conductivity is calculated from the TKN model of Lee and More [4]. The dynamics of the plasma are calculated by solving the fluid equations for conservation of mass, momentum, and energy in a Lagrangian coordinate system using the Medusa code [5]. The thermal conductivity is calculated from the harmonic mean of the classical conductivity and a flux limited free streaming limit. In the present calculation a flux limiting factor of 0.6 is assumed. The equation of state used in our calculation is based on a model developed by one of the authors [6].

The spectral shift of the specularly reflected light is also calculated from the time rate of change in the phase of the reflected light as given by the wavesolver. In this case the spectral shift arises from two factors: the Doppler shift due to expansion of the plasma towards the laser and the nonlinear phase shift due to the changing scale length and refractive index of the profile.

4. Results

Fig. 1 shows the angular dependence of reflectivity for three laser intensities. At an intensity of 2×10^{14} W cm^{-2} the angle of minimum reflectivity is ~30^0 giving an interaction density scale length of ~80nm, assuming a collision frequency at critical density of 0.1 times the angular frequency of the laser [2]. For a lower laser intensity of 4×10^{13} W cm^{-2}, Fig. 1(b), the minimum is shifted to a larger angle of ~50^0, giving a density scalelength of ~15nm. In these measurements the polarization was better than 96%. For s-polarized light the reflectivity is seen to increase monotonically with angle, as expected. For a still lower intensity of 6×10^{12} W cm^{-2}, Fig. 1(c), the angle of minimum reflectivity increases further to ~57^0 as expected since the density profile now becomes more and more step-like. At a much lower intensity of 2×10^9 W/cm^2, at which intensity a 1ps laser pulse does not form a plasma, the measured angular dependence of reflection corresponds to that of an aluminum mirror.

The results of the hydrodynamic simulation for a 1ps gaussian pulse at various intensities are also presented in Fig 1. The value of minimum reflectivity agrees reasonably well with the experimental values for all three intensities. However, the calculations in general predict a higher reflectivity than those observed experimentally, particularly at 0^0 angle of incidence. Factors which could influence this result include the mixture of intensities in the experimental focal spot, the pulse duration and the pulse shape. The effect of these factors has been explored for the case of 2×10^{14} W cm^{-2}. If a gaussian focal distribution is assumed, the calculated reflectivity at 0^0 will change from 80% to 74%. The laser pulses used in our experiment have a duration of 1 - 2 ps. If a 2 ps duration is assumed the reflectivity at 0^0 will change from 80%(1ps, 2×10^{14} W cm^{-2}) to 75% (2ps, 1×10^{14} W cm^{-2}). The pulse shape can also be an important factor. For a 1ps triangular pulse with rise times (time for intensity change from zero to maximum) of 0.2 ps, 1 ps, and 1.8 ps, the reflectives at 0^0 for a laser intensity of 10^{14} W cm^{-2} are 71.1%, 74.5% and 75.4% respectively. Thus, longer rise times lead to higher reflectivities consistent with increased expansion and reduced absorption in the high density plasma above critical density. Such factors can be seen to alter the predicted reflectivity at 0^0 angle of incidence by the order of 10%.

In analysing absorption measurements, it is necessary to determine whether a plasma has been formed due to prepulse on target. We have characterized the formation of preplasma on target as a function of ASE laser energy using a Faraday cup detector and have

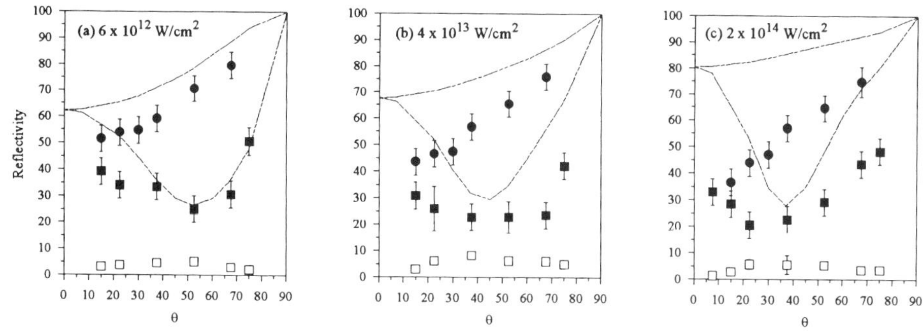

Fig.1 Measured and calculated reflectivity as a function of angle of incidence for aluminum targets at intensities of (a) 6×10^{12} W cm^{-2}, (b) 4×10^{13} W cm^{-2}, and (c) 2×10^{14} W cm^{-2}: circles - s-pol., squares - p-pol, solid symbols - specular reflectivity, open symbols - diffuse reflectivity, dashed lines - simulations.

found a threshold intensity of 10^8 W cm^{-2} for the formation of plasma. For the results presented above we have an intensity contrast ratio of better than 6×10^6 on target. Thus we are below the threshold for any preplasma formation.

The variation of the reflectivity of laser light in the intensity range of 10^{12} to 2×10^{14} W cm^{-2} for 0° and 37° angles of incidence is shown in Fig. 2. The simulation results are also given in the same plots. For 37^0 p-polarized radiation the comparison between simulation and experiment is good. However, for 0^0 angle of incidence and for s-polarized radiation the simulations predict significantly less absorption than measured. For 37^0 p-polarized radiation there is a significant contribution from resonance absorption at the critical density surface which is less sensitive to the conductivity of the plasma while the normal incidence and s-polarized data depends solely on the conductivity of the plasma.

The frequency shifts at two different angles of incidence are shown in Fig 3. Here, the shifts are of the order of two angstroms. The calculated spectral shifts from the simulation model are also shown and are in agreement with the experimentally measured shifts. However, these measurements were taken prior to the absorption measurements

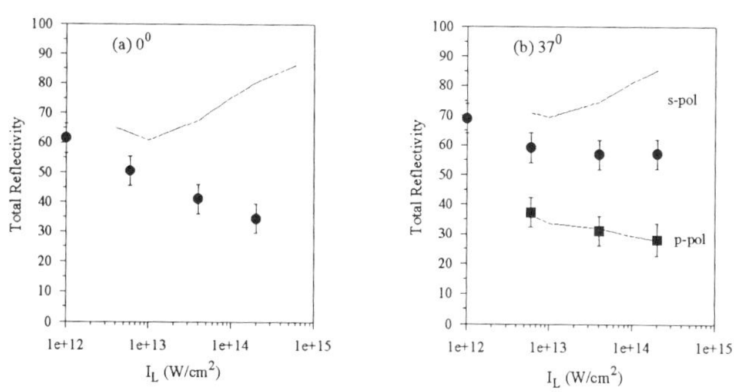

Fig. 2. Total reflectivity from an aluminum target for angles of incidence of (a) 0^0 and (b) 37^0: circles - s-pol., squares - p-pol, and dashed lines - simulation results.

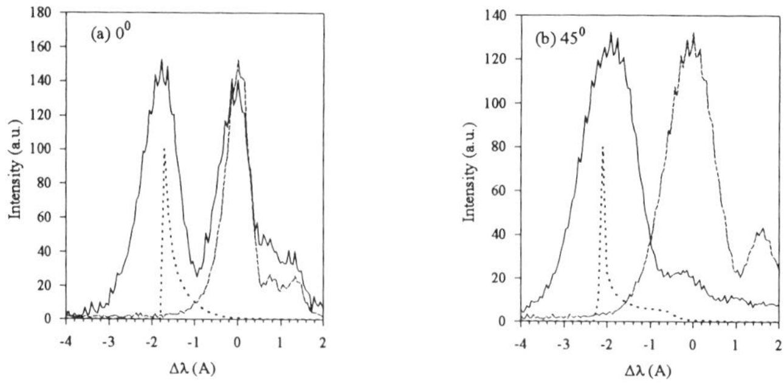

Fig. 3. Wavelength shift of specularly reflected light for (a) $\theta = 0^0$, $I = 2\times10^{14}$ W cm^{-2} and (b) $\theta = 45^0$, $I = 2\times10^{14}$ W cm^{-2}: solid line - experimental measurement (including reference radiation at the incident wavelength in (a) only), dotted line - simulation result and dashed line - spectrum of incident radiation.

discussed earlier and were carried out at higher levels of laser ASE leading to the possible formation of preplasma. This may help explain the agreement between the measured and calculated spectral shift at 0^0 even though the absorbed energy is considerably different in the two cases.

5. Conclusions

To conclude, we obtain absorption at normal incidence of greater than 40% for the intensity range of 6×10^{12} to 2×10^{14} W/cm^2, which for small angles of incidence is higher than the simulation predictions. The observed angle for maximum absorption increases for lower intensities implying that the plasma scale length becomes smaller for smaller intensities as expected. The simulation calculations predict significantly less absorption at 0^0 and agree better at the angle for peak absorption where resonance absorption plays a dominant role. However, further refinements in the modelling calculations to include other factors such as transient ionization will be required in order to carry out more detailed comparisons with the measurements.

References

[1] Milchberg HM, Freeman RR, Davey SC and More RM 1988 *Phys. Rev. Lett* **61** 2364; Sauerbrey R, Fure J, Le Blanc SP, van Wonterghem B, Teuber U and Schäfer FP 1994 *Phys. plasmas*. 1 1635;

[2] Fedosejevs R, Ottmann R, Sigel R, Kühnle G, Szatmári S and Schäfer FP 1990 *Phys. Rev. Lett.* **64** 1250; Fedosejevs R, Ottmann R, Sigel R, Kühnle G, Szatmári S and Schäfer FP 1990 *Appl. Phys. B* **50** 79

[3] Liu X and Umstadter D 1993 *SPIE* **1860** 103; Ng A, Celliers P, Forsman A, More RM, Lee YT, Perrot F, Dharma-wardana MWC and Rinker GA 1994 *Phys. Rev. Lett.* **72** 3351

[4] Lee YT and More RM 1984 *Phys. Fluids* **27** 1273

[5] Christiansen JP, Ashby DETF and Roberts KV 1974 *Comput. Phys. Commun.* 7 271

[6] Havazelet D, Capjack CE, Shvarts D and Marchand R 1989 *Phys. Fluids* **B1** 893

Inst. Phys. Conf. Ser. No 140: Section 9
Paper presented at Laser Interact. with Matter Conf., Oxford, 19–23 September 1994
© *1995 IOP Publishing Ltd*

Investigation of plastic foils with emission and absorption x-ray spectroscopy

István B Földes* and Klaus Eidmann

Max-Planck-Institut für Quantenoptik, D-85748 Garching, Germany
* Permanent address: *KFKI - Research Institute for Particle and Nuclear Physics H-1525 Budapest, POB. 49. Hungary*

Abstract

Radiative heating of plastic foils has been studied by time-resolved soft x-ray absorption and emission spectroscopy. The ASTERIX IV laser beam was converted to x rays in closed Au microcavities which heated the foils and in the mean time it served as a backlighter to absorption spectroscopy. Radiative burnthrough of CH targets was investigated from 1 to nearly 100 eV temperature using both absorption and emission spectroscopy. Reemission of plastic foils of pure CH was compared with F, O and Cl containing plastics. Reemission increases with increasing Z corresponding to the increased continuum radiation.

1. Introduction

High power lasers can produce plasmas with temperatures above 1 keV and densities near to that of solid matter. Laser plasma researches have been motivated mainly by the search for inertially confined fusion plasmas. In recent years applications for studying properties like opacity for the high temperature matter also received interest, mainly because of the astrophysical applications. In order to get rid of the steep density and temperature gradients of laser generated plasmas x-ray heating of the matter can be used. The x-ray heated matter - in contrast to laser plasmas - is free from strong gradients, thus it is more appropriate for investigating high temperature opacities. Also more uniform illumination of the fusion target can be obtained by illuminating it indirectly, i.e. with x rays after a conversion of the laser radiation instead of directly illuminating by the laser beams.

These applications proved to be possible because high conversion of laser radiation to x rays can be obtained in high-Z plasmas[1]. Thermal x-ray sources with temperatures up to nearly 200 eV can thus be obtained[2], therefore radiative heating of matter can be used for (indirectly driven) inertial confinement fusion as well as for studying the properties like opacities of the high temperature matter.

We carried out investigations both by absorption and emission spectroscopy on plastic foils. The choice of these low-Z matters was motivated partly by the simple electronic structure of carbon which allows a comparison with simulations using the SNOP opacity code[3], partly by its applicability in ICF researches.

2. Radiative burnthrough

Radiative heating of plastic foils have been investigated by time-resolved soft x-ray absorption spectroscopy. The third harmonic radiation of the ASTERIX IV. laser beam ($\lambda = 0.44\mu m$) of 450 ps pulse duration and 200 J energy was focused into small, 1 mm Au microcavities. The efficient conversion of laser light to x rays resulted Au wall temperatures higher than 100 eV.

Radiative burnthrough of thin plastic foils was investigated by transmission spectroscopy. The foil to be investigated (1-4 μm parylene) was glued onto a diagnostic hole of the cavity. The time dependence of the transmitted spectra was compared by the emission of the gold wall. The spectrometer was a 5000 l/mm transmission grating combined with a streak camera, thus obtaining 0.07 nm spectral and 30 ps temporal resolution. The cathode of the streak camera did not contain carbon in the investigations of samples containing carbon.

In the case of a thick (3 μm) foil the burnthrough was weak, the initial stage of x-ray heating was studied. Similarly to previous experiments with Be[4] the burnthrough starts near to the cold K-edge. The K-edge will not be so steep, than that of the cold matter any more. It is smoothed and moreover in the beginning it has a red shift. This behaviour could be explained by the thermal smoothing of the cold Fermi-edge of the still degenerate solid matter. The model of Schwanda and Eidmann [4] leads to a higher temperature than was obtained for Be, i.e. T=5±1 eV.

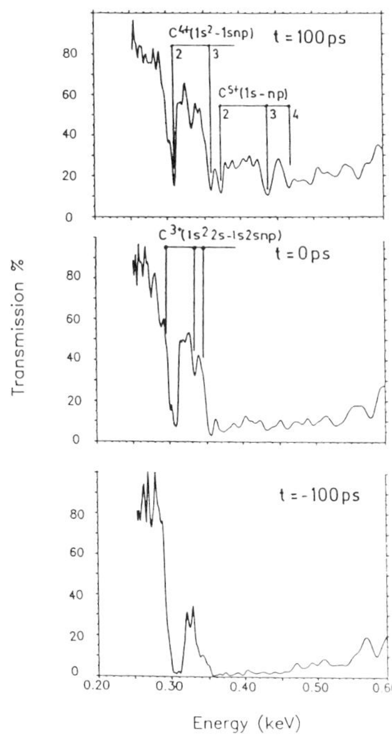

Figure 1. X-ray transmission of 1.9 μm CH foil. Each spectrum is averaged for a 100 ps duration. Time zero is the peak of the x-ray pulse.

Fig. 1 illustrates the time dependence of the strong burnthrough of a thinner, 1.9 μm parylene foil. The transmission here was obtained by normalizing to the Au wall spectra. Note that the transmission of cold CH is practically zero (about 10^{-4}) above the cold K-edge (283.84 eV) and it remains less than 1% below 400 eV x-ray energy. Here again x-ray heating of the material starts near to the K-edge 100 ps before the peak of the heating pulse. As the material is heated up and carbon is becoming strongly ionized, a characteristic transmission window can be observed between the He-like C^{4+} $1s^2 - 1s2p$ and $1s^2 - 1s3p$ absorption lines. The transmission in this window strongly increases reaching more than 60% within the next 200 ps. By this time, i.e. 100 ps after the peak of the heating pulse, the H-like absorption lines also appear, referring to the strong radiative heating of carbon. Calculations of the spectrum with the MULTI hydrocode together with a post-processor [3] are in good agreement with the observed spectrum. The simulations refer to a temperature near to 100 eV in the radiatively heated foil.

3. X-ray reemission

X-ray reemission of radiatively heated carbon was investigated in the same arrangement with a thick CH reemitter foil. The reemitter is situated on the wall on the opposite side of the Hohlraum, thus the reemitted flux was observed by looking through the cavity. There was a 300 ps time window before the plasma filled the cavity thus preventing the observation of the reemitted x rays[5,6]. Before this time however the contrast of the experiment was very good, reemission could well be studied in the Hohlraum configuration. The spectrum is dominated by the H-like emission referring again to a very high temperature. The line spectra arise over a strong continuum radiation which is the consequence of the large layer thickness heated up by the deeply penetrating x-rays. The total reemission of carbon in our range of observation (1-5 nm) ws 5% of the gold reemission.

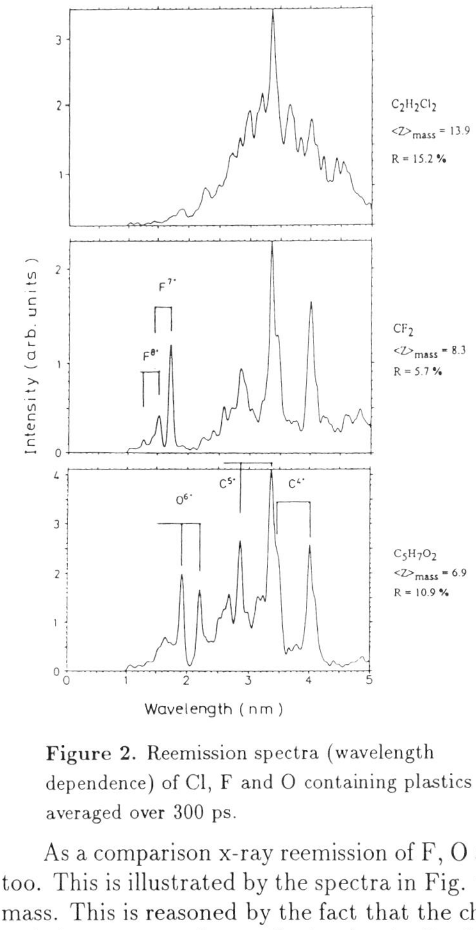

Figure 2. Reemission spectra (wavelength dependence) of Cl, F and O containing plastics averaged over 300 ps.

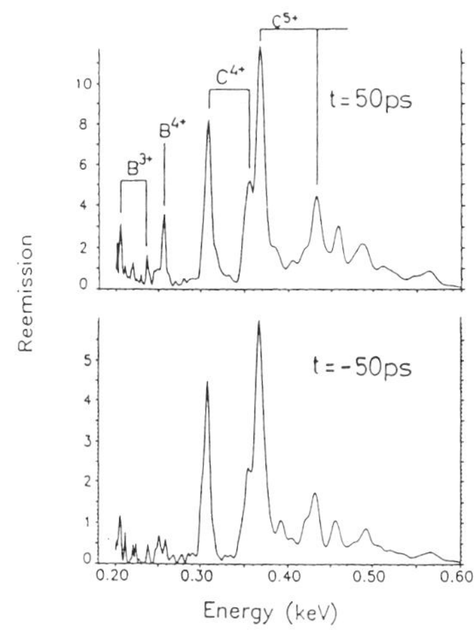

Figure 3. X-ray reemission of B_4C layered on the rear side of 120 $\mu g/cm^2$ carbon. Both spectra are averaged for a for a 100 ps duration.

As a comparison x-ray reemission of F, O and Cl-containing plastics was investigated, too. This is illustrated by the spectra in Fig. 2. Here a Z value is given by averaging over mass. This is reasoned by the fact that the characteristic absorption length of the x rays and the corresponding radiating (optically thick) layer depends on the mass and not on the number of atoms. R is the ratio of the plastic reemission to that of the Au wall. The Cl containing PVDC plastic has the highest Z and - as expected - the highest reemission. In this spectrum the strong carbon Ly-α line is superposed over the weakly resolved Cl spectra which has a strong continuum content. The weaker emitting F and O plastics are still characterized by strong line components, similarly to pure CH foils. H- and He-like F and O lines can be distinguished,too. It is however astonishing that although $< Z >$ is larger for the PTFE (CF_2) the reemission is stronger for the $C_5H_7O_2$. This can be qualitatively explained by noting that the lines of K-shell emission have longer wavelengths for O than for F. The maximal emission of a 100 eV Planckian radiator lies at 2.5 nm, near to the O lines. Thus oxygen will have a stronger continuum radiation than F, the lines of which are situated a factor 2 above the Wien-maximum, thus the continuum content will be lower.

Reemission spectroscopy was also used for the investigation of radiative burnthrough. In this case a B_4C layer was situated below a thin carbon reemitter. The delayed appearance of the boron emission lines refer to the burnthrough time of carbon[5]. If the boron tracer was embedded in a depth of 120 $\mu g/cm^2$ thick carbon, then the boron reemission appears with a delay of 160±30 ps relative to carbon. Boron reemission is immediately dominated by the H-like boron line radiation. This observation is again in good agreement with MULTI simulations [3] which give a temperature of 70 eV in the depth of the boron layer by this time.

References

[1] Mead W C et al. 1988 *Phys. Rev. A* **38** 5275
[2] Sigel R 1987 *Plasma Phys. & Controlled Fusion* **29**, 1261
[3] Eidmann K 1994 *Laser Part. Beams* accepted for publication
[4] Schwanda W and Eidmann K 1992 *Phys. Rev. Lett.* **69**, 3507
[5] Földes I B et al. 1993 *Proc. SPIE* **2015** 179
[6] Földes I B et al., 1994 *Phys Rev. E* **50**, R690

Inst. Phys. Conf. Ser. No 140: Section 9
Paper presented at Laser Interact. with Matter Conf., Oxford, 19–23 September 1994

© *1995 IOP Publishing Ltd*

Calculation of EXAFS in Shock–Compressed Matter

H.–D. Frey and J. Meyer-ter-Vehn

Max-Planck-Institut für Quantenoptik, D-85748 Garching, Germany

Abstract. Extended X–ray absorption fine structure (EXAFS) in shock–compressed matter have been investigated theoretically. EXAFS may be a way to determine the density behind the shock front, giving the second parameter needed for equation–of–state measurements. First experimental data on EXAFS in laser–compressed aluminum were obtained by Hall *et al* [1]. The critical issue is the temperature dependence, since increasing temperature tends to wash out the EXAFS signal. Strong shocks turn the solid probe into strongly coupled plasmas, the regular lattice structure is changed into a plasma fluid with strong ion–ion correlations. Ion–ion potentials are obtained using the Thomas–Fermi model for a quasimolecule, and fluid theory is applied to calculate the correlation function. The photoelectron mean free path, which also reduces the EXAFS signal, is determined from a dielectric approach. Finally, results for dense Al–plasmas are presented, also showing the change along the principle Hugoniot.

1. Extended X–ray absorption fine structure (EXAFS)

EXAFS has become a standard tool for structure analysis of solids [2]. These oscillations in X–ray spectra on the high–energy side of absorption edges are due to interferences of the outgoing photoelectron wave with reflections from neighbour ions, leading to deviations from the smooth, atomic–like absorption coefficient μ_0. Information about the local structure of the solid can be extracted from the dependence of the absorption coefficient μ on the photoelectron energy $E = \hbar^2 k^2/2m_e$.

The standard EXAFS formula for a pair of ions a distance R apart is given by

$$\chi(k) = \frac{\mu - \mu_0}{\mu_0} = \frac{|f(k,R)|}{kR^2}\,\sin\left(2kR + \delta(k,R)\right)\, e^{-2\,R/\lambda_e(k)} \tag{1}$$

([2],[3], for the K–edge). The interference term includes a phase shift due to the potential of the absorber and from the backscattering factor $|f(k,R)|\, e^{i\delta_f(k,R)}$, calculated in the spherical wave approximation [4] from the Thomas–Fermi atomic potential. The attenuation of the photoelectron by inelastic collisions corresponds to a loss of coherence of the electron wave and is included via the mean free path $\lambda_e(k)$.

For disordered systems like dense plasmas, this leads to

$$\bar{\chi}(k) = 4\pi\, n_i \int g(R)\, \frac{|f(k,R)|}{k}\, \sin\left(2\,kR + \delta(k,R)\right)\, e^{-2\,R/\lambda_e(k)}\, dR, \tag{2}$$

with the pair distribution $g(R)$ and the ion number density n_i.

2. Ion–ion interaction and pair correlation

The effective interaction energy $\Phi(R_i)$ of two ions in a plasma medium is determined in a quasimolecule model (see fig.1), calculating the total free energy within the Thomas–Fermi approach. This leads to a short–range effective potential, which depends on density and temperature. Details are given in [5].

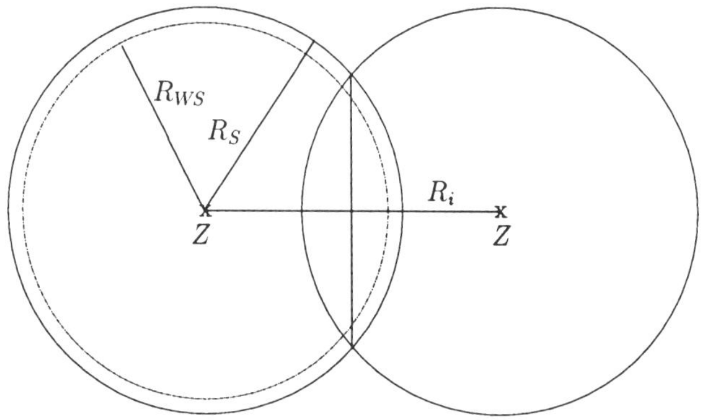

Figure 1. Quasimolecule model with two overlapping spheres, the total volume is adjusted to the Wigner–Seitz volume per ion according to $V_{QM} = 2\,V_{WS} = 2/n_i$ (i.e. $R_S \geq R_{WS}$).

The pair distribution $g(r)$ is calculated from the Ornstein–Zernicke equation

$$h(r) = c(r) + n_i \int c\left(|\vec{r} - \vec{r}\,'|\right)\, h(r')\, d\vec{r}\,', \tag{3}$$

relating the pair correlation $h(r) = g(r) - 1$ and the direct correlation $c(r)$, in combination with the Percus–Yevick approximation

$$c(r) = \left[1 - e^{\Phi(r)/k_B T}\right] g(r). \tag{4}$$

3. Damping due to inelastic electron scattering

The energy dependence of the electron mean free path λ_e also strongly influences the EXAFS signal. The inelastic scattering of the electron in the plasma medium is calculated within dielectric theory [6],[7], using the Lindhard–function for a degenerate electron gas. Results corresponding to dense, low–temperature Al–plasmas — including an explicit calculation of plasmon excitations [5] — are shown in fig.2.

4. Calculated EXAFS: Density and temperature dependence

EXAFS spectra for Al plasmas at twice the solid density ϱ_s are shown in fig.3a for temperatures between 1eV and 20eV. The reduction of the amplitude and range, as well as a shift to higher energies with increasing temperature can be seen.

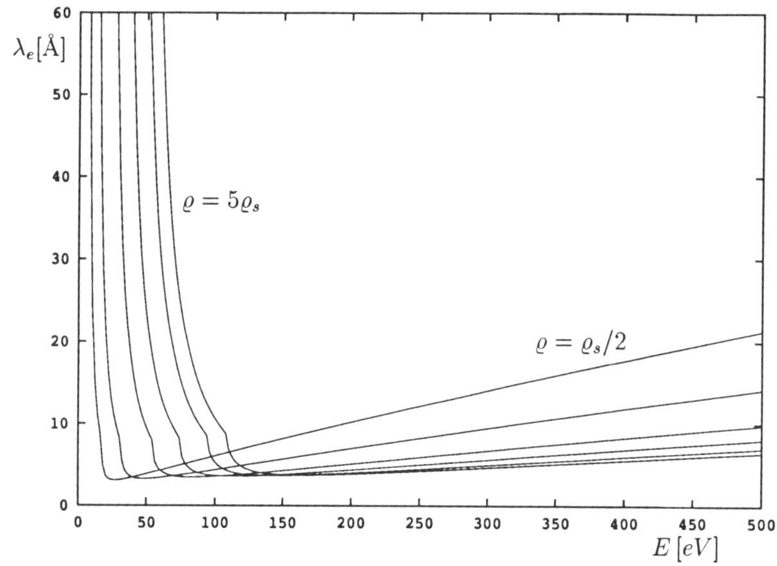

Figure 2. Energy dependency of the electron mean free path in the homogenous electron gas, with electron densities representative for low–temperature Al–plasmas with $\varrho \simeq \varrho_s/2, \varrho_s, 2\varrho_s, 3\varrho_s, 4\varrho_s$ and $5\varrho_s$.

Results for a sequence of states along the principle Hugoniot are given in fig.3b, showing the combined effect of temperature and density variations. A strong shift in the position of the first minimum is visible, which has been used to extract the plasma density. Comparison with fig.3a shows, however, that only combinations of temperature and density can be determined from the extremum positions, a separation of both needs quantitative information about the peak amplitudes as well.

The present results go beyond calculations of Djaoui, Gordon *et al* [8],[9],[7], in particular in the explicit treatment of temperature– and density–dependent ion–ion interaction and pair distributions, and of the plasmon part of the mean free path $\lambda_e(E)$.

References

[1] Hall T A *et al* 1988 *Phys. Rev. Lett.* **60**, 2034
Shiwai B A, Djaoui A, Hall T A, Tallens G J and Rose S J 1992 *Laser Part. Beams* **10**, 41

[2] Koningsberger D C and Prins R (ed.) 1988 *X–Ray Absorption. Principles, Applications, Techniques, of EXAFS, SEXAFS and XANES.* (New York: J.Wiley & Sons)

[3] Stöhr J 1992 *NEXAFS Spectroscopy* (Berlin: Springer)

[4] Müller J E and Schaich W L 1983 *Phys. Rev.* B **27**, 6489

[5] Frey H–D 1994 PhD thesis, Technical University Munich

[6] Ritchie R H 1959 *Phys. Rev.* **114**, 644

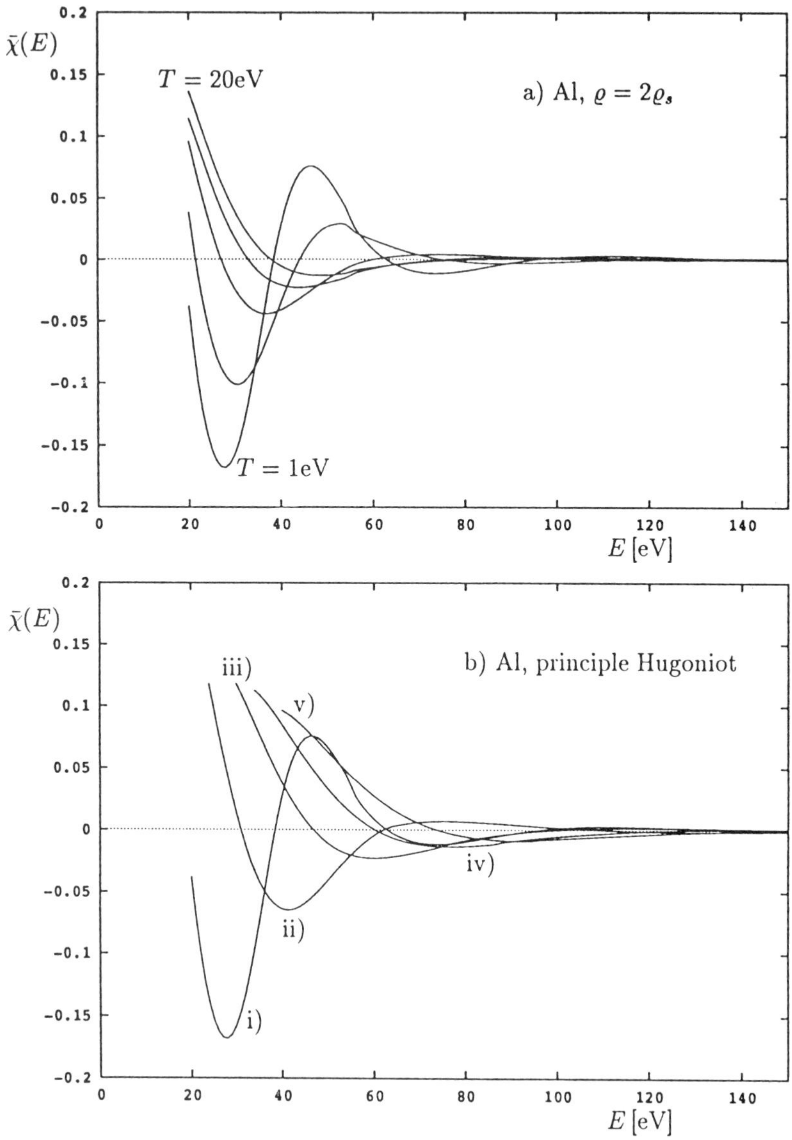

Figure 3. K-edge EXAFS signal $\bar{\chi}(E)$ in Al-plasmas, a) Temperature dependence: $\varrho = 2\varrho_s, T = 1, 2, 5, 10$ and $20\,\mathrm{eV}$; b) principle Hugoniot: i) $\varrho = 2\varrho_s$, ii) $\varrho = 2.5\varrho_s$, iii) $\varrho = 3\varrho_s$, iv) $\varrho = 3.5\varrho_s$, v) $\varrho = 4\varrho_s$.

[7] Gordon F I and Djaoui A 1992 *Phys. Fluids B* **4**, 3985

[8] Djaoui A, Hall T A, Albers R C, Rehr J J and Mustre J 1990 *Laser Part. Beams* **8**, 319

[9] Gordon F I 1993 *Plasma Phys. Contr. Fusion* **35**, 1207

Inst. Phys. Conf. Ser. No 140: Section 9
Paper presented at Laser Interact. with Matter Conf., Oxford, 19–23 September 1994
© 1995 IOP Publishing Ltd

Improved two-moment radiation transport

David Alan Garren†[1,2], Marcel Klapisch‡, John Gardner§ and Denis Colombant||

† Science Applications International Corporation, 1710 Goodridge Dr., P.O. Box 1303, McLean, Virginia 22102, USA

‡ ARTEP, Columbia, Maryland 21045, USA

§ Laboratory for Computational Physics and Fluid Dynamics, Naval Research Laboratory, Washington, DC 20375, USA

|| Plasma Physics Division, Naval Research Laboratory, Washington, DC 20375, USA

Abstract. A two moment model of radiative transport is developed, based on the *ad hoc* assumption that the specific intensity of radiation has the angular dependence of an ellipsoid with the origin lying at one of the foci. This model reduces to the standard diffusion approximation in the limit $\delta \equiv F/cU \ll 1$, with F and U the radiant flux and energy density. However, unlike standard diffusion theory, the present model gives the correct radiative transport in the limit in which all the radiative energy streams in one direction, i.e., $\delta = 1$. This property is crucial in predicting accurately the radiation transport in two and three-dimensional laser-plasma simulations.

1. Introduction

1.1. Radiation transport equation

Computer simulations of laser-plasma interactions require an accurate model for the transport of radiation. The radiation transfer equation [1, 2, 3]

$$\left[\frac{1}{c}\frac{\partial}{\partial t} + \mathbf{\Omega}\cdot\nabla\right] I_\nu(\mathbf{\Omega}) = S_\nu - \kappa'_\nu I_\nu(\mathbf{\Omega}) \tag{1}$$

determines the radiation intensity $I_\nu(\mathbf{\Omega})$ at frequency ν in the direction of the unit vector $\mathbf{\Omega}$. The effective absorption coefficient κ'_ν includes the contribution due to induced emission and S_ν gives the remaining source part of the emission.

[1] Mailing Address: Code 6730, Naval Research Laboratory, Washington, DC 20375, USA
[2] E-mail (INTERNET): GARREN@THIS.NRL.NAVY.MIL.

For a plasma in approximate local thermodynamic equilibrium (LTE), the source is $S_\nu = \kappa'_\nu B_\nu$ and the effective absorption coefficient is $\kappa'_\nu \equiv \kappa_\nu[1 - \exp(-h\nu/kT)]$. Here κ_ν is the standard absorption coefficient, T is the electron temperature, and

$$B_\nu(T) = \frac{2h\nu^3}{c^2} \frac{1}{\exp(h\nu/kT) - 1} \tag{2}$$

is the Planck radiation intensity. For a non-LTE plasma [4], κ'_ν and S_ν in Eq. (1) are modified from their LTE values.

1.2. Moments of radiation transport equation

One method of solving the radiation transport equation (1) is through moments in the radiation direction $\mathbf{\Omega}$. The zeroth and first moments of Eq. (1) yield [2, 3]

$$\frac{\partial U_\nu}{\partial t} + \nabla \cdot \mathbf{F}_\nu = 4\pi S_\nu - c\kappa'_\nu U_\nu \ , \tag{3}$$

$$\frac{1}{c}\frac{\partial \mathbf{F}_\nu}{\partial t} + c\nabla \cdot \vec{\mathbf{P}}_\nu = -\kappa'_\nu \mathbf{F}_\nu \ , \tag{4}$$

for a spherically symmetric source function S_ν. The first three moments of the radiation intensity in Eqs. (3) and (4) are defined by

$$U_\nu \equiv (1/c)\int d\Omega \, I_\nu(\mathbf{\Omega}) \ , \tag{5}$$

$$\mathbf{F}_\nu \equiv \int d\Omega \, \mathbf{\Omega} \, I_\nu(\mathbf{\Omega}) \ , \tag{6}$$

$$\vec{\mathbf{P}}_\nu \equiv (1/c)\int d\Omega \, \mathbf{\Omega}\mathbf{\Omega} \, I_\nu(\mathbf{\Omega}) \ , \tag{7}$$

in terms of the solid angle $d\Omega \equiv \sin\theta \, d\theta \, d\phi$ and the radiation unit vector $\mathbf{\Omega} \equiv \hat{\mathbf{z}}\cos\theta + [\hat{\mathbf{x}}\cos\phi + \hat{\mathbf{y}}\sin\phi]\sin\theta$, with θ and ϕ standard spherical coordinates. The sequence of moments in $\mathbf{\Omega}$ of Eq. (1) does not close since there are always moments through one higher order than there are equations. Therefore, some assumption on either the highest moment or equivalently the form of $I_\nu(\mathbf{\Omega})$ must be made in order to close the system of equations. The closure of the two moment equations (3) and (4) is often expressed in terms of a tensor Eddington factor defined by

$$\vec{\mathbf{f}}_\nu \equiv \vec{\mathbf{P}}_\nu / U_\nu \ . \tag{8}$$

2. Ellipsoidal model for radiation intensity

2.1. Radiative flux reference frame

To close the moment equations (3) and (4) we make the *ad hoc* assumption that the specific intensity of radiation has the angular distribution of an ellipsoid having its origin lying at one of the foci. Mathematically, this has the form

$$I_\nu(\mu) = \frac{(1 - \epsilon_\nu^2)a_\nu}{1 - \epsilon_\nu \mu} \ . \tag{9}$$

with a_ν and ϵ_ν depending upon space $\mathbf{x}$, time t, and spectral frequency ν. The variable $\mu \equiv \cos(\theta)$ is defined so that $\theta = 0$ corresponds to the direction of the spectral flux $\mathbf{F}_\nu$. Since the assumed intensity function is independent of an azimuthal angle ϕ, i.e., $I_\nu = I_\nu(\mu)$, we choose first to define $\mathbf{F}_\nu$ and $\vec{\mathbf{P}}_\nu$ in terms of some coordinate system $(\hat{\mathbf{x}}', \hat{\mathbf{y}}', \hat{\mathbf{z}}')$ that is arbitrary with the restriction that $\hat{\mathbf{z}}'$ lie in the local $\theta = 0$ direction. The angular integrals in Eqs. (5)-(7) then can be expressed in the form

$$U_\nu = \frac{2\pi}{c} \frac{(1-\epsilon_\nu^2)a_\nu}{\epsilon_\nu} \ln\left(\frac{1+\epsilon_\nu}{1-\epsilon_\nu}\right), \tag{10}$$

$$\mathbf{F}_\nu = \hat{\mathbf{z}}'\, 2\pi \frac{(1-\epsilon_\nu^2)a_\nu}{\epsilon_\nu^2} \left[\ln\left(\frac{1+\epsilon_\nu}{1-\epsilon_\nu}\right) - 2\epsilon_\nu\right], \tag{11}$$

$$[\vec{\mathbf{P}}_\nu]_{(\hat{\mathbf{x}}',\hat{\mathbf{y}}',\hat{\mathbf{z}}')} = \begin{pmatrix} P_\nu & 0 & 0 \\ 0 & P_\nu & 0 \\ 0 & 0 & P_\nu \end{pmatrix} - \frac{1}{2} \begin{pmatrix} 3P_\nu - U_\nu & 0 & 0 \\ 0 & 3P_\nu - U_\nu & 0 \\ 0 & 0 & 0 \end{pmatrix}, \tag{12}$$

in terms of a scalar radiation pressure

$$P_\nu = \frac{2\pi}{c} \frac{(1-\epsilon_\nu^2)a_\nu}{\epsilon_\nu^3} \left[\ln\left(\frac{1+\epsilon_\nu}{1-\epsilon_\nu}\right) - 2\epsilon_\nu\right]. \tag{13}$$

The values of radiant energy density U_ν and the magnitude of the radiative flux $F_\nu \equiv \| \mathbf{F}_\nu \|$ in Eqs. (10) and (11) can be expressed equivalently in terms of a_ν and ϵ_ν. The value of ϵ_ν is found from the ratio of Eqs. (10) and (11), i.e.,

$$\delta_\nu \equiv \frac{F_\nu}{cU_\nu} = \frac{1}{\epsilon_\nu} - \frac{2}{\ln\Big((1+\epsilon_\nu)/(1-\epsilon_\nu)\Big)} = \frac{1}{\epsilon_\nu} - \frac{1}{\operatorname{arctanh}(\epsilon_\nu)}, \tag{14}$$

which is a monotonically-increasing function from zero to unity defined over the same range in ϵ_ν. The value of a_ν then determines the effective magnitude of the radiation intensity via Eq. (10). Thus a_ν and ϵ_ν can be found easily from values of U_ν and F_ν.

2.2. Simulation reference frame

The only aspect remaining to be determined is the rotation from the coordinates $(\hat{\mathbf{x}}', \hat{\mathbf{y}}', \hat{\mathbf{z}}')$ of the spectral flux to those of the simulation $(\hat{\mathbf{x}}, \hat{\mathbf{y}}, \hat{\mathbf{z}})$. This is required in order to find an expression for the radiation stress tensor $[\vec{\mathbf{P}}_\nu]_{(\hat{\mathbf{x}},\hat{\mathbf{y}},\hat{\mathbf{z}})}$ in the simulation reference frame in terms of $[\vec{\mathbf{P}}_\nu]_{(\hat{\mathbf{x}}',\hat{\mathbf{y}}',\hat{\mathbf{z}}')}$ of Eq. (12). The radiative flux in the simulation frame is found using the standard Euler angles:

$$[\mathbf{F}_\nu]_{(\hat{\mathbf{x}},\hat{\mathbf{y}},\hat{\mathbf{z}})} = \vec{\mathbf{R}}_y(\theta)\, \vec{\mathbf{R}}_z(\phi)\, [\mathbf{F}_\nu]_{(\hat{\mathbf{x}}',\hat{\mathbf{y}}',\hat{\mathbf{z}}')} \tag{15}$$

with the rotation matrices $\vec{\mathbf{R}}_y(\theta)$ and $\vec{\mathbf{R}}_z(\phi)$ defined by

$$\vec{\mathbf{R}}_y(\theta) = \begin{pmatrix} \cos\theta & 0 & -\sin\theta \\ 0 & 1 & 0 \\ \sin\theta & 0 & \cos\theta \end{pmatrix}, \quad \vec{\mathbf{R}}_z(\phi) = \begin{pmatrix} \cos\phi & -\sin\phi & 0 \\ \sin\phi & \cos\phi & 0 \\ 0 & 0 & 1 \end{pmatrix}. \tag{16}$$

The angles θ and ϕ are determined by the angle of the spectral flux in the simulation frame, i.e., $F_\nu = F_{\nu,x}\hat{\mathbf{x}} + F_{\nu,y}\hat{\mathbf{y}} + F_{\nu,z}\hat{\mathbf{z}}$ via

$$\theta \equiv \arccos\left(\frac{F_{\nu,z}}{\sqrt{F_{\nu,x}^2 + F_{\nu,y}^2 + F_{\nu,z}^2}}\right), \quad \phi \equiv \arctan\left(\frac{F_{\nu,y}}{F_{\nu,x}}\right). \tag{17}$$

The transformation of the radiative tensor in the primed frame $(\hat{\mathbf{x}}', \hat{\mathbf{y}}', \hat{\mathbf{z}}')$ to the simulation frame $(\hat{\mathbf{x}}, \hat{\mathbf{y}}, \hat{\mathbf{z}})$ is found via

$$[\vec{\mathbf{P}}_\nu]_{(\hat{\mathbf{x}},\hat{\mathbf{y}},\hat{\mathbf{z}})} = \vec{\mathbf{R}}_y(\theta)\,\vec{\mathbf{R}}_z(\phi)\,[\vec{\mathbf{P}}_\nu]_{(\hat{\mathbf{x}}',\hat{\mathbf{y}}',\hat{\mathbf{z}}')}\vec{\mathbf{R}}_z^T(\phi)\vec{\mathbf{R}}_y^T(\theta)\ , \tag{18}$$

with the superscript T denoting the matrix transpose. Using Eq. (18) for $\vec{\mathbf{P}}_\nu$ then closes the moment equations (3) and (4).

For spatial regions in which the intensity is nearly isotropic, i.e., $\epsilon_\nu, \delta_\nu \ll 1$, the tensor Eddington factor of Eq. (8) reduces to the optically thick limit with $\vec{\mathbf{f}}_\nu = \vec{\mathbf{I}}/3$. Then $\nabla\cdot\vec{\mathbf{P}}_\nu$ in Eqs. (3) and (4) becomes $\nabla U_\nu/3$, yielding the standard radiative diffusion equations. Equations (3) and (4) also give the correct result for the streaming limit in an optically thin plasma. That is, if all photons are traveling in a single direction, then $\nabla\cdot\vec{\mathbf{P}}_\nu$ becomes ∇U_ν.

3. Discussion

We have modified the radiative diffusion subroutine within a multigroup hydrodynamics code to apply for a tensor Eddington factor given by $\vec{\mathbf{f}}_\nu \equiv [\vec{\mathbf{P}}_\nu]_{(\hat{\mathbf{x}},\hat{\mathbf{y}},\hat{\mathbf{z}})}/U_\nu$. There is very little difference between 1D runs using our new Eddington tensor compared with that of ordinary diffusion theory.

More dramatic differences are expected between the tensor and scalar Eddington factors for 2D and 3D simulations, since this tensor Eddington factor allows for "shadows" that are impossible to obtain with ordinary diffusion theory. We are currently investigating the effect of the shadows on the ablative Rayleigh-Taylor instability in the nonlinear regime. This formulation will not only improve current LTE modeling of laser matter interactions, but it will also be a valid extension of radiative diffusion theory when non-LTE effects are taken into account using the method of Busquet [4].

Acknowledgments

Work supported by the U.S. Department of Energy and the Office of Naval Research. The authors are grateful to Jill Dahlburg for useful suggestions.

References

[1] Zel'dovich Y B and Raizer Y P 1966 *Physics of Shock Waves and High-Temperature Hydrodynamic Phenomena* (New York: Academic) pp 107-72

[2] Minerbo G N 1978 *J. Quant. Spectrosc. Radiat. Transfer* **20** 541-5

[3] Mihalas D and Mihalas B W 1984 *Foundations of Radiation Hydrodynamics* (New York: Oxford) pp 311-6

[4] Busquet M 1993 *Phys Fluids* B **5** 4191

Inst. Phys. Conf. Ser. No 140: Section 9
Paper presented at Laser Interact. with Matter Conf., Oxford, 19–23 September 1994

© *1995 IOP Publishing Ltd*

Low Density Gas and Foam Targets for ICF Long Scale Length Plasma Experiments: Fabrication and Characterization

P L Gobby, M A Mitchell, N E Elliott, M A Salazar, J E Moore, V M Gomez, J C Fernandez and W W Hsing

Los Alamos National Laboratory, Los Alamos, NM 87545, USA

Abstract. Recent ICF experiments performed by the laser-matter interaction group at Los Alamos using the Nova laser to investigate long scale length plasmas required two types of special targets: gas-filled hohlraums and free-standing low density foams, both with densities in the range of 3-5 mg/cc. The mass in each case was provided by hydrogen and carbon - in the form of polymethylpentene for the foam case and 2,2-dimethylpropane (i.e.,neopentane, C_5H_{12}) at 1 atmosphere for the gas case. Dopants of Cl, Ti, Cr and Mn were added to the foams, while Ti and Cr coated carbon fibers were added to the hohlraums - both for isoelectronic x-ray spectroscopic temperature measurements.

1. Introduction

Recently experiments were conducted on long-scale-length plasmas similar to those expected in the National Ignition Facility (NIF). These experiments utilized the smaller, but existing, Nova laser at Lawrence Livermore National Laboratory (LLNL). All NIF ignition designs include either gas-filled hohlraums or low-Z lined hohlraums which would result in a plasma filled hohlraum during the laser pulse. There is concern that the interaction of the incident laser light with the plasma might result in an intolerable fraction of the incident laser light being scattered back out of the hohlraum, thus lowering the deposited energy. This paper deals with the targets used for these experiments.

Los Alamos researchers were primarily interested in plasma densities of 3-5 mg/cc containing only carbon and hydrogen and path lengths of about 2 mm. Densities in this range could be attained in two ways: either with gas or with foam. Gas could yield a homogeneous density but would have to be contained with some sort of laser window. A carbon-hydrogen foam would not require the window, but would be less homogeneous.

Thus two types of targets were requested and delivered: free-standing foam cylinders approximately 2 mm diameter and 2 mm long, and newly designed toroidal hohlraums which would contain gas. Both types eventually also required the inclusion of small quantities of dopant elements to facilitate a temperature measurement.

2. Foam Cylinders

2.1 General Fabrication

The foam cylinders were prepared using an extension of the technique developed by Simandl *et al.* and Williams [1,2]. Only a brief review will be presented here. Polymethylpentene (TPX) was first dissolved in a mixture of 60% durene and 40% napthalene at 180°C. When slowly cooled to approximately 100° C the TPX phase separated from the solvents, thus forming a gel. Continued cooling to room temperature resulted in freezing of the solvents. The solid mass was then broken free from its container and was then ready for further processing.

Significant shrinkage of the solvent/polymer mixture occurred during freezing. Moreover a high density "skin" also formed; these facts collectively prevented casting of the foam cylinders to net shape. Instead the cylinders were machined from the frozen mixture prior to removal of the solvent.

Because of the high vapor pressure of the solvents, the unleached foam had to be stored in a refrigerator (~5°C). Even so, millimeter-sized pieces of the foam would leach completely if left in the refrigerator for a week. Consequently the individual cylindrical pieces were machined rapidly and then returned within five minutes to the refrigerator.

Obviously at 3-5 mg/cc density the foam pieces were extremely fragile and, in our experience, could not withstand any handling after leaching. Therefore the foam cylinders were also mounted in their target configuration prior to leaching.

Carbon fibers were used to mount the foam cylinders. In an effort to minimize the mass in the immediate region of the cylinder, the fibers were not adhesively bonded directly to the fibers; instead each foam cylinder was essentially caged in a superstructure of carbon fibers. (Fig. 1) This superstructure was built around each cylinder as it stood upright on a glass slide prior to leaching. The cylinder was then left out overnight at room temperature during which time enough mass was lost to solvent sublimation that the partially leached foam cylinder could then be lifted by its carbon fiber support. The parts were then transferred to a rough pumping system for a few hours to remove the remainder of the solvents. Characterization for density and dopant concentrations was performed on each foam cylinder only after this mounting and leaching procedure.

2.2 Foam Dopants

Dopants of Cl, Ti, Cr, and Mn were successfully added to these low density foams in amounts of 0.2 - 5.0 weight percent. The techniques used varied, depending on the dopant.

Chlorine was added directly to already leached foam by infiltrating and coating the foams with parylene D ($C_8H_6Cl_2$). Because of the low densities involved a very thin parylene D layer resulted in a relatively large density increase. For example, a TPX foam cylinder originally at 3.5 mg/cc was coated with parylene D to a thickness of only 41 nm (as determined by an *in situ* crystal monitor). Yet the chlorine content was determined to be 4.3 wt. %, and the density increased to 3.9 mg/cc.

Metal dopants were added to the foam by adding soluble metal compounds to the napthalene/durene/TPX during the initial mixing at 180°. For the metals of interest here, these were: chromium (III) chloride tetrahydrofuran complex (1:3), titanium (III) chloride in tetrahydrofuran, and manganese (III) acetate dihydrate in ethanol. In all the metal cases it was very difficult to get loadings above 1 wt. % without the final foam density exceeding 5 mg/cc. This was due to excessive shrinkage that occurred to foams with the higher metal loadings and does not represent a theoretical limit.

2.3 Foam characterization

As previously mentioned the leached foam cylinders were fragile enough that handling had to be minimized. Physical dimensions were determined optically, but the mass could not be reasonably determined via weighing. (Besides not being able to remove the foam from its support structure, the mass of a 3 mg/cc cylinder 2 mm diameter and 2 mm long is less than 20 μg.

The average density of each foam piece was determined using ß-transmission. A Cd^{109} source was used with an aperture of 0.4 mm. Calibration was achieved using various thicknesses of parylene N (poly-para-xylylene). The areal density thus determined was combined with physical dimensions to determine the average density of each piece.

Dopant content in individual foams was determined using x-ray fluorescence. Standards for these measurements were made from thin coatings of parylene D, sputtered titanium/chromium (these two were always used together), and vapor deposited manganese.

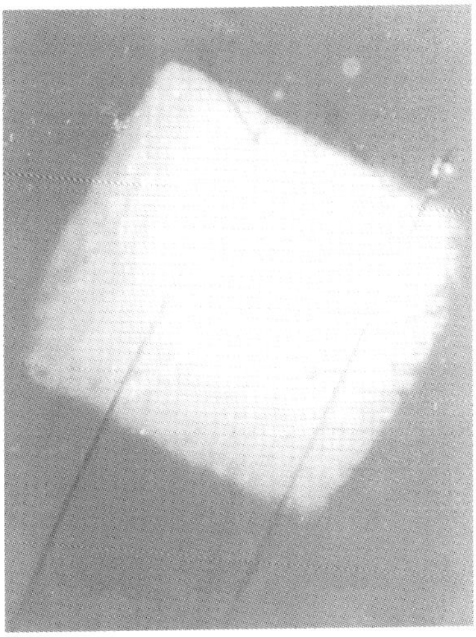

Fig. 1. 3.5 mg/cc polymethylpentene 2 mm diameter x 2 mm long cylinder as mounted with carbon fibers.

Fig. 2. Gas-filled hohlraum of 3.25 mm diameter and silicon nitride window prior to assembly.

3. Gas-filled hohlraums

3.1 Design and fabrication

A special hohlraum was designed for these experiments which increased the laser path-length within the hohlraum, while maintaining the radiation temperature achieved with standard NOVA hohlraums. The 3200 μm diameter toroidal gas-filled hohlraums were constructed with a light entrance hole (LEH) of 1600 μm diameter. Details of the gold hohlraum fabrication procedure are discussed in [3]. Silicon nitride windows were chosen to cover these LEHs and contain the gas. The windows were approximately 250 nm thick. The parent silicon substrate was etched to 1800 μm diameter, while the entire substrate was cut to approximately a 3 mm square. The windows were bonded to the hohlraum using a thixotropic fast-setting epoxy. Fig. 2 shows such a hohlraum and silicon nitride window prior to assembly.

Gas was introduced to the target through a piece of hypodermic tubing, which also served as its support. The Tubing was sealed to a small 5 cc reservoir and valve system bonded to a standard Nova magnetic base. A pressure transducer mounted to the Nova target pylon allowed the target pressure to be monitored up to seconds before shot time.

In our experience the silicon nitride windows could only withstand a pressure differential in one direction - one which pushed the window up against the frame. Pressure differentials which tended to peel the thin window have resulted in window failures. For simplicity we refer to these two cases as positive pressure and negative pressure, respectively. This anisotropy required that the windows were always mounted onto the hohlraum with the silicon substrate to the outside. It also required a filling procedure that avoided subjecting the windows to a negative pressure.

Targets were pre-filled with the gas of interest outside the Nova target chamber. With the target placed inside a separate small chamber with gas line and pressure transducer feedthroughs, the chamber surrounding the target was first evacuated. The interior of the target remained at 1 local atmosphere, and the windows were experiencing positive pressure.

Then the vacuum valve was switched over to the target to remove the air and reduce the pressure difference to zero. The target was then isolated before being filled with ≥1 atmosphere of the desired gas. The special filling chamber was then let back to air, and the pressure difference across the window was reduced.

In most cases the fill pressure has been 1 atmosphere. The choice of gas was used to adjust the desired density. Neopentane was initially used to attain a hydrocarbon density of 3.2 mg/cc, but other gases and mixtures C_5D_{12} and CO_2 have since been used. Sub-atmosphere fills have been avoided up to now, both because the window would experience a negative pressure between filling and subsequent insertion into the Nova target chamber, and also because any leak would cause ambient air to be drawn into the hohlraum during this same time.

3.2 Ti/Cr fibers

In an effort to measure the electron temperature achieved within the hohlraum, 5 µm diameter carbon fibers with thin coatings of titanium and chromium were mounted within the hohlraum. The fibers were coated by vacuum sputtering from a target pressed from 50-50 wt. % titanium and chromium powders.

Uniformity of the coating was assured by rotating the fibers during the deposition. Approximately a dozen fibers were mounted parallel to the axis of rotation at a radius of 2 cm. The deposition flux was at 90° to the rotation axis. It was also necessary that fiber lengths were no greater than 1 cm. Though known for their stiffness, at 5 µm these fibers would easily "wave in the breeze" that occurred during pump down and become entangled with fixturing. This would result in non-uniform coatings.

Fibers thus coated were then characterized and subsequently inserted into the hohlraum through a 50 µm hole previously machined into the hohlraum wall, and manipulated to the desired position. A single small (~100 µm) spot of epoxy both held the fiber in place and sealed the hole.

Characterization of the fibers was carried out using a scanning electron microscope (SEM) with an energy dispersive x-ray analyzer. Typically 1 or 2 fibers from a coating run were sacrificed for this purpose. Fibers were checked for coating uniformity and homogeneity both along and around the fibers. Variations of more than 2-3% only occurred when fibers as long as 2 cm were used. When the lengths were reduced to 1 cm these variations ceased.

Acknowledgment

The authors wish to thank D. Ciarlo of Lawrence Livermore National Laboratory for providing the silicon nitride windows.

References

[1] Simandl R F , Robinson D N , Bolinger W L and Davis W E 1991 *Oak Ridge Y-12 Report* Y-2441.

[2] Williams J E *Eighth Target Fabrication Specialists' Meeting* September 1991, Albuquerque, New Mexico

[3[Foreman LR *et al.*, accepted for publication *Fusion Technology*

Inst. Phys. Conf. Ser. No 140: Section 9
Paper presented at Laser Interact. with Matter Conf., Oxford, 19–23 September 1994

© 1995 US Government

Laboratory High-Energy Astrophysics on Lasers

W. H. Goldstein, D. A. Liedahl, R. S. Walling, M. E. Foord, A. L. Osterheld, and B. G. Wilson

Physics and Space Technology Directorate, Lawrence Livermore National Laboratory, Livermore CA 94550

Abstract. The tremendous range of temperatures and densities spanned by astrophysical plasmas has significant overlap with conditions attainable using high-power laser facilities. These facilities provide an opportunity to create, control, and characterize plasmas in the laboratory that mirror conditions in some of the most important cosmological systems. Moreover, laboratory experiments can enhance astrophysical understanding by focusing on and isolating important physical processes, without neccessarily reproducing the exact conditions of the integral system. Basic study of radiative properties, transport phenomena, thermodynamic response and hydrodynamic evolution in plasmas under properly scaled conditions leads both directly and indirectly to improved models of complex astrophysical systems. In this paper, we will discuss opportunities for current and planned high-power lasers to contribute to the study of high-energy astrophysics.

1. Introduction

The historical role of laboratory astrophysics has been to provide data that could be directly matched to observations. For example, laboratory spectroscopy was originally pursued for the purpose of making elemental identifications through the measurment of wavelength sets that could be directly compared to stellar spectra. While this "database" function continues to have value, the advent of new laboratory plasma sources offers the possibility of a different, richer approach, based on the simulation of aspects of astrophysical systems in the laboratory. These facilities present a novel opportunity to go beyond the acquisiton of basic physical data, and create, control, and characterize plasmas in the laboratory that provide a window on the phenomenology of astrophysical objects.

Simulation experiments may provide direct information on the integral behavior of astrophysical plasmas, or focus on and isolate relevant dynamical processes. In either case, though, they will frequently differ essentially from "database" experiments in requiring the mediation of computer simulations for their interpretation. Astrophysics is unique among the basic physical sciences in that the observable phenomena are often the result of the complex interplay of multiple physical processes, none of which can be modified by the experimenter. Obviously, performing controlled experiment on astronomical objects is beyond our reach. Theory and, more recently, complex computer simulations have been called upon to fill the void left by the absence of controlled experiments. Controllable simulation experiments in the laboratory provide a unique avenue for testing and "benchmarking" the same codes that are applied to astrophysical systems.

In this paper, we discuss in some detail a simulation experiment that can be carried out at currently available laser facilities. The experiment is designed to generate a homogenous volume of photoionized gas, the properties of which can be measured and compared to the predictions of models used to describe X-ray emitting accretion disk plasmas. Experiments aimed at improved understanding of astrophysical opacities, and of the effects of strong magnetic fields are briefly covered to suggest the utility of future high-intensity lasers, including ultra-short pulse lasers and the National Ignition Facility. These and other experiments have also been discussed by Rose in a 1991 review [1].

2. X-ray Photoionized Plasmas

Among the most exciting prospects in astrophysics for the next decade is the opportunity to perform detailed studies of the X-ray emission from accretion-powered X-ray sources. Accretion power results from the conversion of gravitational potential energy into kinetic and radiant energy when material is accreted onto an astronomical object. At sufficiently high accretion rates the total luminosity of some objects can be dominated by the accretion process, as opposed to, say, nuclear fusion. If the object is compact enough, then the energy released near the surface of the star is predominantly in the form of X rays. High accretion rates and compactness characterize three object classes: cataclysmic variables, X-ray binaries, and active galactic nuclei. Interest in these objects is motivated by numerous considerations, but primarily by two: (1) compact objects, particularly neutron stars and black holes, are often difficult to detect, let alone study, except when accreting and (2) the accretion process, now recognized as a fundamental energy-generation mechanism, is not understood.

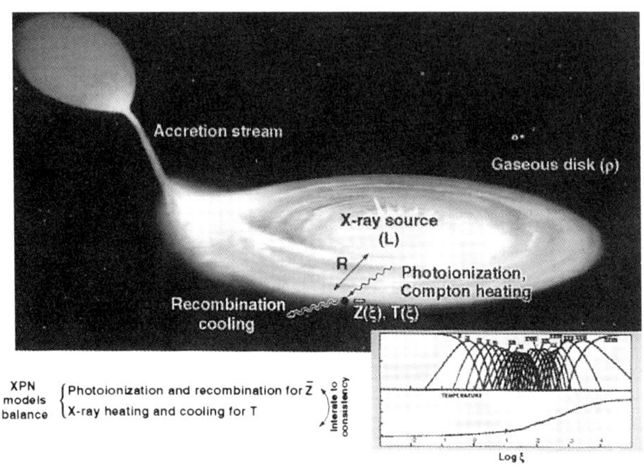

Figure 1. The physics of X-ray photoionized nebulae scales with ionization parameter, ξ=L ρR^2.

The analyses of spectra from accretion-powered X-ray sources, which will comprise a substantial portion of the high-energy astrophysics database, are likely to severely tax the capabilities of the currently available analytical resources. The reason is clear: understanding of atomic physics processes and the matter/radiation interactions in X-ray photoionized plasmas is sketchy.

The essential aspect of an X-ray photoionized plasma is the high degree of overionization relative to a steady-state collision-driven plasma at the same temperature [2]. The ionization state is characterized by the quantity, $\xi = 4\pi F/n$, the ratio of the ionizing radiation flux to the electron density [3]. Whenever ξ exceeds ~10 (cgs), electron impact ionization is less important than photoionization in establishing the charge state distribution. Similarly, the role of electron impact excitation is reduced in comparison to recombination cascades as a population mechanism.

At present, the physical processes of "X-ray nebulae" are modeled by complex computer codes (see, *e.g.*, Ref. 2) which, nevertheless, rely on simplifying assumptions and approximations. For example, the reprocessing medium is assumed to be uniform in density, radiative transfer is assumed to be one-dimensional, and the atomic models are highly simplified for the sake of computational expediency. Given the source luminosity, L, of ionizing radiation, and the density, the models – as shown schematically in Fig. 1 – solve interatively for the ionization balance and electron temperature as a funtion of ξ.

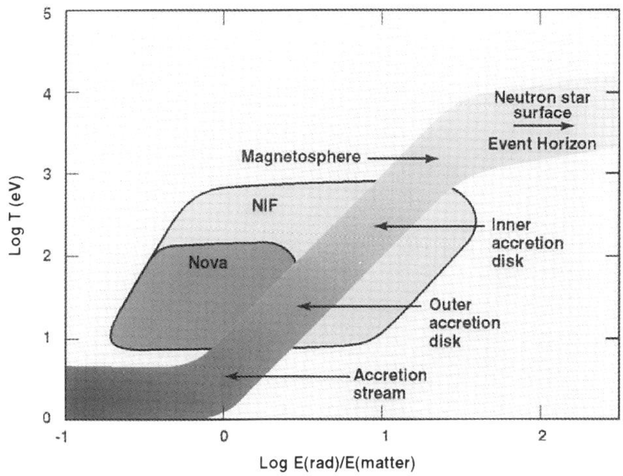

Figure 2. High-power Lasers can radiatively drive laboratory plasmas to conditions simulating accretion powered astrophysical objects.

As shown in Fig. 2, conditions achievable in targets using the Nova laser, and conditions projected for the planned National Ignition Facility (NIF), have excellent overlap with the parameters expected to characterize accretion disk plasmas. In proposed experiments, uniform, photoionized plasmas are created by illuminating a low density gas target with broad-band ionizing radiation from a laser-irradiated high-Z converter target. Preliminary experiments and predictions for gas cells have been made by Carter *et al.* [4], who designed a box-like target in which argon gas is contained by thin-foil windows. This target allows argon pressures from 25 Torr (corresponding to ion densities near 10^{18} cm^{-3}) down to pressures several orders of magnitude lower. Carter also measured a gold converter spectrum on the Nova Two-Beam Facility using .53 μm light in a line-focus geometry at an irradiance of 10^{14} W cm^{-2}. This spectrum has been used to estimate the values of ξ, shown in Table I, achievable at several laser facilities. These values are precisely those expected to obtain in the X-ray emitting regions of accretion disks.

Design calculations for an argon experiment using the 25-Torr gas cell and the measured gold spectrum have predicted ionization into the middle of the L-shell. Then, using absoption and emission spectroscopy, respectively, to measure ionization distribution and temperature, a direct comparison with the predictions of astrophysical simulation codes could be carried out at known – and controllable – density (gas pressure) and ionization parameter (luminosity).

The gas cell design is constrained in that the plasma must have enough time for the photons to ionize the argon ions up to the L shell, and by the rate of propagation into the gas of a shock wave from the window. Simulations show that a 2-ns drive satifies both criteria. Still, the effects of high (i.e., non-coronal) densities, non-stationary ionization kinetics, and optical depth, remain to be evaluated in detail. The first two effects, which are difficult to eliminate in the laboratory experiment, do not map well onto the approximations of the astrophyscial models, but can probably, nevertheless, be minimized or controlled. Optical depth introduces an additional parameter into the models, typically chosen as ρL. While radiation transport experiments in the laboratory have been suggested [1], it's not yet clear whether relevant values of ρL can be attained simultaneously with ξ in gas cell geometries.

TABLE I

	L_{max} (erg/sec)	ρ	R	Log ξ_{max}
Nova 2-Beam (1 ns)	~10^{19}	<10^{18}/cm^3	~1cm	~1
Nova 10-Beam (1 ns)	~10^{20}	"	"	~2
NIF (10 ns)	~10^{21}	"	"	~3

3. Other Experiments

In the field of radiative opacity, the Nova laser has already played an instrumental role in building confidence in complex models of astrophysical phenomena. Experiments there have measured the transmission of X-rays in iron plasmas [5] and essentially verified the opacity predictions of the OPAL code. OPAL opacities had already been shown to resolve discrepencies in stellar pulsation theory [6]. However, it is important to note that experimental confirmation of OPAL has been carried out only at densities ~.01 g/cc, while the major effect on stellar pulsation arises at much lower densities, around .0001 g/cc. It's been shown that, at these low densities, modern opacity codes – which agree at .01 – differ in their predictions [7]. This suggests that laboratory tests at densities below .01 g/cc are important. Unfortunately, signal levels and equilibration are problemmatic under such rarified condition, and a successful experiment may have to await the NIF, where larger plasmas and longer pulse lengths than those at Nova will be available.

A final example illustrates the possible utility of new ultra-short pulse lasers to explore physics relevant to astrophysical systems. It's well known that laser-produced plasmas generate strong magnetic fields, but, until the recent advent of ultrahigh intensity, short pulse lasers, these fields were weak on the scale of competing dynamical effects, and direct localized measurements were neither possible nor essential. New sub-picosecond lasers, are predicted to access a qualitatively new magnetic field strength regime [8] where the fields influence hydrodynamics, energy transport, and atomic structure, and, as a bonus, are pomic XUV transitions, they can probably be directly measured spectroscopically in localized volumes of plasma. redicted simulate in the laboratory those observed or predicted in compact astrophysical objects including cataclysmic variables and white dwarf stars [9]. Because these lasers deposit their energy before appreciable hydrodynamic motion can occur, and because the fields are predicted to be sufficient to produce measureable Zeeman (magnetic field-induced) splitting of atomic XUV transitions, the fields can probably be directly measured spectroscopically in localized volumes of plasma.

Acknowledgments

This work was performed under the auspices of the U. S. Department of Energy by Lawrence Livermore National Laboratory under contract number W-7405-ENG-48.

References

[1] Rose S J, 1991 Laser and Particle Beams **9**, 183.
[2] Kallman T R, and McCray R, 1982 Ap. J Suppl. **50**, 263; Arnaud M, and Rothenflug R, 1985 Astron. Astr. Suppl. Ser. **60**, 425.
[3] Tarter C B, and Salpeter E E, 1969 Ap. J. **156**, 953.
[4] Carter M R, Osterheld A L, Shimkaveg G, Walling, R S and Stewart R E, 1991 Bull. Am. Phys. Soc. **36**, 2353.
[5] Springer P T *et al.*, 1992 Phys. Rev. Lett. **69**, 3735; DaSilva L *et al.*, 1992 Phys. Rev. Lett. **69**, 438
[6] Rogers F J and Iglesias C A, 1992 Astrophys. J. Suppl. **79**, 507; Rogers F J *et al.*, 1986 Astrophys. J. **310**, 723.
[7] Rickert A, 1994, presented at the Sixth International Workshop on the Radiative Properties of Hot Dense Matter, to appear in the J. Quant. Spectrosc. Radiat. Transf.
[8] Sudan R N, 1993 Phys. Rev. Lett. **70**, 3075; Wilks S L *et al.*, 1992 Phys. Rev. Lett. **69**, 1383
[9] Beuermann K *et al.*, 1989 *Accretion-Powered Compact Binaries*, Mauche C W, ed., Proc. of the 11th N. Amer. Workshop on Cataclysmic Variables and Low Mass X-ray Binaries, Santa Fe, Oct. 1989.

Inst. Phys. Conf. Ser. No 140: Section 9
Paper presented at Laser Interact. with Matter Conf., Oxford, 19–23 September 1994

© 1995 IOP Publishing Ltd

Acceleration of matter by radiative plasma pressure

S Yu Gus'kov, S V Mesyats, V B Rosanov

P.N.Lebedev Physics Institute of Russian Academy of Sciences, Leninsky prospect 53,117924 Moscow Russia.

Abstract. Analytical solution of ablation equations for indirect ICF-targets and the expression for hydrodynamic efficiency of shell target compressed by pressure of radiative laser-produced plasma are obtained. It is shown that high efficiency of indirect driven compression (hydrodynamic efficiency 0.1 and shell velocity (3-4)10^7 cm/s) may be attained when using shells with low aspect ratio 10-20, made of Au-foam with initial density (2-3) g/cm^3.

1. Introduction

The problem of an implosion efficiency of indirect driven target, that is the question how to attain the hydrodynamic efficiency more than 0.1 at the final shell velocity (3-4) 10^7 cm/s are considered. Our theoretical research is based on the analytical solution of the problem of a thin shell implosion under radiative laser-produced plasma pressure.

2. Indirect shell target ablation.

The ablation equations for a spherical thin shell [1] are

$$\frac{dM}{dt} = -4\pi R^2 \beta \rho v_s ; \qquad M\frac{d\upsilon}{dt} = 4\pi R^2 (1+\beta^2) P \tag{1}$$

The initial conditions are: $R|_{t=o} = R_o$; $M|_{t=o} = M_o$; $\upsilon|_{t=o} = 0$, where M and R are the mass and the radius of the shell non-evaporated part; υ, the velocity in the direction of center; $M_o = 4\pi R_o^2 \Delta R \rho_o$, the initial mass of layer; ΔR, ρ_o, the thickness and density. Expressions for temperature T, density ρ, pressure P, and sound velocity v_s in the case of a gold target were obtained from the model of equilibrium radiative plasma [2] including the Rosseland mean free pass in form $L(cm) = 10^{-2}\, T(KeV)^{2.5}/\rho(g/cm^3)^{1.5}$.

$$T(KeV) = 0.24\, I_r^{0.36}\, t_r^{0.24} \left(\frac{A}{Z+1}\right)^{0.4}$$

$$\rho(g/cm^3) = 0.057\, I_r^{0.48}\, t^{-0.36} \left(\frac{A}{Z+1}\right)^{0.9}$$

$$P_a(Mbar) = 38\, I_r^{0.84}\, t^{-0.12} \left(\frac{A}{Z+1}\right)^{0.32}$$

$$v_s(cm/s) = 2.6\, 10^7\, I_r^{0.18}\, t^{0.12} \left(\frac{A}{Z+1}\right)^{-0.3}$$

Here and further X- ray intensity and pulse duration I_r and t_r are normalized by 10^{14} W/cm^2 and 10^{-9}s, respectively; A and z are the atomic number and ionization degree of plasma ions; $P_a = (1+\beta^2)P$ is the ablation pressure, β = 1.38 [3].

Solution of equations (1) for a gold target at the final moment of implosion t_c is:

$$u = \frac{v}{v_s} = 1.44 \left(\frac{\alpha}{\bar{\mu}}\right)^{1/2} \qquad (2)$$

$$\mu = \frac{M}{M_o} = 1 - 0.73\,(\alpha\,\bar{\mu})^{1/2} \qquad (3)$$

$$t_c = \left(\frac{\bar{\mu}}{\alpha}\right)^{1/2} \frac{R_o}{\bar{v}_s} \qquad (4)$$

Here $\bar{\mu} = 1+0.086\alpha - [0.086\,\alpha\,(2 + 0.086\,\alpha)]^{1/2}$, $0 \le \alpha \le 4.34$; $\alpha = \frac{\bar{\rho}}{\rho_o}\,\frac{R_o}{\Delta R}$ is the solution parameter; Δ is the shell thikness. Figure 1 shows μ and u as functions of α.

3. Implosion efficiency of an indirect driven target

From the formulas (2)-(4) hydrodynamic efficiency is

$$\eta_h = 0.46\,(\alpha\,\bar{\mu})\,\bar{\mu}^{-2}\left[1 - 0.73\,(\alpha\,\bar{\mu})^{1/2}\right] \qquad (5)$$

The dependence (5), $\eta_h(\alpha)$ has maximum η_{max}= 36% at the parameter α = 1.6 (Figure 2).

Consider conditions of effective indirect driven target implosion and compare the implosion efficiencies of direct (for Nd laser wavelengths) and indirect driven targets.

According to expressions (2) - (3) the final velocity and parameter α of a self-consistent ($t_c = t_r$) target are

$$v(cm/s) \simeq 0.42\, 10^7\, I^{0.46} \left[\frac{R_o / \Delta_o}{\rho_o}\right]^{0.5} \left[\frac{A}{Z+1}\right]^{0.15} \bar{\mu}^{-0.5} \qquad (6)$$

$$\alpha \simeq 1.4\, 10^{-2}\, I^{0.71} \left[\frac{R_o / \Delta_o}{\rho_o}\right]^{1.12} \left[\frac{A}{Z+1}\right]^{0.94} \qquad (7)$$

Fig. 1.

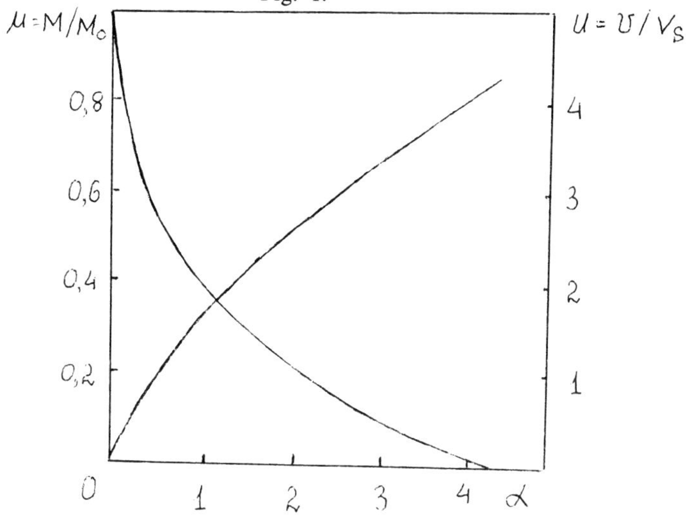

Fig. 2.

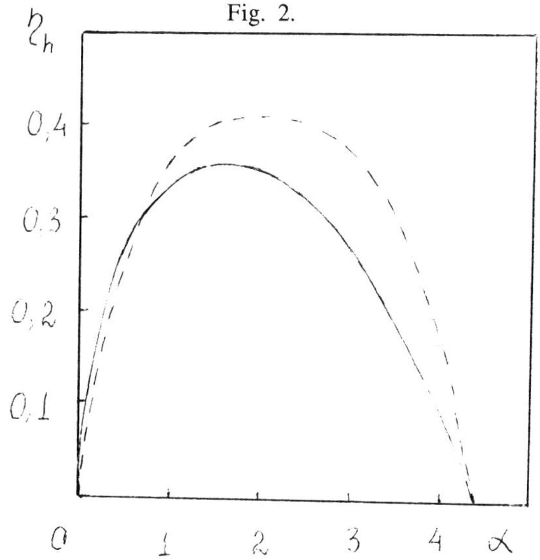

For direct driven implosion the main part of the laser energy is absorbed in the plasma resonance region, therefore, the average corona density is close to the critical density $\rho \simeq 1.7\ 10^{-3}\ (\frac{A}{Z+1})\ \lambda^{-2}$, g /cm^3, and from [1] we have

$$\upsilon(cm/s) \simeq 0.37\ 10^7 I^{0.33} \lambda^{-0.33} \left(\frac{A}{Z+1}\right)^{0.17} \left[\frac{R_o / \Delta_o}{\rho_o}\right]^{0.5} \bar{\mu}^{-0,5} \qquad (8)$$

$$\alpha \simeq 1.7\ 10^{-3}\ \lambda^{-2}\ \left(\frac{A}{Z+1}\right) \left[\frac{R_o / \Delta_o}{\rho_o}\right] \qquad (9)$$

Here $\bar{\mu} \simeq \left[\left(1+\frac{\alpha}{36}\right)^{0.5} - \frac{\alpha}{6}^{0.5} \right]^2$, $0 \leq \alpha \leq 4.5$.

Consider these dependencies for the maximum radiation intensities, corresponding to the direct and indirect driven implosion. Taking into account the physics of laser radiation conversion into X-rays the maximum X-rays intensity is 10^{14} W/cm^2. The maximum intensity of the laser radiation, when the parametric processes of the laser light interaction with plasma are not significant, is connected with the wavelength by the following relation: $I \simeq 10^{14}\ \lambda^{-2}$ W/cm^2.

For indirect targets the value [A / (Z + 1)] is much higher than that for direct ones. Indeed, heavy atoms of the indirect target corona are partially ionized. For example, for gold, at the discussed radiation intensities $Z \simeq 20 - 30$, and, thus, $[A / (Z + 1)] \simeq 7\text{-}9$. On the contrary, direct target corona is completely ionized plasma consisting of light elements, therefore, the relation [A / (Z + 1)] in this case is close to 2. Taking this into account, it is easy to see that for the same values of the parameter $[\ (R_o / \Delta_o) / \rho_o]$ the values of α are significantly higher for indirect targets than for the direct ones (approximately by a factor 30 at laser radiation wavelength $\lambda \simeq 1.06\ \mu m$, and 3 at $\lambda \simeq 0.35\ \mu m$). Such a difference between α values comes from the fact that the radiative corona density significantly exceeds that of the direct target. It is caused by the longer X-rays mean free pass as compared to the laser radiation absorption length. The final velocity of the direct target (for $\lambda = 0.35\ \mu m$)is twice higher than that of the indirect one. This fact due to the direct target corona has high temperature and high sound velocity according with the correlation of X-ray and laser radiation mean free pass mentioned above.

We can see from (6),(7) that the velocities (3-4) 10^7 cm/s of the indirect target can be achieved only at large values of a relative evaporated mass of the shell. The velocity 3 10^7 cm/s for the gold target is achieved at $\alpha \simeq 2.5$ and $1-\mu \simeq 0.83$, the velocity 4 10^7 cm/s, at $\alpha \simeq 3.5$ and $1-\mu \simeq \simeq 0.93$. Besides, the maximum velocity, corresponding to $\alpha = 4.34$ is 4.5 10^7cm/s.

Thus, the indirect way of driving has practically no store to exceed the minimum velocity required by the implosion efficiency. The indirect target can operate in a narrow interval of α-values only: $2,5 \le \alpha \le 4$; the upper limit here corresponds to the hydrodynamic efficiency $\eta_h \simeq 0.1$, the lower, to the velocity 3 10^7 cm/s. On the contrary the direct driving provides a greater possibility to achieve high the velocities of the target shell. Indeed, according formulars (8), (9), for $\lambda = 0.35$ μm, the velocity of 3 10^7cm/s is achieved at $\alpha \simeq 0.17$, and 4 10^7 cm/s, - at $\alpha \simeq 0.3$. The maximum velocity corresponding to the greatest possible $\alpha \simeq 4.5$, is 1.5 10^8 cm/s.

4. Conclusion

For the indirect driven target shell made of solid gold with $\rho_o = 19$ g/cm^3 the values of the aspect ratio are too high: 310 at $\alpha \simeq 2.5$ ($\upsilon \simeq 3$ 10^7 cm/s, $\eta_h \simeq 0.32$) and 420 at $\alpha \simeq 3.5$ ($\upsilon \simeq 4$ 10^7 cm/s, $\eta_h \simeq 0.18$), respectively. When using a combined ablator consisting of two layers: the thin external layer of heavy material with normal density and the internal one of light material - the high values of the aspect ratio can't be notably decreased, due to the requirement of a large relative evaporated mass. To decrease the aspect ratio it is suggested that one should use the gold-foam ablator with the initial density $\sim$ 2-3 g/cm^3. The aspect ratio for such an ablator is 20 at $\alpha = 3.5$, and 15 at $\alpha = 2.5$, respectively.

References

[1] Afanasiev Yu V and Gamaly E G and Gus'kov S Yu 1988 *Laser and particle Beams* 6 1-12

[2] Gus'kov S Yu et al 1993 J. *Sov. Laser Research* **14** 219-223

[3] Afanasiev Yu V and Gus'kov S Yu 1993 *Nuclear Fusion by Inertial Confinement* (Boca Raton: CRC Press) 99-118

Inst. Phys. Conf. Ser. No 140: Section 9
Paper presented at Laser Interact. with Matter Conf., Oxford, 19–23 September 1994

© 1995 IOP Publishing Ltd

ICF target with distributed absorption of laser energy

S Yu Gus'kov[1], V B Rozanov[1], N V Zmitrenko[2]
P.N. Lebedev Physics Institute of Russian Academy of Sciences, Leninsky prospect 53, 117924, Moscow, Russia(1)
Institute for Mathematical Modeling of Russian Academy of Sciences, Miusskaya square 4 125047, Moscow, Russia(2)

Abstract. New approach direct ICF-target concept and design of a new type of ICF-target, which is able to smooth of the nonuniformities of plasma heating under "volume" absorption of laser beams are proposed. The approach is based on the initiation of high-speed electron conductivity processes at the absorption of laser radiation by low-density absorber of light element foam. Numerical simulations show the possibility to reach the thermonuclear gain of 1 at the laser energy of about 200 kJ and the gain of 50 at the laser energy 1MJ for this target.

1. Introduction

In this paper the concept of a direct laser driven target with special absorber of the energy of the laser beams, which ensures smoothing of nonuniformities of energy deposition is proposed. The idea of the "smoothing" absorber is based on feasibility to form "pure" electron heat conductivity waves, that pass ahead hydrodynamic waves, during the total time of laser beams absorption in underdense plasma. Reasoning from the principle of absorber operation we called this target as the Laser Green House.

2. Physics of the Laser Green House Target

The general design of Laser Green House Target is shown on

Figure. It presents the two co-axial shells: an external shell with holes for the laser beams input into the target and an internal shell, which contains a thermonuclear fuel. The space between these shells (absorber) is filled by a low density foam material of light elements (for example, foam plastic) with density close to the critical density of laser-produced plasma. The shells include the layers of a heavy element material to absorb the thermal plasma radiation.

Consider the main physics processes in the absorber of Laser Green House Target.

Time for the creation of the absorber plasma under a "volume" evaporation of a foam matter with macrodensity ρ_*, which occupies the full volume of the absorber, and makes a spherical layer of thickness h, is [1]

$$t_e \simeq (\rho_* / \rho_0)^{1/2}\, h\, [2(\gamma-1)I / (3\gamma-1)\, \rho_*]^{-1/3}$$

here: ρ_0 is the density of the absorber matter in a normal state, for example, $\rho_0 \simeq 1$ g / cm^3 for plastic; γ, the adiabatic index; I, laser radiation intensity. The "volume" evaporation of the foam matter is faster than the "plane" evaporation of the same matter in a normal state and faster than the matter escaping through the introducing holes in factors $(\rho_0 / \rho_*)^{1/2} \gg 1$ and $(\rho_0 / \rho_*)^{2/3} \gg 1$, respectively. For the above noted densities and laser radiation intensity and wavelength $I \simeq 10^{15}$ W /cm^3 and $\lambda \simeq 0.35$ μm the plasma time formation is $t_e \simeq 0.1$ ns, much less than the laser pulse duration needed for ICF target (about a few ns).

The low value of the absorber thickness is limited by size of a region of inverse bremsstrahlung absorption of the laser radiation in undercritical plasma:

$$h_a \simeq 9{,}2\ 10^{-8}\ A^2\ T^{3/2} / z^3\ \lambda^2\ \rho_*^2\ ,\ \text{cm}$$

here: T is the absorber temperature in keV, A and z are atomic number and ionization degree of the absorber atoms, λ in μm. For plasma with critical density $\rho_* = \rho_c = 1{,}6\ 10^{-3}\ A / z\ \lambda^2$ g/cm^3 and for λ = (0.35 - 1.06) μm: $h_a \simeq$ (30 - 100) μm.

The criterion for smoothing of the absorber temperature

Figure
The general design of the Laser Green House Target
I - DT-gas, II - DT-ice, III - high-Z shield,
IV - ablator, V - foam absorber, VI - low-Z,
protective layer, VII - shell.

distribution nonuniformities presents the condition of "pure" heat waves propagation a half of the distances between the absorption region of the neighboring laser beams during the time less than the laser pulse duration. This distance is $\Delta \simeq (4 / k)^{1/2} (1 - \sigma^{1/2}) R$, where $\sigma = kR_L^2/4R^2$ is a relative square of the introducing holes; k, a number of the holes, R and R_L, radii of the target and the hole, respectively.

The "pure" electron conductivity wave is an initial stage of a heat wave of the duration is approximately equal to the time of the electron-ion energy exchange:

$$t_s \simeq 3\ 10^{-12} \left(\frac{A}{Z}\right)^2 \frac{T^{3/2}}{\rho_*},\ s \qquad (1)$$

where: T in keV, ρ_* in g/cm^3.

The self-similar solution for electron conductivity wave give for the temperature smoothing time and temperature:

$$t_\varkappa \simeq 7.4\,10^3 \left(\frac{z\ \rho_*}{(\gamma-1)A}\right) \frac{(1-\sigma^{1/2})^{19/7}(z+4)^{2/7}\ R^{9/7}}{k^{9/14} I^{5/7} \sigma^{5/7}},\ s \qquad (2)$$

$$T_\varkappa \simeq \left[\frac{7}{19}\left(\frac{3}{2\pi}\right)^{2/3} \frac{(I\ S_L)^{2/3}(\ B\ \rho_*)^{1/3}}{\varkappa_0}\right]^{6/19} t^{-2/19},\ keV \quad (3)$$

and the "smoothing" criterion may be written as follows:

$$R \le 3\,10^{-2} \left[\frac{(\gamma - 1)\ A}{z\ \rho_*}\right]^{7/23} \frac{E_L^{7/23}\ k^{9/46}}{(z + 4)^{2/23}(1 - \sigma^{1/2})^{19/23}\ (I\ \sigma)^{2/23}},\ cm$$

In these expressions: E_L is the energy of all the laser beams in J; I in W/cm^3; R in cm.

For $\rho_* \simeq (3\ 10^{-2} - 3\ 10^{-3})\ g / cm^3$, $I \simeq 10^{15}\ W/cm^3$, $\lambda \simeq 0.35$ μm, $E_L = 10^5$ J, k = 6 and $\sigma = 0.1$ the "smoothing" time $t_\varkappa$, according formular (2) is about 0.3 ns and this is much less than the needed laser pulse duration of 3 ns. On other hand, the absorber temperature, according formular (3), is about (1-3) kev and from formular (1) duration of the "pure" heat wave t_s is (0.5 - 8) ns, and this exceeds the "smoothing" time and close to the laser pulse duration.

The "smoothing" criterion gives the upper limit for the radius of the Laser Green House Target. For the above noted laser and absorber parameters this value is 0.18 cm. Thus, there are wide possibilities to optimize the Laser Green House Target parameters, since the radius of a thermonuclear target lies in the interval of (0.07 - 0.1) cm.

This analysis shows the effective smoothing of the nonuniformities of the laser energy deposition in Laser Green House Target by the waves of "pure" electron heat conductivity before hydrodynamic perturbations arise.

Now we are working at 1D and 2D numerical simulations aimed at detailed study into theLaser Green House Target physics. 1D code "DIANA" [2] and 2D code "ATLANT" [3] have been used. 1D simulations of the cryogenic targets with the radius from 0.1 up to 0.2 cm and the absorber thickness of from 0.03 up to 0.08 cm at the laser energy of (0.1 - 1) MJ for λ = (1.06 - 0.35) μm and the pulse duration (2 - 10) ns give the thermonuclear gain of about (2-4) for the laser energy of 0.2 - 0.4 MJ and one of 50 - for the laser energy 1 MJ. The preliminary 2D simulations show theeffective operation of the "smoothing" absorber.

3. Conclusion

Presented theoretical results reveals that the Laser Green House Target is able to provide the uniform distribution of the absorbed laser energy into a target and has the thermonuclear gain. Therefore we believe that the detail studies of the physics of this target and, especially, the physics of the "smoothing" absorber operation are very interesting and important.

References

[1] Afanasiev Yu et al 1988 *Laser and Partical Beams* 6, 1-12

[2] Samarskii A A et al 1983, *Problems of Nuclear Science and Engineering* 2 38-42 (in Russian)

[3] Gamaly E G et al 1988 *Quantum Electronics* 15 1622 - 1632 (in Russian)

Inst. Phys. Conf. Ser. No 140: Section 9
Paper presented at Laser Interact. with Matter Conf., Oxford, 19–23 September 1994

© *1995 IOP Publishing Ltd*

PULSE SHAPING INFLUENCE ON THE TARGETS NEAR BREAK-EVEN POINT

S Yu Gus'kov *, V B Rozanov *, N V Zmitrenko

Inst.Math.Mod. Russian Ac.Sci. 4 Miusskaya Sq. Moscow 125047 RUSSIA
* P.N.Lebedev Phys. Inst. Russian Ac.Sci. 53 Leninski Pr. Moscow 117924 RUSSIA

Abstract. The possibility of pulse shaping for some improvement on laser targets characteristics are discussed. It is shown, that the targets with low ablator aspect ratio (less, than 25) can be ignited due to a moderate pulse shaping. Because the low aspect ratio targets will not be so unstable as high aspect ratio ones, it means, that they can be used in a direct drive break-even experiments.

1. Introduction

It was well known formerly, that a spherical compression of a high aspect ratio thin shell leads to the break-even conditions under absorbed laser energy about 100 kJ [1]. Such kind of the compression of an one-shell cryogenic laser target is associated with a convergence ratio $C_R \geq 20$. But it is found now, that this way requires very high level of the symmetry of a laser irradiation.

For this reason, there were suggested some different ways to improve the symmetry of the irradiation. In particular, it is possible to obtain a sufficiently symmetric action onto a target if one can transform a laser light into a thermal radiation. But this way requires, in general case, more laser energy for an ignition, than a direct way. So, we propose to consider a possibility to use low aspect ratio targets under a direct laser irradiation. An usage of such kind of targets with low aspect ratio allows to reduce C_R and, therefore, to restrict a destroying action of hydrodynamics instabilities. But, at this situation we have to provide a sufficient fuel temperature by any ways. The results of numerical simulations, which are presented in this paper, show, that ignition conditions can be reached by the means of a simple shape laser pulse with a very moderate tailoring of a power.

2. Break-even requrements

The simple direct drive laser target is a micro sphere, which consists of two layers (plastic ablator and liquid DT). It has a central cavity filled by DT-gas under a pressure about a few atmospheres. This target (close to optimal one) for absorbed energy about 100 kJ has the following parameters [1]: initial radius $R_0 \cong 1000\,\mu m$, ablator aspect ratio $A \cong 70$, fuel mass $M_{DT} \cong 20\mu g$. The calculated gain is about 1 - 3. For this target a linear increase of laser power during a period of laser action $\tau = 4ns$ is used and a laser radiation wavelength $\lambda = 1\,\mu m$.

Let us fix this target structure, absorbed energy $E_a = 100$ kJ and the fuel and ablator masses and shall change target radius. In calculations with the different pulse shapes and durations we shall obtain the different values of C_R and gain G. At the same time the profiles of fuel temperature and density are similar, though their average values T and ρ may be quite different.

We can note, that to reach a value $G \geq 1$ we must obtain quite determined values of a fuel area density ρR and a temperature T at the stagnation stage before a burn developes. So, they may be $\rho R \geq 0.3\,g/cm^2$ and $T \geq 3\,keV$ or $\rho R \geq 0.2 g/cm^2$ and $T \geq 5\,keV$ (for target from [1], which was described above). The region of $0.2 < \rho R < 0.3\,g/cm^2$ and $3 < T < 5\,keV$ is a desirable one for gain about 1 - 10. This range of a temperature corresponds the energy of DT in the range $7 < E_{DT} < 12\,kJ$ (for $M_{DT} = 20\,\mu g$).

We can predict the ρR value for different values of C_R and R_0 as $\rho R = \zeta \frac{3M_{DT}}{4\pi} \frac{C_R^2}{R_0^2} \sim \rho^{2/3}$, where a coefficient ζ takes into account density distribution at the stagnation stage. In our calculations $\zeta \simeq 0.4$. Because at the adiabatic process $T \sim \rho^{2/3}$ we obtain for temperature $T \sim \Sigma \frac{C_R^2}{R_0^2}$, where $\Sigma \sim e^S$, S is an entropy of a fuel.

It is seen, that we can reach the same value of ρR, if we reduce the R_0 and value of C_R is reduced too. If we can conserve the fuel entropy, we obtain the same value of T. We can control the entropy level, changing the pulse shape.

If the initial thickness of DT-layer Δ_{DT} sufficiently less than R_0 we can write, that aspect ratio of DT-layer $A_{DT} = \frac{R_0}{\Delta_{DT}} \simeq 4\pi \frac{\rho_0}{M_{DT}} R_0^3$, where ρ_0 is an initial density of a fuel ($\rho_0 \cong 0.2\,g/ccm$). For big values of A it is approximately true, that $A_{DT} \cong \frac{M_a \rho_0}{M_{DT}\rho_{a0}} A$, where M_a is an ablator mass and ρ_{a0} is an ablator initial density. So, if we reduce R_0 the A_{DT} and A decreases as $\sim R_0^3$ and Δ_{DT} increases as R_0^{-2}. For this reason the target with smaller R_0 is more stable, than with greater R_0 ones.

3. Results of computer simulations

We use 1-D Lagrangian code DIANA [2]. It takes into account the main physical processes, which occur in high temperature dense plasma. The radiative losses are included in a form of a bulk emission. Before this code was used to obtain the gain prediction for different kinds of the direct drive targets [3,4].

When we had taken the same fuel mass and reduce R to $700\,\mu m$ and aspect ratio to 25 we have obtained in our calculations with the same laser pulse shape the gains less than one (see line 1 at fig.1). So, we have considered another pulse shape for this low aspect kind of target, so-called two-stage pulse. In this case pulse duration τ is divided into two stages. Both of them characterize by the linear increase of pulse power, but most part of the energy is put into a target during the last short stage. A total pulse duration is τ $(0 \leq t \leq \tau)$, duration of the first stage is τ_1 $(0 \leq t \leq \tau_1)$, and the duration of the last stage is τ_2 $(\tau_1 \leq t \leq \tau)$, so $\tau_1 + \tau_2 = \tau$. The energy deposition during the first stage is E_1 and at the last stage this value is E_2 , so that a total absorbed energy is $E_1 + E_2 = E = 100$ kJ.

When the duration of the last stage is $\tau_2 = 2\,ns$, the results of simulation of such kind of targets are shown at fig.1 (see line 2). In this case an energy deposition of the last stage E_2 decreases with τ from 90 to 80 kJ.

Another example concerns the target with $A = 10$. The numerical results at this case are presented at fig.2. Line 1 corresponds to linear ramped laser pulse and line 2 to two-stage one with $\tau_2 = 1\,ns$. For this line 2 an energy deposition of the last stage reduces with τ from 97.5 to 85 kJ. Value of C_R for this line 2 increases from 10 up to 15 with τ.

4. Conclusion

We can see, that such kind of the pulse shaping essentially improves the processes of compression and burn. We can also to investigate a sensitivity of the results upon laser pulse shape. For line 2 at fig.1 the optimal gain about 10 corresponds to $\tau = 5$ ns ($\tau_2 = 2$ ns, $E_2 = 85$ kJ). If we consider the slight variations of pulse parameters (for example, change pulse duration τ by ± 0.4 ns and energy E by ± 10 kJ) we obtain a reduction of gain on the value about $20 - 40\%$.

The obtained results are correlated with data [5]. Presented gain values correspond to results of moderately shaped pulse in [5]. We also have obtained the gain $G \simeq 20$ for target with $A = 70$, described above, if we had used tailored laser pulse such as $P = P_0\left(\frac{1}{(t-t_0)^2} - \frac{1}{t_0^2}\right)$, $0 \leq t \leq \tau < t_0, t_0 = 1.125\,\tau, \tau = 8$ ns, $E_a = 100$ kJ. This result corresponds to optimized pulse shape from [5].

So, we can conclude, that the pulse shaping is effective for the targets near break-even point and can reduce the dangers, connected with hydrodynamical instabilities. For this reason it may be possible to consider some laser direct drive targets for ignition experiments.

References

[1] Afanasyev Yu V *et al* 1982 *P.N.Lebedev Phys.Inst.Proc.* **134** (Moscow: Nauka) (in Russian) 167-176

[2] Samarskii A A *et al* 1983 *Problems of Nuclear Science & Engineering* (in Russian) **2(13)** 38-42

[3] Basov N G *et al* 1985 *Quant. Electronics* (in Russian) **12** 1289-92

[4] Basov N G *et al* 1988 *Pisma v ZhETF* (in Russian) **48** 245-247

[5] Takabe H *et al* 1991 *ILE Report* **32** 3-12

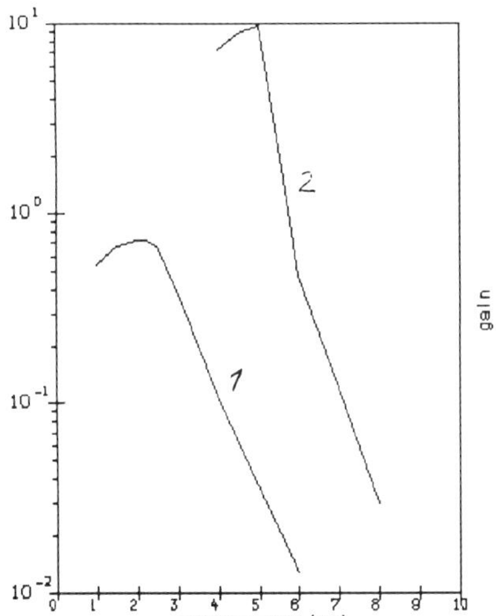

fig.1

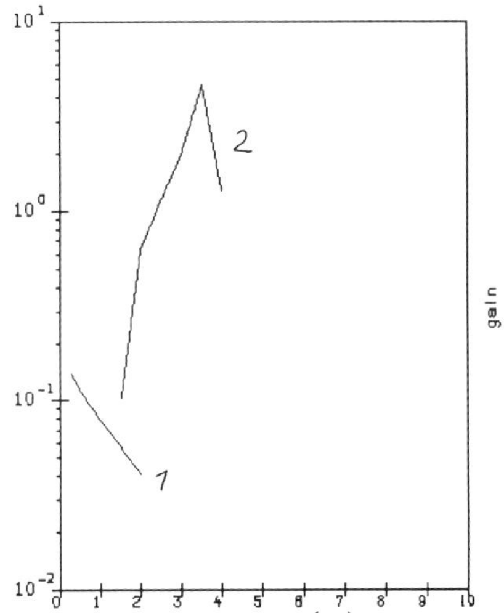

fig.2

Inst. Phys. Conf. Ser. No 140: Section 9
Paper presented at Laser Interact. with Matter Conf., Oxford, 19–23 September 1994
© 1995 IOP Publishing Ltd

The Use of Satellite Line Ratios in the Diagnosis of Plasmas created by the Interaction 12 ps KrF Laser Pulses with Solid Targets

by
T. A. Hall(1), A. Badger(1), A. Djaoui(2), R. Evans(1) and F. B. Rosmej(3)

(1) Physics Department, University of Essex, Colchester, Essex, CO4 3SQ, UK
(2) Rutherford Appleton Laboratory, Didcot, Oxon. OX11 0QX , UK
(3) Institute of Spectroscopy, Russian Academy of Sciences, 142092 Troitzk, Moscow district, Russia.

Abstract

Time resolved and time integrated hydrogen like and helium like emission from the interaction of 12 ps KrF laser pulses with thin aluminium layer targets are presented. The satellites of the resonance lines are used to infer the plasma conditions during the interaction. Modelling of the results has been carried out using the hydrodynamic code MEDUSA and time dependent non-local thermodynamic equilibrium (non-LTE) code to model the satellite intensities . A separate resonance line and satellite line code has been used to predict the detailed shape of the lines. The code results compare favourably with experiment.

1. Introduction

In previous experiments, time integrated emission of Li-like and Be-like satellites in a very thin Al plasma created by a 100 fs full width at half maximum (FWHM) laser has been reported (Mancini 1994, Audebert 1994). In these publications, time dependent calculations of the ionisation distribution and excited state populations were used as a post processor to hydrodynamic simulations in order to calculate synthetic spectra. Agreement with the recorded time integrated spectra was used as the basis for deducing the duration of the emission of the satellites. In another experiment, time resolved spectroscopy was used to measure the density and temperature evolution in the conduction zone of a thick Al plasma created using a 12 ps laser pulse (Riley 1992). From Stark broadened H-like (1s-3p) and He-like ($1s^2$-1s3p) transitions, electron densities of greater than 10^{23} cm^{-3} and temperatures of about 600 eV were inferred near the time of the peak of the laser pulse.

In this work we investigate the diagnostic potential of Li-like and He-like satellites and present results for a thin Al layer (2000 Å) which is irradiated with a 12 ps KrF laser at intensities up to 10^{16} W cm^{-2}. Both time integrated and time resolved spectra are analysed. We show that the time resolved He-like satellites, He-like intercombination line, and Li-like satellites can be used as density and temperature diagnostics during the early stages of the plasma evolution. The rapid changes of the plasma parameters are directly mirrored in the emitted satellite intensities. Time-dependent calculations of line ratios which post-process the hydrodynamic simulation of the laser target interaction are in overall good agreement with the experimental measurements.

2. Experimental set-up

Experiments in which a KrF laser interacted with solid Al targets were carried at the Rutherford Appleton Laboratory. 3J of laser energy was focused using an f/10 calcium fluoride lens on to 2000 Å coatings of Al deposited on a glass microscope slide. The laser pulses were 12 ps FWHM and were Raman shifted in methane to 268 nm wavelength. The ASE prepulse to main pulse power ratio was below 10^{-10} and consequently no plasma was formed before the arrival of the main pulse. By varying the focal spot size from 10 μm to 500 μm, incident laser intensities of from 10^{14} W cm^{-2} to greater than 10^{16} W cm^{-2} were obtained. The focal spot area was imaged using an x-ray pinhole camera. The time integrated spectra from approximately 6 Å to 8 Å were recorded by dispersing the x-rays with an ADP crystal onto a Andor X-ray CCD camera. The time resolved spectrum was recorded using a streaked x-ray spectrograph which consisted of a TlAP crystal dispersing the incoming x-rays onto the entrance slit of a fast x-ray streak camera fitted with a KBr photocathode. The streak rate of the camera was ~ 5 ps/mm and combined with an input slit width of 1 mm gave a temporal resolution of around 5 ps. The spectral resolution of all the spectrometers used was dominated by the source size, which gave $\lambda/\Delta\lambda$~300.

3. Time integrated spectra

A best fit to the shape of the time integrated spectrum near the He-like resonance line W was obtained with the use of the SAT simulation code (Rosmej 1994). This is a full collisional radiative steady-state model which includes in addition to the H-like, He-like and Li-like ground and singly excited states, all 16 doubly excited levels 1s2l2l' in Li-like ions, 20 transitions for $1s^23l$-1s2l'3l'' and 20 transitions for $1s^24l$-1s2l'4l'' Li-like satellites. Figure 1 shows the best fit to a single shot experimental spectrum corresponding to an intensity of approximately 5x 10^{15} W cm^{-2} in the wavelength interval covering the He-like resonance line W, intercombination line Y and the Li-like satellites. The theoretical spectrum was calculated for N_e = 6x 10^{20} cm^{-3} T_e = 260 eV, T_z(He,Li) = 180 eV and convolved with an apparatus function $\lambda/\delta\lambda$ = 400. The different contribution from the W, Y lines the dielectronic capture satellites, inner shell excited satellites are shown.

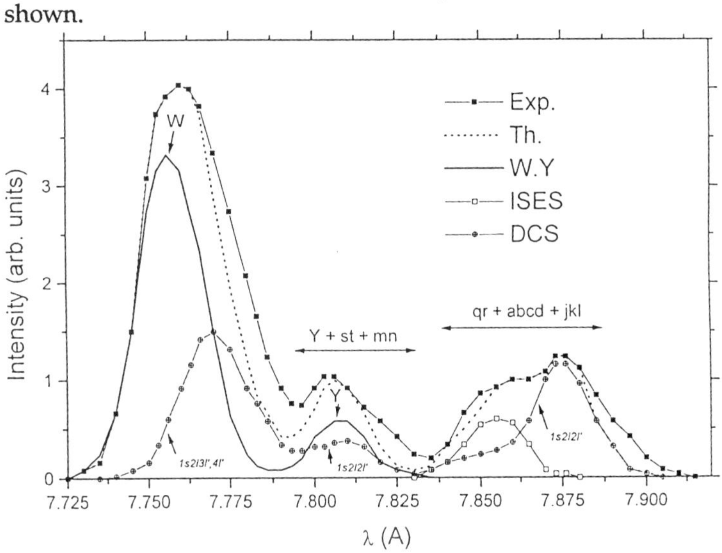

Figure 1. Time integrated experimental spectrum (squares) and best theoretical fit (dotted) curves as calculated for N_e = 6x 10^{20} cm^{-3}, T_e = 260 eV and T_z(He,Li) = 180 eV. The contributions to the theoretical spectrum from the W, Y, ISES and DCS are shown. The contribution to the DCS part from 1s2l2l' and higher order 1s2l3l',4l' are also shown. The Li-like satellite feature under the Y line corresponds to the st and mn component in Gabriel's notation, while the wavelength interval [7.84Å - 7.89Å] correspond to the qr, abcd, jkl components.

There is still some unexplained discrepancy between the theory and experiment on the red wing of the lines (practically all lines show this feature). One reason can be connected with plasma motion but this requires the plasma to flow away from the detector in order for the lines to be red shifted. The Li-like satellites in the interval 7.81Å, 7.8Å have a strong dependence on the ionisation temperature and the assumption of a transient, ionising plasma results in an enhancement of the blue part of the satellites in better agreement with the experiment. The are however other mechanisms such as collisional mixing between the doubly excited levels 1s2l2l' at higher densities, which also have the effect of enhancing the blue wing of the Li-like satellites. The transient nature of the plasma can not therefore be established firmly on the sole analysis of the time integrated spectrum.

4. Time resolved diagnostic based on Li-like and He-like satellites

Experimental resolution limitations and the proximity of many satellites allows us to distinguish two separate features on the long wavelength side of the He-like resonance line W; the Li-like satellites (qr, abcd and jkl) and the blending of He-like intercombination line Y and st, mn Li-like satellites. A third feature of interest is the He-like satellite (excluding 1s2s $^1S_0$ - 2s2p $^1P_0$ and 1s2p $^1P_1$ - 2p^2 $^1S_0$) which occurs on the long wavelength side of the H-like resonance line.

We investigated the possibility to use the wavelength integrated relative intensities of these satellites as a density and temperature diagnostic using the RATION code (Lee 1984). This code is similar to SAT and although with less energy level details, is adequate for this study where the behaviour of individual satellites is not required. The calculated ratio of the Li-like satellites (qr, abcd, jkl) to the He-like satellites has a strong dependence on temperature with a weaker dependence on electron density. The ratio of the Li-like satellites (qr, abcd, jkl) to intercombination line Y plus the st, mn satellites has a strong dependence on density a weaker dependence on temperature. Each one of these ratios is a poor diagnostic when used on its own. However, the two ratios used simultaneously can provide a more reliable diagnostic.

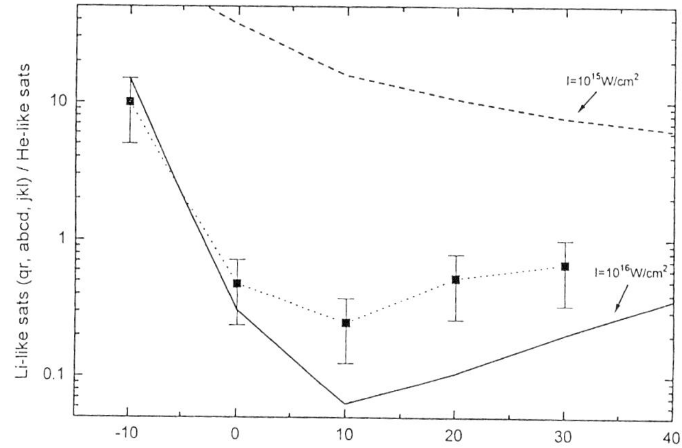

Figure 2. Ratio of the wavelength integrated intensity of the Li-like satellites (qr, abcd, jkl) to the He-like satellites (excluding 1s2s$^1S_0$-2s2p$^1P_1$ and 1s2p$^1P_1$-2p^2 $^1S_0$) as a function of time with respect to the peak of the laser beam as measured experimentally (squares) and as calculated with an incident intensity of 10^{15} W cm^{-2} (full curve) and 10^{16} W cm^{-2} (dashed curve). The experimental error bars are also indicated.

A parameter determination based on these ratios is possible but would still be unable to provide information on the transient nature and spatial gradients of the plasma.

The experiment was simulated with a one dimensional Lagrangian hydrodynamic code which includes time-dependent, non-LTE, average atom model ionisation physics (Djaoui and Rose, 1992). Simulations reveal that the region of maximum Li-like and He-like ion density is characterised by large temperature and density gradients. The plasma evolves very rapidly as a result of heat conduction and hydrodynamic expansion. This rapid change illustrates the need for a calculation of the satellite intensities which takes into account not just the time-dependent character of the problem but also the spatial gradients at a given time. Therefore, we have taken the temperature and density history of a number of computational cells uniformly distributed across the original Al layer from the hydrodynamic simulation as input to FLY, the time-dependent version of the RATION code. FLY then calculates the intensities at each time step and for each cell for transitions of interest. Since the optical depth of the satellites is small, the volume integrated emission is obtained by summing the contribution from each cell. This simplification provides a major advantage compared to the use of resonance lines where optical depth effects on level populations and intensities make the calculation much more complicated (Wark 1994, Djaoui 1994). The ratio of the Li-like satellites to the He-like satellites is shown in figure 2. This ratio is very sensitive to the laser intensity used in the hydrodynamic simulation. The experimental data is bracketed by the two calculations at 10^{15} W cm^{-2} and 10^{16} W cm^{-2}. The agreement is good considering the difficulty in measuring the experimental intensity on target accurately.

5. Conclusions

Our results show that parameter determination in a highly transient plasma using detailed time integrated spectra simulations, can only provide limited information on the plasma state during the emission of the lines under consideration. The inclusion of transient effects through an ionisation temperature different from the electron temperature are suggestive but not conclusive since the same behaviour can also be caused by high density rather than transient effects. The combination of time resolved measurements with detailed hydrodynamic simulation and time dependent calculations of ionisation and excitation are used as the basis for understanding transient plasma behaviour. The calculations are much simpler in the case of satellites since they have small optical depths and experiment and simulation are in overall agreement for incident laser intensities of between 10^{15} W cm^{2} and 10^{16} W cm^{-2}.

References

Audebert P, Geindre J P, Rouse A, Fallies F, Gauthier J C, Mysyrowicz A, Grillon G and Antonetti A 1994 J. Phys. B: At. Mol. Opt. Phys. 27, 3303

Djaoui A and Rose S J 1992 J. Phys. B: At. Mol. Opt. Phys. 25, 2745

Djaoui A, Rose S J and Wark J S 1994 J. Quant. Spectrosc. Radiat. Transfer. 52, in print

Gabriel A H and Paget T M 1972, J. Phys. B: Atom. Molec. Phys., 5, 673

Lee R W, Whitten B L and Stout R E 1984, J. Quant. Spectrosc. Radiat. Transfer. 32, 91

Mancini R C, Audebert P, Geindre J P, Rousse A, Fallies F, Gauthier J C, Mysyrowicz A, Chambaret J P and Antonetti A 1994 J. Phys. B: At. Mol. Opt. Phys., 27, 1671

Rickard G J, Bell A R and Epperlein E M 1989, Phys. Rev. Lett. 62, 2687

Riley D, Gizzi L A, Khattak F Y, Mackinnon A J, Viana S M and Willi O 1992, Phys. Rev. Let. 69, 3739

Rosmej F B 1994 J. Quant. Spectrosc. Radiat. Transfer., 51, 319

Wark J S, Djaoui A, Rose S J, He H, Renner O, Missalla T, Foerster E 1994, Phys. Rev. Lett., 72, 1826

Inst. Phys. Conf. Ser. No 140: Section 9
Paper presented at Laser Interact. with Matter Conf., Oxford, 19–23 September 1994

© *1995 IOP Publishing Ltd*

Mode coupling of the Rayleigh-Taylor unstable surface waves in laser ablation

Susumu Hasegawa , Katsunobu Nishihara

Institute of Laser Engineering , Osaka University , Suita , Osaka 565 , Japan

Abstract . Weakly nonlinear stage of the ablative Rayleigh-Taylor instability has been investigated by perturbation theory . Mode coupling of linear growing waves with wavenumbers k_A and k_B drives new excited waves with wavenumbers $k_A + k_B$, $k_A - k_B$, $2\,k_A$, $2\,k_B$. We have investigated time evolution of the excited waves and found that the ablation effect plays an important role even in nonlinear stage to reduce amplitude of the excited waves . Differences between an ablation surface and a classical contact surface have been clarified .

1 . Introduction

The Rayleigh-Taylor (RT) instability is one of the principal problems inhibiting an achievement of the inertial confinement fusion. The presence of the ablation in the inertial confinement fusion context introduces important modifications to the RT instability evolution that can not be accounted for in the classical RT theory . The principal differences between the ablation surface and the classical contact surface are the presence of mass and heat flows across the surface. These effects can lead to a reduction in a growth rate and even to stabilization at large wave numbers . Takabe et al . [1] found that the linear growth rate $\gamma(k)$ is expressed $\gamma(k)=0.9(kg)^{1/2}-\beta k v_a$ (β=3–4) where v_a is an ablation velocity . If an initial perturbation amplitude is much smaller than its wavelength , the perturbation grows exponentially in time . As the amplitude of the RT instability becomes larger , nonlinear effects become important .

Mode coupling of the RT unstable surface waves has been investigated for the classical contact surface with incompressible fluids [2] . In this paper , we study weakly nonlinear stage of the ablative RT instability by perturbation theory (the 2nd order theory) . As will be shown below , the ablation effects are important even in the weakly nonlinear stage . We assume that there are two linear growing waves with wavenumbers k_A and k_B at the ablation surface , so the mode coupling of the two waves excites new waves with the wavenumber k_0 ($\equiv k_A \pm k_B$, $2\,k_A$, $2\,k_B$) . We investigate the evolution of the excited wave with the wavenumber k_0 . In Sec.II we apply perturbation theory to the basic equations and derive a set of equations up to the 2nd order , and in Sec.III , we investigate the evolution of the excited wave with the wavenumber k_0 and clarify the difference between an ablative RT instability and a classical RT one .

2 . Basic equations

We adopt one-fluid one temperature equations as basic equations [1] . They are assumed to be described in a frame moving with an ablation front , and the equation of motion includes an inertia force $\mathbf{g}$. They are given in the normalized form

$$\frac{\partial}{\partial t}\rho + \nabla\cdot(\rho\mathbf{v}) = 0 \quad , \qquad \rho\,(\frac{\partial}{\partial t}\mathbf{v} + \mathbf{v}\cdot\nabla\mathbf{v}) = -\nabla\rho T + \rho\,\mathbf{g} \quad ,$$

$$\frac{3}{2}(\frac{\partial}{\partial t}T + \mathbf{v}\cdot\nabla T) = -\rho T\nabla\cdot\mathbf{v} + \nabla\cdot(K_0 T^{5/2}\nabla T) \quad ,$$

where ρ,$\mathbf{v}$,T are the density , velocity , and temperature of the ablating plasmas , respectively . They are normalized by the corresponding values at the isothermal sonic point of the stationary state , and we normalize the length by the electron mean free path ℓ_s at the sonic point . We expand the variables up to the 2nd order

$$\mathbf{U} = \mathbf{U_0} + \varepsilon\mathbf{U_1} + \varepsilon^2\mathbf{U_2} + \cdots \quad ,$$

where the column vectors $\mathbf{U_i}$ (i =0,1,2) consist of the i th order variables and are defined as follows

$$\mathbf{U_i} = {}^t(\,\rho_i\,,\,v_{xi}\,,\,v_{yi}\,,\,T_i\,) \quad (\,i=0,1,2\,)\,.$$

Subscript 0 denotes an steady ablation profile , 1 does linear growing mode one , 2 does new excited one . By submitting them into the basic equations , we get the set of equations expressed vector and matrix form as follows

$$\text{0 th} \qquad \mathbf{A}\,\mathbf{U}_0 = \mathbf{0} \quad ,$$

$$\text{1 st} \qquad \frac{\partial}{\partial t}\mathbf{U}_1 + \mathbf{A}\,\mathbf{U}_1 = \mathbf{0} \quad ,$$

$$\text{2 nd} \qquad \frac{\partial}{\partial t}\mathbf{U}_2 + \mathbf{A}\,\mathbf{U}_2 = \mathbf{S}_2 \quad ,$$

where A is the operator matrix . The nonlinear source term $\mathbf{S_2}$ consists of the product of the 1st order variables and excites the new wave with the wavenumber k_0 .

3 . Time evolution of waves excited by mode coupling

We consider a case that there are two linear growing waves whose wavenumbers and linear growth rates are k_A and k_B , γ_A and γ_B , respectively . Mode coupling of the linear growing waves excites new waves whose wavenumbers are $k_A \pm k_B$, 2 k_A , 2 k_B . We solve the initial value problem of the 2nd order equation . Figure 1 shows the time evolution of the 2nd order velocity perturbation (v_{x2}) profile ($k_0 = k_A - k_B = 50$) which is zero initially and excited by the mode coupling of the waves with the wavenumbers $k_A = 160$ and $k_B = 110$. In Fig . 1 , we used the normalized ablation velocity v_a=0.01 and the gravity force normalized by c_s^2/ℓ_s , g=1 , where c_s is the sound speed at 1st sonic point . The position of the ablation front is 0.40625 and that of the 1st sonic one is 0 . As clearly seen in Fig . 1 , in the beginning the 2nd order velocity perturbation is convected towards both the 1st and the 2nd sonic points with the convective velocities ($v_{x0} \pm c_s$) . But later the 2nd order velocity perturbation grows exponentially at the adjacent of the ablation front . The observed convection of the excited perturbation for the ablative RT instability does not exist in the classical surface . In the ablative RT instability , the unstable wave consist of three different modes , i.e. , sound mode , thermal conduction mode and surface mode (entropy mode) , and the velocity perturbation excited by the mode coupling is not an eigen mode of the stationary state . The curl of the 2nd order velocity perturbations is not equal to zero , and the perturbation is shed as vortex .

The solid line in Fig . 2 shows the maximum 2nd order velocity perturbation at the adjacency of the ablation front which is initially equal to zero . In the beginning the 2nd order velocity perturbation spends time in growing exponentially as discussed above . But later the 2nd order velocity perturbation grows exponentially with the growth rate of

$\gamma_A+\gamma_B$. The dotted line in Fig . 2 shows the 2nd order velocity perturbation which initially has a finite value . In the beginning the 2nd order velocity perturbation grows exponentially with the growth rate γ_0 corresponding to the wavenumber k_0 , but later it does with the growth rate $\gamma_A+\gamma_B$. Thus later stage of the RT instability , the amplitude is determined by the mode coupling .

Figure 3 shows the various physical variables at the adjacency of the ablation front when the normalized time (that is $(\gamma_A+\gamma_B)$ x TIME) is equal to 8.6 . The solid line shows the 2nd order velocity , the broken one does the 2nd order mass density , and the dotted one does the 2nd order mass flux . In the ablative RT instability the fluid velocity perturbation is not equal to the time derivative of the surface displacement because of the ablation . The mass flux ($J = \rho$ v) perturbation divided by the stationary mass density gives the time derivative of the local displacement . The absolute value of the 2nd order mass flux has a maximum at the ablation surface .

The ablation effect on the weakly nonlinear stage has been investigated by choosing the ablation velocities $v_a = 0.01$ and 0.02 with the gravity forces g=0.5 and 1 , and the various wavenumbers k_A and k_B that gives approximately maximum linear growth rates . The maximum 2nd order mass flux divided by the product of the 1st order mass fluxes , $J_2 / J_{1,K_A} J_{1,K_B}$, are plotted as functions of the wavenumber k_0=k_A-k_B normalized by the practical target thickness ΔR . The practical target thickness is defined by the distance of the points where the stationary density has 1/e of its peak value . As shown in Fig . 4 , a larger ablation velocity leads to a smaller 2nd order mass flux . It should be also noted that the 2nd order mass flux increases with the increase of the wavenumber k_0 .

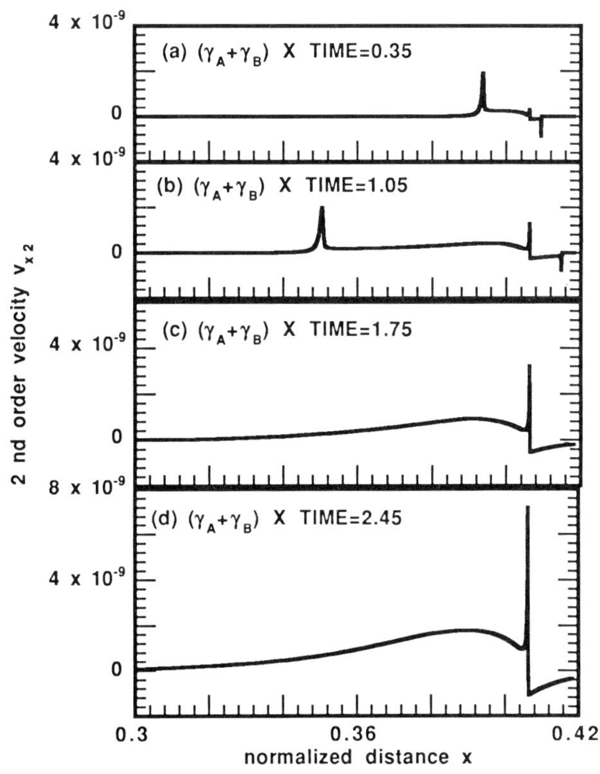

FIG.1 Time evolution of the 2nd order velocity profile

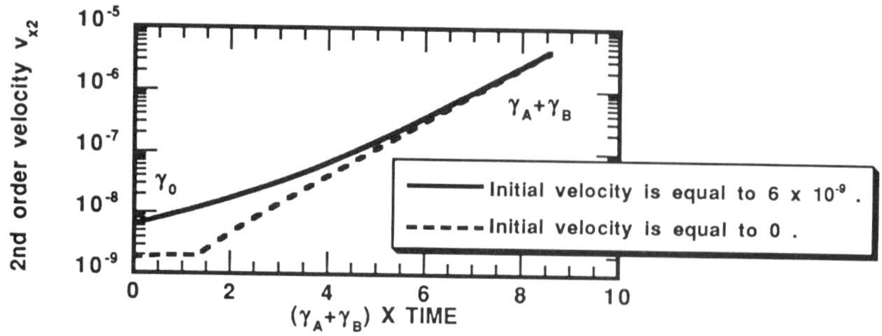

FIG.2 Time evolution of the 2nd order velocity .

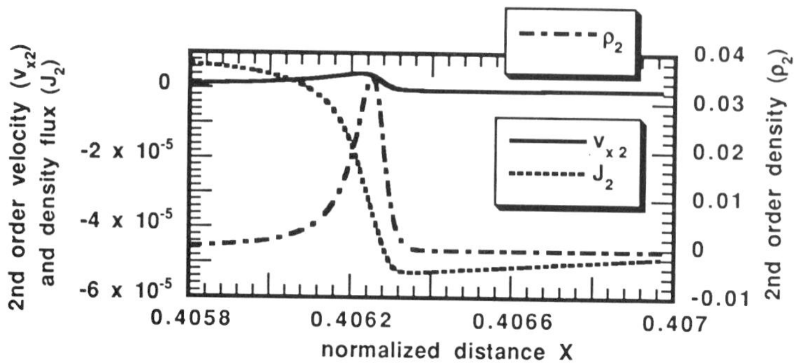

FIG.3 Various physical variables are plotted at the adjacency of the ablation front .

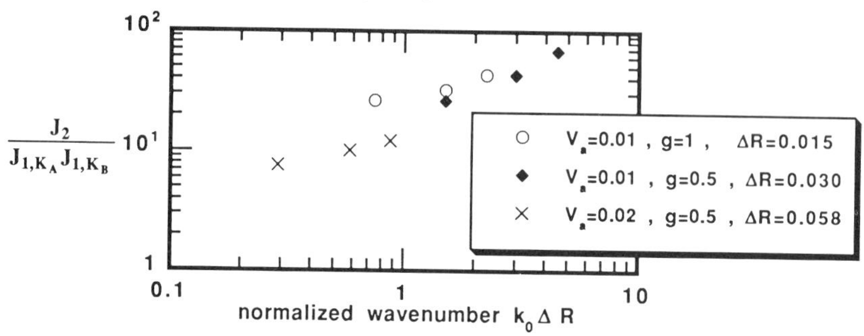

FIG.4 The ratio ($\frac{J_2}{J_{1,K_A} J_{1,K_B}}$) versus the normalized wavenumber $k_0 \Delta R$

References

[1] H.Takabe , L.Montierth , and R.L.Morse , 1983 Phys . Fluids **26** , 2299-307 ; H.Takabe , K.Mima , L.Montierth , and R.L.Morse, 1985 Phys . Fluids **28** , 3676-82

[2] S.W.Haan , 1991 Phys . Fluids **B3** , 2349-55

Inst. Phys. Conf. Ser. No 140: Section 9
Paper presented at Laser Interact. with Matter Conf., Oxford, 19–23 September 1994

© *1995 IOP Publishing Ltd*

LASER PRODUCED HIGHLY CHARGED Ta IONS

H. Haseroth J. Collier A. Kuttenberger K. Langbein
T.R. Sherwood B. Králiková[1] J. Krása[1] L. Láska[1]
K. Mašek[1] M. Pfeifer[1] K. Rohlena[1] J. Skála[1] P. Trenda[1]
P. Parys[2] J. Wołowski[2] E. Woryna[2] J. Farny[3] W. Mróz[4]
I. Roudskoy[5] O. Shamaev[5] B. Sharkov[5] A. Shumshurov[5]
B. A. Bryunetkin[6]

PS-Division, CERN, Geneva, Switzerland
[1]Inst. of Physics, Acad. Sci, Czech Rep., Prague, Czech Rep.
[2]Inst. Plasma Physics and Laser Microfusion, Warsaw, Poland
[3]95445 Bayreuth, Meranierring 28/7,Germany
[4]Inst. Optoelectronics, Military Univ. of Technology, Warsaw, Poland
[5]Inst. of Theoretical and Experimental Physics, Moscow, Russia
[6]NPO VNIIFTRI, Mendeleevo, Moscow Region, Russia

Abstract

Results are presented of investigations on iodine laser interaction with a Ta target. Ion collector, electrostatic ion energy analyzer and Thomson spectrograph were used to determine the ion charge and energy distributions at a long distance ($\sim 200cm$) from the plasma. The highest observed charge state of Ta in the expanding plasma was 53+ . The existence of $Ta46+$ ions was also confirmed inside the plasma corona by X-ray spectroscopy. An electron temperature $T_e > 1.5keV$ was estimated for the experimentally measured mean ion energy $E = 370keV$. The results inspired a revision of a hydrocode modelling the expanding plasma generated by a short (sub nanosecond) laser pulse.

1 Introduction

Laser produced plasma is a potential source of highly charged heavy ions, which are of great importance for large heavy ion accelerator facilities [1]. Even if this idea is not new, it has been revived recently [2] and the laser plasma started to be intensely investigated for this purpose [3, 4].

The laser plasma ion source studies presented up to now were using mainly CO_2 lasers and exceptionally also Nd lasers for interaction with targets of different atomic number Z. Owing to different laser pulse energy and pulse length the power density varied after focusing the beam from $10^{12}W/cm^2$ to $10^{15}W/cm^2$. Fully stripped ions were obtained in case of low- or medium-Z elements including, more recently, for Al [4]. With Z increasing

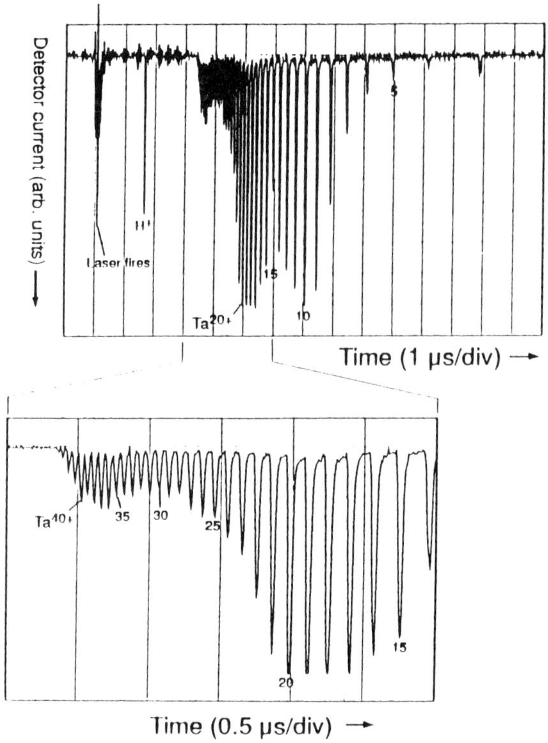

Figure 1: EIA spectra of Ta ions in two different time scales at $207.4cm$ from the target

the production of ions with the highest charge states is more difficult because of the high ionization potential. Moreover, due to the recombination processes the situation in the expanding plasma differs from that in the hot plasma corona. As for the heavy ions, $Ta54+$ were registered by X-ray ion spectroscopy in the plasma corona [5]. The highest charge state of Ta identified so far by the particle diagnostics at a large distance from the target is $Ta26+$ [6].

The theory of charge state dynamics of expanding plasma produced by CO_2 lasers ($\lambda = 10.6\mu m$) and Nd glass lasers ($\lambda = 1.06\mu m$) predicts short wavelength and short pulse lasers to be very efficient in heating the plasma and producing ions of very high charge state [7]. But it also predicts considerable recombination losses of highly charged ions due to the higher plasma density and a rapid cooling of the expanding plasma after the laser pulse cut-off.

2 Experimental

An iodine laser generates a short and intense pulse in the near infrared region ($\lambda = 1.315\mu m$) providing a complementary experimental information to the CO_2 lasers. The

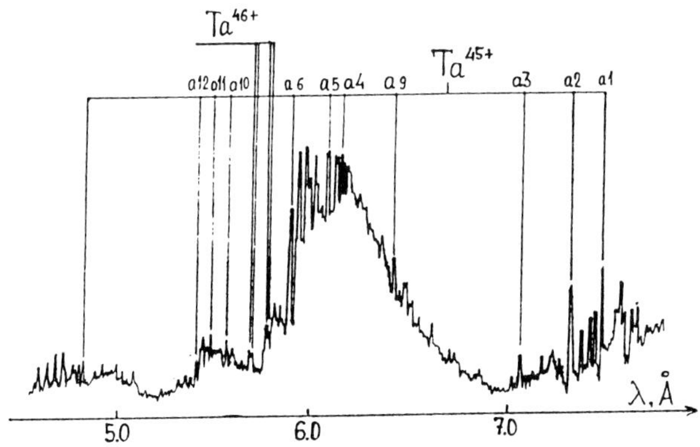

Figure 2: X-ray spectra of $Ta45/46+$ ions from the corona

iodine laser system PERUN in the Institute of Physics in Prague (pulse energy $\sim 50J$, pulse length $350ps$) [8] was used as a driver. The system can also be operated at $2\omega_0$ or $3\omega_0$. The laser beam was focused with $f/2$ optics on a Ta target; the diameter of focal spot was about $100\mu m$. The principal diagnostic technique for studying the Ta plasma were based on the time-of-flight (TOF) method. A cylindrical electrostatic ion energy analyzer (EIA), an ion collector (IC) as well as a Thomson parabola spectrograph (TS) were used [9, 10]. The EIA had a mean radius of $10cm$, the gap between the electrodes $1cm$ and deflection angle $\pi/2$. The IC diameter was $1.5cm$ and the bias voltage $-100V$. The path of flight of the EIA and of the IC were $207.4cm$ and $94cm$, respectively.

The target position with respect to the focus as well as its tilt angle with respect to the laser beam were varied at a fixed laser energy. In our experimental configuration and keeping the laser energy below $32J$ the optimum conditions were found as follows: the focus position $125 - 250\mu m$ in front of the target, the tilt angle about 30 degrees [11]. As an example, ion spectra obtained with the EIA are shown in fig. 1 in two different time scales. The structure of collector signal points to the existence of two ion groups. The faster one, if analysed by the EIA is found to be composed of high charge states of Ta ions and of light impurities ions (H, C, O). Taking into account the relative abundance of ions with a fixed charge state in this fast group, it can be concluded that the number of Ta40+ is about 10^8 per pulse. The peak current density was about $12.8mA/cm^2$. The latest experiments confirm the presence of ions with charge states up to 53+ and with energies at least $4.8MeV$. The same charge states were also produced using the second and third harmonics of the iodine laser frequency. The parabolas from TS are shown in [12]. During the laser interaction with the target the electron density, electron temperature and the laser pulse duration are the parameters controlling the ionization state of the plasma. Using a mixed model given by Busquet [13] and the relation $E = 4(\bar{z}_0 + 1)\ Te$ from [14], an electron temperature $Te \sim 1.7keV$ and average charge state of ions in the focus $\bar{z}_0 \sim 50$ were estimated. Using the relation [7]

$$T_e = 100(P\bar{z}_0/d)^{2/7}\ \ [eV] \tag{1}$$

(P is absorbed laser power, d focal spot diameter) an electron temperature $Te \geq 1.5keV$

for $z_0 \sim 40$ is estimated for our experimental conditions. Measurements of X-ray emission spectra of plasma in the $(4-8)A$ range were performed in order to confront the ion charge states present in the corona with the corpuscular diagnostics results. Two kinds of Bragg crystal spectrographs (flat crystal CsAP, $2d = 26.1A$ for an overview and a high luminosity bent mica for a detailed spectrum) were used. Fig. 2 is showing the spectrum registered by the bent mica spectrograph and its densitogram in the vicinity of $5.9A$. The lines of Co-like $Ta46+$ with the wavelength $5.907A$ (transition $3d-4f$) and the satellite structure next to it are distinguishable. Specifically provided calculation were used for identification of the lines [15].

3 Conclusion

The results of this paper confirm a good efficiency of production of high charge state heavy ions from the laser plasma generated by sub nanosecond pulses in the near infrared region. It was proved that $Ta53+$ ions reach the far expansion zone and, thus, that they survive the recombination loss inevitably occurring in an early stage of the expansion. Rough estimates of the absolute ion numbers support feasibility of ion sources based on such kind of plasma.

References

[1] Haseroth H and Hora H 1991 *Advances of Accelerator Physics and Technologies* (Singapore: World Scientific) p 166

[2] Sherwood T R 1992 *Rev.Sci.Instrum.* **63** 2789

[3] Amidouche Y et al 1992 *Rev.Sci.Instrum.* **63** 2838

[4] Mróz W et al 1994 *Rev.Sci.Instrum.* **63** 1272

[5] Aglitzkiy E V et al 1974 *Kvantovaya Electronika* **1** 2067

[6] Ehler W 1980 *J.Phys.D: Appl.Phys* **13** L29

[7] Roudskoy I 1993 PhD Thesis, ITEP Moscow

[8] Chvojka M et al 1988 *Czech.J.Phys.* **B38** 1337

[9] Denus S et al 1977 *J.Techn.Phys.* **18** 25

[10] Wolowski J et al 1982 *ZhTF* **52** 366

[11] Láska L et al 1994 *Appl.Phys.Lett.* **65** 691

[12] Woryna E et al 1994 at this conference

[13] Busquet M 1982 *Phys.Rev.* **A25** 2302

[14] Farny J 1985 PhD Thesis Military Univ. Technol. Warsaw

[15] Bryunetkin B et al 1994 *Kvantovaya Elektronika* in press

Inst. Phys. Conf. Ser. No 140: Section 9
Paper presented at Laser Interact. with Matter Conf., Oxford, 19–23 September 1994

© *1995 IOP Publishing Ltd*

Investigation of laser imprint and FDD plasma smoothing with the 2D hydrocode POLLUX

M.W.Jones†, M.Desselberger†, J.Edwards‡, M.Dunne† and O.Willi†

† Imperial College of Science, Technology and Medicine, London, UK.

‡ Now at A.W.E. Aldermaston, UK.

Abstract. The effect of laser intensity nonuniformities 'imprinting' on plastic targets in a planar geometry has been simulated. Within this study, the effectiveness of current incoherent beam smoothing techniques such as Induced Spatial Incoherence (ISI), and the use of random phase plates (RPP) was investigated. Finally the use of 'buffer layers' as a smoothing medium in the Foam Direct-Drive scheme (FDD) was studied.

1. Introduction

One of the major problems in direct laser-drive inertial confinement fusion today is that of achieving a sufficiently uniform ablation pressure and thus an uniform target acceleration. Any nonuniformities in the intensity profile of the laser can *imprint* onto the target surface[1] and thus seed hydrodynamic instabilities such as the Richtmyer-Meshkov (RM) and Rayleigh-Taylor (RT) instabilities. This will cause reduced gain, and in the worst case can cause complete target breakup.

This paper computationally investigates the imprinting process, the use of incoherent beam smoothing techniques, RPP and ISI, and also the use of a scheme developed by us[2] which, uses a foam buffer to achieve plasma smoothing of laser intensity nonuniformities. The code that was used for these simulations was a 2D Eulerian hydrocode, POLLUX[3].

2. Planar foils

To illustrate the extent of the problem that occurs with unsmoothed coherent laser beams, an initial set of simulations was performed using a laser which had sinusoidal modulations of period 10μm imposed on it with a peak to valley amplitude of 7:1. This profile was intended to represent a realistically 'dirty' laser while still allowing a simple, quantitative analysis. The laser had a 100ps linear rise to a peak intensity of 10^{14}Wcm^{-2} and was incident on a planar CHO foil, 15μm thick, and 120μm in length. The results at different times can be seen in figure 1. The nonuniformities have imprinted onto the target, and when it subsequently accelerated, the surface nonuniformities grow via the RT instability. It is clear that this foil is almost completely destroyed at the last timestep.

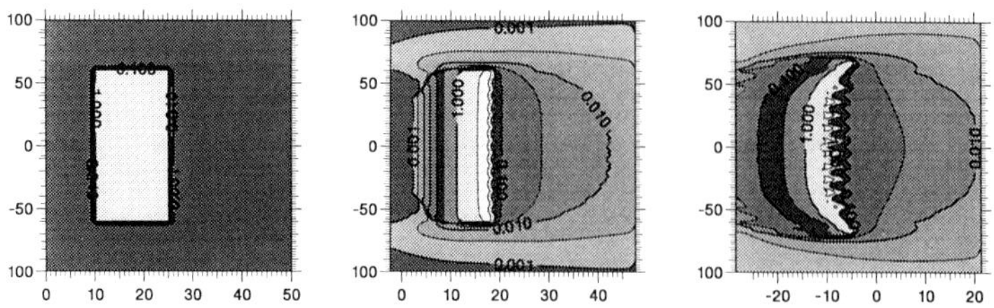

Figure 1. A density contour plot of a laser λ=0.53μm with imposed modulations of λ_{pert}=10μm @ 7:1, incident from the right onto a 15μm CHO foil of solid density 1400 mg/cc. The three plots are at t=0, 250, 750ps after the start of the laser pulse.

3. Beam smoothing

To combat the problem of laser nonuniformities, a number of optical beam smoothing techniques have been proposed and developed. The methods include the use of random phase plates (RPP), induced spatial incoherence (ISI) and smoothing by spectral dispersion (SSD)[4], which all attempt to smooth the laser by destroying its coherence in one form or another. The two methods investigated in this series of simulations are RPP and ISI. The first of these uses phase plates to alter the relative phase of different parts of the spatial profile of the laser, whilst the second in addition allows a temporal reduction in coherence giving a smoother time averaged beam profile.

The density contour plots in figure 2 show some of the results of two sets of simulations, one using an RPP and one using an ISI configuration. The plots are for a much later time (t=1500ps) than the previous figure, showing an improved uniformity at the ablation surface. Each laser had 6 modulation modes imposed upon it, and the coherence time for the ISI case was 2ps. In these simulations, the target was a 25μm CHO foil.

As much of an improvement as these methods are over the simple case, both still imprint the target in the early ($\sim$ 150ps) start up phase. Once a sufficient amount of material has been ablated to form a plasma buffer layer, thermal smoothing can take place, and the laser nonuniformities have a lesser effect on the ablation surface. It is the imprint of the laser at early time which is a problem.

4. FDD scheme

To solve this initial imprint problem, we have recently proposed a scheme called the Foam Direct-Drive (FDD) scheme that uses a low density foam layer on the target. This

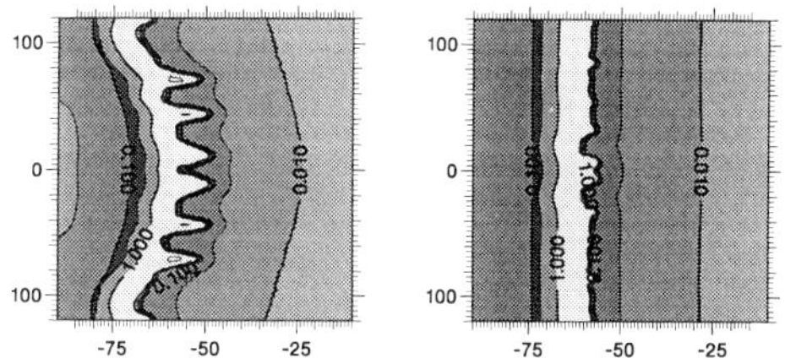

Figure 2. Target: $25\mu m$ of CHO at a solid density of 1400mg/cc. Laser: **a)** RPP case 6 modes, $120\mu m \rightarrow 24\mu m$ temporally coherent, **b)** ISI case with the same mode structure as in a), but a coherence time of 2ps. The two plots are at time t=1500ps after the start of the laser pulse.

foam layer is initially ionized by an x-ray flash so that when the laser is incident on the 'foam' it interacts with a plasma instead of a solid surface. This plasma buffer then thermally smooths the laser nonuniformities[2].

Figure 3 illustrates the amazingly uniform ablation pressures produced using this scheme. The laser conditions were identical to those in figure 1 for the simple foil case, but the target was a $13.9\mu m$ thick CHO foil with a $50\mu m$ foam layer of density 30mg/cc. The target was irradited for 150ps with an x-ray flux of $5 \times 10^{12}\mathrm{Wcm}^{-2}$ before the main laser pulse.

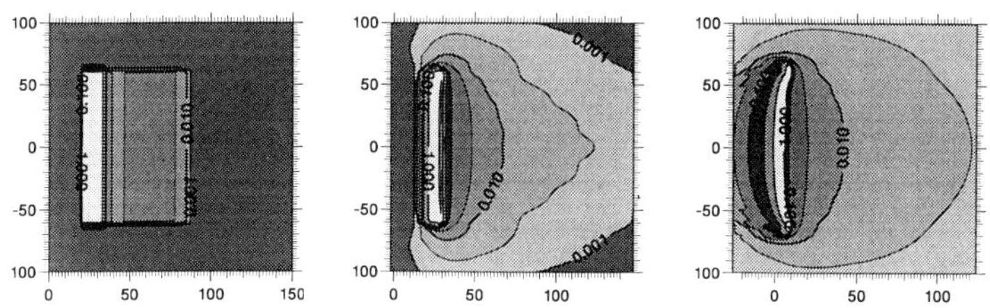

Figure 3. A laser λ=$0.53\mu m$ with imposed modulations of λ_{pert}=$10\mu m$ @ 7:1, interacting with a $13.9\mu m$ CHO foil of solid density 1400 mg/cc and a $50\mu m$ foam layer of density 30mg/cc. The three plots are at t=0, 250, 750ps after the start of the laser pulse.

As a quantitative measure of the smoothing attained in the FDD case, the momentum perturbation of the system was used. This is defined as :

$$\sigma_p = \sqrt{\sum^{\forall i}(p_i - p_{av})^2}$$

where p_i is the momentum for row i integrated from the side of the grid furthest from the laser to the front surface of the foil, and p_{av} is the average value of p_i. To reduce edge effects only the region of the central wavelengths was considered in this calculation.

Figure 4 shows the momentum perturbation of planar foil and FDD scheme as illustrated in figures 1,3. The onset of RT growth for the planar foil case is apparent after 400ps, while for the FDD case, a vastly reduced momentum perturbation is apparent for the full period of the investigation.

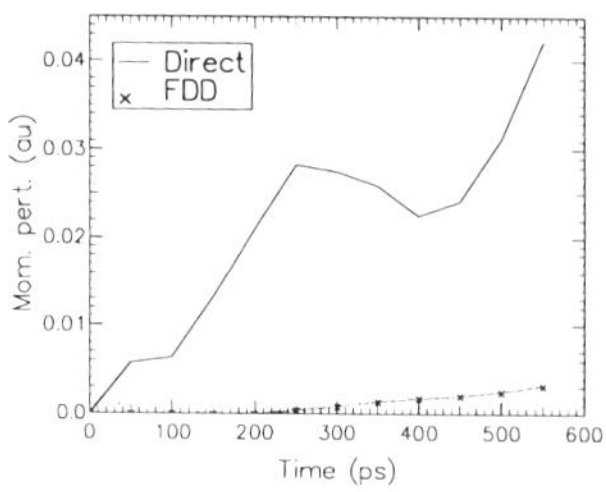

Figure 4. Momentum perturbation in the case of a planar foil, and a Foam-Foil (FDD) configuration

5. Conclusions

While space limitations do not allow a full exposition of all the situations simulated, the above discussion illustrates some of the major results obtained. In particular, the imprinting which still occurs in the ISI, RPP cases and the excellent smoothing that can be obtained with the FDD scheme. The FDD scheme has also now been proven experimentally in a number of cases, and shows great promise for the attainment of uniform implosions in ICF.

References

[1] Desselberger M, Afshar-rad T, Khattak F, Viana S and Willi O 1992 *Phys. Rev. Lett.* **68** 1539

[2] Dunne M, *et al* These proceedings.

[3] Pert G J 1981 *J. Comput. Phys.* **43** 111; 1983 **49** 1

[4] Kato Y, Mima K, Miryanaga N, Aringa S, Kitagawa Y Nakatsuka M and Yamanaka C 1984 *Phys. Rev. Lett.* **53** 1057
Lehmberg R H, Schmitt A J and Bodner S E 1987 *J. Appl. Phys.* **62** 2680
Skupsky S, Short R W, Kessler T, Craxton R S, Letzring S and Squires J M 1989 *J. Appl. Phys.* **66** 3456

[5] Desselberger M, Jones M W, Edwards J, Dunne M and Willi O, *submitted to Phys. Rev. Lett.*

Inst. Phys. Conf. Ser. No 140: Section 9
Paper presented at Laser Interact. with Matter Conf., Oxford, 19–23 September 1994

© 1995 IOP Publishing Ltd

Pressure ionization of dense plasmas in spherical ion-cell model with relativistic corrections

Kenichi Ishikawa* and Thomas Blenski

Institut de Génie Atomique, Département de Physique, Ecole Polytechnique Fédérale de Lausanne, CH 1015 Lausanne, Switzerland

Abstract. We study the continuity of pressure of dense plasmas in case of pressure ionization. Pressure is calculated using a quantum-mechanical stress-tensor pressure formula in the self-consistent field spherical ion-cell model. It appears that in order to preserve the continuity of calculated pressure one must take narrow shape resonances into account properly. We derive also a relativistic pressure formula in the Pauli approximation. The obtained formula is only slightly different from a non-relativistic one. We have found that if relativistic corrections are applied only to bound electron states, it may lead to a discontinuity in the pressure-vs.-density curve even in case of small Z elements. This observation stresses the need of a coherent description of bound and free electrons in the spherical ion-cell models.

1. Introduction

The self-consistent field spherical ion-cell model is often an important part of equation-of-state or opacity calculations of dense plasmas. It is known that a self-consistent field of finite range has continuum resonances accompanying pressure ionization[1]. Several authors[1] have studied this phenomenon and have shown analytically that pressure ionization induces, in principle, no discontinuities in any thermodynamic quantity. R. More discussed[1] this problem and also proposed a quantum mechanical stress-tensor pressure formula for pressure calculations.

In the present study we generalize the quantum-mechanical stress-tensor pressure formula to the case of relativistic Pauli approximation. We report some results of our numerical calculations and discuss the continuity of calculated pressure.

2. Spherical ion-cell model with relativistic corrections

We calculate the electron density by summing over a complete set of bound and free states,

$$n(\mathbf{r}) = \sum_i |\psi_i(\mathbf{r})|^2 f(\varepsilon_i, \mu), \tag{2.1}$$

where, $f(\varepsilon,\mu) = \left[\exp\{(\varepsilon-\mu)/T\}+1\right]^{-1}$ is the Fermi-Dirac function. The wave functions $\psi_i(\mathbf{r}) = (y_{\varepsilon,l}/r)Y_{lm}(\theta,\varphi)$ are calculated from the radial Schrödinger equation:

---

* on leave from Department of Quantum Engineering and Systems Science, Faculty of Engineering, University of Tokyo, Tokyo, Japan

$$\left[-\frac{\hbar^2}{2m}\left(\frac{d^2}{dr^2}-\frac{l(l+1)}{r^2}\right)-eV_{\text{sc}}+V_{\text{mv}}+V_{\text{darwin}}+V_{\text{so}}\right]y(r)=\varepsilon y(r) \quad . \tag{2.2}$$

The self-consistent potential $-eV_{\text{sc}}(\mathbf{r})$ is the sum of the exchange-correlation potential $V_{\text{xc}}(\boldsymbol{r})$ [2] and the electrostatic potential $-eV_{\text{el}}(\mathbf{r})$ determined through the Poisson equation with the boundary condition that it smoothly tends to zero at the radius $r_0=\left[3/(4\pi\rho)\right]^{1/3}$ of the Wigner-Seitz sphere, where ρ is number density of atoms. The chemical potential μ is found from the condition of neutrality of the ion-sphere (the Wigner-Seitz sphere). Eq. (2.2) has three terms for relativistic corrections (Pauli theory[3]) : the mass-velocity term V_{mv}, the Darwin term V_{darwin}, and the spin-orbit term V_{so} [4]. We replace V_{so} by its average value $\overline{V}_{\text{so}}=0$.

The resonances can be found as the wave numbers $k=k_{\text{res}}$ such that

$$d^2\delta_l/dk^2\left(k=k_{\text{res}}\right)=0 \;\text{ and }\; d\delta_l/dk\left(k=k_{\text{res}}\right)>0, \tag{2.3}$$

where δ_l is the phase shift.

3. Pressure formula

Evaluating the radial part of non-relativistic quantum-mechanical stress tensor at $r=r_0$, we get the non-relativistic pressure formula [1,3]:

$$P_{\text{total}}=\frac{\hbar^2}{2m}\sum_{\varepsilon,l}\frac{2(2l+1)}{4\pi}f(\varepsilon,\mu)\left[\left(\frac{d}{dr}\frac{y_{\varepsilon,l}}{r}\right)^2-\frac{y_{\varepsilon,l}}{r}\frac{d^2}{dr^2}\left(\frac{y_{\varepsilon,l}}{r}\right)\right]_{r=r_0}+P_{\text{xc}}(r_0), \tag{3.1}$$

where P_{xc} is the contribution of the exchange-correlation potential.

Similarly, evaluating the radial part of relativistic quantum-mechanical stress-strain tensor [3] derived from the Dirac equation, at $r=r_0$, we obtain the relativistic pressure formula:

$$P_{\text{total}}=c\hbar\sum_s f(\varepsilon_s,\mu)\,\text{Im}\left[\Psi_s^+\hat{\mathbf{r}}\cdot\alpha\frac{\partial\Psi_s}{\partial r}\right]_{r=r_0}+P_{\text{xc}}(r_0), \tag{3.2}$$

where Ψ_s is the Dirac spinor, and $\hat{\boldsymbol{r}}=\boldsymbol{r}/r$. In the present study, we have applied Pauli approximation to the exact formula, Eq. (3.2) and derived the following pressure formula with relativistic corrections:

$$P_{\text{total}}=\frac{\hbar^2}{2m}\sum_{\varepsilon,l}\frac{2(2l+1)}{4\pi}\frac{1}{1+\varepsilon/2E_0}f(\varepsilon,\mu)\left[\left(\frac{d}{dr}\frac{y_{\varepsilon,l}}{r}\right)^2-\frac{y_{\varepsilon,l}}{r}\frac{d^2}{dr^2}\left(\frac{y_{\varepsilon,l}}{r}\right)\right]_{r=r_0}+P_{\text{xc}}(r_0) \tag{3.3}$$

The obtained formula, Eq. (3.3) is different from the non-relativistic formula, Eq. (3.2) only by the factor $\left(1+\varepsilon/2E_0\right)^{-1}$ which is close to unity.

4. Results and discussions

4. 1. Continuity of pressure in pressure ionization of bound levels

Let us take the example of Ar at T = 2.72 eV and at densities around $\rho=4.2\rho_0$ where ρ_0 denotes the solid density (1.8 g/cm^3). In this case we observe the pressure ionization of the 3p level between $\rho=4.20\rho_0$ and $4.21\rho_0$. In Fig. 1 we show the resonance that appears at ρ =

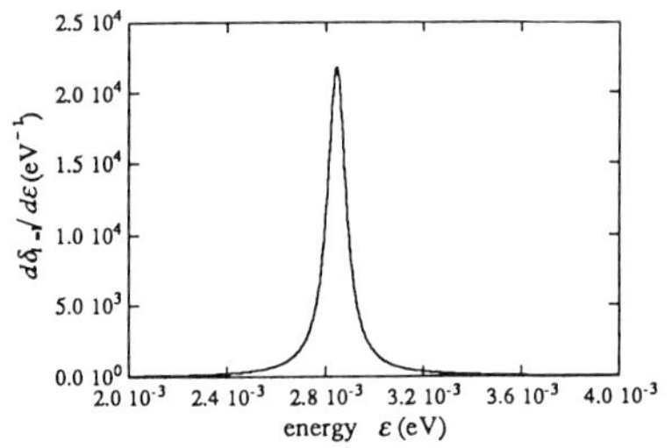

FIG. 1 The resonance of Ar (T = 2.72 eV, ρ = 4.21ρ_0)

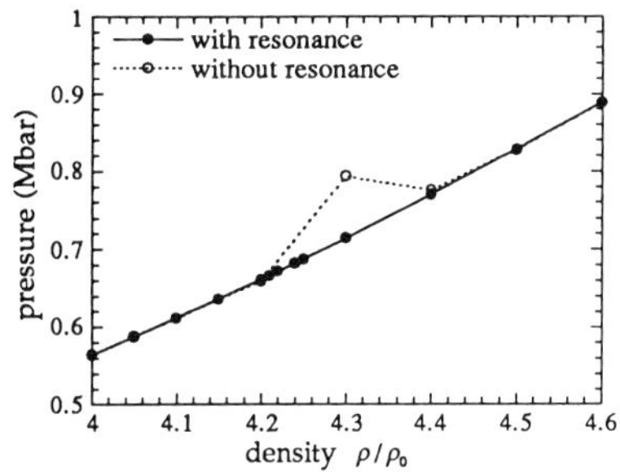

FIG. 2 Pressure vs. density with and without the resonance taken into account

4.21 ρ_0. It has the peak at 2.84×10^{-3} eV and the half-width of 4.5×10^{-5} eV, while the energy spectrum of free electrons extends from 0 to 10 eV. It is therefore clear that resonances like the one shown in Fig. 1 should be searched for at each iteration and that the mesh of integration with respect to energy should be properly chosen in order to take account of their presence.

Fig. 2 shows pressure vs. density of Ar at T = 2.72 eV when the resonance is and is not taken into account. This figure indicates that the calculation with a careful treatment of the resonance gives a completely continuous and smooth curve while the one which does not take the resonance into account leads to a discontinuity at the density of pressure ionization. In Fig. 3 are presented the bound and free electron contributions to the total pressure. The pressure associated with each bound level is always negative, and the contributions of loosely bound levels (especially 3p) are relatively large. This means that when 3p bound electrons are ionized, the bound-electron pressure increases discontinuously as seen in Fig. 3. When the ionization of 3p level occurs, nearly 6 electrons appear in the continuum. When we take the resonance into account properly, this sudden increase of free electron number is absorbed by the resonance. The resonance contribution to the total pressure is negative like that of bound levels, and this manifests itself in the discontinuous decrease in free electron pressure and cancels the abrupt change of bound electron pressure exactly as is shown in Fig. 3. When we do not take the resonance into account, the increase in the number of free electrons appears as increase in the number of high-energy free electrons. This leads to the discontinuous jump of the total pressure as we see in Fig. 2.

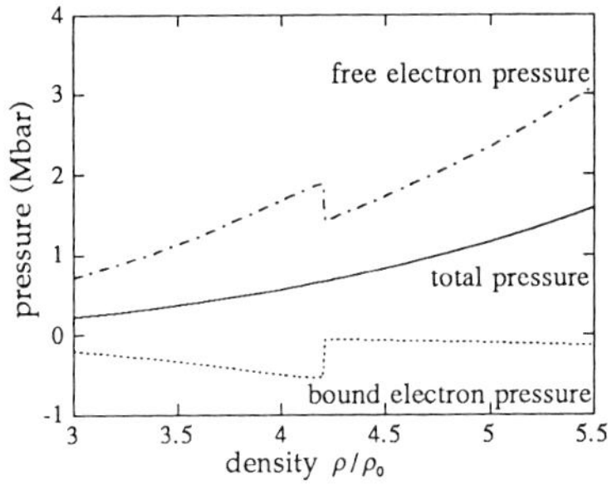

FIG. 3 Bound- and free-electron contributions to the total pressure of Ar (T = 2.72 eV)

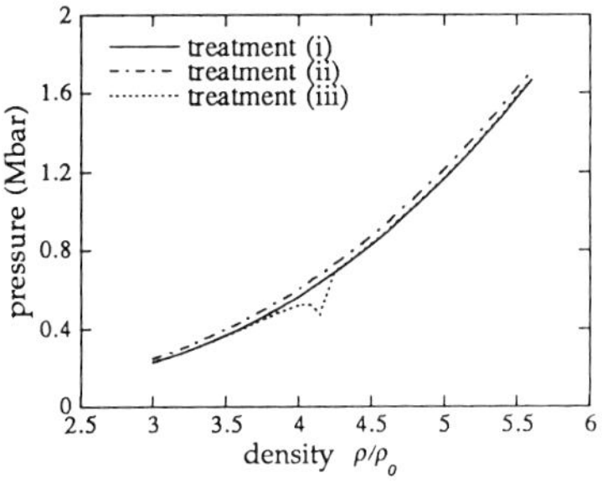

FIG. 4 Pressure vs. density of Ar calculated for the cases (i) - (iii) described in the text

4. 2. Effects of relativistic corrections to bound and free electron states

One may reason that in case of small Z elements only low lying bound levels are affected by relativistic corrections and that the relativistic treatment of bound levels combined with the non-relativistic approximation for free electrons is sufficient. In Fig. 4 we display pressure vs. density of Ar at $T = 2.72$ eV for three calculations: (i) *relativistic treatment of all electrons*, (ii) *non-relativistic treatment of all electrons*, (iii) *relativistic treatment of bound electrons and non-relativistic treatment of free electrons*. We see that the mixed case (iii) leads to the presence of a dip in the pressure curve. We have found that the origin of the dip is as follows. The integral $\int_0^{r_0} y_{3p}^2(r)dr$ for $\rho = 4.05\rho_0$ equals 0.700, 0.634, and 0.701 for the cases (i), (ii) and (iii), respectively. This fact and the neutrality of the ion-sphere lead to a difference in free electron number (2.00 for (i), 2.37 for (ii) and 2.01 for (iii)). Although this difference seems to be small, it concerns the high energy free electrons which have important contributions to the pressure. Thus finally the pressure in the case (iii) is much lower than in the case (ii) and this effect explains the dip seen in Fig. 3. In the case (i), in spite of the fact that the free electron number is lower than in the case (ii), the continuity of pressure is preserved due to the relativistic corrections to free electron states. From Fig. 3 we see that the cases (i) and (iii) give practically the same result except for the dip. Thus for densities where there are no very loosely bound levels, the relativistic corrections to free electrons are nearly negligible.

It is important to note that when one uses exactly the same one-electron Hamiltonian for both bound and free electrons, calculated pressure preserves its continuity (cases (i) and (ii)). This observation stresses the need of a consistent description of bound and free electrons in the self-consistent field spherical ion-cell model.

5. Conclusions

We have studied the pressure ionization and the continuity of plasma pressure calculated with the self-consistent field spherical ion-cell model. It appears that pressure-vs.-density curves are continuous and smooth despite pressure ionization phenomena only when a careful search for narrow resonances is performed at each iteration. We derived a relativistic pressure formula in the Pauli approximation. This formula is only slightly different from the non-relativistic one. Our results indicate that bound and free electrons should be treated in the same manner i.e. both with relativistic corrections or both without them.

6. Acknowledgments

Discussions with F. Grimaldi are gratefully acknowledged. This research has been partially supported by Swiss Federal Office for Science and Education, "Union des Centrales Suisses d'Electricité" and Swiss National Science Fund.

References

[1] More R M 1985 *Adv. Atom. Molec. Phys.* **21** 305-356, and references cited therein
[2] Perrot F and Dharma-wardana M W C 1984 *Phys. Rev. A* **30** 2619-2626
[3] Bethe H A and Salpeter E E 1957 *Quantum Mechanics of One- and Two-Electron Atoms* (Berlin: Springer-Verlag)
[4] Cowan R D 1981 *the Theory of Atomic Structure and Spectra* (Berkeley: University of California Press)

Inst. Phys. Conf. Ser. No 140: Section 9
Paper presented at Laser Interact. with Matter Conf., Oxford, 19–23 September 1994
© *1995 IOP Publishing Ltd*

Transport phenomena of highly dense plasma

A.H. Khalfaoui and D. Bennaceur

Laboratoire Interaction Laser-Matière, 2 Bd Franz Fanon, BP 1017 Alger-Gare 16000, Algeria

Abstract. Transport properties of a system of highly degenerate weakly coupled electrons and strongly coupled semi-classical ions are treated by a quantum collective approach. The model considers electron-ion (e-i) and electron-electron (e-e) collisions in an unified scheme of long and short rang Coulomb interactions. It is shown that the Lorentz ratio of high density plasma may exhibit substantial negative deviation from the ideal Sommerfeld value, due to (e-e) interactions.

1. Introduction

Quantum plasmas with completely degenerate electrons and strongly coupled ions are the subject of great interest in both inertial confinement fusion as well as planetary application [1]. These strongly nonideal plasmas behave very similarly to liquid metals.

Several [2] experimental results on liquid metals showed a departure of the Lorentz ratio L from its ideal value the Sommerfeld number $L_0 = (\pi^2/3)(k_B/e)^2$, defined through the Wiedemann-Franz law as $L = \lambda/\sigma T$, where λ and σ are respectively the thermal and electrical conductivities, T the plasma temperature, k_B the Boltzmann constant and e the electron charge.

The anomalous departure of L/L_0 from unity is due essentially to e-e collisions. This latter effect may contribute inequally to the relaxation times of the thermal and electrical conductivities.

To evaluate the impact of the nonideal effects on transport coefficients, a unification of treatments of short and long-range Coulomb collisions represents a more comprehensive approach to very high-density plasma, for which the shielding distance (Debye lengh) loses its physical meaning as an upper impact parameter.

Density fluctuations for dense plasmas can be respectively split into two approximately independent components associated with the collective and individual particle aspects of the system.

For dense plasmas, considered here, the Debye length is of the same order of magnitude as the interionic distances and the random thermal motion is negligibly small. Hence the collective aspect of the system will be dominant and mainly governs the interaction process in the transport phenomena.

In analogy to the Bloch theory, the role of the longitudinal phonons in the theory of metals is played, in the present model, by the quanta of the plasma oscillations: plasmons and ion sound waves.

The formalism of the procedure used to solve the quantum Boltzmann-Bloch equation will be presented where a variational principle will be used, since the relaxation process will be looked for through the maximization of the entropy of the system.

2. Transport coefficients

The derivation of transport coefficients follows the Kohler variational method [4] where the transport equation is written as:

$$\partial f/\partial t\,\big|_{\text{field}} = \partial f/\partial t\,\big|_{\text{coll}} \quad \Leftrightarrow \quad F = \underline{\boldsymbol{L}}\Phi \tag{1}$$

where $\underline{\boldsymbol{L}}$ is an integral operator which connects the unknown function Φ to the known function of the field F.

$$\partial f/\partial t\,\big|_{\text{field}} = v_x\,(\partial f_0/\partial E)\,[-e\varepsilon - d\xi/dx - (E-\xi)/T\; dT/dx] \tag{2}$$

ε is the electrical field, ξ the Fermi energy, dT/dx is a weak temperature gradient and v_x is the electron velocity in the x-direction.

The collision term as a function of the wave number q of the plasmons and k of the electron is written as:

$$\partial f/\partial t\,\big|_{\text{coll}} = \frac{-\Phi}{8\pi^2 m_s n_s k_B T}\int q^2\, \frac{|U_s(q)|^2}{\omega_s(q)}\, N_q\,\{f_0(k')[1-f_0(k)]\delta(E'-E+\hbar\omega_s) + f_0(k)[1-f_0(k')]\,\delta(E'-E-\hbar\omega_s)\}[1-\Phi(k')/\Phi(k)]d^3k'\;,\quad s=e,i \tag{3}$$

$|U_s(q)|^2$ is the square of the Fourier transform of a shielded Yukawa potential, N_q and $f_0(k)$ are the distribution functions of Bose-Einstein for the plasmons and Fermi-Dirac for the degenerate electrons respectively.
The arbitrary function Φ which is proportional to the electron energy E will be determined by an optimization principle and expanded into a power series of $(E-\xi)$.

Following this procedure the transport coefficients can be written as:

$$\sigma = (e^2/4\pi^3)\,\Sigma_\mu\, c_\mu N_\mu \tag{4}$$

$$\lambda = (1/4\pi^3 T)\,[\Sigma_\mu\, b_\mu B_\mu - (\Sigma_\nu\, c_\nu B_\nu)^2/\,\Sigma_\rho\, c_\rho N_\rho\,] \tag{5}$$

$$N_\mu = \Sigma_\nu\, c_\nu D_{\mu\nu} \tag{6}$$

$$B_\mu = \Sigma_\nu\, b_\nu D_{\mu\nu} = N_{\mu+1} \tag{7}$$

$$D_{\mu\nu} = \frac{1}{(2\pi)^3 k_B T}\iint v_x\,(E-\xi)^\mu\, W(k',k)\,[v_x\,(E-\xi) - v_x'\,(E'-\xi)^\nu\,]d^3k\; d^3k' \tag{8}$$

Here, $D_{\mu\nu}$ is a function of W(k,k') which contains the transition probability P(k',k).

The latter function takes different forms for the absorption and the emission of a plasmon.

3. Wiedemann-Franz law for strongly coupled systems

3.1 Lorentzian approximation

If only the (e-i) collisions are taken into account the Lorentzian ratio as a function of the degree of degeneracy θ ($\theta = k_BT/E_F$, E_F is the Fermi energy) can be written as:

$$L^{ei} = L_0 / [1+r(\theta)] \qquad (9)$$

where $r(\theta)$ is a slowly varying function of order unity.

By solving the Boltzmann-Ziman transport equation taken within a Lorentzian approximation (Leger and Deutsch) [5], investigated several effects on the transport coefficients. Their results showed that inelastic scattering as well as temperature effects produced a negligibly small contributions to the correction of the Lorentz ratio for $\Gamma \leq 50$, ($\Gamma = (Ze)^2/a_ik_BT$ and $a_i = (3Z/4\pi n)^{1/3}$).

The general trend of their findings is shown in Table 1 and compared to the present results , for a fixed density parameter r_s ($r_s = a_e/a_0$, $a_e = (3/4\pi n)^{1/3}$ and a_0 is the Bohr radius).

Table 1: Lorentz ratio in the Lorentzian approximation L^{ei}/L_0 of the present results.

$r_s = 0.5$	Γ	10	30	50	70	90	100
L^{ei}/L_0	present	0.990	0.971	0.951	0.932	0.913	0.893
	LD (ref.5)	1	1	0.925	0.894	0.940	...

3.2 e-e interaction effects

If both (e-e) and (e-i) interactions are taken into account the Lorentz ratio takes the form:

$$L = L^{ei} [(1+\Sigma)/(1+\Lambda)] \qquad (10)$$

where $\Sigma = \sigma_{ei}/\sigma_{ee}$ and $\Lambda = \lambda_{ei}/\lambda_{ee}$.
The correction $(1+\Sigma)/(1+\Lambda)$ can be substantial.

The present results suggesting that, in high-temperature (intermediate coupling) partially degenerate plasma, the relaxation time for electrical transport is significantly larger than it is for thermal transport, thereby showing that (e-e) interactions play an important role in the departure of the Lorentz ratio from its ideal value. Similar conclusions have been reached by Lampe [6].

This specific effect is depected in Fig.1., where the ratio L/L_0 as a function of the coupling strength, is substantially reduced in the intermediate density range ($r_s > 0.5$, i.e. $n \sim 10^{25}$ cm^{-3}).

As the density increases ($r_s \leq 0.1$), the (e-e) collision effect is drastically reduced and L/L_0 tends to unity in the high coupling regime.

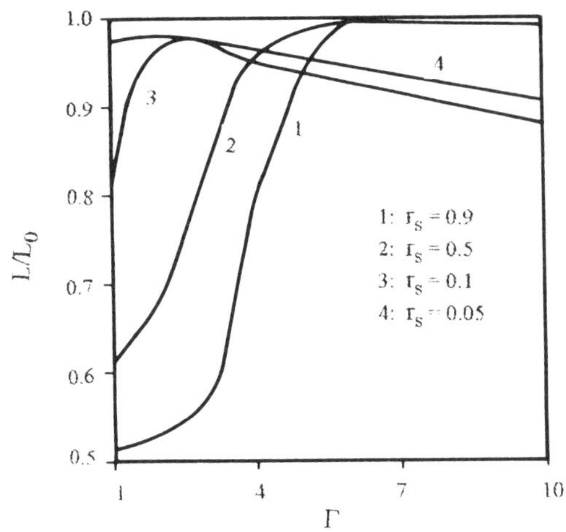

Fig.1: Effects of (e-e) interactions are shown to reduce the Lorentz ratio L/L_0 as a function of the coupling parameter (Γ) for a hydrogen plasma.

References

[1] Flowers E and Itoh N 1976 *Astrophys. J.* **206** 218.
[2] Duggin M J 1969 *Phys. Lett. A* **29** 470.
[3] Ichimaru S 1973 Basic *Principles of Plasma Physics* (Benjamin Reading MA).
[4] Kohler M 1949 *Z. Phys.* **125** 679.
[5] Leger D and Deutsch C 1992 *Phys. Fluids B* **4** 3162.
[6] Lampe M 1968 *Phys. Rev.* **170** 306; 1968 *Phys. Rev.* **174** 276.

Inst. Phys. Conf. Ser. No 140: Section 9
Paper presented at Laser Interact. with Matter Conf., Oxford, 19–23 September 1994

© *1995 IOP Publishing Ltd*

Experimental test of the relative coherence of Equation of States of different materials compressed by Laser Shock Waves

Michel Koenig, Bernard Faral, J. M. Boudenne
Laboratoire pour l'Utilisation des Lasers Intenses, Ecole Polytechnique, 91128 Palaiseau, France

Dimitri Batani, Simone Bossi, Alessandra Benuzzi
Dipartimento di Fisica Università di Milano , Via Celoria 16, 20133 Milano, Italy

Stefano Atzeni
Associazione EURATOM ENEA, C.R.E. Frascati, C.P. 65, 00044 Frascati (Roma), Italy

Mauro Temporal
Laboratori INFN di Legnaro, Padova, Italy

Abstract. We show some preliminary results on a new method developed to test the relative coherence of equation of state table of strongly compressed materials. The method is based on the generation of very high quality shock waves and the use of two steps - two materials targets.

1. Introduction

Studies on Equations of State (EOS) of strongly compressed materials may be realised with shock wave experiments and are very important in astrophysics and in Inertial Confinement Fusion (ICF). In such experiments, the five relevant physical quantities (internal specific energy E, shock velocity D, particle velocity u, shock pressure P, density ρ) are related through the shock and Hugoniot equations:

$$\rho_0 D = \rho\ (D-u)$$
$$p - p_0 = \rho_0 D\ u$$
$$E - E_0 = u^2/2$$

which represent the energy, momentum and mass conservation laws. Hence the measurement of two quantities allows the (absolute) determination of one point on the equation of state of the material f(E,ρ,T)=0.

Stepped targets allow direct "shot to shot" measurements of shock velocity. Of course this method is relevant only if: a) shock pressure is the same on the two side of the step; b) shock pressure, and hence shock velocity, are constant (in time) in the step. The point (a) shows again how it is important to get flat shock fronts as those which may be produced with PZPs [1]. The point (b) may impose some limits to target thickness related to the laser system performance.

Using stepped targets, pressures of the order of 750 Mbars have recently been obtained with the colliding foil technique [2]. In this experiment, a gold foil is accelerated by the radiation emitted from a Hohlraum cavity and collides a gold stepped-target (with a step of known height on the rear side) initially at rest. The induced shock in gold is observed with the use of a streak camera. The transit time of the shock in the gold step is then known, so is the velocity. The corresponding pressure is then inferred from SESAME equation of state data. However, the measurement of a point on the Hugoniot curve is not achieved, since it requires the measurement of two independent parameters of the shock. A recent experiment [3] has shown the possibility to determine on the same laser shot the shock and fluid velocities. Alternatively the determination of the "shock temperature" may be tried with spectroscopic methods [4].

Unfortunately the other parameters are not so easy to be measured (and with the same precision) as shock velocity, and above all the related experiments need very large laser systems in order to be performed. In this paper we present a new method which may overcome such difficulties by testing the relative coherence of equation of state table of strongly compressed materials [5]. The method is based on the generation of very high quality shock waves, i.e. flat shock fronts as those which may be produced with PZPs, and the use of two steps - two materials targets. The scheme of the experiment is shown in fig. 1.

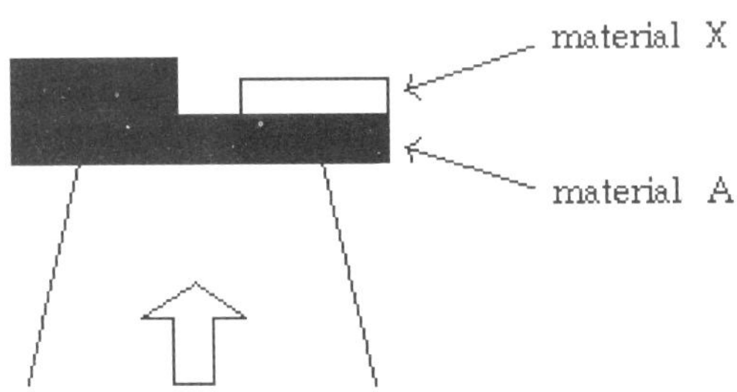

Fig. 1: Scheme of the experimental set-up for the EOS coherence experiment.

The step allows, with a rear face imaging system, to infer the shock velocity for the two materials. A shock pressure P_A is generated in material A by the laser beam and propagates to the rear face of the target on its left side. If we assume the equation of state of the material A to be known, for instance by the SESAME tables, we can then find P_A from the measurement of the shock velocity v_A. On the right side the pressure is changed to a value P_X when it crosses the interface between the two materials. Now the measurement of the shock velocity v_X allows the determination of P_X if also the EOS of X is known. But P_A and P_X are not independent, being related through the impedance mismatch relations and hence we can test the relative coherence of the EOS.

Vice versa if the material A is a "reference" material, whose EOS is supposed to be well known (e.g. Aluminium), and the material X is instead one of which we want to get EOS points, from the determination of v_A and v_X on the same laser shot, we can infer one point on EOS of material X. Hence the method also allows an experimental (relative) determinations of unknown EOS.

2. Experimental results

We begun by verifying that the PZP technique with the use of stepped targets allowed a precise measurement of shock velocity. Fig. 2 shows a streak camera image obtained with such a target. The measured shock velocity is v = 24 µm/ns.

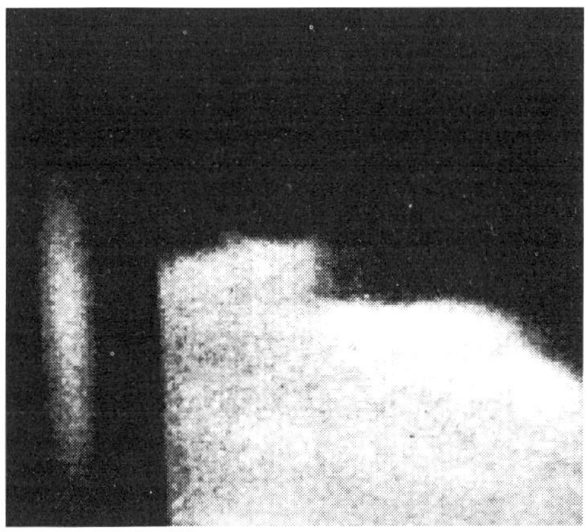

Fig. 2: Rear side shock breakthrough for a Al target with a 7.5 µm basis and a 2.8 µm step. Optical smoothing with PZPs.

Target realisation is very crucial in this kind of experiment. Indeed the space between the two step must be much smaller than the typical focal spot dimensions (in our experiment the focal spot FWHM is 400 µm but the flat region extends over ≈ 180 µm). Moreover, the steps themselves need to be very steep (step gradient lengths only a few microns) in order to achieve a good accuracy in the measurement. Fig. 3 shows a streak camera images obtained with a two steps - two material targets. Measured shock velocities are 20 and 13 µm/ns respectively in Al and in Gold, corresponding to a laser intensity 6 10^{13} W/cm^2. Our preliminary results seem to point out a good coherence of the EOS of Al and Au in this range of laser intensities which correspond to shock pressures of the order of ≥ 10 Mbar in Al. Pressures in Au roughly correspond to the normal increase in pressure (about two times) which is expected for high pressure shocks at the Al-Au interface. However a detailed analysis of the obtained results has not yet been done and the present result is only for information.

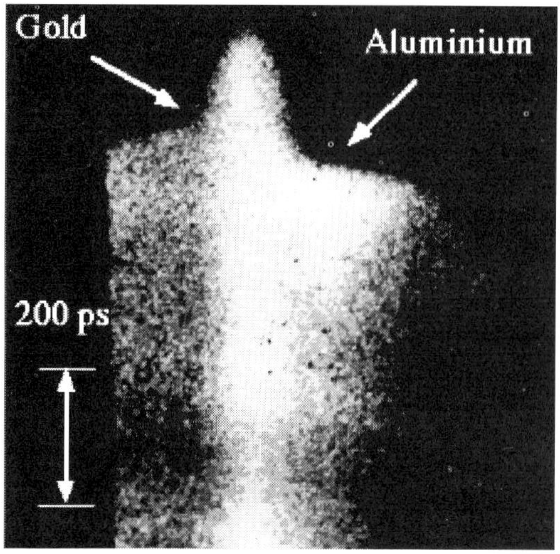

Fig. 2: Rear side shock breakthrough for a Al-Au target with a 7.5 μm basis, a 3.9 μm Al step and a 2.0 μm Au step. The two steps are spaced of about 40 μm. Optical smoothing with PZPs.

Acknowledgements

The targets used in the experiment have been produced at the Laboratoire Cibles, CEA LIMEIL by M.Belingard, F.Zielinski and J.P.Perrine. This work has been supported trough the "Dense Plasma and Laser Compression Physics" network in the framework of the "Human Capital and Mobility" Programme (EEC contract n. CHRX-CT93-0338).

References

[1] Koenig M et al. 1994 *High Pressure Shock Generation and Propagation in Optically Smoothed Laser-Plasmas*, this volume.

[2] Cauble R et al. 1993 Phys.Rev.Lett. **70,** 2102-2105.

[3] Hammel B et al. 1993 *X-ray radiographic measurements of radiation-driven shock and interface motion in solid density materials* Livermore Quarterly report, Vol. 3, N°3.

[4] A.Ng et al. 1985 Phys.Fluids, **28**, 2915-2920.

[5] Koenig M et al. *Mesure de la cohérence des tables d'équation d'état* Rapport Annuel LULI 1993, 82-92.

Inst. Phys. Conf. Ser. No 140: Section 9
Paper presented at Laser Interact. with Matter Conf., Oxford, 19–23 September 1994

© *1995 IOP Publishing Ltd*

High Pressure Shock Generation in Optically Smoothed Laser-Plasmas

Michel Koenig, Bernard Faral, J. M. Boudenne
Laboratoire pour l'Utilisation des Lasers Intenses, Ecole Polytechnique, 91128 Palaiseau, France

Dimitri Batani, Simone Bossi, Alessandra Benuzzi
Dipartimento di Fisica Università di Milano , Via Celoria 16, 20133 Milano, Italy

Stefano Atzeni
Associazione EURATOM ENEA, C.R.E. Frascati, C.P. 65, 00044 Frascati (Roma), Italy

Mauro Temporal
Laboratori INFN di Legnaro, Padova, Italy

Abstract. We show the possibility of obtaining reproducible and quantitative measurements on shock wave dynamics with optically smoothed laser beams. We have used RPPs and Phased Zone Plates as beam smoothing techniques. These lasts allow the production of high quality flat shock waves.

1. Introduction

High pressure shocks are generated as consequence of momentum conservation when lasers are focused on a solid target creating a hot plasma rapidly expanding outwards. Studies on such shocks, their generation and propagation play a fundamental role in Inertial Confinement Fusion (ICF) in order to get almost isentropic compressions of the thermonuclear targets with laser pulse temporal shaping. Shock wave experiments also allow measurements of the Equations of State (EOS) of strongly compressed materials. This is of interest for astrophysics and again for ICF.

In the past, efforts to do quantitative studies have been limited by many experimental problems (target preheating due to long laser wavelengths; strong non-homogenities and hot spots in the beams; small focal spots implying 2D effects). It is known that optical smoothing with the use of Random Phase Plates (RPP) strongly reduces the non-homogenities problem, by breaking the laser beam spatial coherence. In this way it is possible to achieve a good control of the energy deposition. In shock experiments, this allows reproducible results to be obtained. Anyway quantitative results need very high quality shocks, i.e. very flat in space (1-D), and hence we adopted the newly developed technique of Phased Zone Plates (PZP). Here we present some results on shock waves with RPPs and PZPs in a regime of interest for ICF.

2. Experimental set-up

Our experiment is based on the detection of the shock wave arrival on the target rear face which causes a sudden increase of temperature and hence of e.m. emission [1]. We used one of the six beams of LULI ($\lambda = 0{,}53$ µm, $E_{2\omega} \approx 25$ J) focused with a f=50 cm lens in a fairly large focal spot (d≈100 µm) with an intensity $I_L \leq 3\ 10^{14}$ W/cm^2. The temporal behaviour of the laser pulses is gaussian with a FWHM of 600 psec. A 2nd beam is used as temporal fiducial. Targets are simple Al foils of thickness between 2 and 25 µm. The diagnostics are:

- a streak camera working in the visible range with a high temporal resolution (50 psec/mm) coupled to a photographic objective which images the target rear face onto its slit;
- a CCD camera without temporal resolution to image the focal spot;
- a fast diode without spatial resolution to measure the total intensity emitted by the rear face.

3. Experimental results

Fig. 1 shows a typical streak camera image (obtained with a 18 µm Al target); note the time fiducial on the right of the image. The spatial profile of the target rear face emission reproduces the gaussian shape of the focal spot: such shape is reproducible shot to shot with no evidence of localised early shock breakthroughs due to hot spots. This shows that RPPs can assure a good spatial control of laser energy deposition and focal spot shape. Anyway, the gaussian shape of the focal spot makes difficult the use of produced shocks for EOS measurements. Hence we moved to the PZP beam smoothing technique obtaining a top-hat laser intensity distribution [2]. As a result, 2D effects are almost completely eliminated (around the centre of the focal spot) and high quality flat shock waves are generated as otherwise can be produced only by indirect X-ray drive [3].

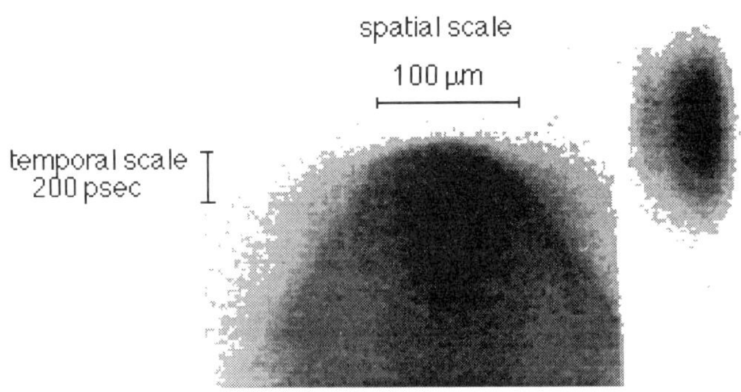

Fig. 1: Rear side shock breakthrough for a 18 µm Al target. Optical smoothing is realised with RPPs giving a measured focal spot diameter FWHM = 106 µm. Time fiducial is on right of the image.

Fig. 2: Rear side shock breakthrough for a 13.8 μm Al target. Optical smoothing is realised with PZPs giving a measured focal spot diameter FWHM = 400 μm. Time fiducial is on left of the image.

With PZPs the focal spot FWHM is 400 μm and the corresponding laser intensity is $I_L \leq 10^{14}$ W/cm^2. Fig. 2 shows a "flat" shock breakthrough obtained with PZPs. The flat region in the picture extends over a ≈ 180 μm length. Pinhole camera images are recorded on each shot which allows a control of the focal spot in the X-ray domain. Unlike the images we got when using RPPs (or usual plasma images) which have a typical gaussian distribution of emissivity, we obtain a clear "flat" emissivity region whose dimensions are ≈ 180 μm, in good agreement with the flat region in the streak camera picture (Fig. 3).

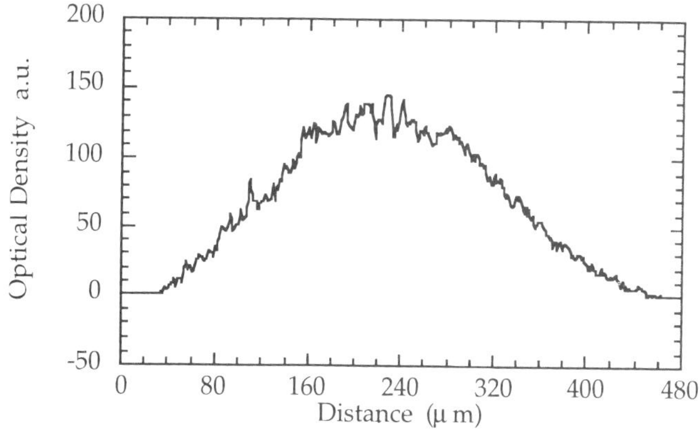

Fig. 3: Pin hole camera image densitometry for a 13.8 μm Al target. Optical smoothing is realized with PZPs giving a measured focal spot diameter FWHM = 400 μm.

To test the quality of the shock waves produced with PZPs we looked at the spatial "flatness" of the shock wave and hence at the temporal "flatness" of its breakthrough. Shock breakthrough showed typical differences in time corresponding to $\Delta t = \pm 7$ ps, or a shock wave whose velocity uniformity is better than ± 4 % across a $\approx$ 180 μm diameter in the sample.

Also, if we take a densitometry of the streak camera image of fig. 2 as a function of time, we see that the rear side luminosity is characterized by a very steep rise followed by a long decay. As observed by Lower et al. [6], this behaviour is typical of shocks when preheating is negligible (no pre-shock signal), but in their case only relatively thick targets were free of such preheating (caused by the hard primary X-rays coming from the cavity), while for us this is true for much thinner targets.

4. Discussion

We obtained good series of reproducible results which are in appreciable agreement with 1D simulation performed with the hydrodynamic code "FILM" developed at Ecole Polytechnique. Typical shock velocities were $v_S \approx 3\ 10^6$ cm/sec = 30 μm/ns with RPPs, and $v_S \approx 24$ μm/ns with PZPs, due to the larger focal spot and lower intensity. Comparisons with FILM simulations, using the SESAME EOS tables, show that these velocities correspond to pressures $P_S \approx 11$ Mbar in Al with PZPs. This result can also be qualitatively obtained with simple analytical model. Indeed in the perfect gas approximation of a plasma, shock velocity and pressure are related by the formula [4]:

$$v_S{}^2 = P/\rho_o$$

where ρ_o=2.7 g/cm^3 is the unperturbed Al density and the shock pressure is given by [4]:

$$P_S\ (\text{Mbar}) = 12.3\ (I_a\ (\text{W/cm}^2)/10^{14})^{2/3}\ \lambda^{-2/3}\ (\mu\text{m})\ (A/2Z)^{1/3}$$

where $I_a \leq I_L$ is the fraction of the laser intensity which is absorbed by the plasma ($I_a \approx 7\ 10^{13}$ W/cm^2 is needed to get $P_S \approx 11$ Mbar). These formulas are valid in the 1D case with constant shock pressure which corresponds to the case of flat-top laser pulses; hence in our case they are useful only as order of magnitude estimates.

Acknowledgements

This work has been supported trough the "Dense Plasma and Laser Compression Physics" network in the framework of the "Human Capital and Mobility" Programme (EEC contract n. CHRX-CT93-0338). A.Benuzzi has participated to the experiment thanks to an ERASMUS grant. S.Bossi got a grant from the French CIES, obtained via the Ecole Polytechnique

References:

[1] Faral B et al. 1993 *Rapport Annuel LULI 1992* 34-44, Ecole Polytechnique, Palaiseau.

[2] Stevenson R M et al. 1994 *Optics Letters* **19** 363-366.

[3] Lower Th et al. 1994 *Phys.Rev.Lett.* **72,** 3186-3189

[4] Fabbro R 1982 *Etude de l'influence de la longueur d'onde de laser sur ...,* These, Université d'Orsay.

Inst. Phys. Conf. Ser. No 140: Section 9
Paper presented at Laser Interact. with Matter Conf., Oxford, 19–23 September 1994

© *1995 IOP Publishing Ltd*

Efficient Raman conversion in Nd:YAP laser through backward stimulated Brillouin scattering in CH_4 gas

Hong Jin Kong, Do Il Chang and Yong Geun Jeon

Department of Physics, Korea Advanced Institute of Science and Technology, 373-1 Kuseong-dong Yuseong-gu Taejon, Korea

Abstract. A noble scheme for efficient Raman laser oscillator to convert the Nd:YAP laser at the wavelength of 1.079μm to the wavelength of 1.575 µm in high pressurized CH_4 gas was studied. By backward stimulated Brillouin scattering(SBS) as a phase conjugated resonator mirror for the fundamental wave at the wavelength of 1.079µm, the Raman laser generation from passively Q-switched Nd:YAP laser has been achieved. We obtained much higher Raman conversion efficiency, 47%, than a conventional Raman laser.

1. Introduction

Stimulated Raman scattering(SRS) in gases provides an efficient method for frequency down conversion of a laser radiation. There have been many works on Raman-shifting the Nd:YAG laser wavelength of 1.06μm to 1.54μm based on SRS in a high pressurized CH_4 gas cell. Because high pressurized CH_4 gas is highly Brillouin active, backward stimulated Brillouin scattering(SBS) in a Raman cell has been a serious limiting factor in the Raman conversion process[1][2].

In this paper we report a noble method[3] for efficient Raman laser oscillator to convert the Nd:YAP laser at the wavelength of 1.079μm to the wavelength of 1.575μm in high pressurized CH_4 gas through the use of backward stimulated Brillouin scattering as a phase conjugated resonator mirror for the fundamental wave.

2. Experiment

In this scheme we used a high reflectivity and a phase conjugation property of

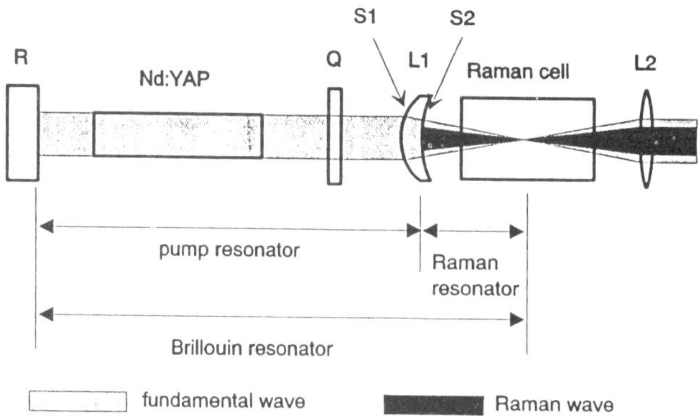

Figure 1 Optical scheme of Raman laser

scheme is similar to the scheme of ref[3] and described in Figure 1. The surface S1 has an anti-reflection for both wavelengths (fundamental wave and Raman wave) and its radius of curvature is $R = 1 - (f_b + d)/n_f$, where n_f is refractance of the lens material for fundamenral wave, f_b is the back focal length of the lens and d is the thickness of the lens. The surface S2 has a dichroic coating whose reflectance is low(6%) for the fumdamental wave and very high(> 99%) for the Raman wave and its radius of curvature of S2 is the same as the back focal length(12cm) of the lens.

In this arrangement, the focusing lens L1 accomplishes complicated functions. S2 of the focusing lens acts as an initial output coupler of a fundamental resonator. Because of the low reflectivity of S2 for the fumdamental wave, most of the backward SBS fundamental wave generated in the Raman cell can be injected into the main oscillator and amplified in the gain medium(Nd:YAP). This feedback and amplification mechanism forms an optical resonator[5] which is represented by Brillouin resonator in Figure 1. The backward Raman beam is reflected by the surface S2 and amplified again.

The rear mirror, R, was a concave mirror with a radius of curvature of 10m. and the optical length of the resonator between the focusing lens and the rear mirror was approximately 50cm. The Nd:YAP rod(ϕ3mm$\times$50mm) was used as an active medium for the fundamental oscillator. The wavelength of the fundamental wave was 1.079μm and the wavelength of Raman wave in CH_4 gas was 1.575μm. The Q-switching filter, Q, was Cr^{4+}:YAG crysrtal. The Stokes output at 1.575μm and the residual fundamental pulse at 1.079μm were collimated by the lens L2(f=15cm). The length of the Raman cell was 15cm. The cell windows were made of fused silica(10 mm thichk) and their surfaces were not coated.

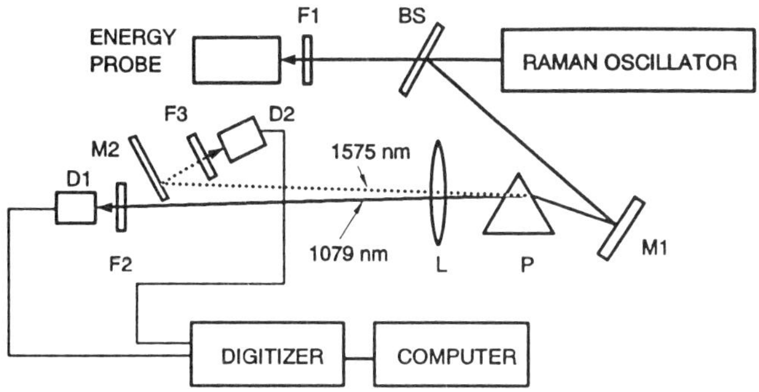

BS, beamsplitter; F1, F2, F3, color filters; M1, M2 mirrors; P, prism; L, lens(f=100cm); D1, D2, photodetectors

Figure 2 Experimental setup for detecting Raman signal

The experiment setup to measure the ouput characteristics of the Raman oscillator is shown Figure 2. The output energy at each wavelength was measured with the pyroelctric energy probe(Laser precision, RJP735). The pulse shape of both wavelengths and output energy were measured simultaneously. The pulse shape at 1.079μm was measured with a silicon photodetector D1(Antel optronics, PS-S5) and the pulse shape at 1.575μm was measured with an InGaAs photodetector D2 (ElectroOptics Technology, ET3000). The output signal of the photodetectors were recorded by the transient digitizer(Tectronix, RTD720) with a temporal resolution of 1ns and transfered to computer by GPIB(IEEE 488).

3. Experimental results.

Figure 3 shows the measured energies of the Stokes output at 1.575μm and the residual fundamental pulse at 1.079μm vs. pressure. The Stokes output genetation started to be generated from 600psi of CH_4 gas and the maximum energy obtained was 16.1mJ at 900psi. The energy of the residual fundamental pulse was about 18.2mJ at this condition, so that the Raman conversion efficiency at 900 psi was nearly 47%. Figure 4 shows the temporal profiles of the Stokes output and the residual fundamental pulse. The Raman output had a pulse width of appromimately 10ns(FWHM). Figure 4 indicates that Raman conversion and stimulated Brillouin scattering cause distortion of shape for the fundamental wave. Raman signal was generated later than the fundamental wave. It is due to the fact that at the first stage the fundamental wave starts to oscillate from the 6% output coupler and then SBS phase conjugation mirror laser, and if the intensity of the fundamental wave becomes higher enough to generate SRS, the Raman wave starts to be produced afterward.

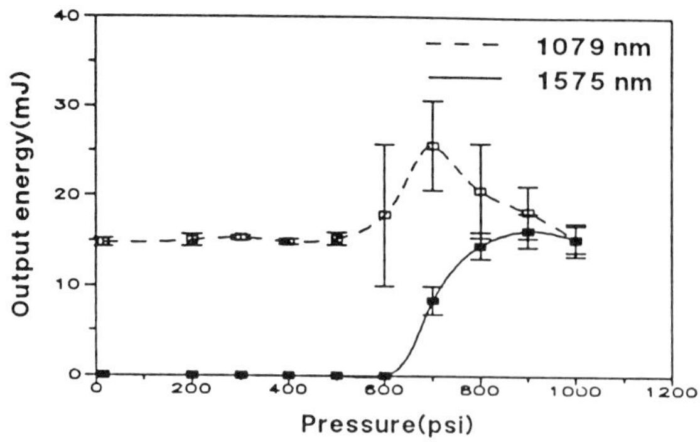

Figure 3 Output energy of Raman laser

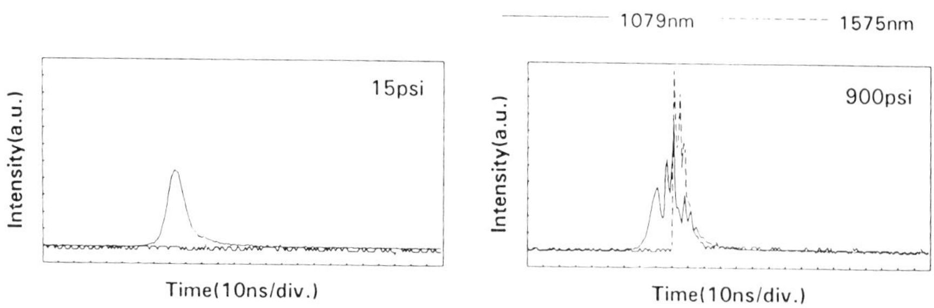

Figure 4 Temporal profiles of Raman and fundamental wave.

4. Conclusion

In summary, we have shown that the backward SBS that occurs in the Raman cell can be used to obtain efficient Raman conversion for Nd:YAP pump oscillator. The Nd:YAP fundamental laser radiation of 1.079μm is downshifted by the stimulated Raman scattering in CH_4 gas to the first Stokes wavelength of 1.575μm. In this configuration, the Raman conversion efficiency was 47% at pressure of 900psi of CH_4 gas similar to the previous work of ref[3].

References

[1] Claudio G Paeazzoli William W Buchman and Robert D Stultz 1988 *IEEE J. Quantum Electron.* **24** 872-880

[2] D C Jones M S Mangir D A Rockwell and J O White 1990 *J. Opt. Soc. Am. B* **7** 2090-2096

[3] H J Kong Y G Jeon and J K Kim *Efficient Raman conversion by utilizing backward stimulated Brillouin scattering Appl. Optics* (to be published)

[4] I D Carr and D C Hanna 1985 *Appl. Phys. B.* **36** 83-92

[5] D Pohl 1967 *Phys. Lett.* **24A** 239-240

Inst. Phys. Conf. Ser. No 140: Section 9
Paper presented at Laser Interact. with Matter Conf., Oxford, 19–23 September 1994
© 1995 IOP Publishing Ltd

High-power Nd:glass laser system in KAIST (Sinmyung I)

Hong Jin Kong, Sang Soo Lee, Ki Gwan Han, Nam Seong Kim, Hyun Soo Kim, Ki Young Um, Jong Rak Park, and Jae Yong Lee

Department of Physics, Korea Advanced Institute of Science and Technology, 373-1 Kuseong-dong Yuseong-gu Taejon, Korea

Abstract. A high-power Nd:glass laser system has been constructed and tested. In this system, we used a Nd:YLF laser as a master oscillator, a 4-pass amplifier for pre-amplification, 5 stages of rod amplifiers, and spatial filtering and image relaying units. The system has demonstrated in excess of 80J(2TW) with 40 psec full width at half-maximum(FWHM) pulse duration. Output gains, energies, and spatial profiles were measured at each amplification stage. A detailed description of the system performance is presented.

1. Introduction

KAIST completed the construction of a one beam Nd:glass laser system named as Sinmyung I capable of delivering 80J(2TW) in 40 psec pulses at system firing rates of 6 shots/hr. Basic features of this system include the adoption of a Nd:YLF laser as a master oscillator, the employment of a 4-pass amplifier for pre-amplification, the use of Nd-doped phosphate glass for rod amplifiers up to 90-mm diameter, and the utilization of spatial filtering and image relaying units.

In this paper, the overall explanation of Sinmyung I system is described in Section 2; the output caracteristic of Sinymung I system is presented in Section 3; finally, the system performance is summarized in Section 4.

2. The Overall explanation of Sinmyung I System

In this section, we introduce the scheme of the TW laser system(Sinmyung I)

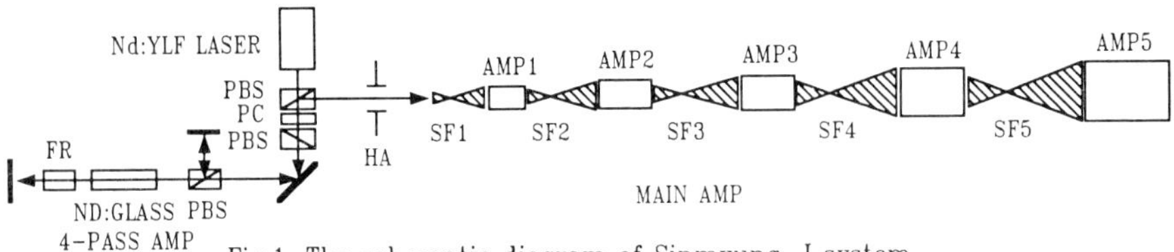

Fig.1. The schematic diagram of Sinmyung I system.
PBS : Polarizing Beam Splitter; PC : Pockels Cell;
FR : Faraday Rotator; HA : Hard Aperture; SF: Spatial Filter

consisting of the master oscillator, the 4-pass amplifier and the 5-stage amplifier. Fig[1] shows the schematic diagram of the laser system. The optical path length was about 40m from the master oscillator to the Amp.5. As shown in Fig[1], a single pulse selected by a Pockels cell propagated into the 4-pass amplifier and then was reflected by a polarizing beam splitter of the Pockels cell because the Pockels cell had no applied voltage when the pulse returned back. This reflected laser pulse was sent to the following 5 amplifier stages.

The 5-stage rod amplifier system consists of two types of cavity. One type(Amp.1 and Amp.2) was a quadruple elliptical cavity with Ag coated reflector and pumped by 4 flashlamps. The other one(Amp.3, Amp.4, and Amp.5) was a circular cavity with a barium sulfate($BaSO_4$) diffuse reflector and pumped by 12 flashlamps. Ellipticities of the reflectors of Amp. 1 and Amp. 2 were 0.62 and 0.67 at which were determined to give the maximum and most uniform pumping efficiency[1]. All amplifiers used phosphate-based laser glass to take adventage of the lower nonlinear index of refraction and higher specific gain than typical silicate-based laser glass[2]. Table[1] lists the parameters of Nd:Glass rods installed in each amplifier cavity. We used the flashlamps with 300mm arc length, 19mm bore diameter and Xe gas of 450 Torr. The lamps was energized by a 450 μsec critically damped pulse-forming network(PFN) consisting of a 200 μF capacitor and a 100 μH inductor. We operated this flashlamp charging energy at about 5% of the explosion energy(24kJ).

Table 1. Parameters of Laser Rods.

Title		Amp1	Amp2	Amp3	Amp4	Amp5
Glass Type		LG750	LG760	LG750	LG760	LHG-8
Size	Diameter(mm)	16	30	40	64	90
	Length(mm)	360	360	360	360	360
Doping % of Nd^{+3}		1.50	1.00	0.60	0.40	0.40
Fluor. Lifetime(μsec)		405	411	428	400	408
Face Angle(degree)		6.0	6.0	6.0	6.0	6.0

3. The Output Characteristic of Sinymung I System

The Nd:YLF laser(Quantronix, Model 4216) was Q-switched and mode locked at 1.053 μm. A single pulse of this output was selected by the Pockels cell. The pulse width of the single pulse was 40 psec(FWHM).

Table 2. Small signal gains of the 5 amplifier stages.

Total pumping energy	Small signal gain		Total pumping energy	Small signal gain		
	Amp. 1	Amp. 2		Amp. 3	Amp. 4	Amp. 5
1.80 kJ	4.6 ± 0.3	2.2 ± 0.5	5.40 kJ	2.1 ± 0.1	1.7 ± 0.1	1.3 ± 0.1
2.45 kJ	8.5 ± 0.4	3.3 ± 0.7	7.35 kJ	3.2 ± 0.2	2.1 ± 0.1	1.4 ± 0.1
3.20 kJ	16 ± 2	6.2 ± 0.6	9.60 kJ	3.9 ± 0.4	2.4 ± 0.1	1.6 ± 0.1
4.05 kJ	29 ± 3	10 ± 1	12.2 kJ	5.4 ± 0.5	2.6 ± 0.3	1.9 ± 0.1

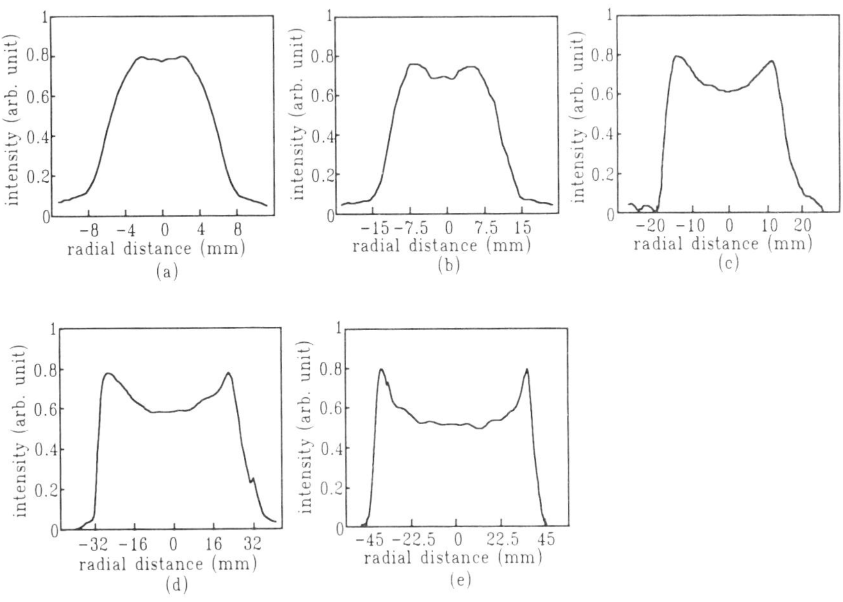

Fig.2. The spatial profiles at the rear surfaces of the 5 amplifier stages. (a) Amp.1; (b) Amp.2; (c) Amp.3; (d) Amp.4; (e) Amp.5

We used the degenerate 4-pass amplifier for a pre-amplifier of Sinmyung I laser system as shown in Fig[1]. The maximum gain of the 4-pass amplifier was 2×10^5 and the merit of this amplifier was a complete compensation for a distortion of a beam polarization due to a thermal birefringence[3].

We measured the single pass gains of the 5 amplifier stages which were listed in Table[2]. The gain of Amp.1 is the largest due to the high Nd-ion doping density. We measured the output energy and the spatial profiles of the Sinmyung I laser system. The output energy was 80J(2TW). The energy of the amplified spontaneous emission(ASE) was less than $10\,\mu$J. In order to operate the 5-stage amplifier system safely at this energy, the output spatial profile of each amplifier should have a good quality without ripples or spikes[4]. We removed unwanted ripples and spikes by installing a spatial filter between amplifiers and used the image relaying technique in which the image of a hard aperture was relayed to the end surfaces of each amplifier rod in order to reduce significantly the deleterious effects of whole beam self-focusing. Fig[2] shows the spatial profiles of an amplified laser beam on the end surface of each amplifier.

4. Conclusion

We constructed and demonstrated the TW level Nd:Glass system(Sinmyung I) consisting of a Nd:YLF laser as a master oscillator, 4-pass amplifier and 5-stage Nd:glass amplifier. The output pulse width of the Nd:YLF laser is 40psec, which was selected from a mode-locked & Q-switched pulse train by a Pockels cell. The maximum gain of the 4-pass amplifier was 2×10^5 and the merit of this amplifier was a complete compensation for a distortion of a beam polarization due to a thermal birefringence, which was considered to be the main feature for Sinmyung I system to produce high laser energy output and no spiking in the spatial mode. Output gains and spatial profiles were measured at each amplifier stage. We obtained the very clean spatial profiles without self-focusing. The Sinmyung I laser system demonstrated in excess of 80J(2TW) with 40psec(FWHM).

References

[1] Kyeong Koo Chi 1993 *Ph. D Thesis,* KAIST in Korea
[2] J Bunkenberg, J Boles and D Brown et.al 1981 *IEEE J. Quantum Electron.*, **QE-17** 1620
[3] Ki Gawn Han 1994 *pre-view for Ph. D Thesis,* KAIST in Korea
[4] David C Brown 1981 *High-Peak-Power Nd:Glass Laser Systems* (New York:Springer-Verlag) ch.7

Inst. Phys. Conf. Ser. No 140: Section 9
Paper presented at Laser Interact. with Matter Conf., Oxford, 19–23 September 1994

© *1995 IOP Publishing Ltd*

SMOOTHING OF ABLATION PRESSURE PROFILE BY CONTROLLING THE POSITION OF CRITICAL SURFACE

B. Králiková J. Krása L. Láska K. Mašek S. Přeučil K. Rohlena J. Skála P. Trenda A. V. Bessarab[1] S. G. Garanin[1] G. A. Kirillov[1] Yu. F. Kiryanov[1] G. G. Kochemasov[1] L. V. Lvov[1] A. B. Ryadov[1] S. A. Sukharev[1] N. A. Suslov[1] A. I. Zaretskii[1]

Institute of Physics of Czech Acad. Sci., Prague 8, Na Slovance 2, 180 40 Czech Republic

[1]Russian Federal Nuclear Centre VNIIEF Arzamas-16, Nizhnii Novgorod Region, Russian Federation

Abstract. A control of the critical surface position by increasing its distance from the ablation surface was attempted using a double pulse regime for plasma formation. A plasma is first formed by a weaker prepulse using the second (red) harmonic of an iodine laser and only then the main, third (blue) harmonic pulse is introduced. It is hoped that the lateral thermal conductivity will help to smooth out the ablation pressure non-uniformuty of a spatial scale of $50 - 100 \mu m$. The experiments with an acceleration of a plane Al foil were performed on the PERUN laser facility ($\lambda = 1.315 \mu m, 0.5 ns$, $E = 40 - 50 J$, conversion to blue ~ 50 %). The effect of the smoothing was evaluated from x-ray pinhole pictures of the focal spot.

1. Introduction

The ablation pressure variations generated by an inhomogeneous illumination of the target surface are in the compression experiments at the root of the onset and subsequent growth of Raighly-Taylor instability with ensuing irregular compression and detrimental mixing of the target material.

Various technics have been invented and applied to cure the illumination inhomogeneities. One class of experiments uses the principle of indirect drive by generating secondary X-ray radiation inside a cavity and letting this radiation to compress the target. The process is very efficient (even if not fully) in smoothing the primary illumination inhomogeneities, but it requires a higher laser energy than for the direct drive. The direct drive experiments try to eliminate the illumination inhomogeneity both by increasing the number of beams and by tinkering at the output laser beam or the beam line to suppress the coherence inherently present in the lasing process.

In this contribution we decided to test the idea of smoothing the ablation pressure inhomogeneity by increasing the distance between the critical and ablation surface and allowing thus the lateral thermal conductivity to wash out the spatial fluctuations occurring on the critical surface due to the uneven illumination. The position of the critical versus ablation surface can be controlled by driving the main heating pulse through a prepulse plasma. The prepulse weakens the density gradient and prolongs the distance the heat deposited at the critical surface must cross to power the ablation. This idea has been tested several times computationally [1, 2, 3]. It was found that the instability growth rate with the prepulse can be several times lower than without one. To illustrate

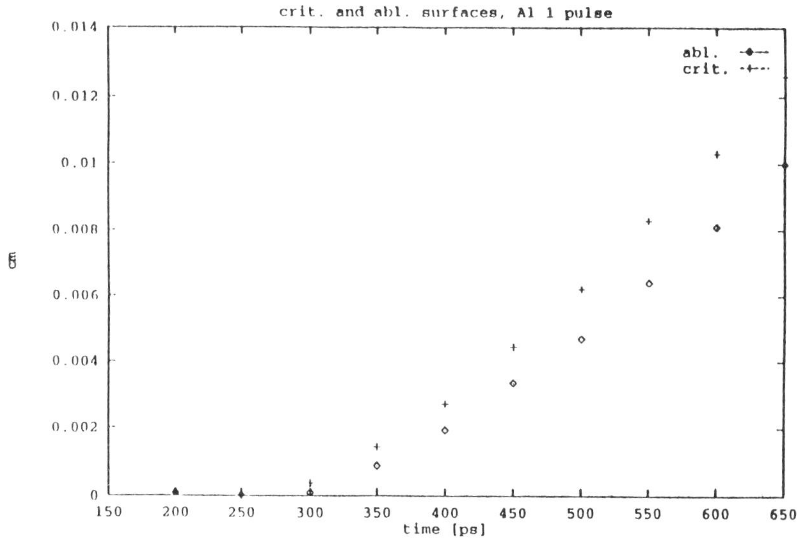

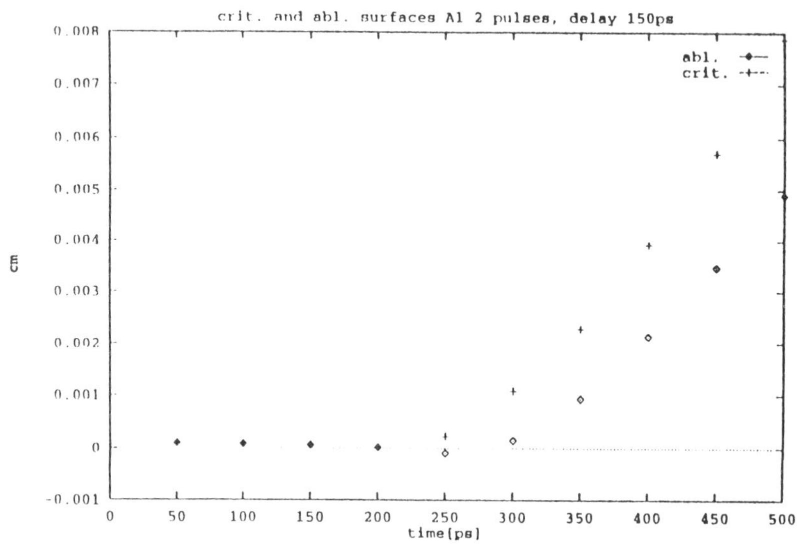

Figure 1: Example of time trajectories of ablation and critical surfaces with and without a prepulse. Iodine laser: $2\omega/3\omega$ peak power ratio 2.5% ($10^{15}W/cm^2$), $300ps$ pulse length, delay between the pulses $150ps$. Target $1\mu m$ Al.

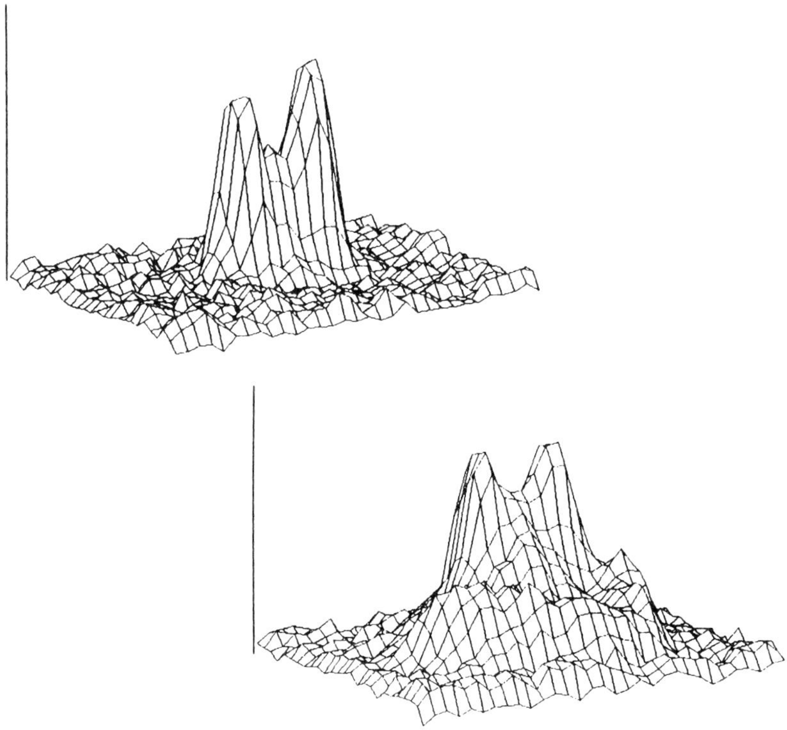

Figure 2: Densitometry of focal image with (right) and without a prepulse. Power ratio 4 %, average power $5.10^{14}W/cm^2$, time delay $1ns$

this idea we applied a 1D-hydrocode (MULTI [4]) to evaluate the time trajectories of the ablation as well as critical surfaces in both the cases fig. 1. It is, however, a common knowledge that a precursor to the main pulse is generally harmful for the laser - target coupling and, hence, if a prepulse were applied it would have to be carefully tuned. We think that this difficulty will be alleviated and the coupling of the heating pulse to the target improved if a thinner precursor plasma is formed. A natural way towards this kind of interaction is to use a two colour pulse. An experiment was thus devised using the converted light of an iodine laser to the second (red light) and third harmonics (blue light) so that the penetration of the heating pulse in the precursor plasma is deeper than if using a double pulse of the same colour.

2. Experimental

To test the idea of smoothing the single beam iodine photodissociation laser installation Perun [5] in the Division of Gas Lasers at the Institute of Physics of Academy of Sciences of the Czech Republic in Prague was used as a driver. The system produces NIR ($1.315\mu m$) pulses $500ps$ long on the first harmonic routinely up to $40J$ in energy.

A conversion to the second and third harmonics by DKDP crystals with ~ 50% efficiency is available. The target chamber is fitted with an aspherical optics which can be adjusted to either of the three harmonics $\omega, 2\omega$ and 3ω of the iodine laser wavelength. The conversion line consists of a converter and of a mixer crystal followed by selective mirrors for the first and second harmonics separation. The crystals operate in the type II - type II conversion and mixing schemes [6], so that the distribution of power between the harmonics can easily be adjusted by tilting them out of resonance angles. The remaining first harmonic power may be used to drive a two frame back lighter. The maximum power available at 3ω is just over $20J$. For the first (trial) experiments the focusing optics was set to 3ω harmonics (blue) and the separated 2ω beam (red) was passing through an auxiliary lens to achieve about $300\mu m$ focus. Optionally a random phase element could be added for further homogenization. The delayed 3ω blue beam was split by a wedge in two slightly diverging beams, which were then focused onto the larger red focal spot to form two separate blue foci of about $50\mu m$ in diameter each. As a target about $1\mu m$ Al layer on a milard foil was used. In this preliminary experiment pinhole cameras were placed at different angles to the optical axis allowing only a passive observation of the focal structure by registering the focal image on an X-ray film. The results of densitometry of the pin hole images of the rear side of the target is seen in fig. 2. In this particular shot both the pulses (red and blue) had nearly the same energy $5J$ and the time delay was $1ns$.

3. Conclusion

The single shot with a two colour pulse is pointing more to an experimental feasibility of the smoothing test rather than proving conclusively that the above mentioned mechanism works. Systematic measurements are planned involving variations in the power ratio as well as time delay between the pulses. We expect the effect to become more pronounced when using overlapping pulses.

References

[1] Gamaly E G et al 1990 *Laser and Particle Beams* **8** 399

[2] Bokov N N et al 1982 *Zh.Prikl.Mat.Tech.Fiz.* **4** 20

[3] Lindl J P 1977 *Lawrence Livermore Laboratory Prepr.* UCRL-79735

[4] Ramis R et al 1987 *Computer Physics Communications* **49** 475

[5] Chvojka M et al 1988 *Czech.J.Phys.* **B38** 1337

[6] Craxton R S 1981 *IEEE J.Q.Electronics* **17** 1771

Paper presented at Laser Interact. with Matter Conf., Oxford, 19–23 September 1994
© *1995 IOP Publishing Ltd*

Effects of plasma physics on capsule implosions in gas-filled hohlraums

E L Lindman, N Delamater, G Magelssen, A Hauer

Los Alamos National Laboratory, Los Alamos, NM 87545 USA

Abstract. Initial experiments on capsule implosions in gas-filled hohlraums have been carried out on the NOVA Laser at Lawrence Livermore National Laboratory. Observed capsule shapes from preliminary experiments are more oblate than predicted. Improvements in modeling required to calculate these experiments and additional experiments are being pursued.

1. Introduction

Capsule-implosion symmetry experiments in gas-filled hohlraums at NOVA are needed to support proposed ignition targets for the National Ignition Facility which are gas-filled to suppress spot motion and help control symmetry. In unlined gold hohlraums the region of the wall heated by the laser (laser hot spot) moves substantially. When this motion is excessive, it interferes with the achievement of the required capsule-irradiation symmetry. This motion can be controlled by placing a thin plastic liner on the inner surface of the hohlraum or by filling it with gas. The plastic liner reduces the wall motion but generates a pressure spike on the axis which destroys the symmetry of the capsule implosion through hydro coupling.

Gas-filled hohlraums are currently preferred because gas fill reduces the wall motion and avoids the hydro-coupling asymmetries. Design calculations suggest that the desired results can be achieved with 1 mg/cm^3 of helium or an equivalent equimolar mixture of helium and hydrogen. When fully ionized this initial gas fill has an electron density of $n_e = 0.033 n_{crit}$. Subsequent motion of the walls and capsule ablater raise the electron density in some regions to values approaching quarter critical.

2. NOVA Experiments

The capsule-implosion-symmetry experiments were carried out in scale-1 hohlraums with standard capsules. A scale-1 hohlraum has a diameter of 1650 μm and lengths between 2100 and 2800 μm. The laser entrance holes (LEH) at the ends of the cylindrical hohlraum have diameters of 75% of the hohlraum diameter. The hohlraums were unlined and filled with 1 atm of either neopentane, propane or methane held in by a 0.6 μm thick Mylar window. Hohlraums with lengths between 2300 and 2700 μm were used with laser pointing from 1125 to 1325 μm such that the laser beams entered at the center of the LEH's.

A "standard" capsule has an inner radius of 220μm and a total wall thickness of 55 μm made up of an inner layer of polystyrene of thickness 3.5 μm, an intermediate layer of PVA of thickness 2.5 μm and an outer layer of CH of thickness 49 μm. 28.5 kJ was delivered to the target using "pulse shape 22" which consists of 1.5 ns foot followed by 1 ns main pulse with a contrast ratio between two and three.

The targets that have been shot are shown in Table 1. The 20 μm thick "symcap" included in Table 1 was a thinner-walled capsule designed to implode faster and thus sample the earlier part of the pulse. It performed as expected.

Table 1: Capsule-shell thicknesses in a gas-fill-vs-pointing array for shots that have been completed.

Gas at 1 atm.	n_e/n_c	ρ [mg/cc]	N-pentane pr. (atm) at same ρ	Capsule at 1125 μm pointing	Capsule at 1225 μm pointing	Capsule at 1325 μm pointing
N-pentane C_5H_{12}	0.126	3.2	1.0	55 μm	55 μm 55 μm	55 μm
Propane C_3H_8	0.077	2.0	0.62	55 μm	20 μm	
Methane CH_4	0.030	0.72	0.22	55 μm		

3. Data and analysis

A typical sequence of images seen by the gated-X-ray-imaging camera is shown in Fig. 1. Time goes from right to left, and the images are separated by about 65 ps. An expanded view of the brightest image at peak compression is shown, and the half-intensity contour used to calculate its eccentricity is drawn in. The cylindrical axis is horizontal and the eccentricity is defined as the length of the vertical semi-major axis divided by that of the horizontal semi-major axis. Using this definition, oblate spheroids ("pancakes") have an eccentricity greater than 1, and prolate spheroids ("sausages") have an eccentricity less than 1. This image has a calculated eccentricity of 1.52.

Experimental and calculated eccentricities are compared in Table 2. The calculations are 2-D cylindrical and include self-consistently the laser deposition, the dynamics of the window that holds the gas in the hohlraum, the gas itself and the wall of the hohlraum as well as the capsule implosion. The energy was deposited using only inverse bremstrahlung for absorption, and no scattering mechanisms were included. The lack of a consistent trend in the experimental data with either pointing or gas density sug-

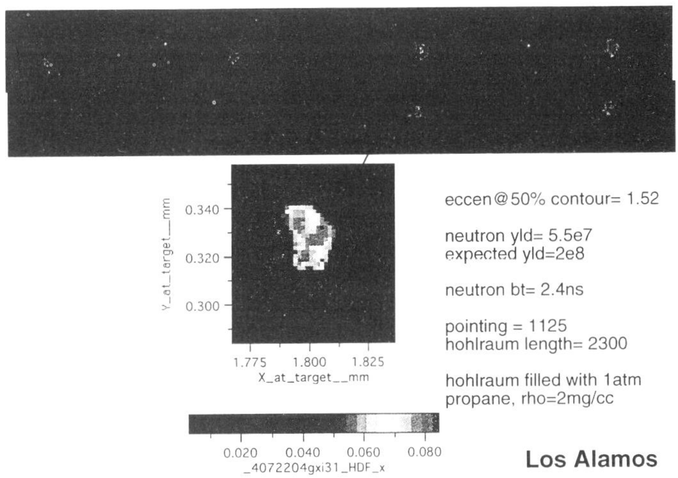

Figure 1: Gated x-ray images of a capsule imploded in propane.

gests that additional experiments with improved procedures and/or diagnostics should be pursued.

The disagreement of the calculated eccentricities with the experimental values is, however, persistent and large enough to suggest that a discrepancy exists between modeling and experiment. A value of $f_e = 0.05$ was used for the electron thermal-flux limiter in these calculations. This choice has been used successfully in the modeling of implosion symmetry in evacuated hohlraums [1]. Since the value of the electron thermal-flux limiter has been chosen in the past to obtain better agreement between theory and experiment, its change is a reasonable thing to try to improve the agreement here.

Table 2: Comparison of observed eccentricities with those calculated with an electron flux limiter of $f_e = 0.05$ and with enhanced electron thermal conductivity in parenthesis.

Gas at 1 atm.	Eccentricities at 1125 μm pointing		Eccentricities at 1225 μm pointing		Eccentricities at 1325 μm pointing	
	Exp.	Theor.	Exp.	Theor.	Exp.	Theor.
N-pentane C_5H_{12}	2.2	0.75 (1.20)	2.2	1.6 (1.8)		2.9 (2.4)
Propane C_3H_8	1.5	0.72 (0.93)				
Methane CH_4	2.1	0.61 (0.92)				
Vacuum No window	0.59	0.67				

Consequently, a series of calculations of the propane experiment was carried out in which the flux limiter was varied between 0.005 and 1.0. A consistent trend was obtained in which the eccentricity rose in value as the flux limiter approached 1. Unfortunately the largest value approached but did not agree with experiment. Since the trend was in the right direction, an additional increase in thermal conduction was included by increasing the diffusion coefficient by a factor of three. Again the calculation moved closer to the experimental values but agreement was not obtained.

The computed eccentricities obtained with a flux limiter of 1.0 and a factor of three enhancement of the diffusion coefficient are given in parenthesis in Table 2. A decrease in the discrepancy is obtained for the other cases as well, but the agreement is still not good.

4. Conclusions

With data from only seven shots, it may be premature to attempt to draw any conclusions. Additional experimental data is required and is currently being pursued. If the discrepancy between theory and experiment, seen in the data shown here, is supported by additional experiments, some changes in the modeling may be required. The inclusion of non-local thermal conduction models [2] may help. But other mechanisms, which redistribute the energy in the hohlraum, should be considered as well. Anomalous absorption which deposits the energy near the LEH is an example. Since only 100 J out of 28 kJ are seen experimentally in hot electrons, however, such mechanisms may be unimportant.

Stimulated Brillouin scatter over a range of angles is another possibility. Energy scattered into a backward cone would remain near the LEH and subsequent transport by thermal conduction or x-rays would make the imploded capsule more "pancake." Approximately 4% of the incident energy is scattered back into the laser optics for each beam in these experiments. The solid angle of one of the f4 beams is approximately .05 steradians. The additional solid angle in which the energy could scatter (4π steradians) is thus 260 times the backscatter solid angle Scattering into 4π at an average intensity of 1/10th the backscatter intensity would remove a significant fraction of the beam.

Acknowledgements

The authors would like to thank R. Berger, L. Powers, L. Suter, B. Wilde, J. Fernandez and W. Hsing for helpful discussions. The targets used in the experiments were fabricated by J. Moore, K. Gifford and M. Salazar. We would like to thank S. Evans, G. Peterson and J. Jimmerson for their help with the diagnostics as well and the NOVA operations staff for their help in carrying out the experiments.

References

[1] Hauer A, Delamater N, Hsing W, Ress D, Suter L, Powers L, Harris D, Landen O, Thiessen R, Phillion D, Amendt P, Magelssen G, Watt R and Hammel B 1993 *Bull. Am. Phys. Soc.* **38** 1884

[2] Luciani J F, Mora P and Virmont J 1983 *Phys. Rev. Lett.* **51** 1664

Inst. Phys. Conf. Ser. No 140: Section 9
Paper presented at Laser Interact. with Matter Conf., Oxford, 19–23 September 1994

© 1995 IOP Publishing Ltd

Expansion characteristics of Optical field ionised channels following irradiation with a 350fs,0.5TW KrF laser pulse.

A. J. Mackinnon, S. Allfrey, M. Borghesi M. W. Jones, O. Willi
The Blackett Laboratory, Imperial College
London SW7 2BZ

Abstract. The interaction of a 350fs, KrF laser pulse with a high density gas jet target has been studied with a time resolved Moiré deflectometer. The diagnostic measured both the spatial uniformity of the plasma immediately after the interaction and the subsequent expansion of the optically ionised density channel. Over the range of gas pressures studied the late time channel expansion agreed well with a simple adiabatic model. The plasma was observed to be very non uniform at high backing gas pressures.

1. INTRODUCTION

The interaction of intense subpicosecond laser pulses with underdense targets can result in plasmas where the optical field of the laser (OFI) rather than collisional effects dominate the ionisation stage. The residual electron temperature is determined, in the absence of other heating mechanisms, by ionisation heating (ATI). This leads to a highly ionised but relatively cold plasma [1].

Plasmas with these properties have been predicted to present a suitable medium for a number of recombination X-ray laser schemes [2]. The simultaneous achievment of high electron densities and low electron temperatures is crucial to the success of the recombination X-ray laser. The mechanisms that can contribute to increasing the electron temperature over the residual ATI level include, inverse bremsstrahlung (IB), Raman heating and ponderomotive heating from beam non uniformities. At low densities the dominant mechanisms are ATI and ponderomotive heating[1], as the density is increased Raman heating and IB become increasingly more important. One of the most effective methods of reducing the contribution of these heating mechanisms is to reduce the pulse length of the ionising laser. For instance Eder et al [2] have predicted that the Raman heating drops from 265eV to 30 eV when the laser pulse width is reduced from 100fs to 50fs.

A number of recent experiments have been carried out to measure electron temperatures in OFI plasmas. Offenberger et al [3] have used Thompson scattering off the interaction laser beam itself to measure the temperatures existing in plasmas produced by the interaction of a 12ps Raman shifted KrF pulse with Helium and Neon gases. Inverse bremsstrahlung was found to be the main heating mechanism. At a slightly shorter pulse length Dunne et al [4] have used the technique described in this paper to measure the temperature of nitrogen plasma after irradiation by a 3ps KrF laser pulse. They measured an electron temperature of 35eV which was again dominated by inverse bremsstrahlung. In contrast to these picosecond measurements Glover et al [5] have recently measured electron temperatures as low as 40eV for 800nm, 2×10^{17}Wcm^{-2} and 150fs laser pulses focused into helium gas. They found that the plasma temperature was independent of ion density over the range of 1-100Torr, consistent with heating arising purely from ATI.

This paper will report on measurements taken with a high pressure gas jet target. The laser pulse length, at 350fs, falls between the picosecond and 100fs regime where there is still a considerable contribution arising from inverse bremsstrahlung and Raman heating.

The influence of beam hot spots on the plasma uniformity is discussed in connection with ionisation defocusing[6] and filamentation [7].

2. **Experimental Arrangement**

The experiment was undertaken on the SPRITE laser at the Rutherford Appleton Laboratory, which was operated in chirped pulse amplification mode. The compression gratings were located under vacuum in the target chamber. Typical operating parameters were a pulse duration 350fs (FWHM), 140-300mJ on target and a 30μm FWHM focal spot (with a F/4 off axis parabolic mirror focusing optic). This gave a peak vacuum intensity of 1.2×10^{17}Wcm^{-2}. The main pulse was superimposed on an ASE pulse of FWHM 20ns and an intensity of 1×10^{9}Wcm^{-2}. The target was a simple solenoid operated gas jet with a 1mm diammeter cylindrical nozzle. The plasma was diagnosed with a temporally independent probe pulse which was split off the main uncompressed heating beam. The duration of the uncompressed beam was less than 10ps. This pulse length was short enough to give a snapshot of the plasma density profile without any fringe distortion arising from plasma motion. A Moiré deflectometer probed the electron density gradient and density profiles at discrete intervals for times up to 2ns after the end of the pulse. A schematic diagram of the experiment is shown in Fig 1.

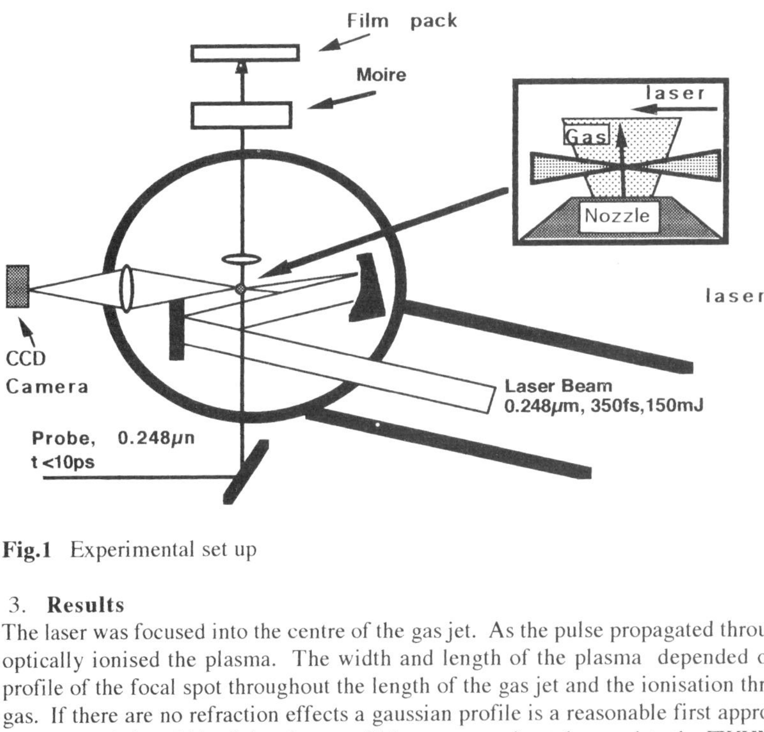

Fig.1 Experimental set up

3. **Results**

The laser was focused into the centre of the gas jet. As the pulse propagated through the gas it optically ionised the plasma. The width and length of the plasma depended on the spatial profile of the focal spot throughout the length of the gas jet and the ionisation threshold of the gas. If there are no refraction effects a gaussian profile is a reasonable first approximation for the laser and the width of the plasma will be approximately equal to the FWHM of the laser spot and the plasma will be approximately 2 Rayleigh lengths long. After the ionisation process the temperature of the ATI heated electrons will be increased by the inverse bremsstrahlung and Raman mechanisms. After the pulse turns off the plasma expands, driving a cylindrical blast wave into the surrounding neutral gas. The Moiré deflectograms in Figs 2 and 3 show how the channel expands between 20ps and 2ns after the interaction for a backing pressure of 10 Bar neon.

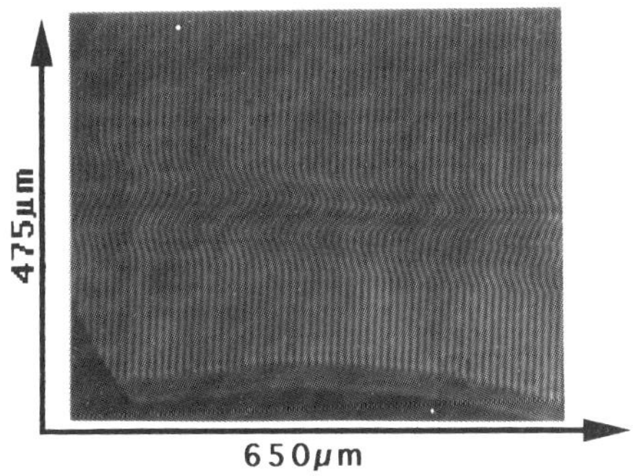

Fig.2 10 Bar backing Pressure, T=+20ps

Fig.3 Channel expansion, T=+ 2ns

The asymptotic expansion characteristics of the density channel are well described by the adiabatic strong shock model developed by Zeldovich and Raizer [8]. From this model, applied in cylindrical co-ordinates, the temporal evolution of the radius of the channel R depends on the ambient gas density ρ, the initial deposited energy E and time t as $R \propto (E/\rho)^{1/4}\, t^{1/2}$

Fig.4 shows how the channel radius increases with time for increasing backing pressure. The straight lines on the graph are fits to the data following an adiabatic $t^{1/2}$ dependence. It can be seen that the lower density data agrees well with the adiabatic model for times greater than 200ps. At a backing pressure of 30 bar the data agrees with the adiabatic model after 1ns.

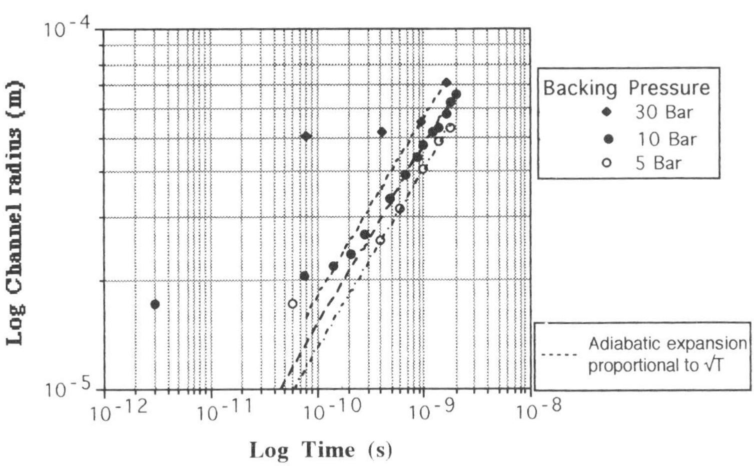

Fig.4 Channel Radius vs time for Neon for increasing backing pressure.

From Fig.4 it can also be seen that as the gas density is increased the asymptotic radius also increased, suggesting that ATI heating, which is independent of density, is not the dominant heating mechanism. The initial electron temperature can be be inferred from the expansion characteristics by modelling the expansion with a 1D hydrocode. Temperatures measured using this technique will be compared to a new analytic inverse bremsstrahlung absorption treatment developed by Pert [9]. This will allow the contribution of ATI, inverse bremstrahlung and Raman heating to the residual electron temperature to be determined.

5. Plasma non-uniformities

Fig.5 shows a deflectogram of the plasma 60ps after the interaction the laser with a backing pressure of 30 bar neon. The deflectogram shows that the plasma is non uniform and has broken up into a large number of channels. These channels were thought to originate from hot spots on the laser beam, which can be seen on the equivalent plane image of the laser beam shown in Fig.6. The non uniformities could simply be density channels resulting from ionisation induced refraction from the laser hot spots. One feature that supports this view can be seen in Fig.5, where the width of the plasma has increased relative to Fig.2. This shows that ionisation defocussing is an important effect in the higher density interaction.

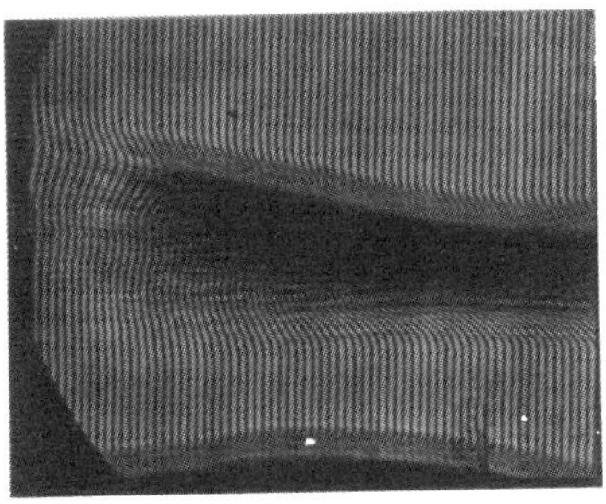

Fig.5 Moiré Deflectogram, 30 bar neon

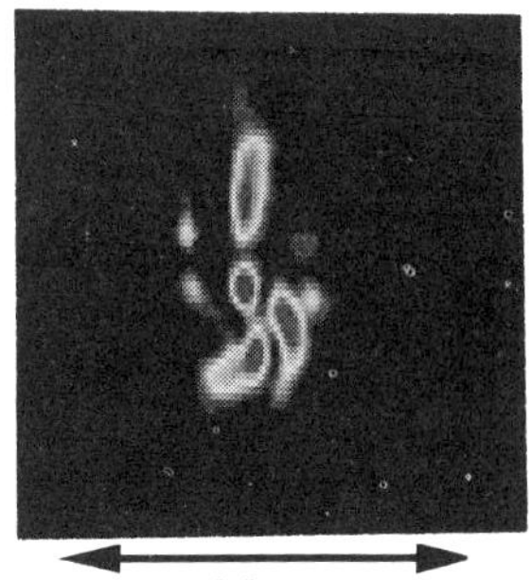

Fig.6 Equivalent plane image of focal spot

An alternative interpretation is that the beam break up arises from filamentation seeded from the beam hot spots. The possible causes of the filamentation are limited by the immobility of the ions, during the laser pulse, to either space charge limited ponderomotive or relativistic mechanisms. Further work is required to deduce the contribution from filamentation to the formation of the plasma non uniformities .

6. Conclusions

The rate of expansion of an OFI produced density channel has been studied for the case of a 350fs pulse interacting with a Neon gas jet target. The late time expansion is observed to be well described by an adiabatic model. The initial deposited energy is also found to increase as the gas density is increased. Further modelling is underway to infer the initial electron temperature from the rate of channel expansion. The plasma was observed to be non uniform at high gas densities. Further analysis and modelling are required to determine whether refraction or filamentation of the hot spots is responsible for the plasma non uniformity .

6. References

1 B.M. Penetrante and J.N.Bardsley, Phys Rev A. 43, 3100, (1992)
2. P.Amendt,D.C.Eder,S.C. Wilks, Phys. Rev. A. 45, 6761 (1992)
3 A. A Offenberger et al Phys.Rev.Lett. 71,3983, (1993)
4 M. Dunne, et al., Phys.Rev.Lett. 72,7, 1994 (1994)
5 T.E. Glover, et al., Phys.Rev.Lett ,73,78,(1994)
6 S.C. Rae Optics Comm. 97 (1993)
7 A.B. Borisov et al., Phys. Rev. Lett. 68, 2309 (1992)
8 Zeldovich and Raizer Physics of Shock Waves.., Academic Press 1966
9 G.J.Pert to be published

Inst. Phys. Conf. Ser. No 140: Section 9
Paper presented at Laser Interact. with Matter Conf., Oxford, 19–23 September 1994
© *1995 IOP Publishing Ltd*

Optical Guiding of 10TW, picosecond laser pulses in preformed density channels at high intensities and densities

A. J. Mackinnon, S. Allfrey, M. Borghesi, A.Iwase, O. Willi
The Blackett Laboratory, Imperial College
London SW7 2BZ
and
F. Walsh
Rutherford Appleton Laboratory,
Chilton,Didcot,Oxon OX11 0QX

Abstract. The interaction of intense laser pulses with high density gas targets has been studied using time resolved optical probing. At high target densities the plasma was observed to break up into small filaments which were refracted at large angles to the propagation direction. It was thought that the refraction was due to ionisation induced defocusing. The use of preformed density channels was found to overcome the refraction and guide the pulse.

1. INTRODUCTION

The propagation of an extremely intense laser pulse through underdense plasmas is an important issue in a number of possible applications. In the recombination X-ray laser scheme [1] a sub-picosecond laser is focused into a neutral gas to create an optically ionised (OFI) plasma. This forms a plasma where the ionisation stage is determined by the intensity of the laser and the electrons are heated predominately by ionisation heating (ATI) [2]. The efficiency and gain of the system are directly affected by propagation issues such as ionisation defocusing and filamentation which limit the length and uniformity of the plasma.

The fast igniter scheme[3] consists of three distinct stages. The first is a conventional laser driven implosion that forms a high density fuel. Then an ultra-intense 100-200ps laser pulse is used to create a density channel leading from the corona to the overdense core region. In the final stage a 1-10ps pulse propagates through the channel until it is absorbed at densities around 100x critical, producing MeV electrons which heat the fuel and initiate thermonuclear burn. This scheme relies on the overdense propagation of the channel forming pulse, which is achieved through a combination of the laser's extremely high radiation pressure and a relativistic lowering of the local plasma frequency. The issue of intensity dissipation in the underdense plasma is also important as any reduction in the laser intensity will reduce the maximum density to which the channel forming pulse can propagate.

A study of the mechanisms affecting the propagation of extremely intense pulses through the underdense plasma is an interesting problem for both these applications. Refraction from density gradients caused by any ionisation will reduce the laser intensity more rapidly than diffraction. This ionisation defocussing mechanism has been studied theoretically by Rae [4] and by Burnett [5], who suggested the use of preformed density channels as a method of overcoming it. The formation of the channel has been studied by Rankin et al [6] theoretically and the guiding of a 100ps pulse down a density channel has been experimentally observed by Milchberg and Freeman [7]. The effect of refraction on the length of the preformed channel has also been recently studied theoretically by Liu and Tripathi [8].

This paper will present experimental results that provided a measurement of the ionisation induced defocusing and studied the use of preformed channels to overcome the refraction.

2. Experimental Arrangement

This experiment was undertaken on the VULCAN laser at the Rutherford Appleton Laboratory, which was operated in chirped pulse amplification mode. The compression gratings were located under vacuum in the target chamber. Typical operating parameters were 15-40J on target in a pulse of temporal width (FWHM) of 1-4ps and of spatial width (FWHM) of around 10 - 20μm (obtained with a F/4 off axis parabolic mirror). This gave peak powers of 5-10TW and a peak intensities of 1-3x10^{18}Wcm^2. The laser was focussed on the boundary of the gas jet target which was a simple solenoid operated gas jet with a 1mm diameter cylindrical nozzle. The plasma was diagnosed with a temporally independent probe pulse which was compressed with a separate set of gratings, the pulse was then frequency doubled using a KDP crystal. The duration of the probe beam was less than 10ps. A Moire deflectometer [9] used the probe to detect density gradients in the plasma. The deflectometer was sensitive to density gradients normal to the probe direction. The preformed channels were formed with co-linear pulses that were either 1% or 10% splits off the interaction beam.

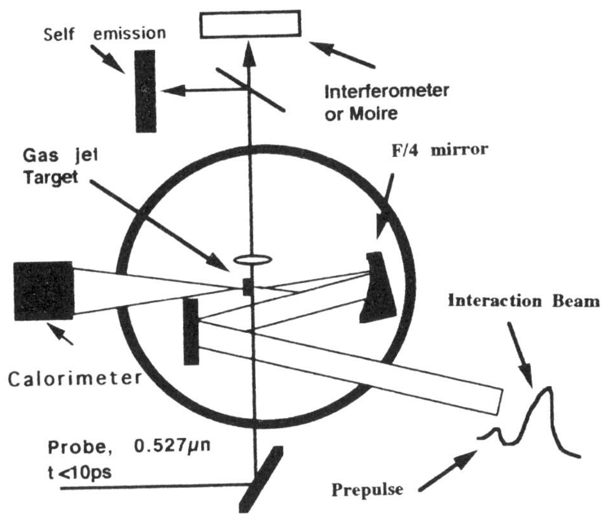

Fig 1. Experimental set up

The space resolved self emission channel was used to detect the time integrated plasma fluorescence at perpendicular to the propagation direction. The film pack was filtered using an interference filter with a centre wavelength of 620nm and a bandwidth of 25nm. The temporal and spatial characteristics of the beam were diagnosed using a single shot auto correlator and an equivalent plane monitor.

3. Results

Figs 2(a),3(a) and 4(a) show Moire deflectograms of the interaction of 6TW pulses with a neon gas jet target at backing pressure of 15 bar. In Fig 2(a) the laser was focused into the neutral gas,Fig 3(a) shows how a density channel has expanded 2.3ns after it was formed with a 1%

prepulse and Fig 4(a) shows how the preformed channel modifies the refraction of the interaction beam. Figs 2(b) and 4(b) show the self emission channels corresponding to 2(a) and 4(a).

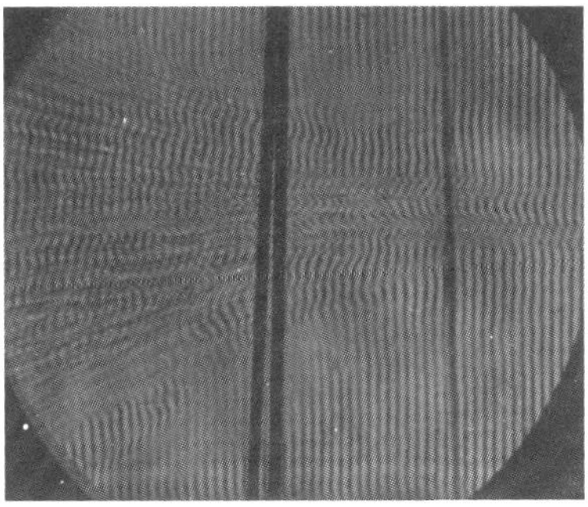

Fig.2(a) No preformed plasma

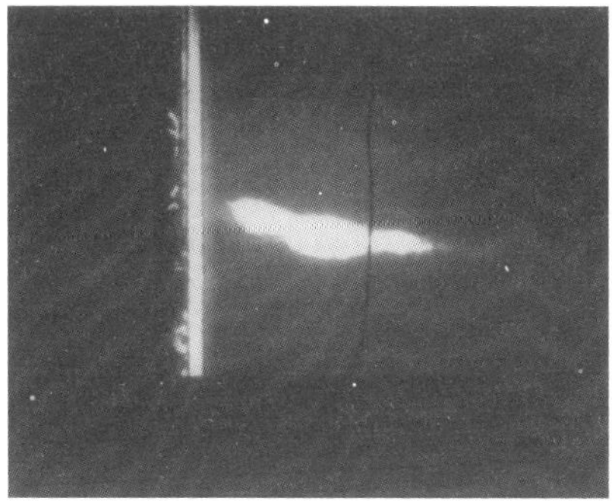

Fig.2(b) Self Emission.

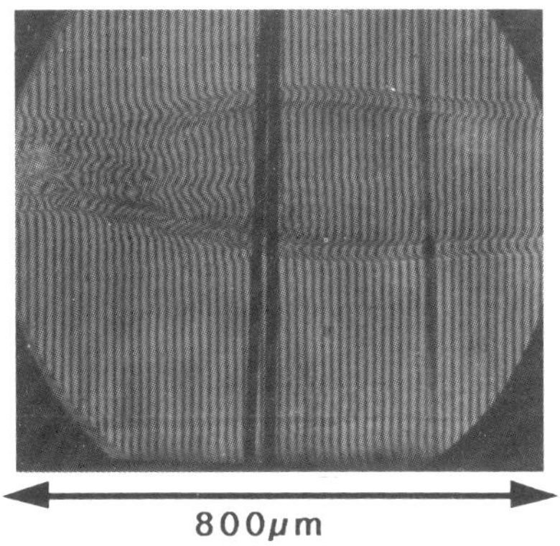

Note : The bar that obscures the centre of the field of view on all the pictures was part of the imaging system(it was the join between the two filters on the film pack),

Fig 3 Preformed plasma only, showing the non uniform channel width along the propagation direction. This deflectogram was taken 2.3ns after the incidence of the prepulse.

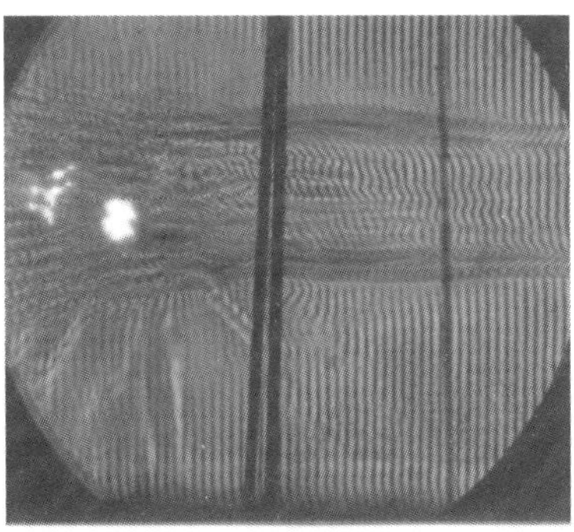

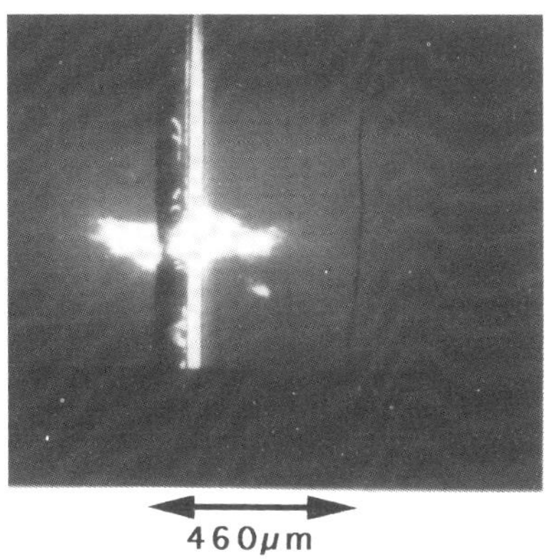

Fig 4 (a) Main pulse focused into density channel. **Fig 4(b)** Self emission channel.

4. **Interpretation**

Fig.2 (a) shows the laser focused directly into the neutral gas jet. From the density gradients visible in this picture it was clear that the laser had broken up into filaments, which were then refracted at large angles to the propagation direction.

The image of the self emission from Fig 2(b) confirmed that as the laser propagated the diameter of the plasma increased rapidly. In the absence of a plasma the diffractive spreading of a laser pulse with a gaussian spatial profile would be given by the Rayleigh length $L_r = \pi a^2/\lambda$. For the laser parameters used in this experiment L_r ranged from 300-600µm and the plasma would extend for approximately 2-5 Rayleigh lengths. The self emission shows that the plasma was around 450µm which was barely more than one Rayleigh length. It is clear from these two pictures that the intensity along the length of the plasma was severely modified by ionisation induced refraction. The length characterising the expansion of the plasma in this case is the defocussing length L_d given by [4,], $L_d = (\lambda/2)N_c/N_e$. L_d was estimated to be around 50µm for our plasma conditions, which was significantly less than a Rayleigh length.

In order to overcome these refractive effects a prepulse was used to form a preformed density channel. Fig 3(a) shows such a channel created by focusing 1% of the interaction pulse into the gas and allowing the plasma produced to expand for 2.3ns. It can be seen that the plasma had a uniform width only over the first 400µm. Regions of high density can also be seen towards the end of the channel. Ionisation induced refraction of the prepulse was thought to be responsible for these effects. Axial non uniformity of the preformed has been attributed to ionisation induced refraction in a recent theoretical paper by Liu [8].

Fig 4(a) shows a deflectogram of the interaction beam focused into the edge of the preformed channel. It can be seen that in the uniform part of the channel the refracted laser light is contained and the pulse is guided. When the channel narrows, part of the pulse manages to escape through the now leaky channel walls while the rest of the light is effectively focused onto the high density unexpanded plasma, with dramatic results. The self emission picture shown in Fig 4(b) connfirms that the laser was guided for approximately 300µm over Fig 2(b). This was equivalent to guiding over at least 6 defocusing lengths.

5. **Conclusions**

Ionisation induced refraction in the interaction of picosecond pulses with dense gas jet targets has been observed using time resolved optical probing. The refraction effects were overcome with the use of a preformed density channel, allowing propagation over many defocusing lengths. Further analysis and modelling is being carried out to give a more detailed insight into the feasibility of the role of preformed channels in the X-ray laser and fast igniter applications.

6. **References**

1. D.C.Eder,P.Amendt,S.Wilks, Phys. Rev. A **45**, 6761 (1992)
2. B.M. Penetrante and J.N. Bardsley, Phys. Rev. A **43**, 3100 (1991)
3. M.Tabak et al., Phys. Plasma.**1**, 1626 (1994)
4. S.C.Rae, Optics. Comm. **97**, 25 (1993)
5. N.H. Burnett and G.D. Enright, IEEE J. Quantum Electron. **QE-26**, 1797 (1990)
6 R. Rankin, C.E.Capjack, N.H. Burnett and P.b.Corkum, Opt. Lett. **16**, 835 (1991)
7 C.G. Durfee III and H.M. Milchberg, Phys. Rev. Lett. **71**, 2409 (1993)
8 C. S. Liu and V.K. Tripathi, Phys. Plasma. **1**, 3100 (1994)
9 0. Kafri, Optics. Lett. **5**. 555 (1980)

Inst. Phys. Conf. Ser. No 140: Section 9
Paper presented at Laser Interact. with Matter Conf., Oxford, 19–23 September 1994
© *1995 IOP Publishing Ltd*

IMPROVEMENTS AND APPLICATIONS OF GAS LENSES.

M.M. Michaelis, M. Kuppen, A. Forbes, N. Lisi, R. Buccellato and A. Prause
Physics Department, University of Natal, Durban, South Africa

Abstract
A brief review of continuous gas lenses is followed by a description of a more recent pulsed gas lens. This shows promise in topics ranging from metal processing to free electron lasers, isotope separation, robotics and laser Q-switching.

1. Gas lens development.

The topic of gas lenses is one that has grown very slowly over the last three decades since their invention at Bell labs in the early 60's [1]. Fig. 1 illustrates some typical CW gas lenses developed by scientists from almost every continent. All but the last two, rely on thermal gradients established by differential heating of the ambient air. A simple "ball park" calculation produces a formula what shows why these lenses are long focal length devices with a narrow field of view [2]:

$$f = \frac{\cot(\sqrt{A}l)}{\sqrt{A}} \quad \text{with} \quad A = \frac{6\times10^{-4}}{r^2}\frac{\Delta T}{T_w}.$$

For a typical gas lens of pipe length $l = 0.2$ m and tube radius $r = 4$ mm with a pipe wall temperature of $T_w = 350$ K and a temperature difference $\Delta T = 50$ K this gives a focal length of 1 m.

Gloge [3] showed with some slightly more complicated calculations that the optic axis of the lens is lowered by gravitationally induced convection to the point where it becomes impractical to use laminar flow pipe lenses with an aperture much greater than 8 mm. Spinning pipe gas lenses however, are able to overcome the effects of gravity [4] and apertures of 4 cm are practical. If larger apertures are required, gas dynamic or pulsed gas lenses have to be considered such as those described in the next section.

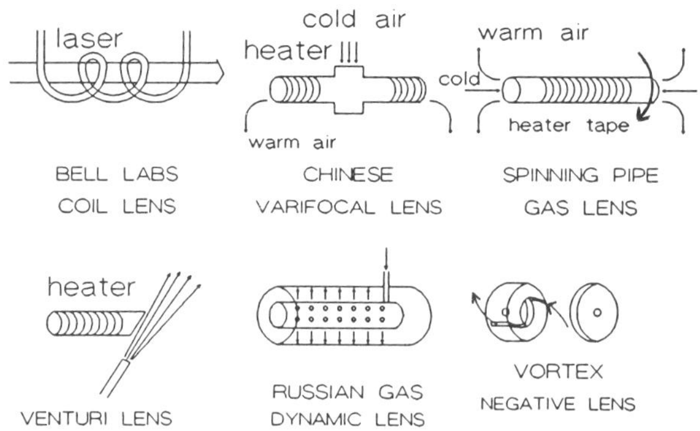

Fig. 1 Some typical continuous wave gas lenses

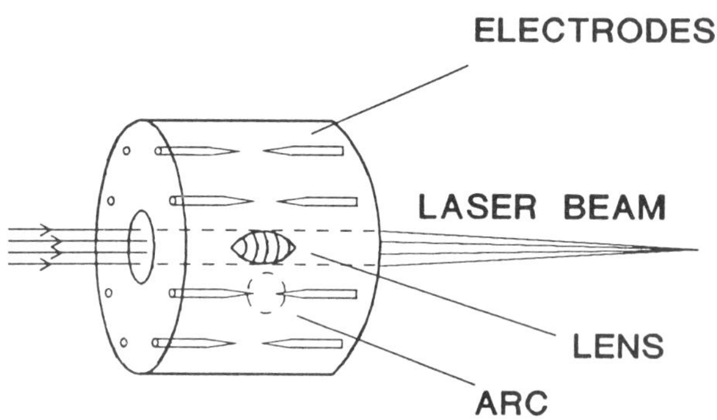

Fig. 2 Schematic of the colliding shock lens

At a series of ECLIM conferences, [5,6,7] we have supported the view that gas lenses could one day become important in laser controlled fusion as a final focusing element able to overcome the problems of high rep-rate, high intensity target illumination, and also capable of protecting delicate optical surfaces from the effects of radiation.

2. Pulsed Gas Lenses.

There are two types of pulsed gas lenses: the decompression gas lens (DGL) and the Colliding Shock Lens (CSL). The DGL is to our knowledge a concept that has only been tested in one dimension. It consists of a cylindrical container, the walls of which can be rapidly opened or even removed. This is accomplished either with fast butterfly valves or as in our 1D experiment, by exploding thin walls made of glass or plastic material. The gas inside the walls is pressurised to several bar and after being released, takes on the density profile required for a GRIN lens [8]. Our experiments [9] served to eliminate some of the initial experimental problems: difficulty of obtaining sub-millisecond venting times; extinction of laser beam due to adiabatic cooling of the air or gas.

The other type of gas lens, the colliding shock lens (fig. 2) [10] is showing greater promise. It consists of a set of spark gaps equispaced on the circumference of a circle. At the centre of the circle colliding shock waves produce a short lived GRIN gas lens. Although it has been shown that the CSL operates close to the diffraction limit, doubts about its practical value were expressed on two fronts. The first was whether it would scale to larger apertures. The argument was that since the refraction is due to the density gradients in the shock front and shock rear, [11,12] the approximate size of the lens would be equal to twice the sum of their thicknesses. This was born out by the fact that our first CSL's were millimetric aperture devices and that initial attempts to scale them by increasing the physical size of the lens were unsuccessful. This is not the place to prolong this discussion save to say that we have recently successfully scaled a larger CSL to a 1 cm optical aperture (f#100) thereby experimentally refuting the shock dimension argument. A simple scaleing law for these lenses has been submitted to J. Appl. Phys.

The second set of doubts over the CSL was as to whether this was a device that could be "rep-rated" at frequencies that would make it useful in the context of high repetition rate pulsed lasers. For want of a suitable power supply, we have only been able to run our lenses at 20 Hz, but there is a strong experimental reason for believing that they could run at tens of kHz: shocks rebounding from the cylindrical wall produce a second focus some 20 μs later.

3. Applications of Gas Lenses.

At previous ECLIMs, we suggested that apart from the obvious mechanical and robotic uses of CW gas lenses, larger aperture pulsed gas lenses might be developed in the context of controlled ICF. We also proposed the use of the CSL as an intra-cavity Q-switching device. Since then we have obtained single and double pulse Q-switching of a ruby laser [13]. We have also demonstrated the versatility of this concept by using an elliptical CSL to generate a line focus (potentially interesting in the context of X-ray lasers). Recently [14] gas lenses have been proposed for the input and output optics of high power Free Electron Lasers. We are currently investigating another high intensity application, laser isotope separation. Finally, gas lenses have been mentioned in the context of Laser Impulse Space Propulsion [9]. This application is ideally suited to these lenses because of their very long focal length and power handling capabilities.

Conclusion.

The topic of gas lenses has spent two decades in obscurity. Having started to overcome limitations of a small field of view and gravitationally induced aberrations, we think these devices may become popular, wherever damage threshold is a problem.

References

(1) D. Marcuse "Light Transmission Optics".
2nd ed. van Nostrand Reinhold, New York (1982).

(2) M.M. Michaelis *et al.* "A gas-lens telescope".
Nature **353**, 547-548 (1991).

(3) D. Gloge "Deformation of Gas Lenses by Gravity".
The Bell System Technical Journal **46**, 357-365 (1967).

(4) M. Notcutt *et al.* "Spinning Pipe Gas Lens".
Optics and Laser Technology **20**, 243-250 (1988).

(5) M.M. Michaelis *et al.* "Gas Lens Laser Produced Plasma".
19th ECLIM. Madrid 1988.

(6) M.M. Michaelis *et al.* "Gas Lens Applications".
20th ECLIM. Schliersee 1990.

(7) M.M. Michaelis *et al.* "Applications of the Colliding Shock Lens".
21st ECLIM. Paris 1993.

(8) R. Buccellato Ph.D. Thesis, University of Natal, Durban (1994).

(9) R. Buccellato *et al.* "Pulsed Gas Lenses".
Optics and Laser Technology **25** (4), 247-250 (1993).

(10) R. Buccellato *et al.* "Colliding Shock Lens".
Optics Communications **101**, 350-355 (1993).

(11) M. Bacon *et al.* "Refractive fringe diagnostics of sparks in air".
Journal of Applied Physics **66** (3), 1075-1079 (1989).

(12) M.M. Michaelis *et al.* "Refractive fringe diagnostic of spherical shocks".
Optics and Laser Technology **23** (5), 283-288 (1991).

(13) N. Lisi *et al.* "The Colliding Shock Lens as an
Intracavity Q-switch Element".
Applied Optics, in press, (1994).

(14) W.H. Christiansen "Gas Optics Applicable to the Free Electron Laser".
Seventh International Symposium on Gas Flow and Chemical Lasers.
Vienna 1988.
SPIE **1031**, 474

Inst. Phys. Conf. Ser. No 140: Section 9
Paper presented at Laser Interact. with Matter Conf., Oxford, 19–23 September 1994

© *1995 IOP Publishing Ltd*

RADIATION-HARDENING of FINAL OPTICS for an ICF REACTOR[§]

G.H. Miley, S.G. DelMedico,[†] M. Petra, and O. Barnouin[†]

Fusion Studies Laboratory • University of Illinois
100 NEL, 103 South Goodwin Avenue
Urbana, IL 61801-2984 U.S.A.

Abstract

One approach to radiation hardening of the final optics for a laser-driven inertial confinement fusion (ICF) reactor is to employ a fused silica optic held at an elevated temperature to cause continuous thermal annealing of color centers. A design of this type using fused silica wedges as the final optical element, arranged to deflect the laser beam out of line-of-sight of the target, has been proposed by workers at Lawrence Livermore National Laboratory.[1] In the present paper, an experimental program designed to study thermal annealing using a pulsed research fission reactor for irradiations is described. Preliminary results suggest that a possible approach is to hold the fused silica wedges at a modest temperature (~300 °C) during operation, but to apply periodic 20-30-minute "pulses" of high-temperature (~600 °C) annealing.

1. Introduction

The final optics for a laser-driven fusion reactor must survive a variety of intense radiation, including fast neutrons, x-rays, and target debris. The 14-MeV neutron irradiation represents the most difficult component to protect against, since it is not possible to design an optically transmissive shielding material for these neutrons. This problem has been examined in some detail by J. Woodworth et al.[1] Based on a review of the possible use of various types of optics, including multi-layer dielectric mirrors and grazing incident metal mirrors, they concluded that the use of refractive optics is potentially most attractive. One concept, illustrated in Fig. 1., uses a set of fused silica wedges to deflect the beam out of the line-of-sight of the target. A combination of a fast shutter and x-ray absorbing gas is used to stop other radiation and debris in front of the fused silica wedges. The basic issue then, is to develop a radiation-hardened wedge design with a reasonable lifetime against neutron-induced reductions in transmission.

Assuming fused silica were used for the wedges, one approach for radiation hardening[1,2] involves the use of continuous *in situ* thermal annealing. While it appears that the induced decrease in transmission can be handled by this method, long-term effects on fused silica of operation at elevated temperatures in an intense radiation environment are not known. In addition, it is not clear that *in situ* annealing can completely overcome transient effects during the very high dose rate encountered in ICF. Most data to date for such annealing are concerned with steady-state irradiations. Thus, we are currently studying this problem in some detail, using the University of Illinois' pulsed TRIGA reactor to provide the irradiation.

----

§ Portions of this work were supported under Contract DE-FG02-94ER81849.
† Rockford Technology Associates, Inc. Champaign, Illinois 61820-5579

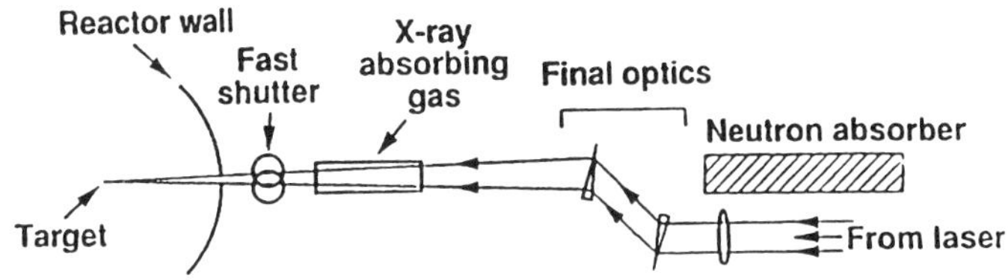

Figure 1. Schematic of the final optics concept employing wedge deflectors.

2. *In Situ* Annealing Estimates

The goal for this work is to develop radiation-hardened optics for operation in a hypothetical 1000-MW electric ICF power plant,[1] where silica wedges, placed 40 m from the target, would receive a fluence of 10^{12} n/cm^2 pulse (corresponding to 1.6×10^{20} n/cm^2/year or 3.2×10^{11} rads/year), based on available steady-state irradiation data. *In situ* annealing characteristics for this case show that annealing at ~400 °C is very effective in removing color center transmission losses from fused silica. (See Fig. 2.) After a modest radiation dose, considerable darkening of the glass at shorter wavelengths occurs, but most can be removed with several hours' annealing. Using this and other similar data, it is concluded that for the reference case, as shown in Fig. 3, *in situ* annealing during operation at ~325 °C would maintain 99% transmission at 354 nm.

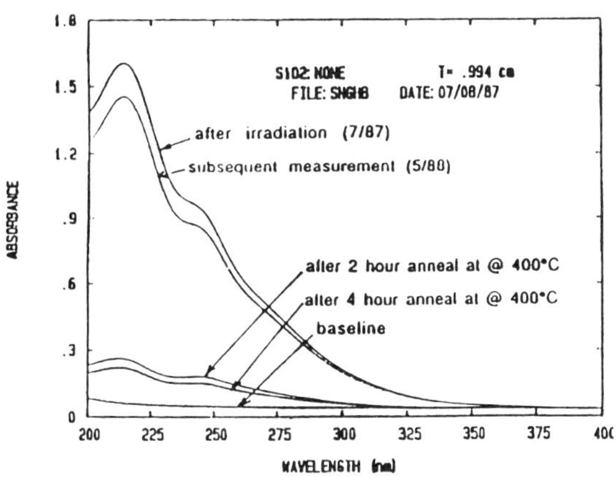

Figure 2. Color center absorption in fused silica from 14-MeV neutron irradiation.

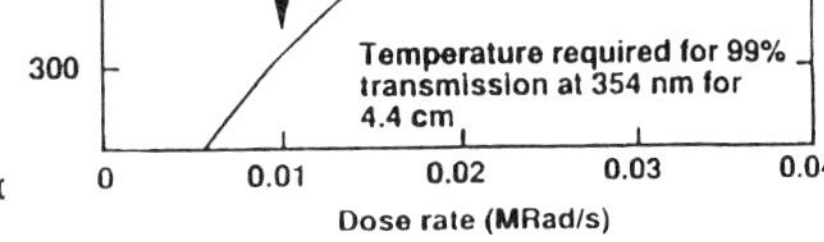

Figure 3. Operating temperature required to maintain 99% transmissivity through 4.4-cm section of a 30-cm aperture Brewster angle silica wedge, as a function of distance to target (upper scale) and dose rate (lower scale).

The annealing temperature strongly depends upon the radiation intensity on the optics, hence on the location of the fused silica wedges. Thus, if the wedges were moved in closer to obtain better focusing, the annealing temperature would have to be increased. At distances greater than the reference 40 m, lower annealing temperatures can be employed. Unfortunately, the database used for these projections is extremely limited. Thus far, samples have only been irradiated with neutron doses of up to 1.5×10^{16} n/cm^2, corresponding to just one hour of baseline reactor operation.[3] More data is needed to confidently extrapolate these results to the expected doses of 1.6×10^{20} n/cm^2/year. At present, a 14-MeV neutron source capable of carrying out such long-term irradiations is not available. The U. of Illinois' TRIGA provides a means of simulating absorbed doses equivalent to one year's operation of an ICF reactor, but this is achieved with a mixture of lower-energy (~<1 MeV) neutrons and gamma rays. However, if the results are compared on the basis of comparable actual absorbed dosage, using an effective dose equivalent, this still provides an approximate simulation of ICF damage results. In other words, the key absorption centers in fused silica of interest for ICF operation are the E' and B_2 bands (at 215 and 246 nm, respectively), and these bands are also dominant in the TRIGA reactor irradiations, although the energy efficiency for their formation is different. Thus if a proper dose equivalent for color center formation is employed, the information obtained can be interpreted in terms of 14-MeV effects with fair confidence.

3. Initial Results

Transmission changes in the high dose-rate field during reactor pulses were measured on several different materials, including fused silica (Corning 7940), sapphire, CsI (single crystal), spinel, ALON™, and Pyrex™ (Corning 7740). Studies were done at various wavelengths and temperatures, and the recovery of optical transmission during and after pulsed irradiation was measured. As illustrated in Fig. 4, these measurements typically show rapidly increasing radiation-induced absorption during a pulse, followed by decreasing absorption in the "tail" of the reactor pulse due to the rapid decay of the shorter-lived "transient" color centers. Transmissivity correspondingly shows a rapid decline and then recovery during the pulse. Long-lived ("permanent") color centers remain, however, long after cessation of the pulse. Clearly, such effects must be reduced or avoided for satisfactory operation of the final optics in an ICF reactor.

The absorption coefficient is found to be a sensitive function of temperature of the sample during irradiation. The peak absorption, as well as the recovery time, was significantly reduced by use of elevated temperatures. A typical example is given in Fig. 4 for a Pyrex™ sample (a) and for a fused silica sample (b) held at various temperatures during the reactor pulse (c). The increase in temperature supplies the electrons trapped at color centers with enough energy to to escape the defect trap. Lower temperatures are sufficient to temporarily increase transmission, but the sample darkens more quickly when irradiation resumes. This effect is attributed to the fact that lower temperatures may remove trapped electrons but still allow the build-up of radiation-induced defects. Such defects leave the material subject to rapid build-up of color centers when irradiation resumes. It is necessary to raise the sample temperature to approximately one-half of the melting point in order to continuously anneal out these defects. Higher temperature annealing can effectively remove radiation-induced defects; however, the higher temperatures needed appear to be too high for long-term applications, due to possible physical deformation effects. Thus we are investigating an alternate approach;

namely the use of lower temperature *in situ* annealing, combined with brief periodic (<30 minutes) application of high temperatures. For ICF, *in situ* temperatures of ~300 °C could be combined with high-temperature treatments at 600-700 °C.

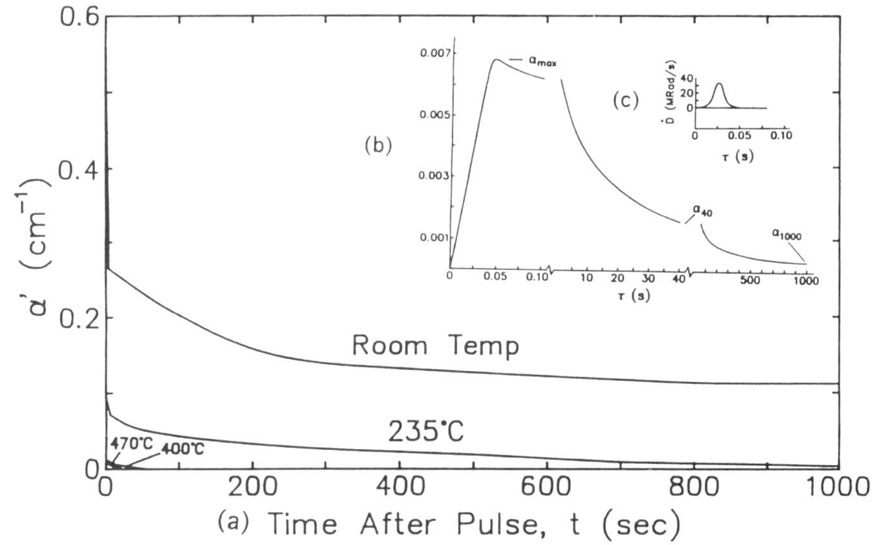

Figure 4. (a) Change in absorption coefficient of Pyrex™ at 633 nm during and after a reactor pulse; (b) change in absorption coefficient of fused silica at 325 nm during and after a reactor pulse; (c) dose rate during the reactor pulse

4.0 Conclusion

In summary, these data indicate the potential benefit of holding the final optical components of an ICF reactor at an elevated temperature. However, to avoid physical damage to the optics over long periods of operation, temperatures of ~300 °C appear desiarble. Periodic high-temperature annealing, combined with continuous lower-temperature *in situ* annealing, provides an alternative approach that could be attractive, but this approach requires more study.

5.0 References

[1] J.G. Woodworth, et al., 1992 *14th Symposium on Fusion Engineering*, IEEE Conf. Proceedings, 895.

[2] G.H. Miley, O. Barnouin, et al., 1991 *Fusion Engr. and Design*, 18, 341-348.

[3] G.H. Miley, et al., 1991 "Workshop on Radiation Effects on Diagnostics," Princeton Plasma Physics Laboratory.

Inst. Phys. Conf. Ser. No 140: Section 9
Paper presented at Laser Interact. with Matter Conf., Oxford, 19–23 September 1994

© *1995 IOP Publishing Ltd*

Laser-X-ray Conversion

L.S.Mkhitarjan, G.G.Kochemasov

Russian Federal Nuclear Centre - All Russia Scientific Research Institute of the Experimental Physics (VNIIEF), Arzamas-16, Nizhniy Novgorod Region, Russia.

Abstract. An analytical model of laser corona evolution dynamics allowing numerical evaluation of X-ray radiation generation from the corona has been proposed. It is shown that for elements with high Z the main portion of radiation is due to the line emission. The coefficient of laser radiation conversion to X-rays shows a small dependence on the target material opacity in plane geometry whereas in spherical geometry this dependence is sufficiently pronounced.

Many scientists believe, that fusion target implosion symmetry is a problem that can be most easily solved by laser radiation conversion to X-rays. With this approach, what becomes of critical importance for laser fusion (LF) research is the conversion efficiency and X-ray energy characteristics inside the cavity containing a fusion target. Therefore, it would be of interest to look consistently at the in-cavity phenomena responsible for radiation field generation and thus define the main field characteristics.

There are publications [1-3] with the attempts like these which involve some rather free assumptions, and thus make it impossible to do reliable conclusions as to the quantitative aspect of the problem. We mean this paper to consider laser radiation conversion to X-rays for plane case (open geometry) and the details of radiation field generation inside confined space.

When a wall is illuminated by a laser pulse, there develop two X-ray generation sources significantly different in their properties.

Firstly, this is plasma produced directly by laser radiation (L-corona). While heat is transferred in L-corona by electrons, and light absorption occurs due to inverse bremsstrahlung, the plasma represents a hot (~1 keV or higher temperature with laser intensity $\sim 10^{14}$ W/cm^2) and low-density (about critical density ρ_c) formation.

And, secondly, plasma produced by X-rays emitted by L-corona (R-corona). Heat conduction in this plasma occures due to radiation transfer, so its temperature is substantially lower and density much higher than for L-corona. The latter situation makes L-corona dynamics virtually independent of whether L-corona is generated immediately at the wall or at the "tail" of R-corona. Therefore, the characterisation of laser corona can be made irrespective of R-corona considerations.

L-corona generates primary X-radiation, and for this radiation it may be considered as transparent (due to high temperature and density nonuniformities). And therefore, this radiation has a considerably nonequilibrium spectrum.

R-corona is what partly converts this radiation into equilibrium spectrum. The X-ray emission fluxes resulting in this way are illustrated by Fig. 1.

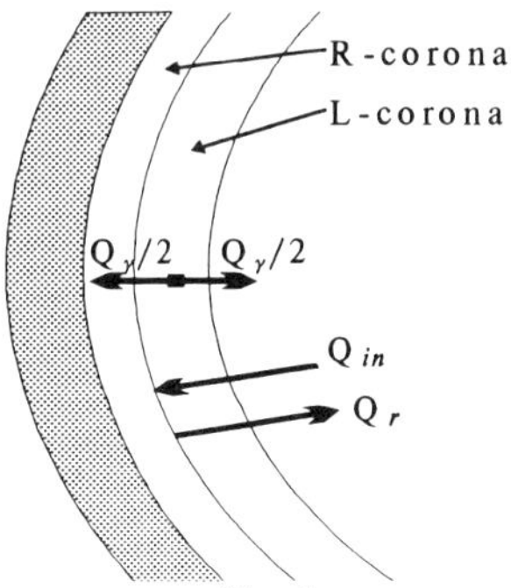

Fig. 1

L-corona emits X-rays of $Q_\gamma(t)$ power. While this corona has small thickness with respect to the inside space, one half of this radiation goes inside the cavity, and the other contributes to the X-ray flux incident upon R-corona. However, the radiation from R-corona Q_r itself fills the cavity and also contributes to the incident flux Q_{in}.

Inside subsonic region one can consider L-corona to be quasisteady. Temperature profile for this region with good accuracy is $T = T_s\left(\frac{ax}{\Delta}\right)^{2/5}$, where $a = \frac{5}{14}\left(\frac{c_V}{c_T}+1\right)\left(1+\sqrt{1+56B\left(\frac{c_V}{c_T}+1\right)^{-2}}\right)$; $B = \frac{2\kappa}{7}\frac{A_\gamma T_s}{c_T^3}$; $\Delta = \frac{2\kappa}{7}\frac{T_s^2}{\rho_s c_T^{3/2}}$; κ - is the electronic heat conductivity $\left(Q_e = -\frac{2\kappa}{7}\frac{\partial T^{7/2}}{\partial x}\right)$; A_γ -plasma emissivity (plasma radiation loss $\propto A_\gamma \rho^2\sqrt{T}$); c_V - plasma specific heat; T_s, ρ_s -are temperature and density at sonic point respectively.

Then, by using isothermal expansion approximation for supersonic region, we will obtain the energy emitted by L-corona:

$$Q_\gamma = A_\gamma \rho_s^2\sqrt{T_s}\,\frac{85}{12}\frac{\Delta}{a}\left(1+\frac{6av_st}{85\Delta}\right); \quad v_s = \sqrt{c_T T_s}.$$

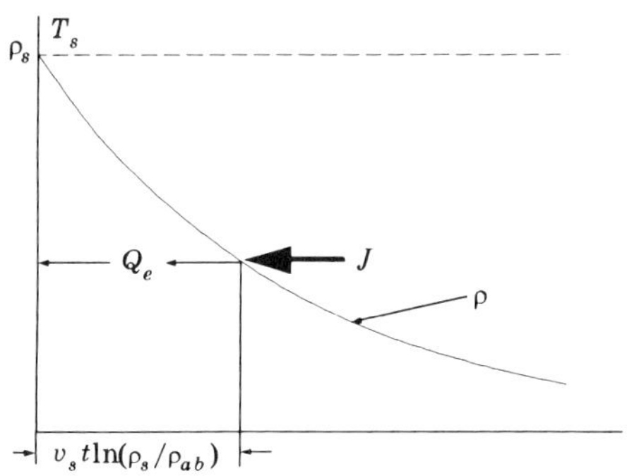

Fig. 2

To get the L-corona gasdynamics characteristics (T_s, ρ_s), we take its approximated profile (see Fig. 2), where $\rho_{ab} = \min\left[\rho_c; \left(\frac{2T_s}{k_T c_T^{1/2} t}\right)^{1/2}\right]$; k_T - is the inverse bremsstrahlung coefficient $J(x) = J_0\exp\left(-\int_x^\infty k_T\frac{\rho^2}{T^{3/2}}dx\right)$.

Based on the energy balance and assuming that electronic heat conduction flux Q_e is only absorbed at the wave front, we have the following set of equations:

$$\rho_s v_s^3\left(\frac{5}{2}+\frac{c_V}{c_T}\right)\left(1+\frac{b\Delta}{av_st}\right) = J_a(1-\eta_0)$$

$$\frac{2\kappa}{7}\frac{T_s^{7/2}}{v_s + \ln\left(\frac{\rho_s}{\rho_{ab}}\right)} = J_a - \rho_s v_s^3\left[1+\left(\frac{5}{2}+\frac{c_V}{c_T}\right)\frac{b\Delta}{av_st}\right]; \qquad (1)$$

$$b = \left[\frac{c_V}{c_T}\left(\frac{2}{5}+\frac{\pi}{16}\right)+\frac{1}{5}-\frac{\pi}{32}\right]\left(1+\frac{2c_V}{5c_T}\right)^{-1};\ \eta_0 \equiv \frac{Q_\gamma}{J} = \frac{\mathcal{L}}{\mathcal{L}+1};$$

$$\mathcal{L} = \frac{85B}{12a\left(\frac{5}{2}+\frac{c_V}{c_T}\right)}\left(1+\frac{6av_s t}{85\Delta}\right)\left(1+\frac{b\Delta}{av_s t}\right)^{-1}$$

J_a - is absorbed laser intensity.

These equation can be easily solved iteratively, thus allowing to find T_s, ρ_s and η_0 (i.e. Q_γ).

To calculate the conversion ratio, we only have to know R-corona radiation. For this purpose, we will also make use of isothermal expansion approximation. We will describe the R-corona heat transfer with quasisteady equation by adding into it a distributed source term representing the absorption of nonequilibrium spectrum Q_γ. Further, we will assume the spectral free path length to be of the form $l(\omega, T, \rho) = \frac{l_\omega(T)}{\rho^k}$, and Rosseland mean-free-path length to be equal to $l = l_0\frac{T^n}{\rho^k}$.

Then we have the following set of equations:

$$\left(\frac{5}{2}+\frac{c_V}{c_T}\right)\rho_f v^3 = Q_\gamma\frac{1+\beta}{2} - (1-\beta)\frac{\sigma c T_r^4}{4};\quad v^2 = c_T T_r,$$

$$\frac{\sigma c T_r^4}{3} = Q_\gamma\left(S_f - \bar{S}_\gamma\right)\frac{1+\beta}{2} - \frac{\sigma c T_r^4}{4}(1-\beta)S_f, \tag{2}$$

$$S_f = \frac{\rho_f^k}{l_0 T_r^n}\frac{vt}{k};\quad \bar{S}_\gamma = \frac{1}{2Q_\gamma l}\int_0^\infty l_\omega(T)Q_\omega d\omega,$$

where Q_ω is the radiation spectrum of L-corona $\left(\int_0^\infty Q_\omega d\omega = Q_\gamma\right)$, σ - is Stefan-Boltzman constant, c - is light velocity, β=1-S_h/S, S - is the total area of the cavity walls, S_h - total area of holes in the cavity walls.

Clearly, spectral composition of L-corona emission and details of the R-corona opacity spectral dependence may significantly influence the albedo value of R-corona, particularly where the condition S_f>>1 is satisfied with the allowance being not large.

In practice, $l(\omega,T,\rho)$ depends on photo absorption and line absorption (high and medium Z), therefore S_f and S_γ have to be found through numerical integration.

We will illustrate the method by solving eqs. (1) and (2) with simulated free-path length $l(\omega, T, \rho) = \lambda\omega^p T^m/\rho^k$ for p=k=3/2, m=1/2. Then, $l \approx \frac{15}{4\pi^4}\frac{\lambda T^{m+p}}{\rho^k}\Gamma(5+p) \equiv l_0\frac{T^{m+p}}{\rho^k}$.

While our consideration is only valid for cases when S_f>>1 we take l_0 such that there would be, e.g., $S_f \approx 4$ for $T \approx 0.2$ keV and $\rho \approx 0.5$ g/cm^3. Then, $l_0 = 8\cdot 10^{-3}$ and for spectrum $Q_\omega = Q_\gamma \exp\left(-h\omega/T_\gamma\right)/T_r$ we have $S_\gamma = 0.06(T_r/T_\gamma)^{3/2}$.

Value A_γ, which determined mainly by lines emission, we have taken according to [4], and loss of energy due to ionisation (to calculate c_V) according to [5]. With these values for gold, we have obtained the following results for laser pulse width t=1 ns.

	$A_\gamma = A_\gamma$ [4]			$A_\gamma = 0.2A_\gamma$ [4]	$A_\gamma = A_\gamma$ [4] $l_0' = 2l_0$
λ [μm]	1	1	0.35	1	1
J_a [10^{14} W/cm^2]	1	10	10	1	1
T_γ [keV]	1.21	5.65	3.5	1.72	1.21
Q_γ [10^{14} W/cm^2]	0.94	7.7	8.2	0.8	0.94
Open geometry (β=0)					
T_r [keV]	0.13	0.21	0.23	0.12	0.124
Q_r [10^{14} W/cm^2]	0.28	2.15	2.63	0.21	0.24
$\eta = \left(\frac{Q_\gamma}{2} + Q_r\right) \Big/ J$	*0.75*	0.6	0.67	0.61	*0.71*
Closed geometry (β=1)					
T_r [keV]	0.24	0.425	0.45	0.222	0.215
Q_r [10^{14} W/cm^2]	3.3	33.5	42	2.47	2.19
$\alpha = \frac{Q_r}{Q_\gamma + Q_r}$	0.78	0.81	0.84	0.76	0.7
$\frac{Q_\gamma}{2} + Q_r$ [10^{14} W/cm^2]	*3.8*	37	46	2.9	*2.7*

The table demonstrates the influence of the wavelength upon the conversion ratio and its dependence on the R-corona opacity and L-corona emissivity.

What is most noticeable is that η dependence on opacity is extremely weak for plane case.

On the contrary, in closed space the X-ray flux emitted by the walls is significantly more sensitive to R-corona opacity, than for plane case.

Conclusions

1. In plane case for high-Z materials, more than half of absorbed laser energy is converted into X-rays.
2. X-ray emission by plane target is weakly dependent on the target opacity.
3. For closed geometry, R-corona albedo for high-Z materials may be as high as 0.7-0.8.
4. X-ray emission from cavity walls is highly sensitive to the walls opacity.

References

[1] R.Sigel and et al. *Phys.Fluids*, **B2**(1), 199 (1990)

[2] Zhang Jan. *Laser Interactions and Related Plasma Phenomena*, vol. **8**, H.Hora and G.H.Miley (eds), p. 205 (1988)

[3] S.Yu.Gus'kov, M.Kolanovsky. *Journal of Soviet Laser Reaserch*, **13**(5), 390 (1992)

[4] D.E.Post and et al. *Atomic Date and Nuclear Data Tables*, **20**, (1977)

[5] Thomas A.Carlson and et al. *Atomic Date*, **2**, 63 (1970)

Inst. Phys. Conf. Ser. No 140: Section 9
Paper presented at Laser Interact. with Matter Conf., Oxford, 19–23 September 1994

© *1995 IOP Publishing Ltd*

Study of Cannonball target implosions by x-ray spectroscopic measurements

H Nishimura, H Shiraga, T Endo[1)], K Fujita, H Takabe, Y Kato, S Nakai, E Förster[2)], and I Uschmann[2)]

Institute of Laser Engineering, Osaka University, Suita, Osaka 565 Japan
1) Faculty of Engineering, Nagoya University, Nagoya 464-01, Japan
2) Max-Plank-Arbeitsgruppe Röntgenoptik an der Friedrich-Schiller-Universität Jena D-O 07743 Jena, Germany

Abstract. Fusion pellet implosion by laser-generated x rays was investigated by means of time-integrating spectroscopic measurements. Deuterium fuel was seeded with a small amount of Ar in order to determine the electron temperature and the density of the compressed fuel, respectively, from the emission intensity ratio and the broadening of Lyβ (Ar^{17+} 1s-3p) and Heβ (Ar^{16+} $1s^2$-1s3p) lines. Monochromatic images of these lines were also taken with toroidally bent crystals for the first time to provide temperature distribution of the compressed fusion pellets. By comparing with one-dimensional (1D) fluid dynamic simulations, observed results were examined in terms of shell disintegration at the stagnation phase.

1. Introduction

One of the current issues in inertial confinement fusion research is to understand the influence of fluid instabilities which degrade spherical convergence of the pellet and thus may result in discrepancy of neutron yields between the experiments and 1D-simulations [1]. The purpose of this study is to derive temperatures and densities of the compressed core from the x-ray spectra of the Ar dopant by the help of spectrum analysis code RATION[2] and to compare the results with 1D fluid simulation for understanding of implosion behavior. As long as emission intensities of these lines increase with increase in density and temperature as a result of compression, the observed time-integrated spectra may represent the plasma condition at close to the instant at which 1D spherical convergence starts being degraded.

2. Experiments

Gekko XII facility was used for the implosion experiments. Experimental conditions adopted were the same as were used before [1]. Mean x-ray flux on the fusion capsule estimated by the illumination code [3] is 1.1×10^{14} W/cm^2 (corresponding to the radiation temperature of 182 eV).

The fusion capsule was set at the center of a gold cylinder cavity. The inner radius and the length of the cylinder were respectively 800 and 1600 μm and the thickness was 10 μm. Fuel capsule consisted of three different layers : The inner-most layer was polyvinyl alcohol of 1.2 μm-thick. The middle layer was parylene of typically 6 μm-thick and the outer-most layer was polytetrafluoroethylene as an x-ray absorber. The thickness of the outer coating was varied from 2 to 14 μm to control implosion behavior from the stagnation-free mode to the high compression mode. The inner radius of the capsule was 155 ± 5 μm. Surface roughness and thickness of each capsule were investigated by using optical interference microscopes and a scanning electron microscope. Spectral decomposition of the surface roughness was made for the analysis of implosion stability. The capsule was filled with a mixture gas of deuterium (10 ± 1.5 at) and argon (0.04 ± 0.01 at, corresponding to 0.2 ± 0.05 at %). Estimated optical depths of the Lyβ and Heβ lines are 0.3 - 0.5.

X-ray emission from the Ar dopant was measured by two different instruments: One is a time-integrating and 1D space-resolving x-ray spectrograph. The other was a time-integration but 2D space-resolving x-ray monochromatic microscope by the use of toroidally bent Si and Ge crystals [4]. Spectral window of the microscope was adjusted to be 10-20 eV to cover broadened Ar lines. Spatial resolution was checked to be better than 10 μm.

An x-ray streak camera coupled to a slit camera was simultaneously used to measure the implosion time from the Ar emission and the glass pusher emission. For this particular use the $CF_{1,3}$ coated GMBs were additionally prepared. Yield of DD neutrons was measured by using absolutely calibrated neutron detectors.

3. Experimental results and analysis

Experimentally observed x-ray spectra were analyzed by using the RATION code. First, electron density of the compressed fuel was obtained from the width of the Stark-broadened Lyβ and Heβ lines. Resultant electron densities are plotted in Fig.1 as a function of the initial areal mass density of the capsule. The electron density increases with increase in the capsule areal density $\rho\Delta R$ up to 2.66 mg/cm^2, but decreases beyond this value. The maximum density, namely the maximum fuel density attained under the current condition, is about 7 × 10^{23} cm^{-3}. This implies that the radial convergence ratio (= R_i / R_f, ratio of the initial to final fuel radii) of 11 has been attained unless higher Z materials from the capsule gave notable influences on the measurement. Symbolic dots and lines seen in the figure represent the volume-averaged electron density obtained from 1D simulation.

The electron temperature was obtained from the line intensity ratio of Lyβ to Heβ. In RATION, the effect of photo-excitation and ionization is considered by the use of escape factors [8]. This treatment is of particular importance due to the fact that the optically thick ines such as Lyα (Ar^{+17} 1s-2p) and Heα (Ar^{+16} 1s^2-1s2p) possess common ground states respectively with Lyβ and Heβ and thus electronic populations of these ground states are considerably influenced by the radiative processes.

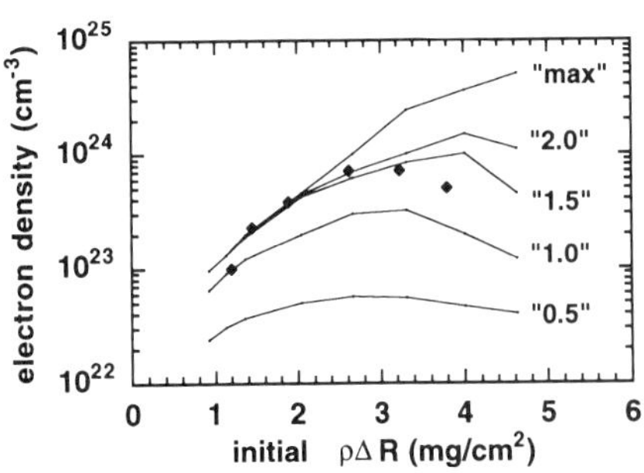

Fig.1 Electron density of the compressed fuel derived from the spectral widths of Heβ line and from ILESTA-1D simulations as a function of of initial areal density of the capsule.

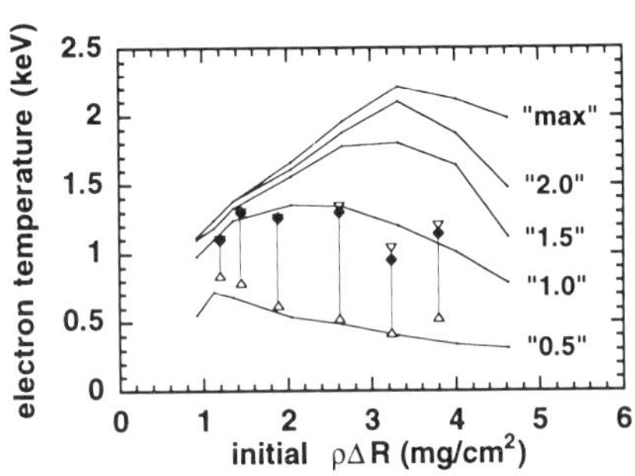

Fig.2 Electron temperature of the compressed fuel obtained from the line ratio of Lyβ to Heβ and from ILESTA-1D simulations Three differnt temperatures are shown for the effective plasma size of 0.01(▽), 1(◆), and 20 μm(△).

To obtain the escape factor first we need an effective plasma size. In the presence of temperature and density gradients, the electronic population and the spectral shape may be so different between neighboring plasmas that the energy coupling among these plasmas via radiative transport will be weaker than that for the homogeneous plasma case. This means that if we adopt straight-forwardly the radius of compressed fuel as the effective plasma size, the optical depth will be over estimated. Consequently, temperatures were obtained assuming optically thin plasma size of 1 μm as a representative size deduced from the similar indirect drive experiments [5]. For comparison temperatures for a completely optically thin case and those for the experimental radius of 20 μm were obtained. Results are shown in Fig.2. Expect for the largest plasma case shown with open triangles (△), the obtained temperatures show a similar trend to the electron density when the areal mass density of the capsule is increased. The highest electron temperature achieved is about 1.3 keV for the same areal mass density as provides the maximum electron density.

Figures 3 (a) and (b) show the monochromatic images of Ar Lyβ and Heβ lines obtained with the microscope. Temperature distribution of the fuel was derived from the intensity ratio of these images as shown in Fig.3 (c). Four cooler spots at the core center (which may be caused illumination asymmetry) and relatively higher temperature in the surrounding region are seen. It is confirmed that the space averaged temperature obtained from this observation is consistent with that from the above measurement.

4. Comparison with simulation

Experimentally obtained electron density and temperature were compared with one dimensional fluid calculation with ILESTA-1D [6,7]. To simulate the experiments, a concentric double-shell structure was adopted: A fuel capsule is surrounded by a gold spherical shell and laser irradiates the inner surface of the outer shell. One unique radius of the converter was determined so that the calculated implosion times coincide with the implosion times observed for various capsules of the experiments. A good coincidence was obtained for 500 μm radius.

According to the behavior of the shock wave propagating through the compressed fuel, we shall define each specific time as "0.5" when the converging shock wave first reaches to the target center and "1.0" when the reflection shock encounters the pusher-fuel contact-

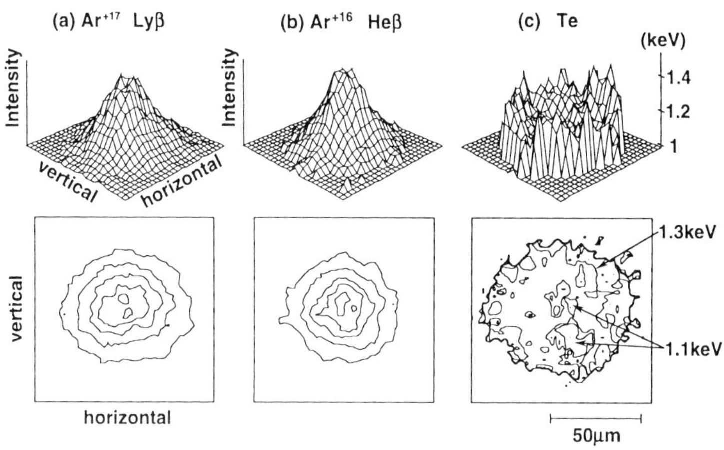

Fig.3 Monochromatic image of the Lyβ (a) and Heβ (b) lines of Ar dopant in the deuterium fuel. (c) Temperature distribution of the core derived from intensity ratio of the images.

surface and starts decelerating the pusher moving inward. In the same manner, the labels "1.5" and "2.0" represent the times for the second bounding of the shock wave and the second pusher deceleration. The level "max" does not represent a specific time, but the peak density or teh peak temperature attained in each implosion.

Resultant volume averaged density and temperature at each labeled time are plotted in Figs.1 and 2. Experimentally obtained electron densities are between the curves of "1.0" and "1.5" and the electron temperatures are between "0.5" and "1.0". A similar trend is seen in the comparison of the experimentally obtained DD neutron yields with simulations. Experimental data points fit well with the curve of the time "1.0" except for the heavier targets of $\rho \Delta R$ larger than around 3 mg/cm^2.

5. Discussion and conclusion

Growth of perturbations on the outer surfaces of the capsule due to the Richtmyer-Meshkov (RM) instability [8] and Rayleigh-Tailor (RT) instability [9] were calculated using a linear analysis with semi-empirical dispersion relation of the instabilities [10]. Parameters for the analysis were taken from the ILESAT-1D simulations.

Calculated perturbation amplitudes on outer surface of the capsule are smaller than pusher thickness, suggesting no occurence of shell breaking at least in the accelerationphase. Since the capsule radius remained constant in the experiment, the thinner wall capsule seems to result in faster growth of the RT instability due to its higher aspect ratio (defined by a ratio of capsule radius divided by capsule thickness). However comparison of the experimentally obtained density, temperature, and neutrons with simulations show better agreement with respective maximum values for the thinner wall capsule. In the simulation the pusher wall is initallythin enough to suffer from preheating by harder x-ray component of the drive and the shell becomes thicker in later time. Under such a high adiabat condition implosion mode is more likely to the stagnation-free, exploding pusher type. Then in-flight aspect ratio of the shell becomes low, ablation velocity large, and density gladient scale length long. As a result shell compression would be stabilized. For better understanding, however, there still remain basic uncertainties in the temperature and density derivations arising from time- and space-integrated measurements.

Acknowledgements

The authors sincerely acknowledge R.W.Lee for offering the RATION code.This work was in part supported by Scientific Research Foundation of Ministry of Education, Science and Culture of Japan

References

[1] For example, Nishimura H et al.1993 Int. Conf. Plasma Physics and Controlled Nuclear Fusion Research, 1992 (IAEA proc. 14th Int.Conf. Würzrburg, 1992) Vol. 3, pp97-103

[2] Lee R W, Whitten B L, and Strout R E, 1984, J.Q.R.S.T. **32,** 91.

[3] Nakamura M, et al., 1992 Laser and Part. Beams, **10**, 421.

[4] Uschmann I, et al., to appear in Rev. Scientific Instruments Jan. 1995.

[5] .Keane C J, et al., 1993 Phys.Fluids, **B 5**, 3328.

[6] Takabe H, et al., 1988 Phys. Fluids **31**, 2884.

[7] Takabe H and Nishikawa T, 1994 J. Q.S. R. T **51**, 379.

[8] Richtmyer R D, 1960 Commun. Pure Appl. Math. **13**, 297; Meshkov E E, 1969 Izv. Akad. Nauk SSSR Mekh. Zhidk. Gaza **4**, 101

[9] Lord Rayleigh, 1950 Scientific Papers II, 200, Cambridge, England, 1900; Sir Geoffrey Tailor, Proc. R. Sco. London **A 201**, 192.

[10] Takabe H, et al., 1985 Phys. Fluids **28**, 3676; Taback M, et al., 1990 Phys. Fluids **B2**, 1007.

Inst. Phys. Conf. Ser. No 140: Section 9
Paper presented at Laser Interact. with Matter Conf., Oxford, 19–23 September 1994
© *1995 IOP Publishing Ltd*

Kinetic simulation of DT ignition and burn in ICF targets

A Oparin, S Anisimov and J Meyer-ter-Vehn
Max-Planck-Institut für Quantenoptik, D-85748 Garching, Germany

Abstract. The transport of fast charged particles during ignition and burn of DT plasma in ICF targets is treated as Brownian motion, solving a Langevin equation for a number of macroparticles. The metod is equivalent to solving the Fokker-Planck equation. In this work the method is applied to target configurations of actual interest.

1. Motivation

Inertial Confinement Fusion is approaching ignition of DT targets within the next generation of high-power lasers [1]. The problem lies in the tiny masses ($M_{DT} \sim 1\,mg$) and sizes ($R \leq 50\,\mu m$). Laser fusion is probing the lower threshold for DT ignition. Here, we test the envisioned ignition parameters $T_{ign} \geq 5\,keV$, $(\rho R)_{ign} \geq 0.3\,g/cm^2$ and the burn rate formula $\phi \simeq H_f/(H_B + H_f)$, where $H_f = \rho R$ is the confinement parameter and the burn parameter $H_B \simeq 7\,g/cm^2$, on basis of *stochastic transport theory.* Within the present model, it is found that these conditions are too optimistic; the product $(\rho R)_{ign} T_{ign}$ should be 50 % larger for stable ignition and a factor 2 larger for saturating burn.

2. Stochastic Transport Method

The transport of charged particles during ignition and burn of DT plasma is treated as Brownian motion, solving the Langevin equation

$$\frac{dv_i^\alpha}{dt} = h_i^\alpha(\vec{v}^\alpha) + g_{ik}^\alpha(\vec{v}^\alpha)\xi_k(t), \tag{1}$$

where the functions for drift $h_i^\alpha(\vec{v}^\alpha)$ and diffusion $g_{ik}^\alpha(\vec{v}^\alpha)$ are discussed below and $\vec{\xi}(t)$ with $\langle\xi_i(t)\rangle = 0$, $\langle\xi_i(t)\xi_k(t+\tau)\rangle = \delta_{ik}\delta(\tau)$ describes "white noise"; α denotes different particle species and $i, k = 1, 2, 3$. The brackets imply stochastic averaging. Solving equation (1) for a number of particles is equivalent to solve the Kolmogorov equation

$$\frac{\partial f_\alpha(\vec{v},t)}{\partial t} = -\frac{\partial}{\partial v_i}[A_i^\alpha(\vec{v})f_\alpha] + \frac{1}{2}\frac{\partial^2}{\partial v_i \partial v_j}[B_{ij}^\alpha(\vec{v})f_\alpha], \tag{2}$$

$$A_i^\alpha(\vec{v}) = h_i^\alpha(\vec{v}) + \frac{1}{2}g_{ik}^\alpha(\vec{v})\frac{\partial g_{jk}^\alpha(\vec{v})}{\partial v_j}, \qquad B_{ij}^\alpha(\vec{v}) = g_{ik}^\alpha(\vec{v})g_{jk}^\alpha(\vec{v}).$$

for the probability density f_α, and the functions $h_i^\alpha(\vec{v})$ and $g_{ik}^\alpha(\vec{v})$ can be obtained by comparison with the Fokker-Plank equation for plasma with collisions. Explicit expressions are derived by Ivanov and Shvets in [2] for different important cases.

The stochastic equations (1), written for each macroparticle m ($1 < m < M_\alpha, M_\alpha$ - the number of particles of sort α) with the coordinate equations

$$\frac{d\vec{v}^\alpha_m}{dt} = \vec{h}^\alpha(\vec{v}^\alpha_m) + G^\alpha(\vec{v}^\alpha_m)\vec{\xi}, \qquad \frac{d\vec{x}^\alpha_m}{dt} = \vec{v}^\alpha_m \tag{3}$$

and the initial conditions $\vec{x}^\alpha_m|_{t=0} = \vec{x}^\alpha_{0m}, \vec{v}^\alpha_m|_{t=0} = \vec{v}^\alpha_{0m}$ form the system describing the dynamics of the particle ensemble with the distribution function satisfying the Fokker-Plank equation. The velocity equation (3) is solved using an Eulerian scheme (index α and M suppressed)

$$\vec{v}^{n+0.5} = \vec{v}^n + \vec{h}(\vec{v}^n, t^n)\frac{\Delta t}{2} + G(\vec{v}^n, t^n)[\Delta\vec{\eta}]_{0.5} \tag{4}$$

$$\vec{v}^{n+1} = \vec{v}^{n+0.5} + \vec{h}(\vec{v}^{n+0.5}, t^{n+0.5})\Delta t + G(\vec{v}^{n+0.5}, t^{n+0.5})\Delta\vec{\eta} \tag{5}$$

where $\Delta\vec{\eta}$ is the increment of the Wiener process in finite time Δt:

$$\Delta\eta_k = \int_{t^n}^{t^{n+1}} \xi_k(t)dt = \gamma^n_k\sqrt{\Delta t}, \qquad [\Delta\vec{\eta}]_{0.5} = \frac{\vec{\gamma}^n\sqrt{\Delta t}}{2},$$

and γ^n_k is a normally distributed random quantity with zero average and unit dispersion.

3. Ignition and burn calculation

Stochastic 3D transport of ${}^4He$, ${}^3He$, T and p is solved together 1D two-temperature hydrodynamics and multigroup radiation transport [3]. Three thermonuclear reactions ($D+T, D+D, D+{}^3He$) are taken into account. In each temporal step Δt, the trajectories of typically 1500 macroparticles are calculated, which represent the true particles burned in time Δt in the target volume. Then the hydrodynamic quantities are advanced.

In Fig. 1 equatorial projections of simulated trajectories are presented for a typical homogeneous case $\rho R = 1\,g/cm^2$ and $T_e = T_i = 20\,keV$. Trajectories with an arrow in the middle correspond to macroparticles stopped in the target, otherwise they escape.

Figure 1. Illustrative macroparticle trajectories in target.

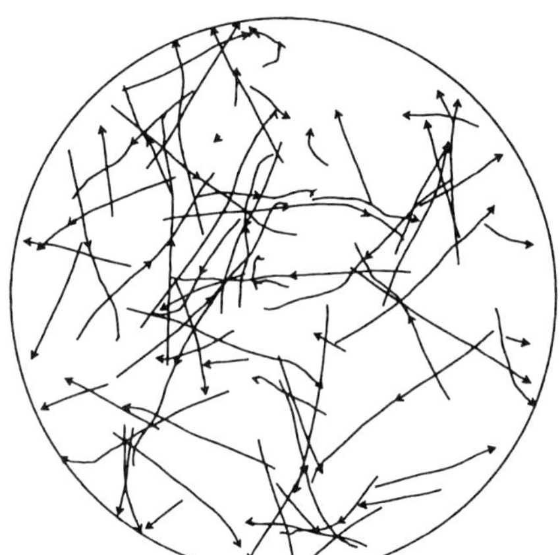

Ignition and burn of uniformly compressed DT spheres (so called "volume ignition") was studied first. Initial configurations with masses $M_{DT} = 10^{-6} - 10^{-2}\,g$, confinement parameters $H_f = \rho R = 0.2 - 3.0\,g/cm^2$ and tempertures $T = T_i = T_e = 4 - 10\,keV$ are considered; all other parameters like densities ρ, radii R, etc. are calculated from M_{DT},

H_f and T. Results are given in Fig. 2 in terms of burn efficiency (i.e. fraction of DT fuel burned) versus H_f for $M_{DT} = 1\,mg$ and $T = 10\,keV$.

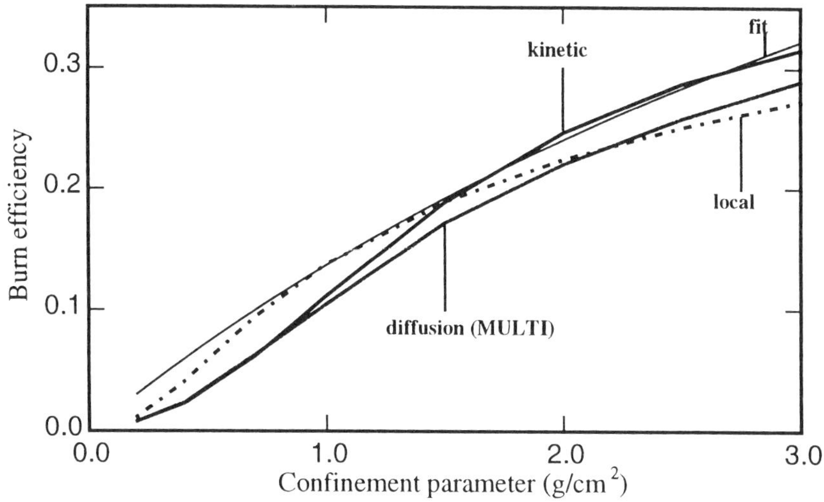

Figure 2. Ignition of uniformly compressed microspheres.

Results of the present work (thick solid curve) can be represented by the simple formula (thin solid curve) for the burn efficiency $\phi = H_f/(H_B + H_f)$ for $H_f > 1.5\,g/cm^2$ and $H_B = 6.31$; similar agreement is obtained for lower temperatures, at $7\,keV$ with $H_B = 6.85\,g/cm^2$ and at $5\,keV$ with $H_B = 8.16\,g/cm^2$. In Fig. 2, the present results are also compared with other burn calculations based on local α-deposition and one-group diffusion transport of α-particles [4].

It is interesting to see that, for large enough H_f, particle transport produces more burn than α-diffusion and even local α-deposition. However, for $H_f < 1.5\,g/cm^2$, local deposition overestimates the burn efficiency and transport considerably reduces the burn.

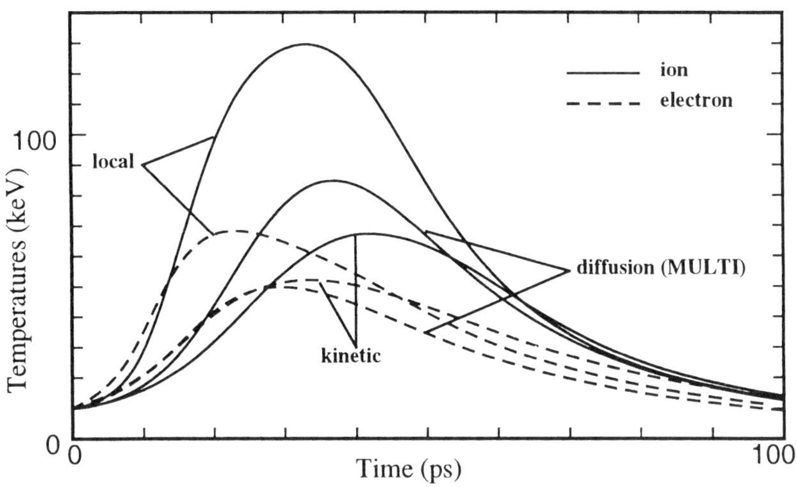

Figure 3. Temperature evolution during DT burn.

The temporal evolution of ion and electron temperatures are plotted in Fig. 3 for a uniform configuration with $M_f = 1\,mg$, $T = 10\,keV$ and $H_f = 1.5\,g/cm^2$.

It is seen that local α-deposition strongly overestimates the central burn temperature; kinetic transport gives peak temperatures even lower than diffusive transport. Electron temperatures peak earlier than ion temperatures due to more efficient energy transfer to electrons during ignition when the plasma temperature is still low.

The DT fuel energy needed for ignition can be considerably reduced (by a factor of 10 and more) by igniting a central portion of the fuel only (the spark) and letting burn propagate into surrounding cold fuel. In Fig. 4, ignition and burn of such spark configurations are plotted in terms of trajectories in a plane spanned by the confinement parameter $H_s = \int \rho dR$ and the average ion temperature T_s of the spark; the dots mark different initially isobaric configurations with uniform spark and cold fuel regions (10 % of the mass in the spark).

The dashed lines given by $H_s T_s = 2.5 - 3.0\,keV \cdot g/cm^2$ mark the ignition threshold. It is seen that burn, indicated by a strong temperature excursion, only develops for initial configurations at and above this ignition line. Note that the increase of H_s is due to spark recompression by heating the cold compressed fuel. The values $T_s = 5\,keV$ and $H_s = 0.3\,g/cm^2$, often used as ignition parameters, do not lead to ignition in the present calculation.

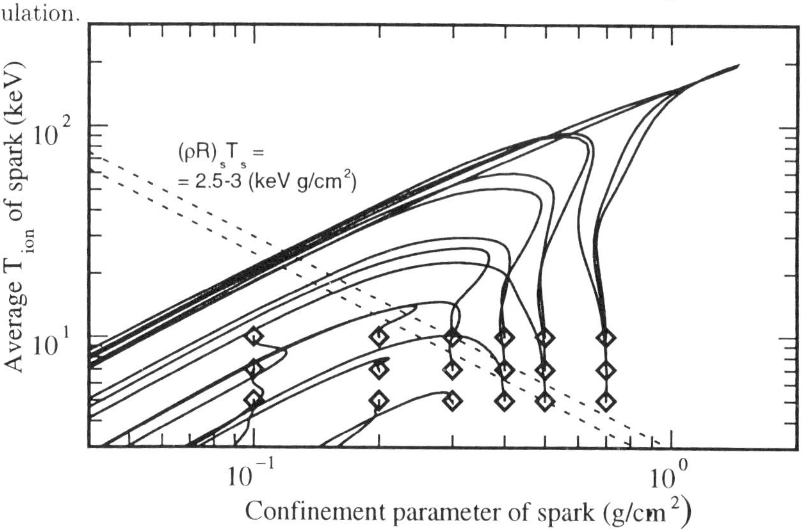

Figure 4. Burn trajectories for spark ignition.

References

[1] Lindl J D, McCrory R L and Campbell E M Sept.1992 *Physics Today*

[2] Ivanov M F and Shvets V F 1980 *Zh. Vychisl. Mat. i Mat. Fiz.* **20** 682
Ivanov M F and Shvets V F 1980 *Sov. Phys. Tech. Phys.* **25** 647
Ivanov M F and Shvets V F 1978 *Sov. Phys. Dokl.* **23** 130
Ivanov M F and Shvets V F 1979 *Chislen. Met. Mekh. Splosh. Sred* **10** 64

[3] Anisimov S I and Oparin A M 1993 *JETP Lett.* **57** 634

[4] Ramis R, Schmalz R and Meyer-ter-Vehn J 1988 *Comp. Phys. Comm.* **49** 475

Inst. Phys. Conf. Ser. No 140: Section 9
Paper presented at Laser Interact. with Matter Conf., Oxford, 19–23 September 1994

© *1995 IOP Publishing Ltd*

Metallic hydrogen by laser compression

A Oparin and J Meyer-ter-Vehn

Max-Planck-Institut für Quantenoptik, D-85748 Garching, Germany

Abstract. We propose an experiment for metallic hydrogen production by a kilojoule laser pulse.

1. Metallic hydrogen

Solid hydrogen is a molecular *hcp* crystal with density $0.088\,g/cm^3$. It is an insulator, transparent for optical light. At tenfold compression and pressures of $2-3\,Mbar$ pressure, the molecular crystal is expected to transform into a monatomic *bcc* structure which forms a metal consisting of protons and electrons only. It is the simplest metal possible and therefore of outstanding basic interest. Attempts to make metallic hydrogen have been unsuccessful, so far. Static compression in diamond anvil cells [1] and also dynamic compression using two-stage gas guns [2] have reached maximum pressures up to $2\,Mbar$, but the expected transition has not yet been found experimentally.

2. Uniform laser compression demonstrated at MPQ/ILE

High power lasers allow to generate multi-megabar pressures in solids. However, direct irradiation of a surface leads to rippled shock fronts due to nonuniformities in the laser spot. Very uniform shock fronts could be produced recently at MPQ/ILE, using indirect illumination of the shock sample by thermal radiation produced in laser driven cavities [3]. This technique offers new possibilities for dynamic high-pressure experiments, including solid hydrogen compression to the metallic state. An explicit way to do this is described in the present work.

3. Multiple shock compression

A basic difficulty is that the solid hydrogen has to be compressed about ten times at low temperature. A single shock leads to $4-5$ fold compression at the maximum and then heats the sample excessively. Low entropy compression to high densities can be achieved, however, by a sequence of relatively weak shocks. In [4] Aoki and Meyer-ter-Vehn reported about numerical simulation of cryogenic hydrogen compression by cylindrically imploding heavy gold liner with velocity 1 km/s. The metallic state was achieved at a temperature about 1000 K. Nevertheless the planar geometry may be preferable for experiments. An attractive option is to use two planar flyers impacting symmetrically from two sides on the hydrogen sample.

4. Simulation of symmetric flyer impact

The symmetric impact of two gold flyers, 100 μm thick and initially at 2.5 km/s velocity, on a plane layer of 200 μm solid hydrogen has been simulated, based on the SESAME EOS. The parameters have been chosen such as to reach the metallic phase. This phase lives for about 20 ns, and the temperature is about 2000 K, low enough for the metallic phase to exist. The thickness of hydrogen in the metallic state is about 15 μm and is almost minimal for diagnostic. The perspective density plot in Figure 1 clearly shows the sequence of shocks during the compression period. We can also see that the metallic hydrogen when achieved is very homogeneous. The first shock, the most important for sample heating, needs about 20 ns to reach the midplane, and the total compression time is 50 ns. Such long times are needed in order to keep the strength of the first shock low. These times are long compared to typical 1 ns pulse times of existing high power lasers.

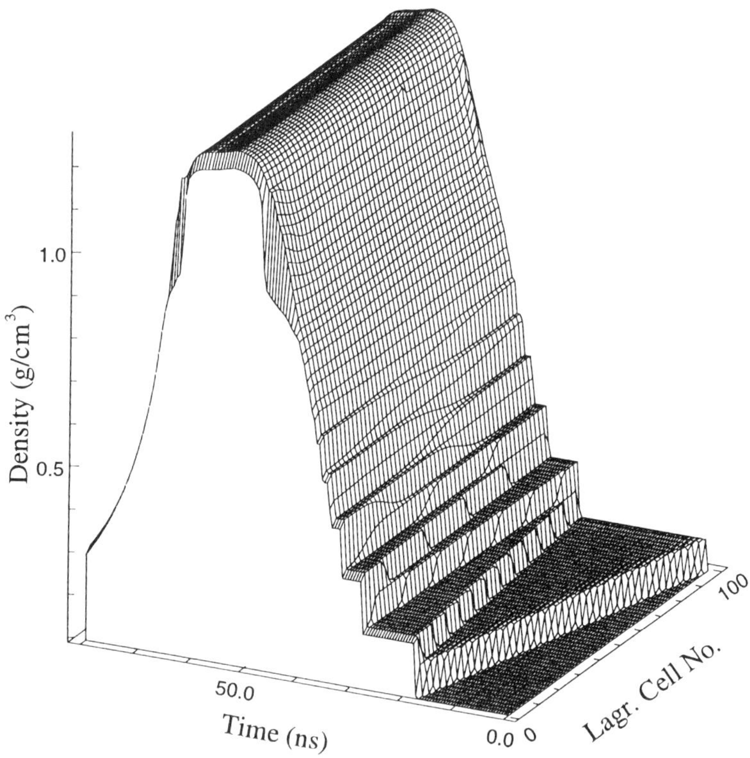

Figure 1. Perspective plot of density evolution.

5. How to compress slowly (50 ns) with a short laser pulse (1 ns)

The problem is to transform the pulse energy into kinetic energy of a massive flyer layer moving at relatively low velocity. The solution is to cover the flyer plate by a thick plastic cushion. The corresponding calculation is displayed in terms of flow diagrams in Figures 2 and 3. The thickness of the plastic layer is 700 μm. It is assumed that the laser pulse is incident on a gold converter foil (50 μm) from the inner side. A laser pulse

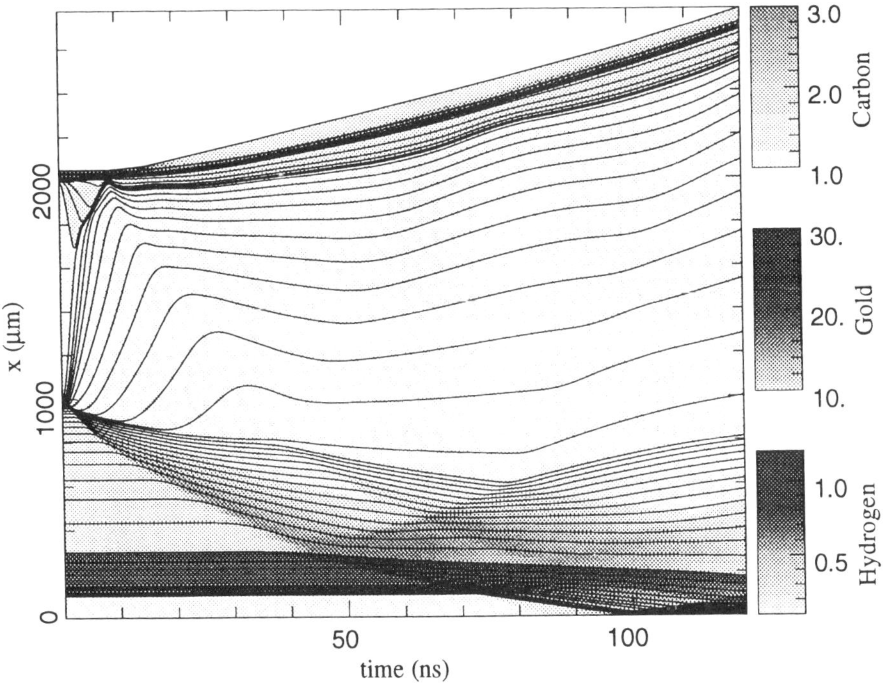

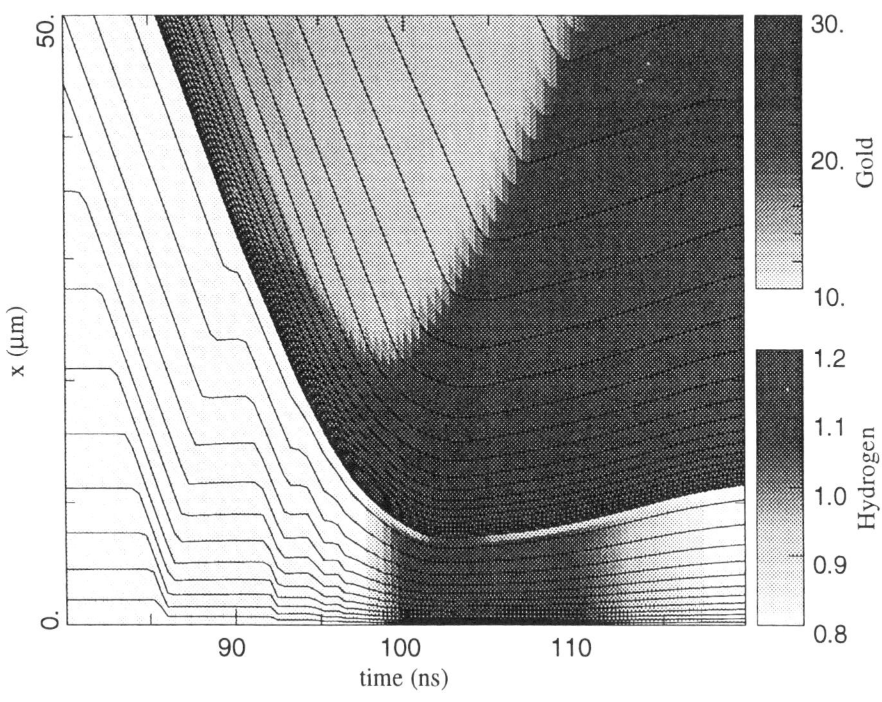

Figures 2 and 3. Flow diagrams with a density (g/cm^3) distribution.

with $1\,ns$, $30\,TW/cm^2$, $30\,kJ/cm^2$ is used to produce a $1\,ns$ X-ray pulse which drives a shock into the plastic impulsively. Its pressure is high initially ($6\,Mbar$), but decays rapidly when it passes through the plastic and the flyer layer, setting them into moderate motion. It is sufficient to compress the hydrogen to the metallic state. The Figure 3 represents a zoom on the high compression hydrogen region of Figure 2. The strength of the first shock in hydrogen is 30 kbar and the final hydrogen temperature is still low (about 3000 K).

About 80 % of the laser energy are transferred to the plastic, and 0.7 % are finally found in the hydrogen layer at maximum compression. It is important to note that only energies of order 7 Joule have to be invested to compress hydrogen samples 0.1 mm thick and 1 mm in diameter. Kilojoule pulses available from existing laser facilities should therefore be sufficient to reach the metallic state.

6. Proposed experimental configuration

A possible experimental configuration is sketched in Figure 4. The stack of plastic ablator and gold liner described above is mounted on the radiation cavity used in the MPQ/ILE shock experiments. This configuration is doubled and applied symmetrically from two sides to a disk-shaped solid hydrogen sample in the centre. Also the compressed configuration is illustrated. The fronts of the compression waves running toward the hydrogen will have a slightly convex shape due to losses in radial direction, and therefore highest compression is expected to occur first at the centre. Part of the hydrogen will be squeezed out radially. Optical observation of the compressed material is possible within the plane of the disk. Besides optical diagnostics, electric conductivity measurements may be possible, using the two gold liners as electrodes.

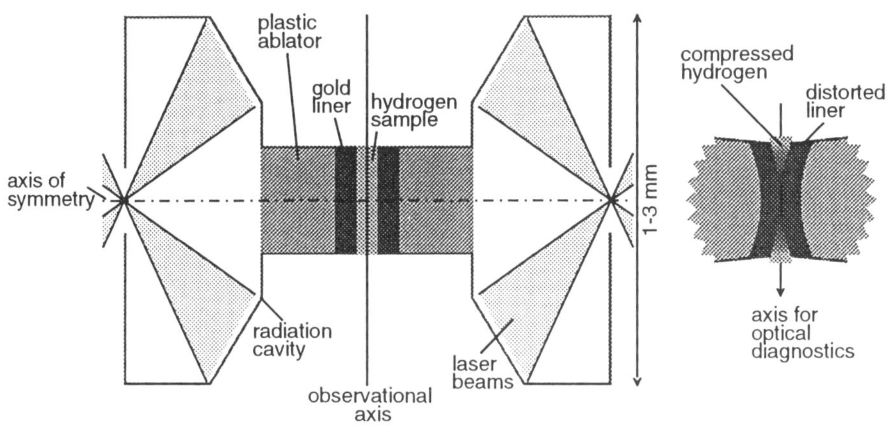

Figure 4. Proposed experimental configuration.

References

[1] Mao H and Hemley R 1994 *Rev. Mod. Phys.* **66** 671

[2] Weir S T, Nellis W J and Mitchell A C (in press) *High Pressure Science and Technology-1993* eds. Schmidt S C *et al* (American Institut of Physics)

[3] Löwer Th *et al* 1994 *Phys. Rev. Letters* **72** 3186

[4] Aoki T and Meyer-ter-Vehn 1994 *Phys. Plasmas* **1** 1962

Inst. Phys. Conf. Ser. No 140: Section 9
Paper presented at Laser Interact. with Matter Conf., Oxford, 19–23 September 1994
© 1995 IOP Publishing Ltd

The Ion Model of Matter

N.Y. Orlov

Institute for Mathematical Modelling, 4-A, Miusskaya Sq., 125047, Moscow, Russia.

Abstract. The most interesting results of quantum mechanical model of hot and dense matter based on the self- consistent field method for statistical ensemble of plasma ions are presented. Comparisons with experimental data are shown. The quantum mechanical explanation of experimental observed phenomenon of "enlightment" is given.

1. Introduction

In order to solve the scientific and technological problems connected with controlled nuclear fusion, exploration of space and other questions we have to know the heat conduction coefficients, the light absorption coefficients, Rosseland or Plank means and other plasma characteristics. Some theoretical models of hot and dense matter have been created in the past for the calculation of such characteristics. The well-known Hartree-Fock-Slater model (HFS) does not give accurate results as the temperature decreases. The Detail Configuration Account (DCA) [1] has used the Hartree-Fock equations for separate ions in plasma and the Saha method for ion concentrations. This approach meets great difficulties in the case of heavy chemical elements [2]. Moreover, it may be used in the case of ideal or weak nonideal plasma only.

Therefore another theoretical model of matter has been proposed [3] in 1982. It was called "The ion model of matter"(IM). The system of self-consistent field equations, which may be obtained by using of W. Kohn and L. J. Sham method [4], was solved for statistical ensemble of atoms and ions of plasma in contrast to HFS and DCA methods. Certainly, such system is more complicated in comparison with previous models. But it was solved at first for "pure" matter which contain only atoms and ions of alone chemical element and then for more complicated case of mixture of chemical elements [5]. The relativistic effects were taken into account [6,7], new variation approach to description of electron energy structure was developed [8]. The comparison of its results with a lot of experimental data shows a good agreement [8,9]. And what is more this model gives the theoretical explanations for some phenomena which were observed in experimental investigations but which have not had such explanations [10]. It gives a possibility to perform reliable quantum mechanical calculations for nonideal plasma into inaccesible before region of low enough temperatures.

Rather like theoretical model has been proposed by F. Perrot [11] in 1982.

2. The ion model of matter

The system of equations of ion model of matter may be obtained by using of W. Kohn and L. J. Sham method [4]. It includes the self-consistent field equations (the Schrodinger equations [3] or the Hartree-Fock equations [12]) for bounded electrons for each plasma ion, the Gibbs distribution for ion concentrations and the electroneutrality condition. The approximate form for exchange interaction is used in the case of Schrodinger equations. We'll consider of this case.

This way seems to be more difficulty than DCA. But one circumstance makes easy this problem. The fact is that the ion concentrations differ from each other greatly. If we found the method for dividing of statistical ensemble into two parts (the abundant species ions and the rare species ions respectively) before quantum mechanical solution of whole problem, we would evaluate the characteristics of abundant species ions immediately and for rare species ions we would use the theory of perturbations.

Such method has been found [3,9]. The numerous calculations show that most abundant species ions correspond to electron configurations nearest to Fermi distribution. More rare species ions correspond to its one-electron exitations, still more rare species ions correspond to two-electron exitations etc.

For each ion from first group (group of "base ions") we write down the Schrodinger equations

$$\hat{H}_j \Psi^j_\alpha(q) = E^j_\alpha \Psi^j_\alpha(q), \tag{1}$$

Ψ^j_α(q) - one-electron wave function, q - aggregate of $\vec{r}$ and spin σ. By writing it in the form $\Psi_\alpha(\vec{r}) = \frac{1}{r} R_{El}(r) Y_{lm}(\theta,\varphi)$, we can write down for bounded electrons

$$-\frac{1}{2} R^{j\prime\prime}_{nl} + \left[-V_j(r) + \frac{l(l+1)}{2r^2}\right] R^j_{nl} = E^j_{nl} R^j_{nl} \tag{2}$$

with the boundary conditions $R^j_{nl}(0) = 0;\ R^j_{nl}(r_0) = 0$ and standard condition $\int_0^{r_0} R^{j2}_{nl}\, dr = 1$ (j is the number of ion with electron state occupation numbers N^j_{nl}),

$$r_0 = \left[\frac{3}{4\pi}\frac{1}{N_A}\right]^{1/3} \frac{1}{a_0}\left[\frac{A}{\tilde{\rho}}\right]^{1/3}, \tag{3}$$

(r_0 - the ion sphere radius, $a_0 = 5.292 \cdot 10^{-9}\,cm$, $N_A = 6.02 \cdot 10^{23}$ - Avogadro number, $\tilde{\rho}$ - density gm/cm^3),

$V_j(r) = V^j_d(r) + V^j_{ex}(r)$, where $V^j_d(r)$ and $V^j_{ex}(r)$ are "direct" and "exchange" potentials respectively. The approximate form for $V_{ex}(r)$ was proposed in [13]. The electron density may be written as a summ of densities of bounded and free electrons $\rho_j(r) = \rho_{j1}(r) + \rho_{j2}(r)$. For $\rho_{j2}(r)$ the quasiclassical approximation is used [3].

Then we impose the electroneutrality condition upon the ion group $\sum W_j N_j = Z$,

$$N_j = \sum_k N^j_k = \sum_{nl} N^j_{nl} + 4\pi \int_0^{r_0} \rho_{j2}(r) r^2\, dr \tag{4}$$

and W_j is given by formula

$$W_j = Cg_j \exp\left[-\frac{E_j - \mu \sum_k N_k^j}{T}\right] \quad (5)$$

E_j - total energy of ion, T - temperature of matter, μ - chemical potential (we use atomic system of units, g_i - statistical weight).

By solving this set of equations by iteration method we'll find the electron energy levels for each ion from first group and evaluate the total energy.

Now we'll consider the ions from second group (the rare species ions). In order to take into account such ions we can apply the theory of perturbations. Now we can construct the different theories of perturbations for different "base ions". They differs from each other by collection of one-electron energy levels and by collection of one-electron wave functions. Therefore we can apply each of these theories for evaluation of characteristics of nearest to considered "base ion" electron configurations only [9].

3. The results of evaluations

The comparison with experimental absorption spectrum is given at Fig. 1. It shows the experimental light absorption coefficient [14] (continuous line) in aluminium for T = 18 ev., $\rho = 0.05 g / cm^3$ and ion model result (dotted line) [9].

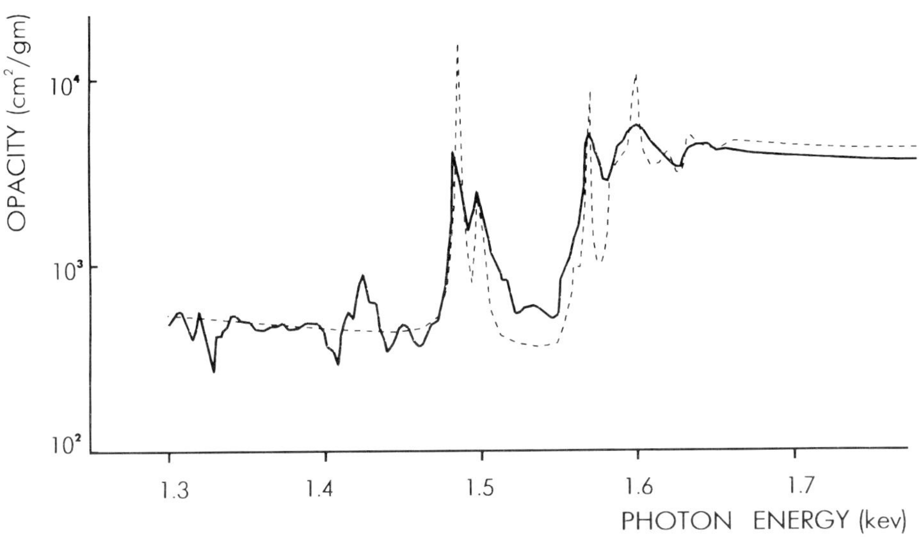

Fig. 1

We have compared our results with experimental data connected to mesurements of the light absorption coefficients æ (cm^{-1}) depending on plasma density (n_a - atom concentration $1/cm^3$) in argon at temperature $T = 2 \cdot 10^4 K^0$ and wave length $\lambda = 5800 \mathring{A}$. The dotted line at the Fig. 2 presents our theoretical results and the continuos line presents the experimental results [15]. We can see that at low densities the light absorption coefficient increases with the increasing of density. But at high enough density its behaviour is changed. At one time the theoretical explanation of this experimental results was absent. To give the quantum mechanical explanation of this phenomenon we have calculated the absorption spectrum. Figure 3 shows the bound - free absorption spectrum at lg n_a = 19.3 (—) and at lg n_a = 20.4 (- - -). The pointer at the picture shows the frequency of measurements corresponding the wave length $\lambda = 5800\,\overset{o}{A}$. It is seen that spectrum changes with density increasing and cross section decreases for mentioned wave length.

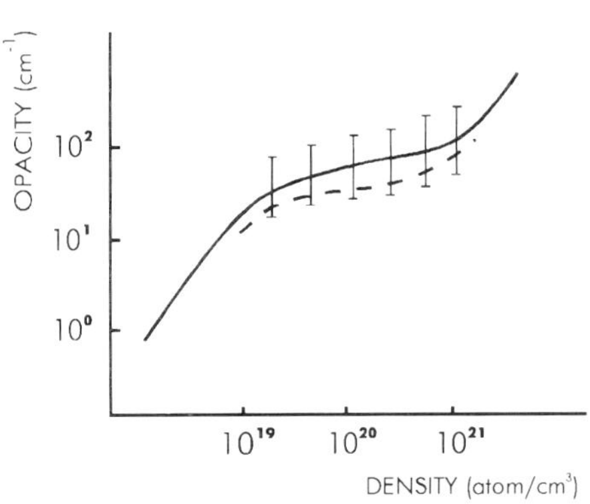

Fig. 2

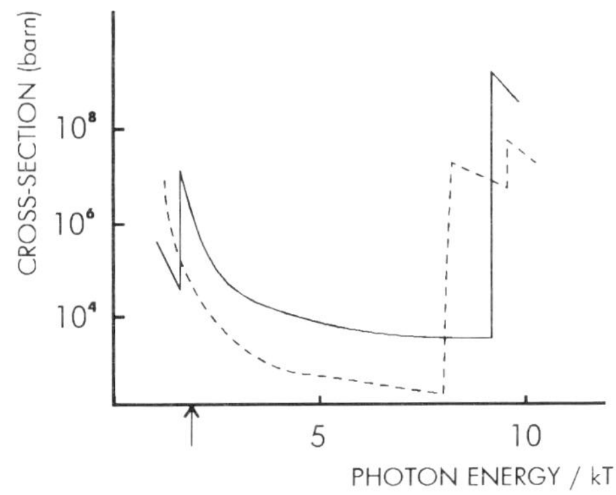

Fig. 3

REFERENCES

[1] D. Salzman et al. *Phys. Rev.* A.**32**, 3627 (1985).
[2] B.F. Rozsnyai. *J. Quant. Spectrosc. Radiat. Transfer.* **27**, 211 (1982).
[3] A.F. Nikiforov et al. *Preprint Inst. Appl. Mathem. The USSR Academy of Sciences.*, N **144**, 1982.
[4] W. Kohn and L.J. Sham, *Phys. Rev.* A**140**, 1133 (1965)
[5] N.Y. Orlov. *USSR Comput. Math. Mathem. Phys.* **27**, 1058 (1987).
[6] N.Y. Orlov. *Differential Equations.* **20**, 1253 (1984).
[7] N.Y. Orlov. *Mathem Modelling.* **2**, 33 (1990).
[8] N.Y. Orlov. *Differential Equations.* **27**, 1237 (1991).
[9] N.Y. Orlov. *Mathem. Modelling.* **4**, 19 (1992).
[10] A.F. Nikiforov, et al. *Vopr. Atomnoi Nauki i Tehniki*, **4(6)**, 36 (1979).
[11] F. Perrot, *Phys. Rev.* A**25**, 489 (1982).
[12] N.Y. Orlov. *Mathem. Modelling.* **5**, 3 (1993).
[13] A.F. Nikiforov et al. *Vopr. Atomnoi Nauki i Tehniki.* **4(6)**, 16 (1979).
[14] S.J. Davidson et al. *Appl. Phys. Lett.* **52**, 847 (1988).
[15] V.E. Fortov et al. *J. Eksp. Teor. Fis.* **76**, 141 (1979).

Inst. Phys. Conf. Ser. No 140: Section 9
Paper presented at Laser Interact. with Matter Conf., Oxford, 19–23 September 1994

© 1995 IOP Publishing Ltd

Investigations of Nd-laser interaction with medium-Z targets

P Parys L Ryć J Wołowski E Woryna E Zielińska

Institute of Plasma Physics and Laser Microfusion, 00-908 Warsaw, 23 Hery St., Poland

Abstract. Measurements of the energy distribution of different ion species at Nd:glass laser irradiance of about 10^{14} *W/cm²* on Al, Fe, Cu, and Pt targets are presented. The abundance of ion species and the average charge state of Fe and Cu plasmas as a function of ion energy are determined at distance about of 230 *cm* from the targets. Also, the influence of the target position versus the focal spot position on the X-ray radiation, reflection of laser light, and ion production are studied.

1. Introduction

Our investigations show that the laser-produced plasma parameters depend clearly on the position of the laser beam focus with respect to the target surface. The most intense X-ray emission in the applied experimental system is obtained for the focus distance of 2 to 3 *mm* in front of the target and inside the target, while the highest ion emission for the distance of about 1 *mm* in front of and inside the target. When the laser beam is focused precisely on the target surface, both these quantities take their minimum values at maximum reflectivity of the laser beam on the plasma. Hence, for such a position of the focus, the absorption of the laser radiation in the plasma is lower and the amount of the plasma produced is less also because of the minimum dimensions of the focus.

2. Experimental set-up

Schematic diagram of the experimental set-up is shown in Fig. 1. Targets (*T*) were irradiated with Nd:glass laser fundamental pulses. The laser pulse energy was up to 15 *J* in about 1.2 *ns*. The laser beam was focused with an f/2 aspherical lens controlled in the laser beam axisdirection with the accuracy of 20 *μm*. Targets were located in the geometric centre of the target chamber (*CH*) normally to the laser beam axis. 3 and 6 *μm* thick Al foils, 12 *μm* thick Al foils covered with 0.05 *μm* Cu layer, and Fe, Cu, and Pt slabs were used as targets. An electrostatic ion-energy analyzer (*EIA*) [1] located at an angle of 45 degrees, and at a distance of 230.2 *cm* from the target was used to measure the energy distributions of ion species. Three ion collectors (*S45,S55,S90*) located at 45, 55, and 90 degrees angles, and at the distances of 77.9, 75.3, and 46.1 *cm*, respectively, were used for angular distribution measurements of plasma expansion. The X-ray measurements were performed using an X-ray spectrometer (*XRS*) with a flat ADP crystal (0.68-0.83 *nm* wavelength range) and a CCD linear array consisting of 1728 elements, a set of four semiconductor detectors - BPYP 44 photodiodes (*SD4*), filtered with 300, 600, 900, and 1200 *μm* thick Be foils with 1/e cut-off energies of

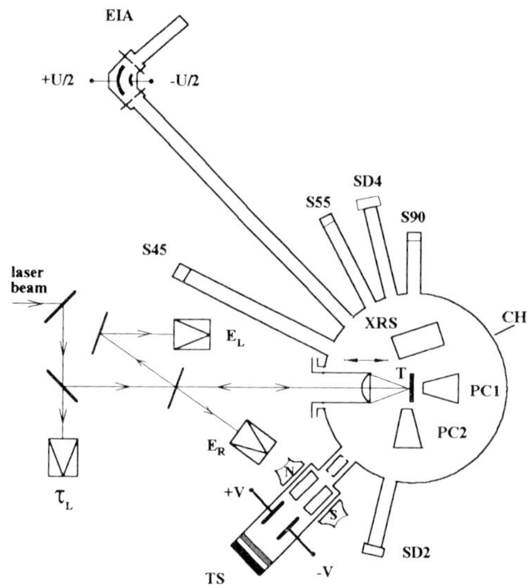

Fig. 1. Schematic diagram of the experimental arrangement

3.1, 3.8, 4.3, and 4.7 *keV*, respectively, a set of two semiconductor detectors - BPYP 41 photodiodes (*SD2*), filtered with 6 μm thick Al and 10 μm thick Be foils, an X-ray vacuum diode with an Al cathode filtered with 5 μm thick policarbonate foil with maximum sensitivity below 1 *keV*, and two pinhole cameras (*PC1,PC2*). Temporal shapes of the laser and X-ray pulses were recorded with a SCD 1000 Tektronix digitizer. The laser energy and the energy reflected from the plasma were measured with a Rj 7620 energy ratiometer. Ion current pulses from the ion collector and the EIA were recorded with a TDS 450 Tektronix digitizing oscilloscope. The pressure was about 4×10^{-6} *Torr* inside the *CH*, and $(6\div8)\times10^{-6}$ *Torr* inside the EIA.

3. Experimental results

The influence of the target position versus the laser beam focal spot position on ion and X-ray emission was investigated using a 3 μm thick Al foils as targets. The target - focusing lens distance was varied by moving the lens with regard to the targets to keep the measurement geometry. Sizes of focal spots, soft and hard X-ray intensities, the laser energy reflected from the plasma, andionization state of the plasma were investigated as a function of the focusing lens position. The minimum diameter of the focal spot was estimated to be about 100 μm (lens position x = 4.80 *mm*) on the basis of the X-ray images obtained with the *PC1*, Fig. 2. The soft X-ray measurements shows the occurrence of two maxima (at x = 3.40 and 6.30 *mm* with a dip at x = 4.80 *mm*, Fig. 3a. A similar behavior is observed for ion emission. Fig. 3c shows the ion current of Al^{3+} ions with the energy of 3 *keV*. This dependence has its minimum also at x = 4.80 *mm*. Simultaneously, the harder X-ray intensity reaches its maximum at x = 4.80 *mm*. The ratio of the reflected to the incident laser energy, E_R/E_L, as a function of focusing lens position is shown in Fig. 3b. For example, Fig. 3c shows the ion current of Al^{3+} ions with the energy of 3 *keV*. This dependence has also its minimum at x = 4.80 *mm*.

Spectral distributions of soft X-ray emitted from Al, Cu+Al, and Cu targets are registered. The electron temperature, T_e, and the electron concentration, n_e, of Al plasma were estimated

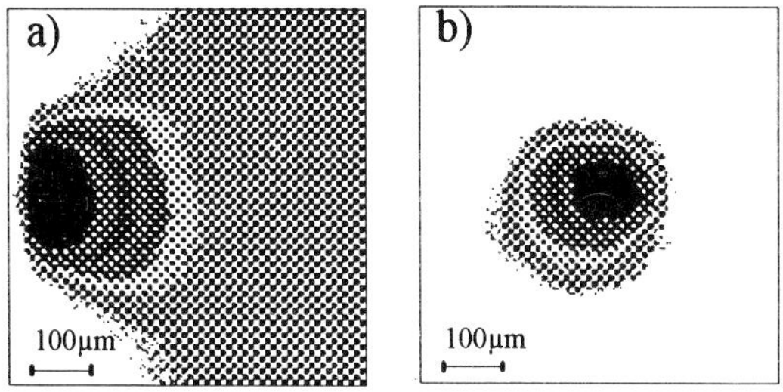

Fig. 2. Side (a) and back (b) view of X-ray images of the minimum focal spot

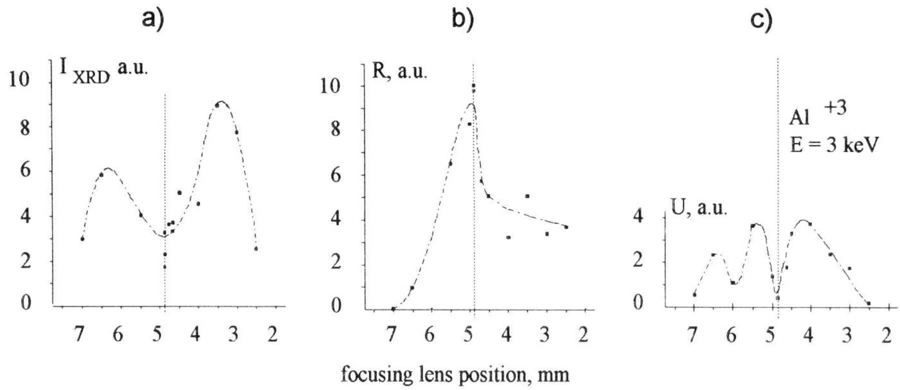

Fig. 3. Soft X-ray (a), reflectivity (b), and ion current (c) versus focusing lens position

on the basis of the intensity ratio of the (j,k) satellites and Al^{11+} resonance lines. $T_e = 390\ eV$ and $n_e = (1 \div 1.5)\times 10^{20}\ cm^{-3}$ were obtained. The electron temperature and the average charge state of Al plasma in the focal spot $<z_0>$, estimated from ion collector signals, were in the ranges of $T_e = 225 \div 500\ eV$, and $<z_0> = 10.7 \div 11.2$, respectively.

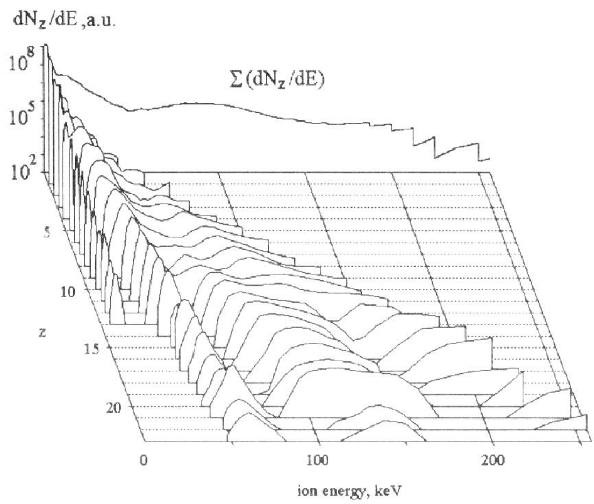

Fig. 4. Ion spectrum from the EIA and the energy distribution of Fe ion species

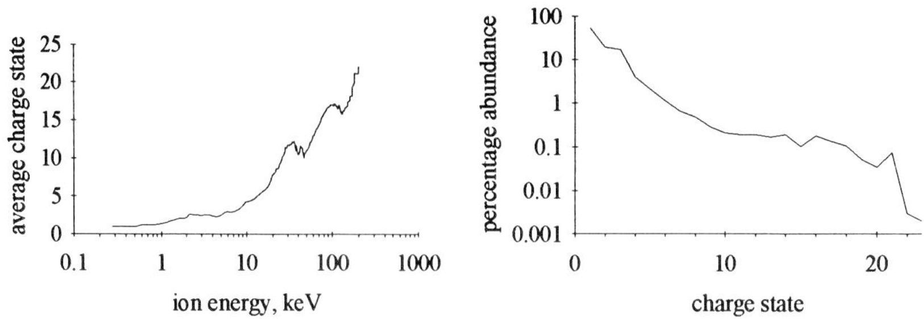

Fig. 5. Average charge state vs. ion energy and percentage abundance of Fe plasma

The Al, Fe, Cu, and Pt ion species spectra with different energy to charge state ratio were measured by varying the deflection voltage U/2 of the EIA from shot to shot. Fig. 5 shows the ion spectrum of Fe ions obtained from the EIA. On the basis of them, the energy distribution of ion species (Fig. 4), the percentage abundance, and the average charge state of plasmas as a function of ion energy were determined, Fig. 5. The maximum registered charge state and the maximum kinetic energy of ions were as follows: 12+ for Al targets, 24+ and about 200 *keV* for Fe targets, 18+ and about 100 *keV* for Cu targets, and 19+ for Pt targets. X-ray spectrometric measurements shows the existence of Cu^{21+} ions in the focal spot area, while the maximum registered charge state of Cu obtained by means of the EIA was only 18+.

4. Conclusions

At the IPPLM, within the program on the studies of medium- and high-Z laser-produced plasma, investigations of the dependence of such plasma properties on the way of the laser beam focusing on the target surface were performed. Placing the target in various distances from the laser beam focus influences not only the power density in the focus but also the amount and the temperature of the colder plasma produced outside the focus volume. Parameters of this plasma depend strongly on the atomic number of the target material [2]. The work presents results of investigations of properties of the laser-produced plasma of various atomic numbers (Al, Fe, Cu, and Pt) obtained from measurements of parameters of ion fluxes and X-rays emitted from the plasma for various positions of the focus with respect to the target. The dependence of the laser radiation reflectivity on the focus position with respect to the target was also measured. As a result of those investigations, a clear dependence of the above mentioned quantities on the way of the laser beam focusing on the target was observed. On the basis of the results obtained, the plasma temperature and its charge state as well as ion energy distributions far away from the target were determined. An important result of the work is showing the need for optimizing the way of the laser beam focusing on the target in investigations concerning laser-plasma sources of X-rays and ion sources.

The X-ray diagnostics used in the experiments described make it possible to monitor the operation of the laser system, to estimate the laser focus spot, and to determine the main plasma parameters additionally to ion diagnostics.. The obtained results are consistent with the ones given in [3].

References

[1] Denus S et al 1977 J Tech Phys **18**(1) 25.
[2] Mróz W et al 1992 Laser and Particle Beams **10** 689
[3] Amiranoff F et al 1982 J. Physique **43** 1037

Inst. Phys. Conf. Ser. No 140: Section 9
Paper presented at Laser Interact. with Matter Conf., Oxford, 19–23 September 1994

© *1995 IOP Publishing Ltd*

Studies of ion fluxes emitted normally to the target surface from various Z-number targets

P Parys J Wołowski E Woryna J Farny[a)] W Mróz[b)]

Institute of Plasma Physics and Laser Microfusion, 00-908 Warsaw , 23 Hery St., Poland
a) 95445 Bayreuth, Meranierring 28/7, Germany
b) Institute of Optoelectronics, Military Univ of Technology, 00-908 Warsaw, Poland

Abstract. An aspherical lens - ellipsoidal mirror illumination system was used in the experiments. Results on highly charged ions from Al, Mo, Ta, and Pb targets in laser-produced plasma are presented. A cylindrical electrostatic ion-energy analyzer and an ion collector were used in the measurements. The influence of focusing conditions and a tilt angle on emission of highly charged Ta ions was studied. The maximum charge state of 42+ with energy at least 1.0 *MeV* have been recorded. Energy distribution of ion species, average charge state, and percentage abundance of the expanding plasma were determined at the distance of 323.5 *cm* from targets. The Nd:glass laser used in the experiments delivers up to 15 *J* to the target in 1.2 *ns* (FWHM) pulses.

1. Introduction

The subject of optimization of laboratory laser-plasma sources of multiply charged ions are the maximum flux and the charge of ions emitted from a laser-produced plasma. Previous investigations ([1,2]) suggested the maximum ion emission in the direction perpendicular to the target surface illuminated with a laser. For the purpose of precise investigation of parameters of multiply charged ion fluxes emitted perpendicularly to the target and parallel to the laser beam axis, an aspherical lens - ellipsoidal mirror illumination system was applied. Moreover, the influence of the target tilt angle with respect to the laser beam axis as well the beam's focus position with respect to the target surface on the flux, the energy, and the maximum charge state of ions emitted from the plasma was investigated. Considerable influence of the way of the laser beam focusing on the expanding ion parameters is shown in [3]. The results obtained demonstrate the possibility of getting high degrees of ionization and high maximum energies of ions (for Ta: $z = 42$, $E = 1.0$ *MeV*) at optimum experimental conditions, far away from the laser-heated target (above 3 *m*). These results require more detailed theoretical and modeling analysis of the laser-produced plasma expansion. Results of preliminary calculations, performed taking into account the two plasma temperatures (thermal plasma and colder plasma [4]), are consistent with experimental results at more detailed analysis of the influence of recombination on the expanding plasma parameters.

2. Experimental arrangement

The laser beam from a Nd:glass laser with the output energy E_L up to 15 *J* in $\tau_L = 1.2$ *ns* was focused on target surfaces to the minimum focal spot diameter of about 120 *μm* by means of an aspherical lens - ellipsoidal mirror illumination system, Fig. 1. Thick Al, Mo, Ta, and Pb strips were used as targets. They were not especially cleaned before laser irradiation. The ions from laser-produced plasmas were studied experimentally using mainly time-of-flight techniques. Both a cylindrical electrostatic ion-energy analyzer (EIA) and an ion collector (IC) [5] were used to measure ion fluxes emitted normally from targets and parallel to the

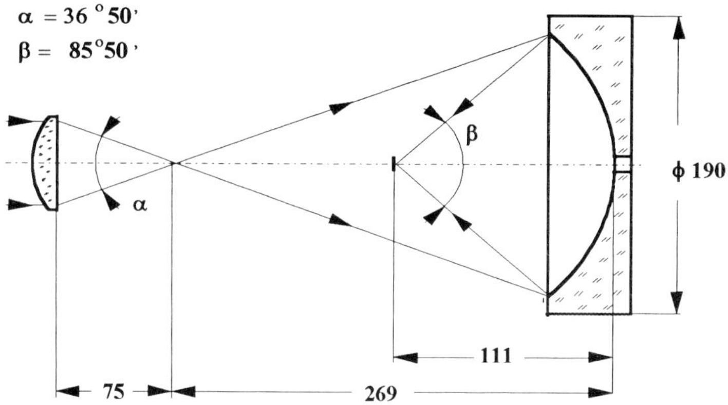

Fig. 1. Scheme of the aspherical lens - ellipsoidal mirror illumination system.

focused beam axis. The path of flight of the EIA and the IC were 323.5 *cm* and 191.5 *cm*, respectively. Ion collector pulses and ion spectra from the EIA were registered with a DSA 601 Tektronix digitizing signal analyzer. The temporal shape of laser pulses was recorded by the SCD 1000 Tektronix oscilloscope, and the incident laser energy was measured with a Rj 7620 energy ratiometer. Both the EIA and the IC were evacuated to $(3\div5)\times10^{-6}$ *Torr*.

3. Experimental results

Most of the experiments were performed in condition of sharp focusing the laser beam onto target surfaces (power intensity 6×10^{13} W/cm^2). Typical Ta ion spectrum from the EIA and the ion collector signal are presented in Figs. 2a and 2b, respectively). The energy distribution of Ta ion species, Fig. 3, were obtained on the basis of about 50 laser shots in wide range of deflection voltage U/2 (from 0.05 to 4 *kV*) of the EIA. Fig. 4 shows the charge state - energy diagram of Ta plasma. The average charge state of Ta plasma as a function of ion energy is shown in Fig. 5, and the percentage abundance of Ta ion species in Fig. 6. The secondary ion-electron emission coefficient was not taken into account during Ta data treatment. The maximum registered charge state and energy of Ta ions were $z = 27$ and $E = 650$ *keV*, respectively. For Mo targets the maximum charge state was $z = 32$.

However, it was determined that the best focal position of the laser beam was about 0.5 *mm* inside the target, Fig. 7, when the maximum obtained charge state was $z = 42$, and the maximum ion energy was at least 1.0 *MeV*. Also the influence of a tilt angle of Ta target with respect to the laser beam was studied. At the tilt angle $\varphi = 15$ degrees the maximum registered charge state was decreased from 21+ to 15+ (Fig. 7) and the amplitude of the collector signal was decreased of about 3 times.

In the experiments performed with Al targets, the maximum registered charge state and the maximum energy of ions were $z = 13$ and $E = 200$ *keV*, respectively. The energy distribution of Al ion species (Fig. 9), and the average charge state as a function of ion energy (Fig. 10) were determined taking into account the secondary ion-electron emission coefficient given by Cano [6]. The electron temperature of Al plasmas was estimated to be $T_e = 320\div350$ *eV* from ion collector signals, while from X-ray spectrometry measurements it was $T_e = 370\pm30$ *eV* [7].

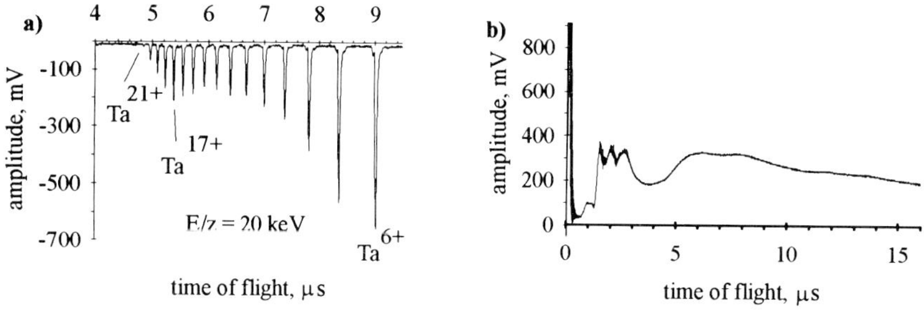

Fig. 2. Ta ion spectra from the EIA (a) and the ion collector signal (b).

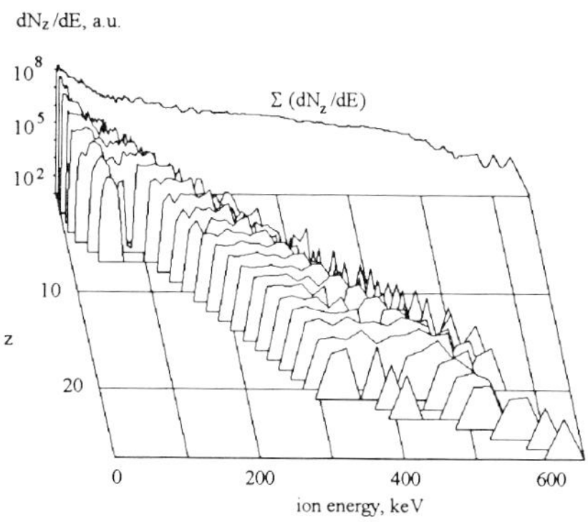

Fig. 3. Energy distribution of Ta ion species.

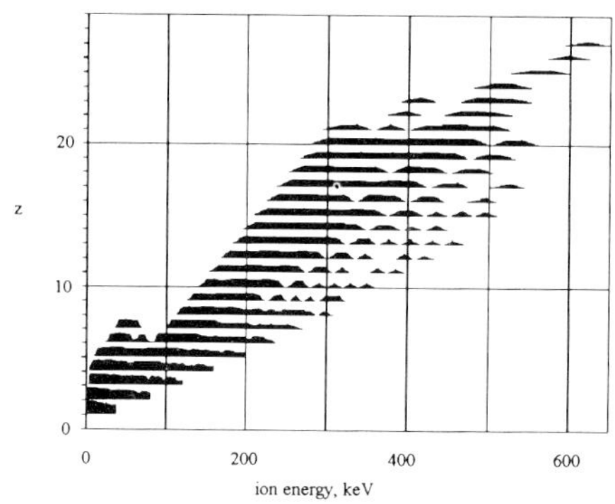

Fig. 4. Charge state - energy diagram of Ta plasma.

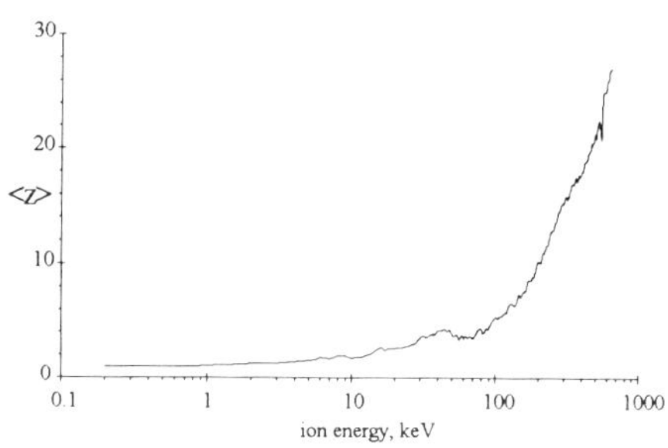

Fig. 5. Average charge state of Ta plasma as a function of ion energy.

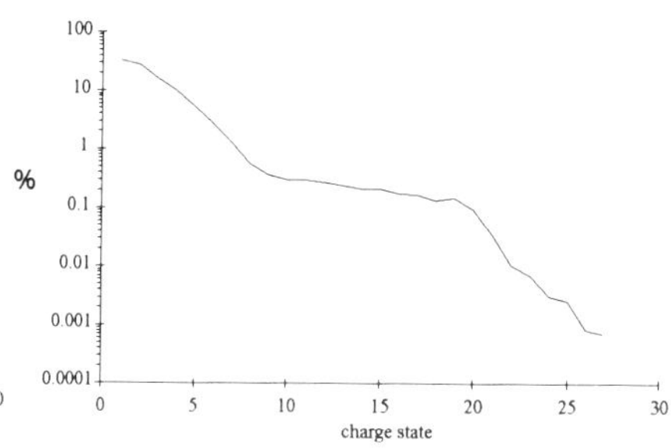

Fig. 6. Percentage abundance of Ta plasma.

The measurements performed at the focus position of 0.5 *mm* inside the Pb targets show the maximum charge state of 34+, and the maximum energy of about 1.1 *MeV*.

4. Conclusions

Highly charged ions have tendency to propagate perpendicular to a target.

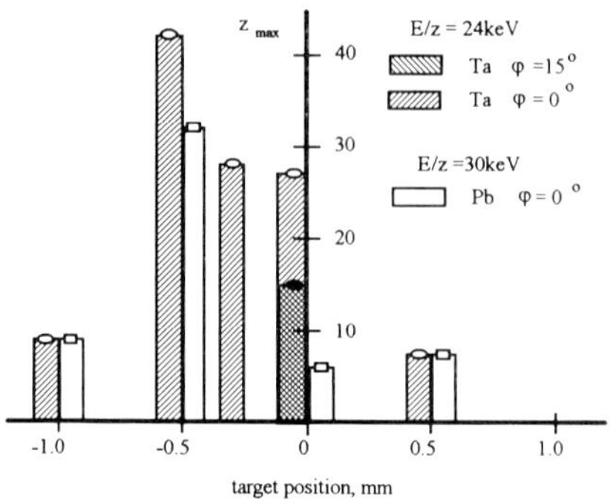

Fig. 7. Influence of the focal spot position and the tilt angle of target on maximum charge state of ions.

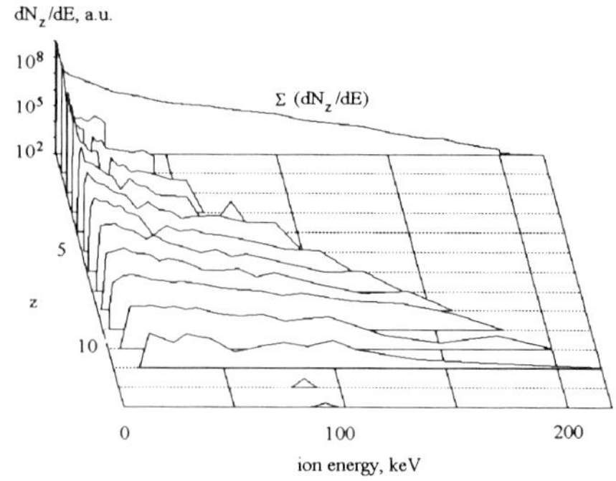

Fig. 8. Energy distribution of Al ions species

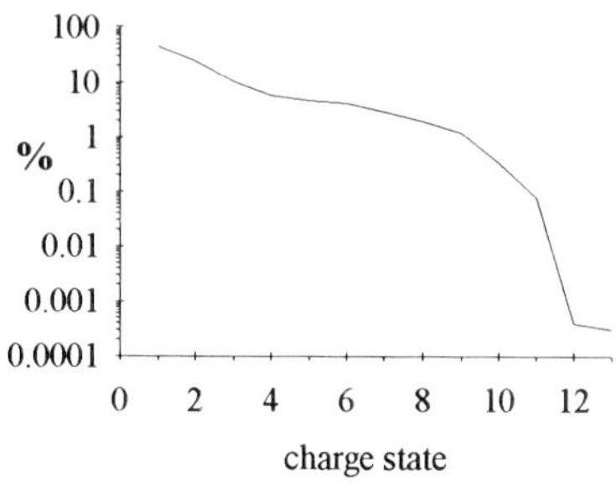

Fig. 9. Percentage abundance of Al plasma.

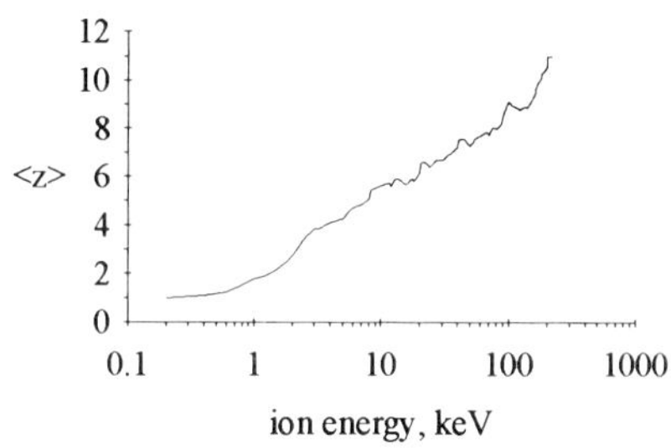

Fig. 10. Average charge state of Al plasma as a function of ion energy.

The optimum focal spot position was 0.5 *mm* inside Ta and Pb targets.
Maximum registered charge state and energy of the ions at various focusing conditions were as follows:

Al targets: 13+ and 0.2 *MeV*, sharp focusing,
Mo targets: 32+, sharp focusing,
Ta targets: 27+ and 0.65 *MeV*, sharp focusing,
42+ and 1.0 *MeV*, focus 0.5 *mm* inside the targets,
Pb targets: 34+ and 1.1 *MeV*, focus 0.5 *mm* inside the targets.

Theoretical modeling of ion pulses with two temperature exponential initial density profile and careful treatment of recombination processes is in agreement with results of experimental measurements.

References

[1] Lewis C L S et al 1982 J Phys D: Appl Phys **15** 69
[2] Basov N G et al 1987 Zh E T F **92**(4) 1299
[3] Parys P et al 1994 this conference
[4] Mróz W et al 1992 Laser and Particle Beams **10**(4) 689
[5] Denus S et al 1977 J Tech Phys **18**(1) 25
[6] Cano L G 1973 J Appl Phys 44(12) 5293
[7] Dyakin V M et al 1994 this conference

Inst. Phys. Conf. Ser. No 140: Section 9
Paper presented at Laser Interact. with Matter Conf., Oxford, 19–23 September 1994
© *1995 IOP Publishing Ltd*

Inverse-bremsstrahlung absorption of strong laser fields in dense plasmas

Susanne Pfalzner†[1], and Paul Gibbon‡

† Gesellschaft für Schwerionenforschung, 64220 Darmstadt, Germany

‡ C.E.A.-Saclay, DRECAM/SPAM, 91191 Gif-sur-Yvette, France

Abstract. Calculation of inverse bremsstrahlung absorption in the important high-density/high-intensity regime has to include two effects: the collisionality of the strongly-coupled plasma and the non-linear effects due to the intensity of the laser field. Standard perturbational treatments are inadequate to describe this situation; instead, a newly developed code will be presented which calculates collision rates by direct computation of particle-particle interactions. This so-called 'tree-code' is much faster than similar molecular dynamics codes because of its $N \log N$ computation time. It will be shown how the tree method combined with standard non-equilibrium techniques can be used to calculate inverse bremsstrahlung in this regime.

1. Introduction

Irradiation of pellet surfaces in inertial confinement and short-pulse laser interaction with solid targets can produce plasmas with densities up to $\sim 10^{23}\text{cm}^{-3}$. An important absorption process in such plasmas produced by lasers of wavelength $\lambda < \lambda_{Nd} = 1.06\mu m$ is inverse bremsstrahlung. A self-consistent theoretical treatment of inverse-bremsstrahlung in the high-density/high-intensity regime must include two effects:

- collisionality of the strongly-coupled plasma
- non-linear effects due to the intensity of the laser field

Standard perturbational treatments which assume either weak-coupling or small quiver velocities are inadequate to describe this situation. On the other hand, the usual plasma simulation techniques such as fluid, Particle-in-Cell and particle-collision (or Fokker-Planck) codes can not describe the collisionality self-consistently.

[1] E-mail : PFALZNER@GSI.DE

The most straightforward computational method to treat the collisions correctly is to integrate the particle orbits using a direct force summation including all particle-particle pairs [7]. However,the required computation time is proportional N^2, which becomes prohibitively expensive for more than a few hundred particles.

Furukawa et. al.[3] have studied absorption using a P^3M code, in which the macroscopic field determined by PIC is corrected by calculating forces between nearest neighbours directly [4]. The main drawback with P^3M is that it cannot cope well with long-wavelength density perturbations, nor with highly non-linear systems in which the particles tend to cluster. The latter situation may require such large nearest-neighbour sums that the N^2 component dominates, thus cancelling the gain in efficiency made by using PIC for the distant particles.

In the context of gravitational many-body problems, a hierarchical tree algorithm has been recently developed with a computation time proportional to $N log N$ [5, 6]. For dense plasmas a similar tree code has been developed for the simulation of dense plasmas where special attention is required for the Coulomb multipole expansion - which must cater for charges of opposite sign - and to the boundary conditions imposed by the physics.

2. Computational method: Hierarchical tree code

The motivation for using a hierarchical tree code for collisional plasma simulation is that it has two advantages over other methods (eg: Fokker-Planck, P^3M)

- an exact description of collisions, including many body encounters and large angle scattering
- a gridless treatment which permits arbitrary geometry and high density contrasts

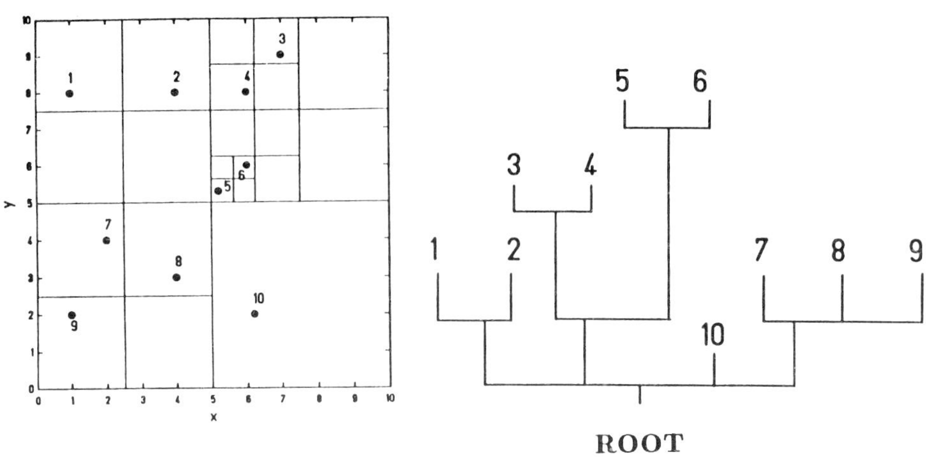

Figure 1. Space division process and resulting tree structure

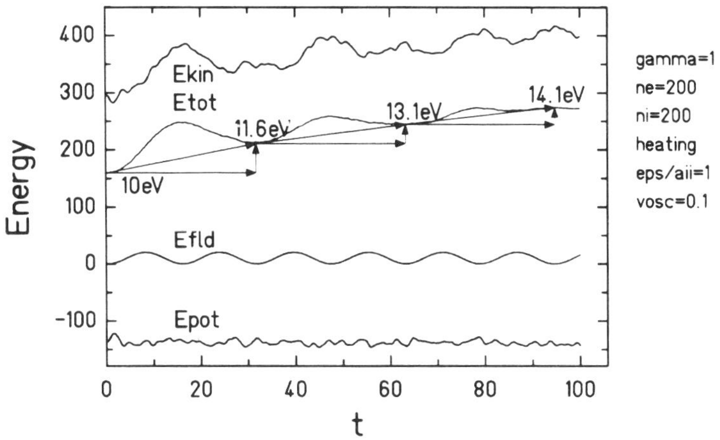

Figure 2. Temporal development of kinetic, potential, field and total energy of a dense plasma with applied sinusoidal external electric field

The most time consuming part of a dynamic N-body code is the force calculation. Summing all direct particle-particle interactions involves N(N-1)/2 $\sim$ N^2 operations. The basic idea of a hierarchical tree construction was first described by Barnes and Hut [5]. The difference with standard N-body codes is to sum the 1/r-forces of immediate neighbours directly and then take coarser groupings of particles at larger distances. A important feature of this code is that the space is not divided by a regular grid, instead the entire space recursively subdivided. The resulting data structure is know as the 'tree'. An example is given in fig.1, which shows the space division and the resulting tree structure for a given particle distribution (for simplicity shows a 2-dimensional picture although the code is really 3-dimensional). The force is evaluated for every individual particle starting at the root and working towards the leaves. The size s of the current cluster is compared with the distance d to the current particle $s/d \leq \theta$ where θ is a tolerance parameter. Having so calculated the force the positions and velocities can be up dated and for the next time step the new tree can be built again. For more detailed information about the numerical method see [1], [2] and references therein.

3. Preliminary results and future investigations

The inverse bremsstrahlung of a dense plasma is calculated by adding an external electric field. Therefore the force which each individual particle feels contains an additional component due to the laser, $F_{field} = v_{osc} sin(\omega t)$, with v_{osc} as the amplitude of the quiver velocity due to the laser field and ω the laser frequency. Fig.2 shows the kinetic, potential, field and total energy of a typical simulation of such a situation. Due to inverse bremsstrahlung the plasma heats up and the absorption of the laser energy by the plasma can be measured by the increase of the total energy.

Although this heating process is realistic it makes the calculation of the absorption rate difficult, because each data point is basically at a different temperature. Ideally one

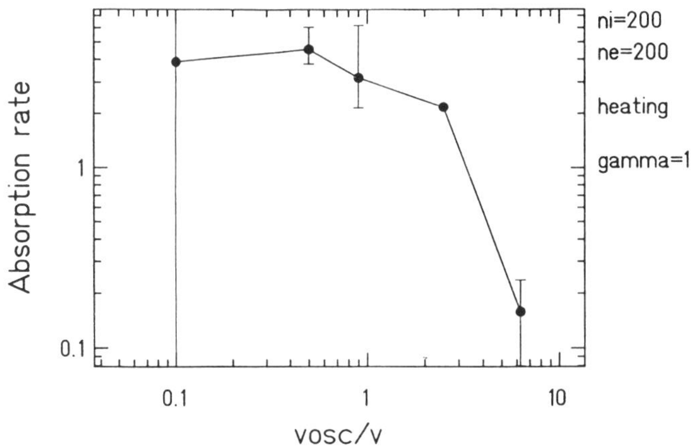

Figure 3. Absorption ratio as function of $v_{osc}/v_{thermal}$

would like the system to stay isothermal. One way of doing this is to rescale the whole system to the initial temperature after each laser cycle or apply an external constraint[8].

Fig.3 shows the rate at which the laser energy is absorbed by the plasma through inverse bremsstrahlung as a function of the ratio $v_{osc}/v_{thermal}$. The result shows generally the trend that is found in less dense plasmas: for low $v_{osc}/v_{thermal}$ the absorption rate is nearly constant, but as soon as $v_{osc}/v_{thermal} \simeq 1$ the absorption rate falls rapidly ($\sim 1/v_{osc}^3$). As fig.3 shows these results contain a rather large error for several reasons. The number of simulation particles has been rather small and the time over which the sample was taken was only a few laser cycles. The goal of future work will be to simulate a few thousand electrons and ions over about 20 laser cycles.

References

[1] Pfalzner S and Gibbon P 1994 *Comp.Phys.Comm* **79** 24

[2] Pfalzner S and Gibbon P 1995 *Many-body tree methods in Physics* (Cambridge University Press)

[3] Furukawa h and nishihara k 1990 *Phys. Rev.* A **42** 3532

[4] Eastwood J W and Hockney 1981 *Computer Simulation Using Particles* (New York: McGraw-Hill)

[5] Barnes J and Hut P 1986 *Nature* **324** 521 **79** 24

[6] Hernquist L 1988 *Comp.Phys.Comm* **48** 107

[7] Hansen J P and mcdonald i r 1981 *Phys. Rev.* A 23 2041

[8] Allen M P and Tildesley D J 1987 *Computer Simulation of Liquids* (Oxford University Press)

Inst. Phys. Conf. Ser. No 140: Section 9
Paper presented at Laser Interact. with Matter Conf., Oxford, 19–23 September 1994

© 1995 IOP Publishing Ltd

NEO-LISP

Deflecting Near-Earth Objects using High Average Power, Repetitively Pulsed Lasers

C. R. Phipps

Chemical Science and Technology Division
Mail Stop E543, Los Alamos National Laboratory
Los Alamos, New Mexico, 87545, USA

and

M. M. Michaelis

Department of Physics, Faculty of Science
University of Natal
Durban 4001, South Africa

Abstract

Several kinds of Near-Earth objects exist for which one would like to cause modest orbit perturbations, but which are inaccessible to normal means of interception because of their number, distance or the lack of early warning. For these objects, LISP (Laser Impulse Space Propulsion) is an appropriate technique for rapidly applying the required mechanical impulse from a ground-based station. In order of increasing laser energy required, examples are:
1) Repositioning specially prepared geosynchronous satellites for an enhanced lifetime,
2) Causing selected items of space junk to re-enter and burn up in the atmosphere on a computed trajectory, and 3) Safely deflecting Earth-directed comet nuclei and earth-crossing asteroids (ECA's) a few tens of meters in size (the most hazardous size). We will discuss each of these problems in turn and show that each application is best matched by its own matrix of LISP laser pulse width, pulse repetition rate, wavelength and average power. The latter ranges from 100W to 5GW for the cases considered. We will also discuss means of achieving the active beam phase error correction during passage through the atmosphere and very large exit pupil in the optical system which are required in each of these cases.

1. Near-Earth Objects (NEOs)

For the purposes of this paper, NEOs include comets and Earth-crossing asteroids (ECAs), plus manmade satellites. The former pose a threat spectrum ranging from overwhelming, if improbable, disaster for life on Earth, through 1.5 -10-cm size objects, now sufficiently numerous to pose a significant threat to Space Station Freedom, to micron-size dust which gradually degrades solar cells and sensors on satellites in low-Earth orbit (LEO). The latter are considered from two aspects: those in geosynchronous orbit (GEO) which we wish to keep "on station", i.e., located above a fixed spot on Earth for as long as possible at minimum cost, and the million or so pieces of manmade detritus from 35 years of space exploration which we would like to bring down. All of these cases, except for NEOs smaller than about 1 cm, can be dealt with using state-of-the-art detection systems and variations of the LISP concept.

2. Summary of the LISP Concept

Laser Impulse Space Propulsion (LISP) [Phipps and Michaelis 1994] is hardly the first proposal of photon propulsion [see Sänger 1956, Marx 1966, Möckel 1972a, b and 1975, Kantrowitz 1972], which actually predates lasers. The principal contributions of our work have been to revive interest in the subject, to show how modern laser and optical technology makes possible applications which were not viable in earlier times, and to emphasize the importance of using pulsed rather than cw lasers for most applications.

In the LISP concept, a repetitvely-pulsed laser transmits a high-quality beam to the propelled object's surface. The object then propels itself by the reaction force produced by the laser -driven ablation jet. The repetitive pulse format is crucial to permit using optimized laser-plasma interaction parameters, with laser target irradiation parameters I τ and λ remaining free parameters for each problem. The mass ablated from the object plays the rôle of mission fuel. The correct laser parameters to heat the ablation jet to the required temperature for the desired I_{sp} in each application can be calculated [Phipps and Michaelis 1994]. Given a criterion such as minimum cost per pound lifted into low Earth orbit (LEO), optimum values of C_m, Q* and I_{sp} result for each problem. Given other parameters such as laser range z, it is usually found that an optimum set (I, λ, τ) also results. The tradeoff between per-pulse energy W and laser repetition rate f is decided by laser and optics technology as well as by the requirements for stable atmospheric beam propagation. For most applications, a ground-based laser is used.

3. Momentum Coupling and Specific Impulse

For opaque materials in vacuum irradiated by pulsed lasers at or above plasma threshold intensity, C_m is given within a factor of 2 by [see Phipps, *et al.* 1988]

$$C_m = 3.95\ \mu_A{}^{0.44}/[Z^{0.38}(Z+1)^{0.19}(I\lambda\sqrt{\tau})^{0.25}] \quad \text{dyne-s/J.} \quad [1]$$

while

$$C_m Q^* = v_E = g I_{sp} \quad \text{cm/s} \quad [2]$$

and

$$g C_m I_{sp} = C_m{}^2 Q^* = 2{\cdot}10^7\ \eta_{AB} \quad [3]$$

by inspection. The two elements of the pairs (C_m , Q*) and (C_m , I_{sp}) are not independent, but increasing one decreases the other. For orbit transfers, the laser energy cost C per unit mass delivered is

$$C = Q^* \left\{ \frac{1-\exp\left[\Delta v/v_E\right]}{\exp\left[\Delta v/v_E\right]} \right\} \quad \text{J/g} \quad [4]$$

Table 1: No one type of laser is best for all scenarios

LISP variant	Definition	Laser Power	Optimum λ	Governs λ	Laser Location
LISK	GEO Station-keeping	100 W	530 nm	Range & atmospheric transmission	Earth surface
LEGO-LISP	LEO to GEO orbit transfers	10 kW	800 nm	Laser diode parameters and fiber loss	On board
LO-LISP	LEO re-boost	100 kW	4 μm	Energy cost & atmospheric transmission	Earth surface
LEO-LISP	Direct launch, to LEO	2.4 GW	4 μm	Energy cost & atmospheric transmission	Earth surface
NEO-LISP	Near-Earth-Object deflection	4.8 GW	4 μm	Range, energy cost and SRS	Earth surface
LISK-BROOM	Clearing Space Junk	20 kW	530 nm	Range, object size	Earth surface

Minimizing C gives the solution

$$C_m = 3.18{\cdot}10^7 \eta_{AB}/\Delta v \quad \text{dyne-s/J.} \quad [5]$$

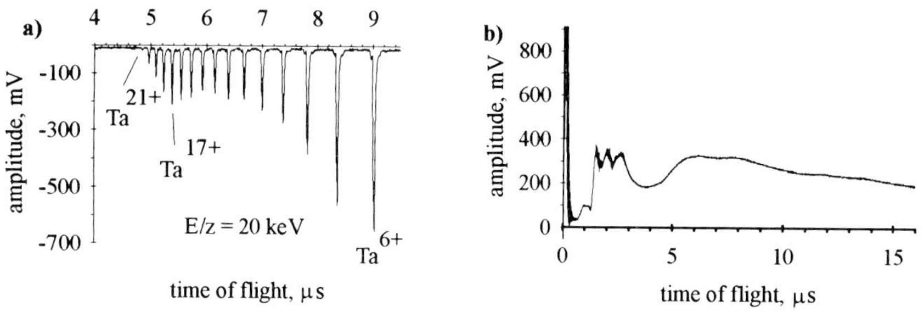

Fig. 2. Ta ion spectra from the EIA (a) and the ion collector signal (b).

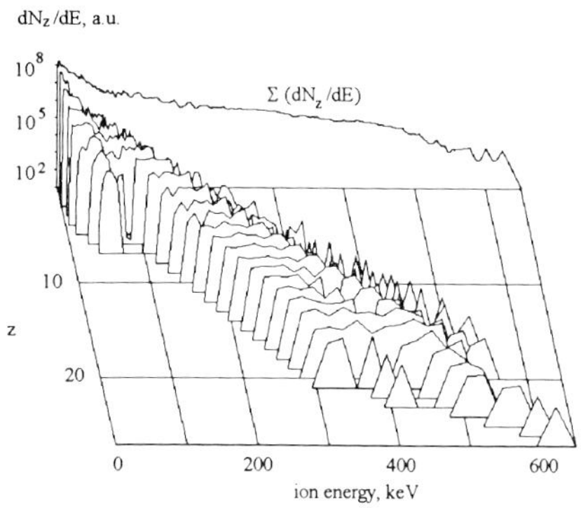

Fig. 3. Energy distribution of Ta ion species.

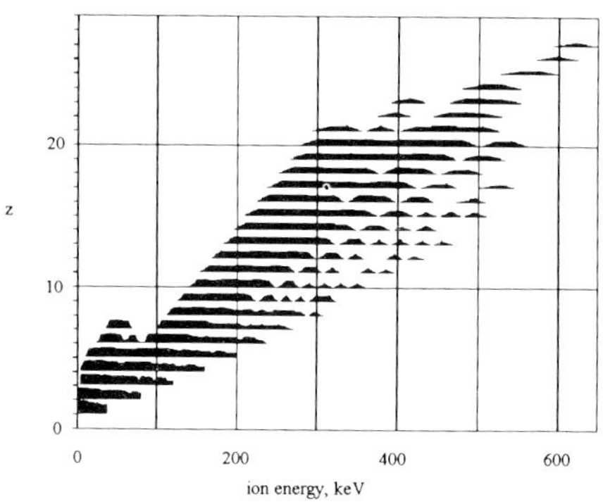

Fig. 4. Charge state - energy diagram of Ta plasma.

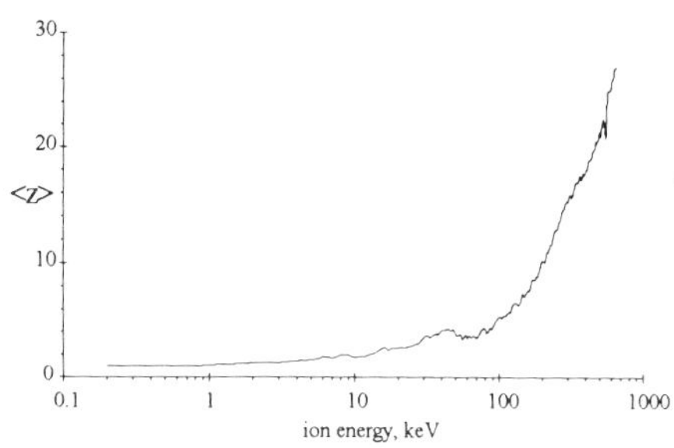

Fig. 5. Average charge state of Ta plasma as a function of ion energy.

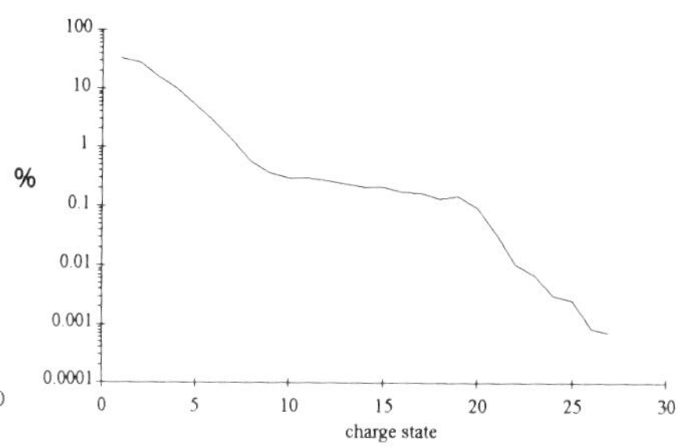

Fig. 6. Percentage abundance of Ta plasma.

The measurements performed at the focus position of 0.5 *mm* inside the Pb targets show the maximum charge state of 34+, and the maximum energy of about 1.1 *MeV*.

4. Conclusions

Highly charged ions have tendency to propagate perpendicular to a target.

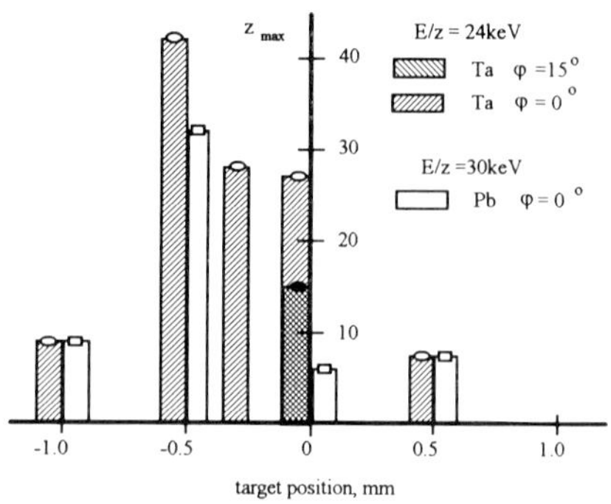

Fig. 7. Influence of the focal spot position and the tilt angle of target on maximum charge state of ions.

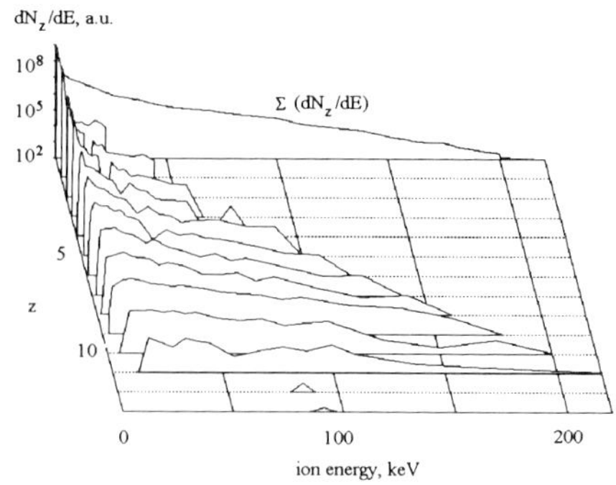

Fig. 8. Energy distribution of Al ions species

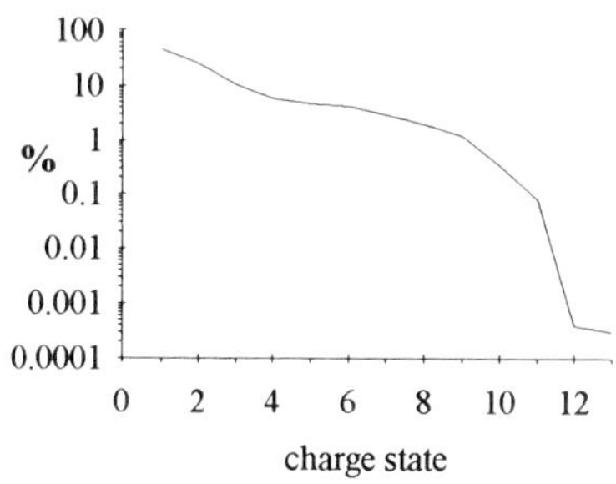

Fig. 9. Percentage abundance of Al plasma.

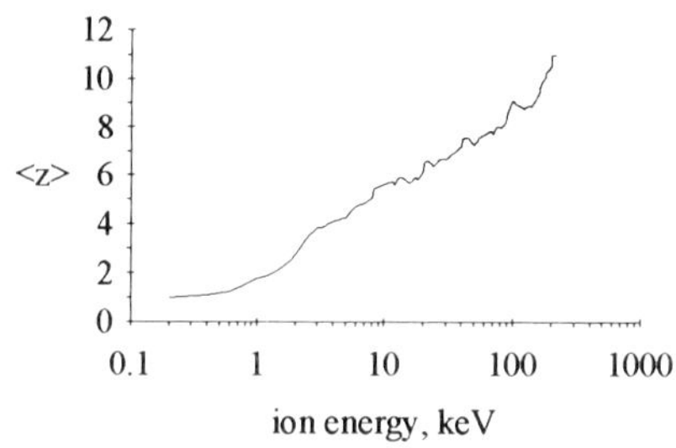

Fig. 10. Average charge state of Al plasma as a function of ion energy.

The optimum focal spot position was 0.5 *mm* inside Ta and Pb targets.
Maximum registered charge state and energy of the ions at various focusing conditions were as follows:

Al targets: 13+ and 0.2 *MeV*, sharp focusing,
Mo targets: 32+, sharp focusing,
Ta targets: 27+ and 0.65 *MeV*, sharp focusing,
42+ and 1.0 *MeV*, focus 0.5 *mm* inside the targets,
Pb targets: 34+ and 1.1 *MeV*, focus 0.5 *mm* inside the targets.

Theoretical modeling of ion pulses with two temperature exponential initial density profile and careful treatment of recombination processes is in agreement with results of experimental measurements.

References

[1] Lewis C L S et al 1982 J Phys D: Appl Phys **15** 69
[2] Basov N G et al 1987 Zh E T F **92**(4) 1299
[3] Parys P et al 1994 this conference
[4] Mróz W et al 1992 Laser and Particle Beams **10**(4) 689
[5] Denus S et al 1977 J Tech Phys **18**(1) 25
[6] Cano L G 1973 J Appl Phys 44(12) 5293
[7] Dyakin V M et al 1994 this conference

Symbols

A_s	target irradiation area (cm^2)	m_o	initial mass of the object (f)
C	energy cost per unit mass delivered (J/g)	n	total number of laser pulses
C_m	momentum coupling coeff.=$\delta p/\Phi$ (dyne-s/J)	$\aleph$	laser beam quality factor
D	laser beam diameter at launch (cm)	δp	momentum fluence = $\delta m\ v_E$ (dyne-s/cm^2)
d_s	laser beam diameter at target (cm)	P	laser power (W) = W/Δt
η_{AB}	ablation efficiency = $(1/2)\delta m v_E^2/10^7\Phi$	Q*	specific ablation energy = $\Phi/\delta m$
f	laser repetition rate (s^{-1})	R	Earth-centric object orbit radius (cm)
Φ	target-incident laser fluence (J/cm^2)	Δt	mission time
g	standard gravity (980 cm/s^2)	τ	laser pulse duration (s)
h	altitude (cm)	Δv	target velocity increment (cm/s)
G	gravitational constant	v	object velocity
I	laser intensity on target =Φ/τ (W/cm^2)	v_E	effective exhaust velocity (cm/s)
I_{sp}	specific impulse = v_E/g (s)	W	laser pulse energy = $\int \Phi dA_s$ (J)
λ	laser wavelength (cm)	ω	laser energy per unit target mass (J/g)
δm	target ablated mass fluence (g/cm^2)	z	range to target (cm)
m_A	atomic mass number of ablation plasma	Z	ionization state of ablation plasma
M	mass of the Earth (g)		

References

CRAMER, B. AND BOGERT, P. 1992 NASA internal report "Status of micrometeoroid and orbital debris protection"

FABBRO, R., *et al.*1991 *J. Appl. Phys.* **68** pp. 775-784

FLURY, W. AND MCKNIGHT, D, 1993 *Adv. Spce Res.* **13** pp. 299-309

HILLS, J., 1992 Los Alamos internal report LA-UR-92-2321

HO, C., PRIEDHORSKY, W AND BARON, M 1993 *SPIE* **1951** pp. 67 *et seq*

KANTROWITZ, A. 1986 *Proc. SDIO/DARPA Workshop on Laser Propulsion,* J. T. Kare, ed. LLNL Conf-860778, Livermore CA, p. 1

MÖCKEL, W.E. 1972a *J. Spacecraft and Rockets* **9**, p. 863

MÖCKEL, W.E. 1972b *J. Spacecraft and Rockets* **9**, p. 942

MÖCKEL, W.E. 1975 *J. Spacecraft and Rockets* **12**, p. 700

PHIPPS, C. R. *et al.* 1988 *J. Appl. Phys.* **64**, pp. 1083-96

PHIPPS, C. R.., *et al.* 1990 *J. Laser and Particle Beams* **8** pp. 281-297

PHIPPS, C. R. AND DREYFUS, R. W. 1993 Chapter 4 in *Laser Microprobe Analysis*, A. Vertes, R. Gijbels and F. Adams, eds., John Wiley, NY, pp,. 369 *et seq.*

PHIPPS, C.R. AND MICHAELIS, M. M. 1994 *J. Laser and Particle Beams* **12** no. 1, pp. 23-54

SÄNGER, E. 1956 *Aero Digest,* p. 68

SKOURDOULIS, C. D. AND KOSMIDIS, C. E. 1992, *SPIE* **1810**, p. 646

SKOURDOULIS, C. D., *et al.* 1994, paper B-IX/P17, European Materials Research Society Spring Meeting, Strasburg

Inst. Phys. Conf. Ser. No 140: Section 9
Paper presented at Laser Interact. with Matter Conf., Oxford, 19–23 September 1994
© *1995 IOP Publishing Ltd*

Symmetric implosion of indirect-driven ICF target. A MULTI-2D numerical simulation

J. Ramirez†, R. Ramis†, and J. Meyer-ter-Vehn‡
†E.T.S.I.Aeronáuticos, Univ. Politécnica, 28040 Madrid, Spain
‡Max-Planck-Institut für Quantenoptik, D-8046, Germany

Abstract

The radiation-driven hydrodynamics of a spherical capsule inside a cylindrical target is analyzed by using Multi-2D Code. The multilayered capsule (DT-gas, DT-solid and C-based ablator) is compressed quite symmetrically ($\pm 1.5\%$ peak to peak radial deviation in DT-ablator interface at half compression) up to one fourth the original diameter, through axial deposition of $2 \times 6MJ$ modulated ion beams in axial cylindrical converters; high-Z stopping walls are used to prevent direct radiation from entering the capsule. Higher compression ratios are expected in subsequent numerical simulations if direct radiation is effectively avoided to the capsule.

I. Introduction

In reaching fusion conditions by imploding a spherical capsule through high power laser or ion beams the spherical symmetry of implosion is known to be a critical point. Since direct illumination of the capsule (allowed rms deviation 1-2%) [1-2] is difficult to achieve, indirect illumination has been widely considered in the last years; the beam is then used to heat a small cavity where the capsule is embedded; radiative thermalization leads to a fast symmetrization of the radiation field and the capsule is expected to implode fairly symmetrically, driven by thermal X-rays.

Different schemes for indirect-driven inertial confinement fusion targets have been proposed in the literature [3-5]. Focussing on implosion symmetry and related parameters: geometry, input energy vs. time, etc., we have simulated several of these schemes by using Multi-2D Code [6] (see also a separate contribution by R. Ramis et al. to these proceedings) and have come to the design discussed in this paper.

II. The target

In summary, the target consists of a central DT-gas sphere ($4mm$ dia.) covered by a first DT-solid layer ($0.3mm$ thick), a second C-based ablator layer ($0.3mm$ thick), and located at the center of a cylindrical lead casing ($0.5mm$ thick, $13mm$ o.d.) with flat ends; twin modulated ion beams ($6MJ$ each) are axially applied to radially restricted cylindrical converters ($2mm$ axial diamension, $4mm$ o.d., Berilium) at each end; high-Z stopping walls (0.25mm thick, 8.2mm o.d., lead) prevent direct radiation from entering the capsule and a low density CH-based filling material ($35mg/cm^3$) in the cavity provides for some mechanical strength to the system.

III. The numerical simulation

When applying the modulated pulses ($100TW \times 5ns$ prepulse, $40TW \times 25ns$ plateau and $500TW \times 10ns$ main pulse each) the target follows the evolution shown in Fig. 1 for selected instant times (note that only one quadrant of a meridional section is shown).

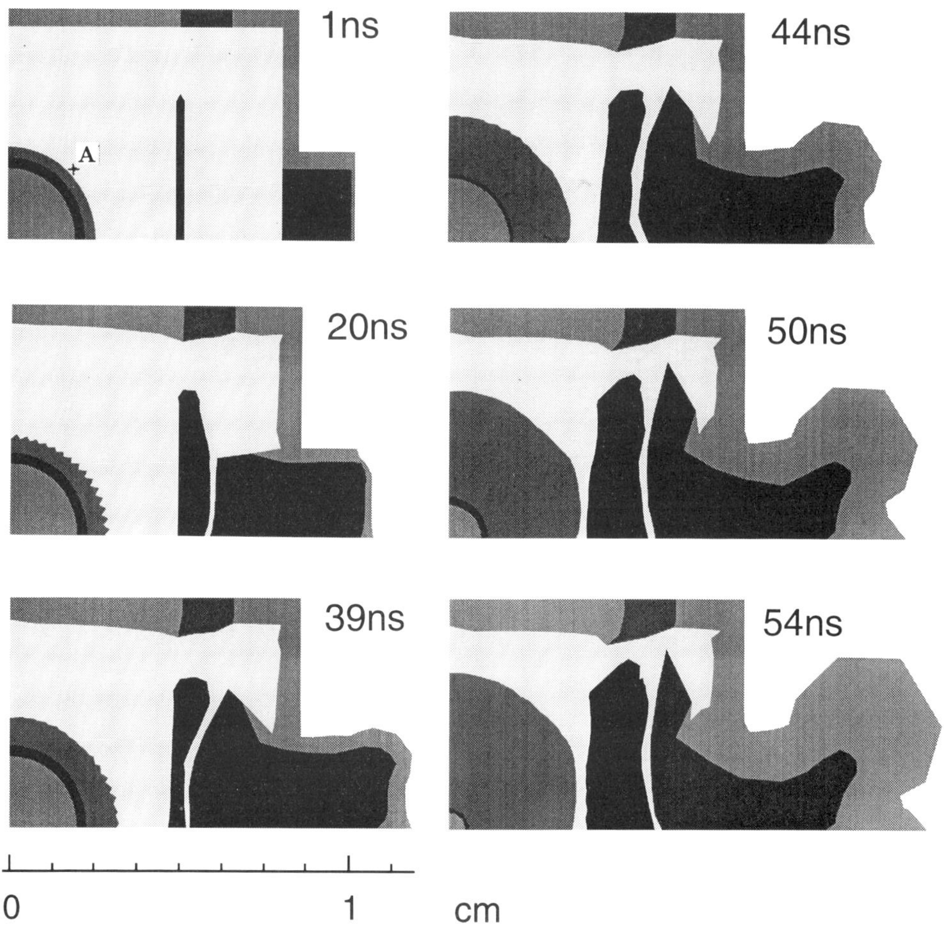

Fig. 1 Radiation-driven hydrodynamics at selected times 1, 20, 39, 44, 50, and 54 ns

Both converters move initially towards the stopping walls and finally enter the cavity but direct radiation to the capsule is avoided at any time; ablative process takes place not only on C-based ablator but also on stopping walls and inner casing-surfaces thus changing the size of the annular window through which radiation energy enters the cavity like a fluid (see Fig. 2).

We must note a highly distorted mesh (not shown) along the process but a good implosion symmetry of the capsule; a detailed plot of DT-solid/C-ablator interface at

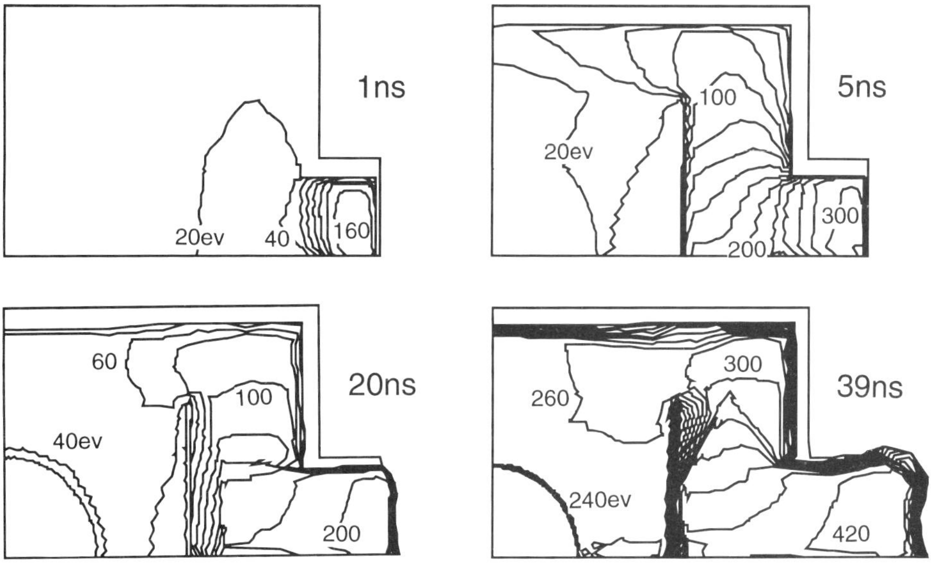

Fig. 2 Radiation Temperature contour lines (20*ev* each) at 1, 5, 20, and 39 *ns*.

different times is shown in a separate contribution to these proceedings (by J. Meyer-ter-Vehn et al.); a detailed analysis of this interface at half compression (1.15*mm* radius at 50*ns*) shows ±1.5% peak to peak radial deviation. At final calculation (54*ns*) DT-solid has been compressed to one fourth (in diameter); nevertheless, main pressure shock has not yet reached the center of the capsule, there is still some non burnt ablator material (about 17%) and higher compresion ratios are expected if the calculation proceeds a few *ns* more, after solving some present numerical problems. Some Rayleigh-Taylor instability is starting to develop at DT-gas/DT-solid interface, however, at this time.

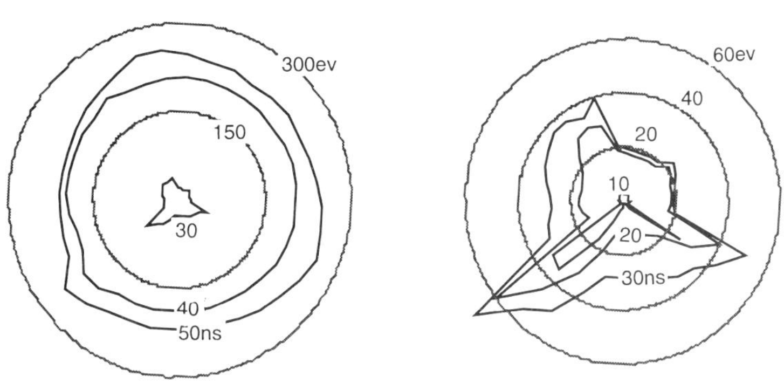

Fig. 3 Radiation field at point A (2*mm* axial, 2*mm* radial coordinates).

To visualize the symmetrization process of the radiation field in the cavity we show in Fig. 3 some polar plots of the radiation energy seen from a point near the capsule (see point A in Fig. 1) at different times; obviously, at early times direct radiation coming from the stopping wall-casing window is dominant but later, particularly when main pulse is active, the radiation field appears quite symmetric. We must also note the dark rectangle on the casing (Fig. 1), just above the stopping wall, representing a reduced oppacity (to one half) area and helping to control the $l = 4$ asymmetry in the radiation.

IV. Conclusions

This numerical simulation shows that a fairly symmetric implosion of a fuel capsule can be obtained by indirect illumination through ion beam deposition on two opposite converters. The inner geometry of the cavity changes substantially during the process because the radiation induced hydrodynamics and this change must be taken into account in any realistic calculation. The effective screening of the capsule against direct illumination is a critical point.

The hydrodynamic efficiency of the target (still very low, aprox. 1.5%) is expected to increase by a proper selection of relevant parameters such as opacities, densities, materials, etc. besides target geometry and pulse shaping.

References

[1] McCrory R L and Verdon C P in *Inertial Confinement Fusion* (Proc. Course and Workshop in Varenna, 1988) Editrice Compositori, Bologna (1989) 83

[2] Atzeni S *Europhys. Lett.* **11** (1990) 639; *Laser Part. Beams* **9** (1991) 233

[3] Murakami M and Meyer-ter-Vehn J 1991 *Nuclear Fusion* **31** 1315

[4] Kang K M, Lutz K J, Maruhn J A and Tahir N A 1993 *Il Nuovo Cimento* **106 A** 1825

[5] Meyer-ter-Vehn J, Basko M, Ramis R and Rickert A 1993 *Il Nuovo Cimento* **106 A** 1883

[6] Ramis R and Meyer-ter-Vehn J 1992, *MULTI-2D A Computer Code for Two-Dimensional Radiation Hydrodynamics* Report MPQ174.

Inst. Phys. Conf. Ser. No 140: Section 9
Paper presented at Laser Interact. with Matter Conf., Oxford, 19–23 September 1994
© *1995 IOP Publishing Ltd*

Second Harmonic Generation in Plasma Generated by High Intensity Picosecond Laser Pulses

Jean Roussie, Jacques Briand, Michel Armengaud, Alain Gomes and Yves Quemener.

IRSAMC-GILMT, URA 277 du CNRS, Université Paul Sabatier, Toulouse France.

Second harmonic time resolved spectrum in Aluminium plasma created by high-intensity picosecond laser pulse was obtained using the P102 laser facility in the CEA's study centre of Limeil (France). The laser we used was a glass-neodyme picosecond one delivering pulses whose power is about 10 TW. The prepulse and postpulse were suppressed using the plasma-mirror technique. These experiments allowed to study the behaviour of Langmuir plasmons generated by decay instability in such plasmas. The duration of plasmons along times range as long as 100 ps, the laser pulse lasting only 1 ps, allowed their free behaviour study in the absence of pump wave. The obtained spectrum present a similar double particularity: a broadening spectral line in which narrow spectral lines develop persisting after the vanishing of the broadening one. This broadening line may be associated to the collapse of Langmuir plasmons, the narrow lines may be associated to an emission of solitons from the collapsing plasmons. This emission of solitons may be interpreted as an energy loss for the collapse, leading to a duration much larger than theoretical predictions give.

INTRODUCTION

In the past few years, development of high-intensity picosecond laser pulses has opened new investigation fields in high energies physics and ultra short phenomenon. These lasers, using Chirp Pulse Amplification (CPA) are able to produce pulses shorter than 1 picosecond with energy higher than 10 Joules corresponding to a power higher than 10 TW and a focused irradiation in the range of 10^{17} W/cm^2 [1][2]. The experiences describe below were performed in collaboration with the C.E.A. study Centre of Limeil-Valenton using the P102 laser facility which presents the characteristics above mentioned. The chirp of the pulse delivered by the oscillator was obtained in an optical fibre. After temporal compression using a pair of grating, the main pulse duration was 1 ps, and he carried 90% of the laser energy, the remaining energy being carried by a pedestal which duration ranges over 100 picosecond [3].

Previous works about second-harmonic generation in nanosecond laser-matter experiments have shown that lateral second harmonic was produced by diffusion of the laser light on plasmons generated by decay instabilities (Parametric Decay Instability and Electronic Decay Instability) [4][5][6]. The temporal behaviour of the second harmonic spectra follows the

behaviour of the diffusing plasmons. It has been shown that at high irradiations ($I>10^{14}$W/cm^2) the obtained spectra were characteristic of strong Langmuir turbulence [4][5] and collapse of plasmons, theoretically predicted by Zakharov in 1972 [7]. The duration of collapse, in contradiction with theoretical predictions, were found to be about few tens picosecond, in agreement with ionospheric measurements [8][9].
For this reason the development of high intensity picosecond laser with pulse duration much shorter than collapse the duration is of great interest for studying strong Langmuir turbulence. As a matter of fact, they allow the instability study in the absence of pump wave, leading to the free behaviour of strongly turbulent plasmons.

EXPERIMENTAL SET-UP

Experiments were performed using a CPA laser presenting a pedestal which power was 10^6 times lower than the main pulse's one, corresponding to an irradiation of 10^{17} W/cm^2 on the target and this leading to pre-plasma creation. We suppressed this prepulse plasma by using plasma-mirror technique [10][11]. It consists of interposing a glass blade between the focusing lens and the target at a convenient distance of the lens in order that the prepulse

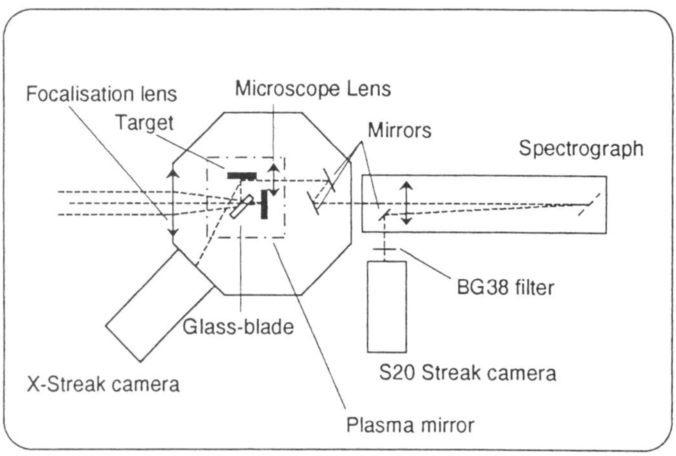

Figure 1: Experimental Set-Up

passes through it, meanwhile the main pulse creates a plasma on it. This plasma reflects the laser beam which is focused on the target. As a matter of fact, during the main pulse duration the plasma's expansion length in front of the glass-blade is small compared to the laser's wavelength and the plasma acts as a plane mirror ($L \approx c_s \tau \approx 0.1 \mu m$ where c_s stands for the sound speed and τ for the main pulse duration). The postpulse is, the on the glass-plasma being in expansion, partially absorbed and partially reflected, and then doesn't create a plasma on the target. It acts as a 'lighthouse' illuminating the plasma and particularly the plasmons, diffusing on them and thus generating second harmonic emission.

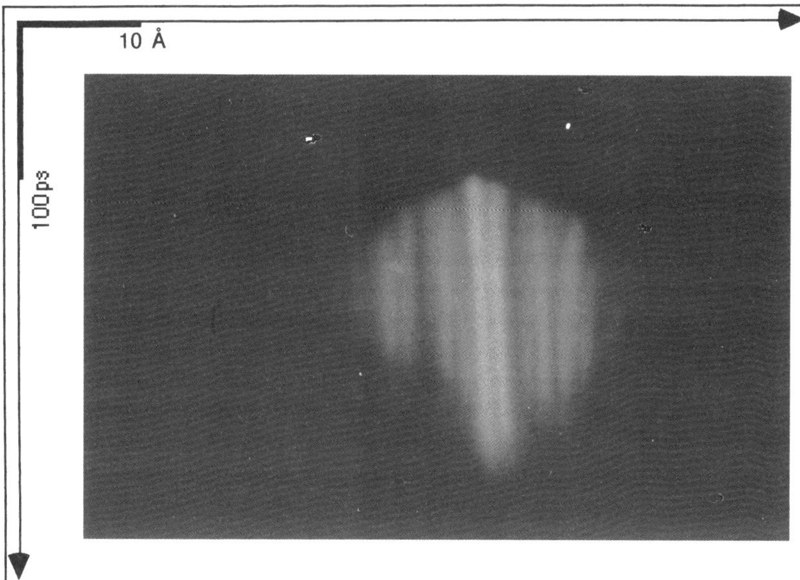

Figure 2: Second harmonic spectrum

To obtain the second harmonic spectrum, the plasma was imaged by a microscope lens on the entrance slit of a spectrograph using two lenses of 500 mm focus length and a 2500 lines per mm holographic grating. The spectrally stretched image of the plasma was then sent through a BG38 filter on the entrance slit of a S20 Thomson streak camera [fig 1]. The temporal resolution of the system including the imprecision due to the delay between two extreme rays on the grating is about one picosecond, which is small in comparison with the duration of the observed phenomena, as it will be seen later.

RESULTS

A typical obtained spectrum is presented in the figure 2. All the obtained spectra present the same characteristics: A main broadening line in which narrow spectral lines develop, persisting after the vanishing of the main line. The evolution of the main line last few tens to a hundred picosecond, the narrow lines persist a hundred picosecond more.

Owing to the conservation laws for the diffusion:

$$\begin{cases} \vec{K} = \vec{k_e} + \vec{k_o} \\ \Omega = \omega_e + \omega_o \end{cases}$$

(where $(\Omega, \vec{K})$,$(\omega_o, \vec{k_o})$ and $(\omega_e, \vec{k_e})$ stand respectively for the pulsation and the wave vector of the diffused light, the incident laser light and the plasmons), lateral second harmonic spectrum results from the diffusion of the incident laser light on plasmons in the vicinity of the critical surface [4][5][6].

The obtained spectra present similitudes with the lateral second harmonic spectra obtained in nanosecond regime [4][5].

In both cases, the main lines are broadening lines whose evolution stands along time ranges of few tens to hundred picoseconds. Unlike in nanosecond regime, the broadening line doesn't shift to the red in picosecond regime in absence of Doppler effect. These lines may be associated, like in nanosecond regime, to Langmuir Collapse [4]. The narrow spectral lines don't appear in lateral second harmonic spectra in nanosecond regime. These lines are characteristic of stationary solitonic structures generated from the collapsing plasmons. Calculations are running to understand the effect of such solitonic emission on collapse dynamics. This solitonic emission, if confirmed, would be an energy loss for the collapsing plasmons and could be responsible of the stretching of the collapse duration compared to theoretical predictions.

CONCLUSIONS

Spectra of the second harmonic were obtained in picosecond regime. These spectra confirm the results obtained previously about duration of plasmons evolution in ultra strong Langmuir turbulence regime toward collapse. Moreover, in picosecond regime, solitons seem to start from the collapsing plasmons. These solitons, carrying a part of the collapsing plasmons energy, might be responsible of the slowing down of the collapse phenomenon in comparison with the theoretically predicted duration (few tens picosecond against one picosecond)

Acknowledgements

We would like to acknowledge the searchers of the CEA-Limeil and particularly Mr Jolas, Louis-Jacquet and Mme Gary for their useful help during the experiments.

REFERENCES

[1] P. Maine & al. 1987 *IEEE J. Quantum Electronics* 24,2 p398

[2] P. Maine & al. 1988 *Proc of the SPIE*

[3] data C.E.A.

[4] J. Briand & al. 1990 Phy. Fluids B 2, 160-5

[5] F. Dahmani & al. 1991 Phy. Fluids B 3, 2558-67

[6] S. Abdelli & al. 1992 Laser Particle Beams 10, 629-37

[7] V.E. Zakharov 1972 Soviet Physics JETP 35, 908-14

[8] P.Y. Cheung & A.Y. Wong 1985 Phy. Fluids 28, 1538

[9] P.Y. Cheung & A.Y. Wong 1985 Phy. Rev. Let. 55, 1880

[10] H.C. Kapteyn & al. 1990 Springer Series Chemical Phys. 53, 122

[11] D.M. Gold & al. 1991 SPIE 1413, 41

Inst. Phys. Conf. Ser. No 140: Section 9
Paper presented at Laser Interact. with Matter Conf., Oxford, 19–23 September 1994
© 1995 IOP Publishing Ltd

Liquid Hydrogen Layers in Plastic Capsules of Large Diameter†

*J. Sanchez,T. Bernat, E. Mapoles, W. Giedt

**J. Sater, D. Bittner, R. Bieri, M. Monsler

*University of California, Lawrence Livermore National Laboratory, P.O. Box 808, Livermore, CA 94550

**W. J. Schafer Associates, 303 Lindbergh Avenue, Livermore, CA 94550

Abstract The inner surface of the fuel layer in laser fusion targets must be very smooth prior to implosion, to prevent Raleigh-Taylor instabilities from mixing cold fuel into the hot spot. If a liquid hydrogen fuel layer can be stabilized, surface tension will rapidly damp out fluctuations of the liquid-vapor interface. This may provide a sufficiently smooth surface to prevent mixing. We report experimental data on thick (approximately 50 μm) liquid fuel layers supported using the thermal gradient method. We are studying mixtures of H2:D2, inside of 2 mm ID plastic capsules. A variety of oscillations in the liquid as well as stable liquid surface configurations have been observed.

1. Introduction

Targets for Inertial Confinement Fusion (ICF) require very high uniformity layers of liquid or solid deuterium-tritium (DT) fuel. High uniformity solid DT layers have been obtained using the "beta layering" technique[1]. However, the surface finish naturally achieved by the ice layer through this process has an rms. surface roughness 2 - 3 times larger than the maximum desirable rms. surface roughness for a National Ignition Facility (NIF) scale target. The fuel's inner surface roughness causes Raleigh-Taylor instabilities which mix cold fuel into the hot spot preventing efficient fuel ignition. Liquid layers, if achievable, would be smoother due to surface tension effects.

In this article, we present results of recent experiments on liquid H2-D2 layers supported using the thermal gradient[2] technique. Hydrogen-deuterium (HD) mixtures are used because they are much easier to handle in experiments without concern for radioactivity and they are physically equivalent to DT mixtures for studying the thermal gradient layering process.

In the thermal gradient technique, a positive (hot on top) temperature gradient is externally impressed on the capsule containing the H2-D2 mixture. This temperature gradient concentrates the component with the higher vapor pressure (H2) in the colder region (towards the bottom) leaving the top of the capsule with a higher concentration of D2. The difference in the surface tension of the two components produces a surface tension gradient that pulls the liquid towards the area of higher surface tension (D2). These surface tension driven flows counteract the gravitational slumping of the liquid layer.

2. Experiments

2.1. Cryogenic test cell description

The static gradient experiment presented here was accomplished with a 2.2 mm outer diameter (OD) hand blown glass capsule with an average wall thickness of ~ 15 μm. For the dynamic gradient experiment, a 3 mm OD glass shell constructed from two 0.5 mm thick hemishells was used. The capsules for our experiments were fitted with a 0.35 mm diameter fill tube which is used to fill the capsules in-situ after they are cooled below the dew point of the H2-D2 mixture.

The capsule was positioned in the center of a cylindrical cavity 6 mm high and 15 mm in diameter, containing helium gas to cool the capsule and provide the medium for establishing the thermal gradient. The top and bottom of the cavity had calibrated germanium resistance thermometers and heaters mounted on them. The top assembly was vacuum insulated except for the side facing the capsule to minimize the heat leaked outside to the test chamber.

The helium gas chamber, which retains the helium heat exchange gas is cooled by a variable temperature cryostat whose temperature could be controlled between 8.0 K and 40 K. Optical windows in the cryostat allow direct observation of the sample.

For all test cases, the capsule fill consisted of a 50-50 mixture of H2 and D2 transferred to the capsule thorough a metering valve. Once a sufficient amount of liquid was collected, the temperature of the test chamber was allowed to cool below the freezing point of the hydrogen mixture while maintaining the desired average temperature between the gradient plates constant. The cooling continued, until a solid hydrogen ice plug was formed in the fill tube outside the cylindrical cavity, sealing in the fill.

Figure 1 shows the camera setup used to image the test capsule. The resolution of this arrangement is limited by the CCD resolution and the need to frame the OD of the test capsule (the inner diameter (ID) does not have good contrast), to a value of ~ 3 microns per pixel.

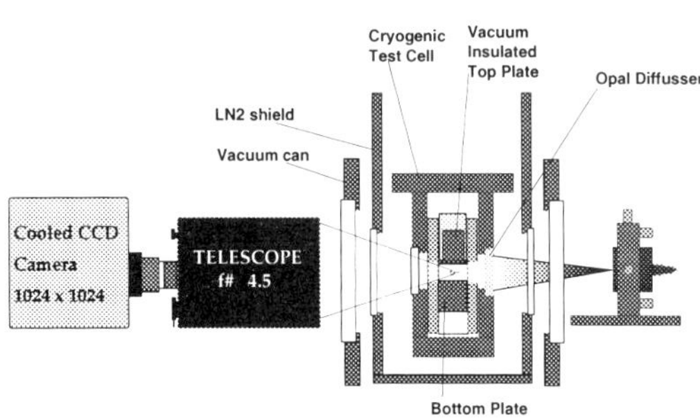

Figure 1

2.2. Data acquisition procedure

The layer thickness data was acquired using two different techniques: first with a static thermal gradient applied; and second, with a dynamic gradient generation technique.

In the static thermal gradient technique, the temperature gradient was increased in small steps of 0.1 to 0.5 K/cm, allowing enough time between steps for the temperature to stabilize. The sequence started with a small negative gradient (hot on bottom), and proceeded to the largest positive gradient attainable. A full frame picture of the capsule was acquired at the beginning of the sequence, and subsequently whenever a specified gradient change was completed. The maximum gradient attainable by this method was 18 K/cm at an average temperature of 21 K.

With the dynamic gradient generation technique, the helium gas chamber is allowed to cool to near the lowest achievable temperature, generally between 7 and 9 Kelvin. Power is then applied to the heater in the vacuum insulated top gradient plate by a controlled voltage source causing its temperature to increase rapidly. With this arrangement temperature gradients in excess of 29 K/cm were achieved. For this experiment, the CCD was formatted to acquire 1024 by 2 or 4 pixel frames aligned vertically with the center of the capsule for a timed "streak" record of the behavior of the central (thickest) portion of the liquid layer as both the average temperature and the gradient were rapidly changing.

3. Layer Thickness and Symmetry

3.1. Thin shell with static gradient

The static gradient test data was collected at an average capsule temperature of ~18 Kelvin. A picture was recorded at increments of 1.6 Kelvin per cm. in the varying thermal gradient. The layer thickness was measured using a profile of the image generated at ~ 45 degrees from the vertical to minimize the effect of the obscuration as well as that of the fill tube which is located in the horizontal plane. Figure 2 shows a series of profiles thus measured that were assembled into a time history similar to a "streak" record. This "streak" record shows that the layer thickness at the top of the capsule increases roughly by the same amount that the bottom layer thickness decreases as the temperature gradient is increased from ~ - 4 Kelvin per cm. to ~ 15 Kelvin per cm.

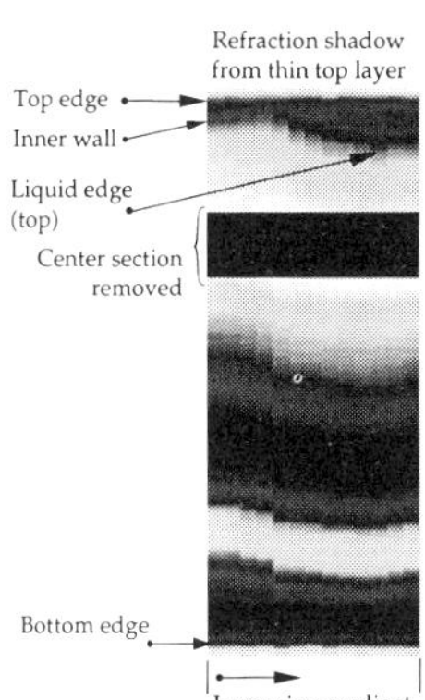

Left, figure 2 shows streak record of layer thickness versus time

Right, figure 3 shows a plot of layer thickness vs. thermal gradient

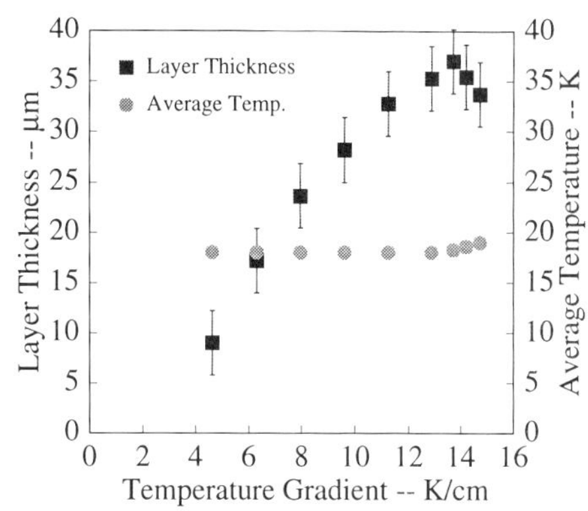

Figure 3 shows the measured top layer thickness versus the thermal gradient. The layer thickness increases smoothly with the temperature gradient until the average temperature starts to increase on the last three data points from about 18 to 19 Kelvin. This increase in the average temperature causes the liquid layer thickness to decrease by evaporation even though the thermal gradient is still increasing.

3.2. Thick shell with dynamic gradient

Figure 4 shows a typical "streak" obtained using the dynamic gradient generation technique. The record starts with hydrogen ice randomly distributed over the capsule, being thickest at the bottom where the liquid was collected before freezing. The capsule's average temperature at the start of the record was ~8.8 Kelvin. As the top plate temperature increases, the first 26 seconds of the record show the melting of the hydrogen ice as it progresses from the top of the capsule towards the bottom.

The melting process continues through the onset of oscillations in the liquid which start when the average temperature of the capsule reaches ~16.3 Kelvin, until the liquid oscillations are fully developed, at ~16.8 Kelvin. These temperatures are very close to the melting temperature of 16.5 Kelvin expected for a 50:50 mixture of H2-D2.

Figure 5 shows the measured layer thickness (after stabilization) for several of the data-sets. An unexpectedly large decrease in thickness occurs with a small increase of the average temperature even though the temperature gradient is still increasing. This large thickness change seems incompatible with previous data and may indicate the existence of very low frequency oscillatory modes in the liquid excited by the dynamics of the experiment.

3.3. Thermal gradient generation by light absorption

A new technique is being developed to generate the required thermal gradient by absorption of light of the proper wavelength within the capsule wall. This laser generated thermal gradient technique opens a new set of control parameters for the liquid layer by using spatially and temporally modulation of the heat source. Figure 6 shows layer thickness supported with this technique.

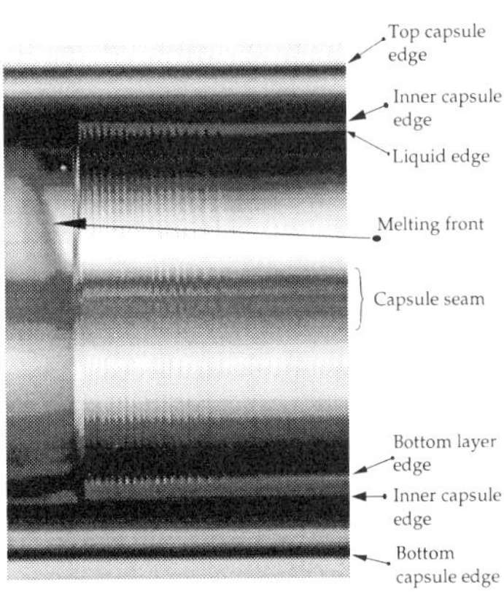

Left, figure 4 shows streak record of dynamic thermal gradient experiment

Right, figure 5 shows layer thickness measured after stabilization

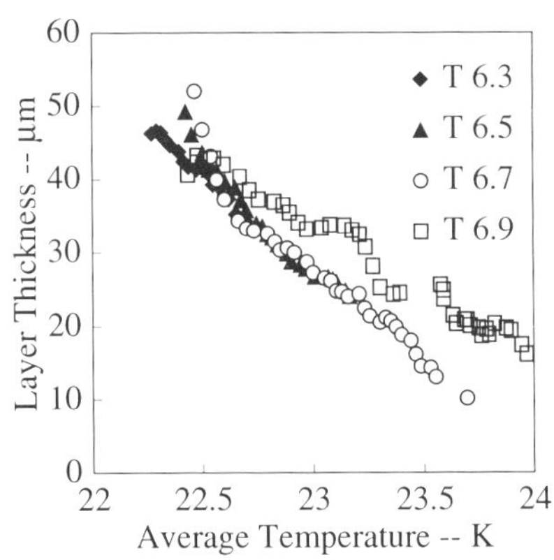

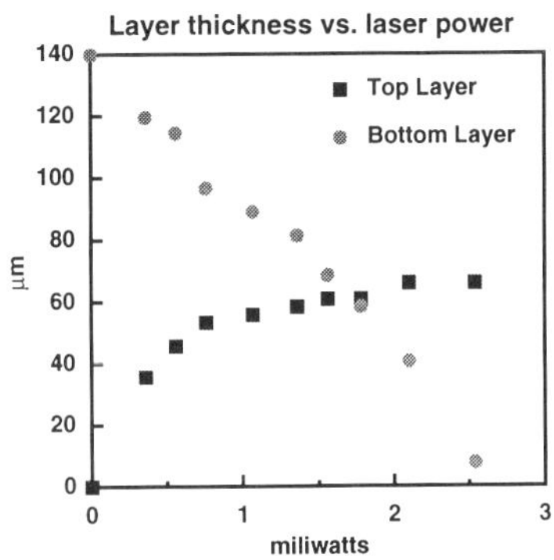

Left, figure 6 shows laser power required for thick liquid layers

Right, figure 7 shows data correlation and simple model fit

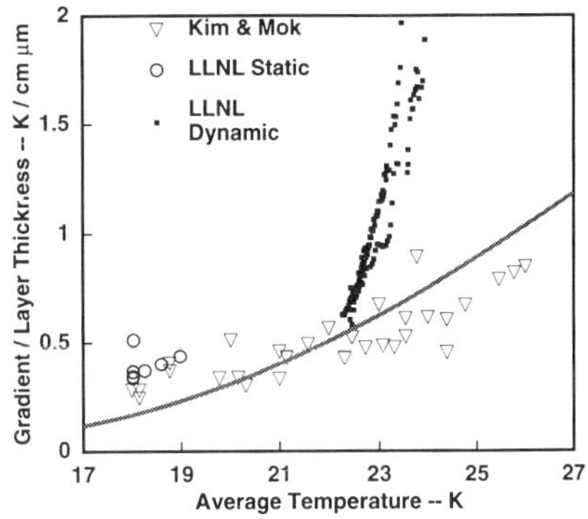

4. Comparison of the data

Much of the liquid layer thickness data obtained to date varies not only as a function of the thermal gradient, but also as a function of the capsule average temperature. Figure 7 shows a plot of the thermal gradient divided by the liquid layer thickness as a function of the capsule's average temperature. The data on this figure represents the static gradient technique, the dynamic thermal gradient technique, and previous experiments on thin centered liquid layers at the U. of Ill_2. For the static and the dynamic thermal gradient data, the levitated layer thickness was used in this plot, while the thin layer data from K. Kim is for a centered layer. The trend line shown was calculated using a simple analytical model developed by M. Monsler to capture the most important aspects of the layer thickness correlation. The model uses an assumed value for $d\sigma/dT$ which implies that a component separation of a few percent is necessary to generate the required surface tension gradient. The dynamic gradient experiment data starts near the trend set by the static data, but quickly diverges from it as the layer becomes thinner. This behavior may be caused by unobserved low frequency oscillation modes in the liquid layer thickness excited by the dynamically impressed thermal gradient.

5. Conclusion

Results of recent experiments on liquid H2-D2 layers supported using thermal gradients demonstrated that ~50 µm thick liquid layers can be successfully levitated with thermal gradients of reasonable magnitudes. A new laser generated thermal gradient technique opens possibilities of better control of layer geometry by spatial-temporal modulation of the heat source.

6. References

1. T. P. Bernat, E. R. Mapoles, and J. J. Sanchez, "Temperature- and Age-Dependence of Redistribution Rates of Frozen Deuterium-Tritium," ICF Quarterly Report Vol. 3 no. 3, page 57. (1991).
2. K. Kim, L. C. Mok, M. J. Erlenborn, and T. P. Bernat, "Non contact thermal gradient method for fabrication of uniform cryogenic inertial fusion target," J. Vac. Sci. Technol. A. Vol. 1 no. 2, page 1196. (1985)

Inst. Phys. Conf. Ser. No 140: Section 9
Paper presented at Laser Interact. with Matter Conf., Oxford, 19–23 September 1994
© 1995 IOP Publishing Ltd

A new possibility of charge-transfer pumping for xuv by means of laser-produced plasma and laser-evaporated gas

I.F. Shaikhislamov, Yu.P. Zakharov

Institute of Laser Physics,630090,Novosibirsk,Russia

Abstract. Despite the recent success of electron-collision and recombination pumped XUV-ray laser schemes there is strong motivation to investigate other excitation methods as well. In the 70-s and 80-s along with this schemes a charge-transfer pumping of high-Z laser-produced plasmas on neutral gas as a source of XUV-laser radiation was proposed and investigated. This scheme has two main advantages. Charge-transfer pumping populates preferentially ion' losing levels due to its quasi-resonant character and principally there is a possibility to pump all laser-produced ions. However, the experiments have showed that in the simple outline with uniform gas atmosphere an effective mixing between high-Z plasma ions and gas particles can't be achieved. In this report a proposition aimed to overcome this problem is briefly presented.

1. Introduction

A charge-transfer pumping scheme has been proposed for the first time by Vinogradov A.V. and Sobel'man I.N. in [1]. According to theoretical estimations the gain $KL > 5$ can be achieved in the range $\lambda > 100\mathring{A}$ if gas and ion concentrations are: $Ng\,Ni > 10^{17}cm^3$. In the range $\lambda > 10\mathring{A}$ charge-transfer pumping can provide short-time sources of XUV radiation in the form of few lines and with efficiency 1%.

Experimentally it has been tested in a series of works by Elton R.C. et al [2]. In the experiments with laser-produced plasma expanding into background gas an inversion of population for various ions ($C^{+4}, C^{+5}, N^{+4}, N^{+5}, Li^{+2}\ldots$) has been demonstrated. The levels of preferential population well accorded to predicted by charge-transfer mechanism. Nevertheless, this inversion didn't show influence from gas pressure (in the range $10^{16} < Ng < 10^{17}cm^{-3}$) and quantitatively was no more than 10^{-3} of maximum possible value. Moreover, to authors' explanation this inversion was due to charge-transfer of ions on neutral atoms of the same expanding plasma (for example $C^{+6} + C^{0}$) and a role of background gas was to create a mixture of this particles.

The reason of such situation has been clarifies in other experiments (for example [3]) and especially in numerical simulations [4]. At first millimeters from the target the background gas is very rapidly ionized by hot electrons and photons. When this processes fall down due to adiabatic cooling of expanding plasma then another process switches on - a blast wave. When gas concentration is high enough ($Ng > 3 \cdot 10^{16}\ cm^{-3}$) the blast wave forms fully after first few millimeters from the target and outrun the plasma front. As the result gas is ionized by blast wave before contact with high-Z ions and mixing between them is entirely prevented. Therefore, under high and uniform gas concentrations this mentioned processes greatly reduce the intensity of charge-transfer between plasma and gas over all distances from the target.

2. A new scheme of charge-transfer pumping

Thus, having reveal the main obstacles on the way of charge- transfer pumping we can try to construct conditions when this obstacles are removed. It is obvious that instead of uniform gas atmosphere a compact gas cloud can be used. If there is a gap between gas cloud and plasma target of order of few millimeters up to one centimeter the plasma will have a time to cool down before reaching gas and won't ionize it from the very start. The blast wave formation also takes some time (for energy accumulation). Therefore, during some time a pure charge-transfer process between plasma and gas under high concentrations will be realized. To create compact gas cloud a low-intensity laser irradiation of the second target is proposed. This scheme has two side-line advantages. Firstly, the problem of absorption of generated emission by dense neutral medium will be solved and secondly, the possibility opens to create gas from various and very exotic materials.

Gas cloud has to have sharp boundary, so as to prevent too fast blast wave formation. There is an empirical condition under which the significant blast wave isn't generated:

$$\int Ng\, dl < 3 \cdot 10^{16} cm^{-2}$$

where $'l'$ is a path of plasma in gas atmosphere. Then, if gas cloud has an exponential profile $Ng \approx \exp(-X/L)$, with a scale L and we wish to prevent blast wave' formation up to point where $Ng > 3 \cdot 10^{16}\ cm^{-3}$, then the condition $L < 1$ mm has to be satisfied. A possibility of creating of gas clouds with such parameters by low-intensity laser irradiation has been experimentally demonstrated for example in [5].

As the first step it is now planned to demonstrate an effectiveness of this scheme of charge-transfer pumping in the gain experiment with the mixture of C^{+4} ions and atoms of Indium. C^{+4} ions can be easily obtained in large amounts because of its low potential of ionization($64.5\,ev$). Using the atoms of low ionization potential (as for example $In, 5.8\,ev$) allows to pump the level $n = 4$ of C^{+3} ion, and losing at $\lambda = 1200\mathring{A}$ (4-3 transition) becomes possible:

$$C^{+4} + In^0 \rightarrow C^{+3}\ (n=4) + In^+ \rightarrow C^{+3}\ (n=3) + In^+ + Photon(1200\mathring{A})$$

A gain as high as $10cm^{-1}$ can be achieved even for the densities of ions and neutrals about $10^{16}\,cm^{-3}$, whereas densities up to $10^{17}\,cm^{-3}$ will be used.In the figure below a sketch of proposed experiment is shown.

For measurement of gas cloud density we suggest to apply a miniature piezoelectric probe of high sensitivity and response time. The probe has undergone testing on the effects of back-scattered CO_2-radiation from the target.

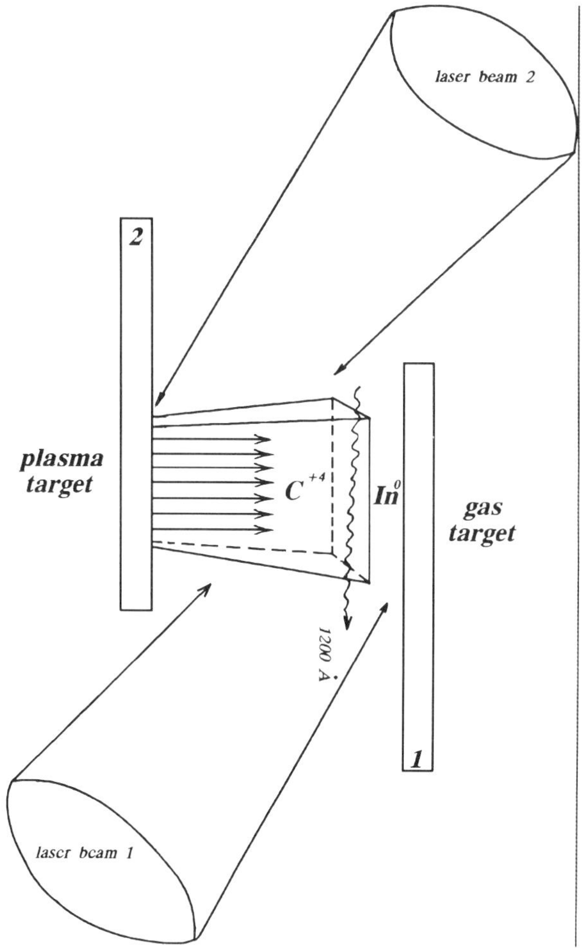
laser beam 2
2
plasma target
C +4
In0
gas target
1200 Å
1
laser beam 1

The proposed experiment has not only practical but also fundamental interest. The principal possibility and features of charge-transfer under high concentrations of reagents has not been explored yet. This can be of interest for example in such astrophysical phenomenon as Supernova shell' luminosity under the impact with compact hydrogen clouds [6].

References

[1] Vinogradov A V and Sobel'man I N 1972 *J. of Theor. Phys. (Russian)* **a6** 2113

[2] Dixon R H and Elton R C 1977 *Phys.Rev.Lett.* **38** 1072.

[3] Basov I G et al 1982*Itogi nauki i tehniki (Russian)* **26**

[4] Giuliani J and Mulbrandon M 1989 *Phys.Fluids B1* **7** 1463.

[5] Knyazev B A et al 1986 *J. of Theor. Physics (Russian)* **56** 1319.

[6] Wise W M and Sarazin L C 1989 *The Astrophysical J.* **345** 384.

Inst. Phys. Conf. Ser. No 140: Section 9
Paper presented at Laser Interact. with Matter Conf., Oxford, 19–23 September 1994
© 1995 IOP Publishing Ltd

Numerical Simulation of Optically Ionised Plasma Recombination X-ray Lasers

Y F Shao *+, S J Rose *⊕, and A Djaoui *

*Rutherford Appleton Laboratory, Chilton, Didcot, Oxon OX11 0QX, UK
⊕Department of Physics and Space Science, University of Birmingham, Birmingham B15 2TT, UK
\+ Institute of Applied Physics and Computational Mathematics, P. O. Box 8009, Beijing 100088, P R China

Abstract. A non-LTE average atom model which has been modified to include tunnel ionisation and ATI heating is presented, and the model is applied to study the optically ionised Li-like Ne recombination x-ray laser. Results from the calculation show that high gain in Li-like Ne at 98Å can be produced with a 100fsec duration, 0.248 μm laser of intensity $2\times10^{17}\ Wcm^{-2}$ and the gas density of $10^{-3}\ gcm^{-3}$. The duration of x-ray lasing is about 1.5psec.

1. Introduction

Recent advances in the technology of ultrashort pulse, high intensity lasers provide a possible new approach [1] to recombination x-ray lasers. A neon gas target is directly ionised by means of tunnel ionisation in an intense optical field to produce a He-like Ne plasma with a low electron temperature. The plasma rapidly recombines through three-body recombination to the excited states of Li-like Ne to produce population inversion and gain. Compared with the conventional X-ray lasing schemes, this new scheme has a very high energy efficiency[2] and is a potential candidate for a "table top" x-ray laser. Some numerical modelling of this scheme has already been presented[1,2,3,7]. In this work, we use a non-LTE average atom model to calculate the gain which is modified to include tunnel ionisation and above threshold ionisation (ATI) heating .

2. Model

In the original average-atom model [4], we construct differential equations for the time-evolution of the average population of the different principal quantum shells $P_n(t)$ by considering the average rate of populating and depopulating a particular shell. The following processes are considered: collisional excitation and ionisation (by free electrons in the plasma), their inverse processes: (collisional deexcitation and three-body recombination), and the spontaneous radiative rates. When a ultrashort pulse, high intensity laser interacts with a gas target (neon) the rates for collisional processes can be exceeded by the optical ionisation rates. In order to simulate this situation, we modify the original model to include tunnel ionisation and ATI heating.

Tunnel ionisation is a very non-linear process [5] because the rate rises very rapidly above a threshold intensity[6]. We use the tunnel ionisation theory of Ammosov, Delone, and Krainov [8] who show that ionisation rate out of shell n can be approximated by

$$W_n = 1.61\omega_{a.u.}\frac{Z_n^{\,2}}{n_{eff}^{\,4.5}}\left[10.87\frac{Z_n^{\,3}}{n_{eff}^{\,4}}\frac{E_{a.u.}}{E}\right]^{2n_{eff}-1.5}\times\exp\left[-\frac{2}{3}\frac{Z_n^{\,3}}{n_{eff}^{\,3}}\frac{E_{a.u.}}{E}\right]\quad s^{-1}\qquad(1)$$

where $\omega_{a.u.}$ is the atomic unit of frequency, E is the instantaneous laser electric field, $E_{a.u.}$ is the atomic field strength and an effective principal quantum number n_{eff} is defined as $n_{eff} = Z_n(13.6/U_i)^{1/2}$. In our model, U_i is approximated as the appropriate average ionisation energy and Z_n is approximated by the average ionisation.

When a short pulse high intensity laser interacts with a plasma the major heating mechanisms, which should be considered, include non-linear inverse bremsstrahlung, above threshold ionisation and stimulated Raman scattering. When the laser pulse $\tau_{FWHM} \leq 50\,fs$, electron heating from inverse bremsstrahlung is small [5], and relatively little electron heating from SRS is expected [7]. In the present calculations, we take a Gaussian laser pulse at 0.248 μm wavelength and 100 fs FWHM duration, and the ATI heating is considered as the only heating mechanism in our model. So the temperature as calculated is a lower limit to the true temperature when other heating processes are included. The formula for calculating the average ATI energy was taken directly from [5]. When we calculate the electron temperature, we assume that all free electrons come from the ground state and the ATI energy is considered as a heating source term in the free electron energy balance.

The Li-like ground states $1s^2 2l$ and the excited states $1s^2 3l$ and $1s^2 4l$ are denoted α, β and γ respectively. The shell-averaged gain between β and α and between γ and β is calculated by reconstructing the population of states α, β and γ from the average-atom shell populations, $P_n(t)$, using the method of Djaoui and Rose [4]. A Doppler line width is assumed $(T_i = T_e)$.

3. Results of Simulation

In the calculation we take a Gaussian laser pulse at 0.248 μm of wavelength and 100 fsec FWHM duration with a peak at 10^{-13} sec. From Fig. 1 (a), we find that neutral Ne gas can be ionised to He-like Ne when the laser intensity is higher than $2\times10^{17}\,Wcm^{-2}$. Comparison with the result in [1], shows good agreement. Fig. 1 (b) shows that the higher the laser intensity, the more residual energy will be produced. In computing the gain profiles, we choose the peak intensity of $2\times10^{17}\,Wcm^{-2}$ and fix the density at $10^{-3}\,gcm^{-3}$ throughout the calculation. The time-histories of the reduced fractional populations of α, β and γ, and the fraction of He-like Ne are plotted in Fig. 2 (a). The shell-average gain profiles of the Li-like Ne lines are shown in Fig. 2 (b). The shell-averaged gain on the $\gamma \rightarrow \beta$ transition appears earlier in time than for $\beta \rightarrow \alpha$ and the duration of gain for $\beta \rightarrow \alpha$ is about 1.5 psec. The maximum shell averaged $\beta \rightarrow \alpha$ gain coefficient is about 460 cm^{-1}. Comparing with the results in [2], the gain profiles are similar, but the values of gain are different. For the $3d_{5/2} \rightarrow 2p_{3/2}$ fine-structure transition in Li-like Ne at 98Å, the gain coefficient is approximately the half of the shell averaged value. Also according to Ref. [9], the Stark width is a few times large than the Doppler width in Li-like Neon for the $3d_{5/2} \rightarrow 2p_{3/2}$ transition for an electron temperature and density of $25eV$ and $2.5 \times 10^{20} cm^{-3}$ respectively. The maximum gain of the $3d_{5/2} \rightarrow 2p_{3/2}$ fine- structure transition in Li-like Ne at 98 Å would thereby be reduced to approximately $100 cm^{-1}$ in fairly good agreement with the results in [2].

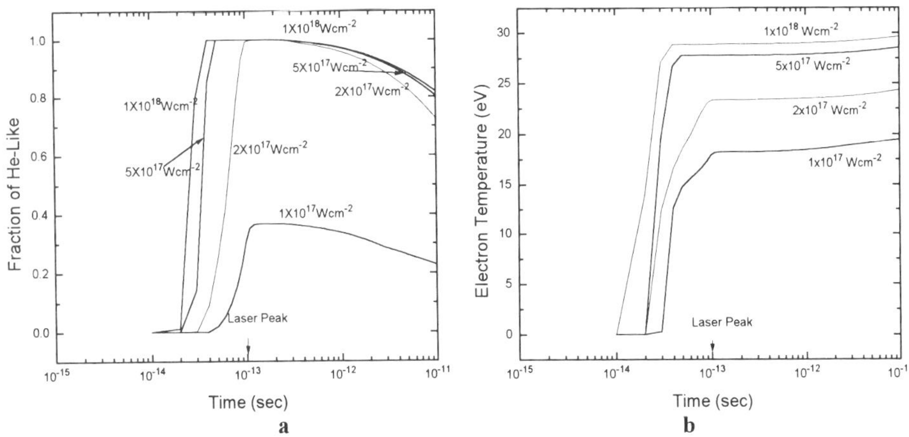

Fig. 1: The time evolution of (a) the fraction of He-like Ne and (b) the electron temperature

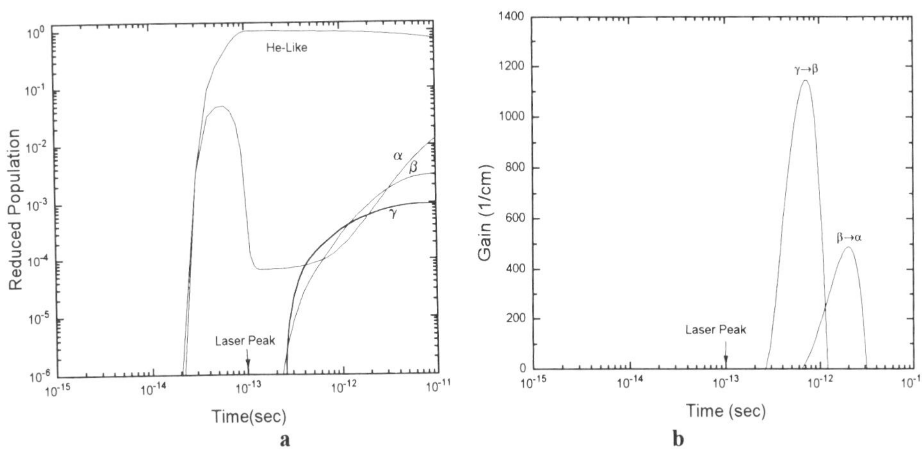

Fig. 2: (a) The time evolution of (a) the reduced populations (fraction / degeneracy) of He-like Ne and the excited states of Li-like Ne α, β and γ, and (b) the gain profiles. (The Li-like ground states $1s^2 2l$ and the excited states $1s^2 3l$ and $1s^2 4l$ are denoted α, β and γ respectively.)

Positive gain is only found when the laser intensity is higher than $2\times10^{17}\,Wcm^{-2}$. However increasing the intensity to $1\times10^{18}\,Wcm^{-2}$ results in a decrease in the gain to three quarters of the value at 2×10^{17}. Further increase in intensity results in a further decrease in the gain. This behaviour occurs because for low intensities, ionisation to He-like Ne does not occur, yet for higher intensities, a higher electron temperature results and the free electrons do not so rapidly recombine through the excited states of Li-like Ne to produce a population inversion and gain.

4. Conclusion

The optically ionised plasma recombination X-ray lasers scheme requires that the laser directly ionises neon gas to He-like Ne by tunnel ionisation to produce a low electron temperature. The scheme is very sensitive to laser intensity: in our calculations, we show that the gain for lasing to the ground state can be produced in a fairly narrow range of laser intensity. The maximum gain coefficient of the $3d_{5/2} \rightarrow 2p_{3/2}$ fine-structure transition in Li-like Neon at 98Å is predicted to be about $100cm^{-1}$ and the duration of lasing is about 1.5ps for a 100fsec duration, $0.248\,\mu m$ wavelength laser-driver of intensity $2\times10^{17}\,Wcm^{-2}$ and material density of $10^{-3}\,gcm^{-3}$.

References

[1] D. C. Eder, P. Amendt and S. C. Wilks
'Table-Top X-ray Lasing Using A High-Intensity Short-Pulse Driver'
Internal Report of LLNL 1991.

[2] P. Amendt, D. C. Eder and S. C. Wilks, *Phys. Rev. Lett.* **66**, 2589(1991)

[3] D. C. Eder, P. Amendt et al to be published in *Physics of Plasma*

[4] A. Djaoui and S. J. Rose, *J. Phys. B: At. Mol. Opt. Phys* **25**, 2745 (1992)

[5] A. Djaoui and A. A. Offenberger, *Physical Review* **E 50** (1994)

[6] B. M. Penetrante and J. N. Bardsley, *Physical Review* **A 43**, 3100(1991)

[7] D. C. Eder, P. Amendt, and S. C. Wilks, *Physical Review* **A 45**, 6761(1992)

[8] M. V. Ammosov, N. B. Delone, and V. P. Krainov, Zh. Eksp. Teor. Fiz. **91** 2008 (1986) [Sov. Phys.-JETP **64**, 1191 (1986)].

[9] P. Amendt, D. C. Eder, S. C. Wilks, M. J Dunning, and C. J. Keane, Proceedings of SPIE's International Symposium on High Power Lasers, Los Angeles, CA. January 20-25, 1991.

Inst. Phys. Conf. Ser. No 140: Section 9
Paper presented at Laser Interact. with Matter Conf., Oxford, 19–23 September 1994
© 1995 IOP Publishing Ltd

Core Corona Coupling in Laser Irradiated Vacuum Insulated Plasma Target

D.P. Singh, I. Deha* and M. Vaselli

Istituto di Fisica Atomica e Molecolare - C.N.R., Via del Giardino 7, 56127 Pisa (ITALY)
*Permanent Address: USTHB Institut de Physique, B.P. 132 Bab Ezzouar-Algiers (ALGERIA)

Abstract. When a spherical plasma target is irradiated uniformly and symmetrically from all sides by intense laser beams, the hot electrons generated at plasma critical layer via laser absorption tend to form a hot cloud. A vacuum layer is introduced in the target to reduce the hot electron flux penetrating radially inward to the center of the target. Considering the self-regulation model of core-corona coupling, the dependence of this coupling on laser wavelength, the width of the vacuum layer and other plasma parameters is studied analytically. These calculations indicate that the use of vacuum layer inside the target has a strong effect on heat transfer efficiency from hot corona to the plasma core.

1. Introduction

The preheating of internal core of the plasma target by superthermal electrons produced by intense laser beam at the plasma resonance layer poses a serious problem in target compression in inertial confinement fusion. Several ways of shielding the target from these long range hot electrons have been proposed such as, use of high Z layer, anomalous return current-resistance, seeded magnetic fields etc. An efficient way to eliminate adverse effects of hot electrons is to create a vacuum gap between the plasma core and the plasma critical layer where such electrons are produced, which would retard or prevent the penetration of the hot electron flux into the plasma core. Experiments [1] carried out with vacuum gap in the target also confirm the reasonable character of the approach and also suggest that the adverse effects of support structure can be minimised to a tolerance level. It has been observed that even a small vacuum gap of about 200

μm can reduce the hot electron flux by more than three order of magnitude. The advantages of creating a vacuum gap is to retard/prevent the flow of hot electrons crossing it and in this way, to reduce the mass of the target. The spherical plasma target with a vacuum gap is irradiated symmetrically from all sides by intense laser beams. Assuming that most of the laser energy is absorbed at plasma critical layer, a number of hot electrons are generated there which drive the ablation of target. These hot electrons penetrate inside the target up to depth of their mean free path producing a hot electron corona. However, many electrons are stopped due to the space charge created by vacuum gap. Considering this fact in view, self-regulation model of core-corona coupling is developed to study the heat transfer efficiency from the hot corona to the dense core.

2. Theoretical Analysis

In steady state case of laser irradiated spherical target, core-corona coupling involves two processes occurring simultaneously which govern the heat transport to the target core which are: energy transfer from hot electrons to cold electrons in core-corona overlapping region, and thermalization of core-corona overlapping region with whole core. Following Langmuir [2] the effective density (n_h) of hot electrons crossing the vacuum gap (d) is

$$n_h = \frac{k_B T_h n_{cr}}{\left[(k_B T_h)^{\frac{1}{2}} + d\,(2\,\pi\, e\, n_{cr})^{\frac{1}{2}}\right]^2}\,, \qquad (1)$$

where k_B, e, n_{cr} and T_h are Boltzman constant, electronic charge, plasma critical density and hot electron temperature respectively. These hot electrons passing through the vacuum layer form a hot electron corona surrounding the plasma dense core driving the target ablation. The expression for the energy equipartition time τ_{eq} in core-corona overlapping zone can be given as [3],

$$\tau_{eq} = \frac{3}{4}\sqrt{\frac{m}{\pi}}\frac{(k_B T)^{3/2}}{e^4 n_h \ln \Delta}\,, \qquad (2)$$

T (= $(T_h+T_c)/2$) is the mean electron temperature in the plasma ablation region. The self-regulation condition [4] (the width of the plasma ablation is of the order of mean free path of hot electrons) of the corona may be given as

$$\left(\frac{r_c}{r_e}\right)^2 = \frac{1}{\left(1 - \frac{d}{r_c}\right)^2} + \frac{2\,\lambda_h}{r_c}\frac{n_h}{n_{cr}} \quad , \tag{3}$$

r_c is the radius of plasma critical layer. The hot electron mean free path at plasma critical layer (λ_{hc}) may be expressed as

$$\lambda_{hc} = \left[k_B T\right]^2 / \left[\pi e^4 (Z+1) n_{cr} r_c \ln \Delta\right] \quad , \tag{4}$$

Cooling rate of the corona is

$$S_0 = \frac{3}{4}\frac{k_B(T_h - T_c) n_{cr} r_c^3}{\tau_{eq}}\left[2\,\ln\left(1-\frac{d}{r_c}\right) + \ln\left[\frac{1}{\left(1-\frac{d}{r_c}\right)^2}\right] + \frac{2\lambda_h}{r_c}\frac{n_h}{n_{cr}}\right] \quad , \tag{5}$$

Mass of the whole core may be written as

$$M = \frac{m_i\, n_{cr} r_c^3}{3}\left[1 + \ln\frac{\rho}{\rho_0} + 3\,\ln\left(1 - \frac{d}{r_c}\right)\right] \quad , \tag{6}$$

(ρ/ρ_0) is the ratio of plasma core density to that of the external region. The expression for core-corona coupling may be obtained from the energy balance condition between the core and the corona as

$$\tau_c = \frac{2}{3}\frac{\left[1 + \ln\frac{\rho}{\rho_0} + 3\,\ln\left(1 - \frac{d}{r_c}\right)\right]}{\left[2\,\ln\left(1 - \frac{d}{r_c}\right) + \ln\left(\frac{1}{\left(1 - \frac{d}{r_c}\right)^2}\right) + \frac{2\,\lambda_h}{r_c}\frac{n_h}{n_{cr}}\right]}\tau_{eq} \tag{7}$$

which describes the general dependence of core-corona coupling on the vacuum gap, laser wavelength and plasma parameters. When $d = 0$ aforementioned expression reduces to that given by Kidder and Zink [4] and is represented by (τ_{c0}). The respective results on the efficiency of vacuum insulated target over normal target (without vacuum gap) are obtained for the set of parameters: $4\pi M = 48\mu g$, $Z = 1$, $\rho/\rho o = 10^4$.

3. Conclusion

Figure 1 displays the variation of relative core - corona coupling

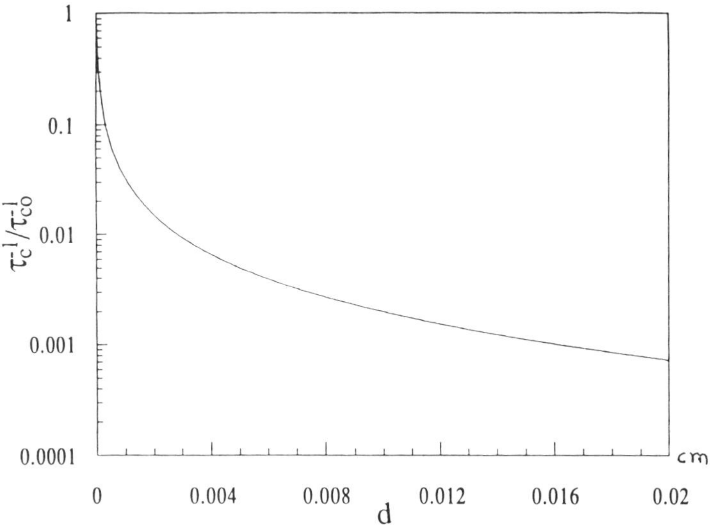

Fig 1. Plot of ($\tau_c{}^{-1}/\tau_{c0}{}^{-1}$) versus vacuum gap (d) for $\lambda_{\mu m}$=10.6 and T = 50 keV.

efficiency ($\tau_c{}^{-1}/\tau_{c0}{}^{-1}$) of vacuum insulated target with the vacuum gap for CO_2 laser wavelength. It is seen that even a small gap of about 100 μm is able to reduce the coupling by more than three orders of magnitude. Furthermore, the relevant calculations for Nd laser wavelength indicate that this reduction of core corona coupling enhances additionally by about two orders of magnitude for the same set of parameters. In general laser wavelength scaling of ($\tau_c{}^{-1}/\tau_{c0}{}^{-1}$) shows a strong increase in lower wavelength regime ($\lambda \leq 1$ μm) and seems to nearly saturate for higher wavelengths ($\lambda \geq 10$ μm). However, the presence of vacuum gap does not modify the qualitative dependence of core-corona coupling efficiency with mean electron temperature in the plasma ablation region.

References

[1] Lee K. et al., 1979, *Nuclear Fusion* **19,** 1447.

[2] Langmuir I., 1923, *Phys. Rev.* **21,** 419.

[3] Spitzer L. Jr., 1962, *Physics of Fully Ionized Gases*, 2nd Edn. Interscience, New York.

[4] Kidder R. and Zink J. W.,1972, *Nuclear Fusion* **12,** 325; Singh, D. P. et al. 1989 *Laser and Particle Beams,* **7,** part 1, 111.

Inst. Phys. Conf. Ser. No 140: Section 9
Paper presented at Laser Interact. with Matter Conf., Oxford, 19–23 September 1994

© *1995 IOP Publishing Ltd*

Intermittency and chaos in stimulated Raman scattering

M M Škorić and M S Jovanović

Vinča Institute of Nucl. Sciences, P.O. Box 522, 11001 Belgrade, Yugoslavia

Abstract. Transition from coherent to strongly nonlinear incoherent regime, or spatio–temporal chaos, is studied for saturated SRS in typical laser fusion conditions. Plasma is assumed bounded, uniform and weakly collisional. The corresponding set of the coupled differential equations with nonlinear detuning of EPW taken into account, is solved numerically in space-time with nonzero source boundary conditions. Through the variation of laser and plasma parameters, in particular, by increasing the pump strength, the system is driven toward the chaotic state. A route via steady state, periodic and quasiperiodic regimes, with an intermittent transition to an extensive chaos is revealed. Distinctive features, such as spiky–burst-like waveforms, broad spectra and incoherent dynamics are observed, which are qualitatively consistent with results obtained by other authors, based on more complex models.

With increasing evidence that stimulated Raman scattering (SRS) often becomes absolute in laser fusion plasma experiments [1,2], the problem of the nonlinear saturation of this process has come to a spotlight in many experimental and numerical studies. Since SRS nonlinearly couples an incident laser light (pump) and scattered radiation with electron plasma waves (EPW's), the physics of the excited waves, in particular EPW's, appears a key factor determining the course of nonlinear saturation of the SRS process in the model of bounded, spatially uniform, weakly collisional plasma we are considering. We write the coupled equations describing spatio-temporal evolution of the three waves in the SRS process as follows [3,4]:

$$\frac{\partial a_0}{\partial \tau} + V_0 \frac{\partial a_0}{\partial \xi} = -a_1 a_2, \qquad \frac{\partial a_1}{\partial \tau} - V_1 \frac{\partial a_1}{\partial \xi} = a_0 a_2^*,$$
$$\frac{\partial a_2}{\partial \tau} + V_2 \frac{\partial a_2}{\partial \xi} + \Gamma a_2 - i\delta \left|a_2\right|^2 a_2 = \beta_0^2 a_0 a_1^*, \tag{1}$$

where a_0, a_1 and a_2 are normalized, slowly-varying amplitudes of the pump, backscattered and EPW, respectively, τ and ξ – normalized time and space co-ordinate, Γ – EPW linear damping rate and $\beta_0 = v_{osc}/c$ – electron quiver velocity in the pump wave electric field. A wave resonance is assumed, given by the perfect matching conditions $\omega_0 = \omega_1 + \omega_2$ and $\mathbf{k}_0 = \mathbf{k}_1 + \mathbf{k}_2$, fulfilled in a physical space.

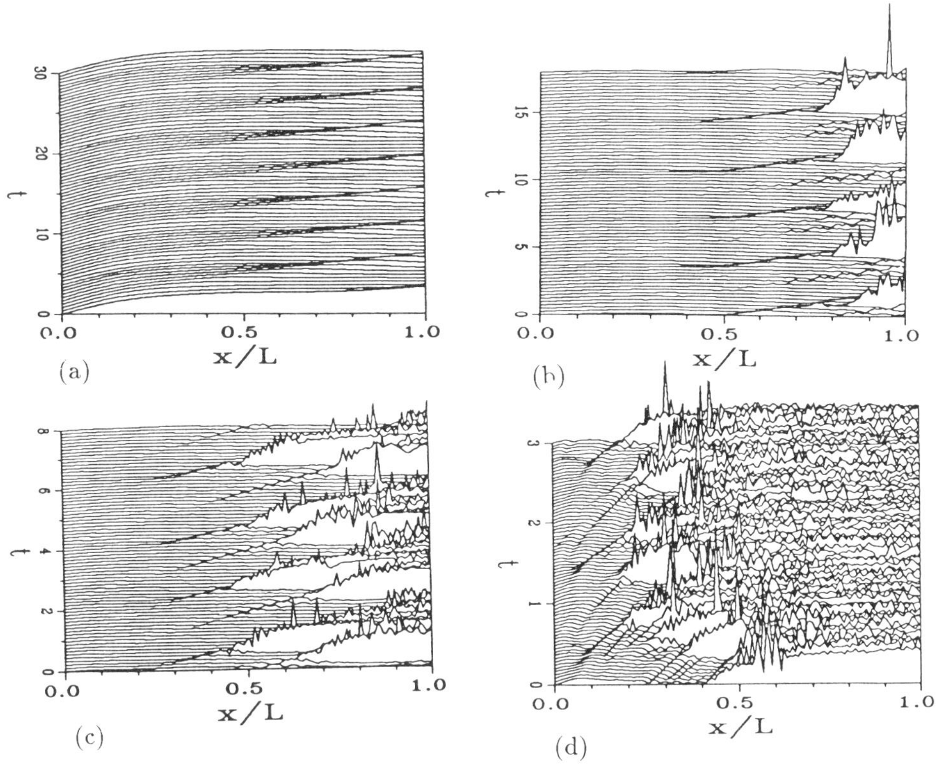

Figure 1: Space-time patterns of electron plasma wave in stimulated Raman backscattering for the values of the pump strength parameter β_0: a) 0.025, b) 0.03, c) 0.04, d) 0.1.

A cubic self-modal nonlinear term is added in the EPW equation in order to account for possible nonlinear phase detuning (shift) due to ponderomotive effects or relativistic correction to electron mass in large amplitude EPW [5-7].

Plasma parameters in simulations were: plasma length $L = 100\,c/\omega_0$, density $n_0 = 0.1\,n_{cr}$, electron temperature $T_e = 1\,keV$, EPW damping rate $\nu_{epw} = 10^{-5}\omega_{pe}$ and the initial electromagnetic noise $\varepsilon_1 = 0.01$, indicating the relative noise level with respect to the pump wave. The pump parameter β_0 was varied in the interval (10^{-3}–10^{-1}) resulting in different types of SRS dynamical behaviour. For low β_0-values the dynamics is seemingly regular-like (simple periodicity, quasiperiodicity). With increasing β_0 spatio-temporal patterns obtain complex forms, and the system attains chaotic-like features. We focus our attention on the spatio-temporal evolution of EPW, because it exhibits the most interesting and most exciting features. Moreover, all the richness of SRS physics originates from the EPW nonlinearity; thus, the dynamics changes are most emphasized in the EPW spatio-temporal patterns.

As described in the previous paper by these authors, SRS can, under appropriate conditions [3,4], transit from a regular regime (periodic cycle) to high-order chaotic state recognizable by unpredictable (irreproducible) evolution and burst-like behaviour

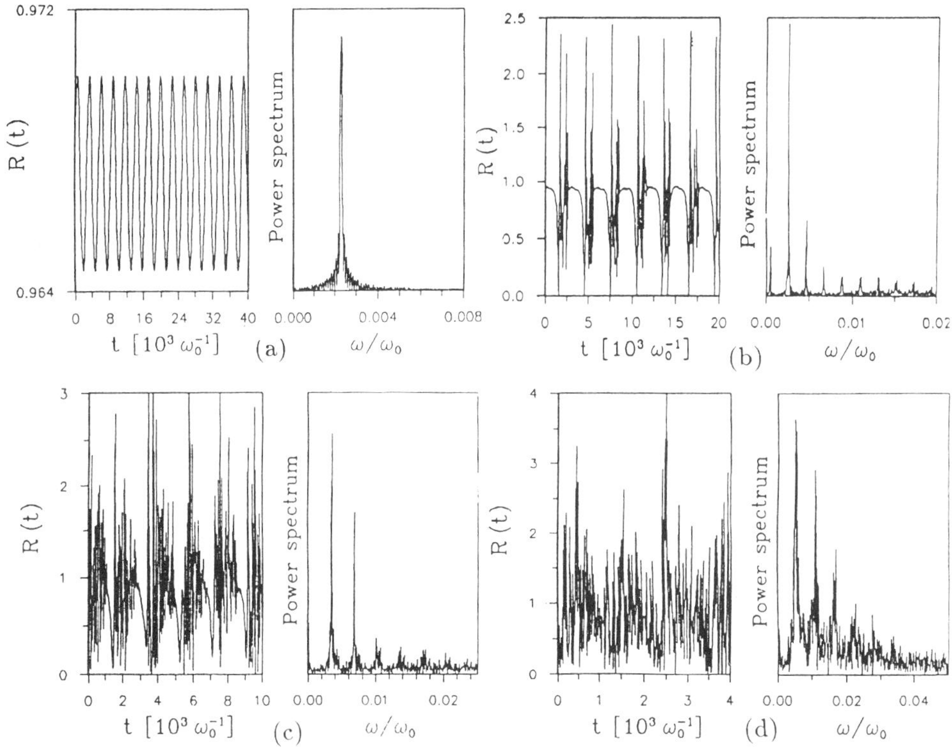

Figure 2: SRS reflectivity vs. time (1) and corresponding power spectra (2) for β_0-values from Fig. 1.

of relevant physical quantities. A route to chaos, passing through different dynamical regimes and leading to a developed spatio-temporal chaos is revealed. These transitions are illustrated in Fig. 1, which shows space-time patterns of the EPW, for the pump values: $\beta_0 = 0.025$, 0.03, 0.04 and 0.1, respectively. After regular oscillations in space-time, visible in Fig. 1a, we observe in Fig. 1b a physical situation that spatial coherence in the EPW profile is still, predominantly, preserved in time, apart from the occasional bursts in temporal evolution. Such type of behaviour, in which periodic or quasiperiodic oscillations are interrupted by chaotic bursts, generic to our model, is called *intermittency* in the terminology of studies of nonlinear dynamics and chaos, and is widely observed at the onset of turbulence in many hydrodynamic flows.

More complex features of spatio-temporal evolution of EPW's are revealed by further increasing of the pump strength. The regularity in its dynamics is replaced by nonlinear incoherency and burst-like aperiodicity. For $\beta_0 = 0.04$ (Fig. 1c) turbulent area occupies approximately up to 50% of the plasma space, and this percentage becomes considerable larger for greater β_0 (e.g. $\beta_0 = 0.1$, Fig. 1d). The case represented in Fig. 1c is named *spatio-temporal intermittency*, while Fig. 1d demonstrates the state of a *spatio-temporal chaos* [8,9]. The latter figure displays the coexistence of patches of turbulence immersed in the rest of structure, still in the laminar state.

In order to shed more light at the route to chaos via intermittency of SRS in a plasma, temporal dependence of *reflectivity* R (fraction of input energy that returned from the plasma via the backscattered wave) and *power spectrum* of the scattered wave at the entrance point $x = 0$ is demonstrated in Fig. 2. There is only one peak in the power spectrum for the case $\beta_0 = 0.025$, while the next value ($\beta_0 = 0.03$, Fig. 2b) produces the spectrum which is more rich in characteristic frequencies. Although, the latter spectrum still contains a cascade of distinct peaks; the broad-band noise appears indicating a loss of coherence.

With the strengthening of the pump, dynamics becomes more complex and power spectra exhibit ever more frequencies. For the case of developed chaos (Fig. 2d) the frequency spectrum is practically continual, which means that various modes of oscillations are present, so that the evolution of the process is unpredictable (chaotic).

The simulation results, partially presented in the present paper, exhibit a sort of a qualitative agreement with recent experiments [1,2] and numerical simulations [10,11]. The main advantage of our model is a physical simplicity and transparence, with respect to more complex nonlinear models of SRS, based on Zakharov-Maxwell equations [12,13]. An investigation of relation to spectral data on plasma wave properties obtained in more recent SRS experiments [2] is in progress.

References

[1] Drake R P *et al.* 1988 *Phys. Rev. Lett.* **60** 1018
[2] Villeneuve D M *et al.* 1993 *Phys. Rev. Lett.* **71** 368
[3] Škorić M M and Jovanović M S 1994 in *Laser Interaction and Related Phenomena vol. 11* (ed. by Miley G H and Hora H, AIP, New York)
[4] Škorić M M and Jovanović M S 1994 to appear in *Dynamical Systems and Chaos* (ed. by Aoki N, Pergamon Press)
[5] Salomaa R R E and Karttunen S J 1986 *Phys. Scripta* **33** 370
[6] Bingham R 1990 *Phys. Scripta* **T30** 24
[7] Kruer W L 1990 *Phys. Scripta* **T30** 5
[8] Hohenberg P C and Shraiman B I 1993 *Physica* **D37** 109
[9] Cross M C and Hohenberg P C 1990 *Rev. Mod. Phys.* **T30** 5
[10] Bezzerides B *et al.* 1993 *Phys. Rev. Lett.* **70** 2569.
[11] Bonnaud G *et al.* 1992 *Phys. Fluids* **B4** 423
[12] Rozmus W *et al.* 1987 *Phys. Fluids* **30** 2181
[13] Kolber T *et al.* 1993 *Phys. Fluids* **B5** 138

Inst. Phys. Conf. Ser. No 140: Section 9
Paper presented at Laser Interact. with Matter Conf., Oxford, 19–23 September 1994
© *1995 IOP Publishing Ltd*

Comparison of shock breakout uniformity from a solid target using Phased Zone Plate arrays and Random Phase Plates.

R M Stevenson , A M Evans

AWE Plc, Aldermaston, Reading, Berkshire RG7 4PR U.K.

Abstract. The uniformity of the shock breakout from the rear surface of a directly illuminated solid aluminium target was used to determine the improvement in shock planarity from laser beam smoothing using Phased Zone Plate arrays (PZP) or Random Phase Plates (RPP). The results produced using a PZP show a four fold improvement in intensity distribution variations with an almost top-hat profile.

1. Introduction

The spatial intensity distribution of high power laser systems used for laser plasma interaction studies and ICF research plays a critical role in determining beam-plasma coupling. In general, targets must be placed in the quasi-near field for relatively large area illumination and in this region the inherent coherence of the beam produces a deeply modulated distribution which can seed non-linear plasma and hydrodynamic instabilities .

The random phase plate was developed by Kato *et al* [1] to scramble the phase of the laser beam with the resultant improvement in the intensity profile due to the superposition of many small beamlets. However, this technique is inefficient at producing top-hat intensity profiles. The use of a phased zone plate array by Stevenson *et al* [2] has been shown to produce a near top hat intensity profile with high efficiency (~82%).

In this paper we present streak camera measurements of the uniformity of the shock breakout from the rear surface of a solid aluminium target which was directly illuminated using 1) a defocused laser beam, 2) a Phased Zone Plate Array or 3) a Random Phase Plate.

2. Experiment

The shock breakout measurements rely on the fact that a shock wave strongly heats the target material and can produce temperatures of ~5ev (50,000 K). At these temperatures the shocked material is highly luminous and the shock arrival at a surface can be detected optically. Because of the high conductivity of a metal, its surface remains opaque until the shock wave is within a very short distance of the rear surface. The shock wave traverses this distance within a few picoseconds. Hence, the luminosity of the target surface increases to a high value almost instantaneously as the shock wave reaches it. Variations in the measured shock breakout time can be directly related to non uniformities in the incident laser intensity profile.

The diagnostic equipment consisted of a pair of optical streak cameras with orthogonal slits which view the rear surface of the 35μm (+/- 0.5μm) thick aluminium target. The cameras had a temporal resolution of 13ps and a spatial resolution of 10μm. Each of the optical beam lines had a spectral window of 600-800 nm selected by an OG590 filter combined with the spectral response of the cathode of the streak camera. The target was illuminated by one heating arm of the HELEN laser system at AWE which produces approximately 250J of second harmonic light (526nm) in a 300ps pulse. The laser system is configured to produce an approximately flat top near field intensity profile.

3. Results

A series of shots were fired with and without the PZP's and RPP's in order to evaluate their performance. In all figures time runs from left to right. The result shown in figure 1a is the streaked shock breakout for a 500μm defocused laser spot and the corresponding tracking of the shock front (figure 1b). This shows typical low spatial frequency variations and asymmetries that are present in all high power laser beams. Figures 2a & 2b correspond to the use of a Random Phase plate to produce a 500μm spot. This result shows the characteristic sinc^2 profile produced by these devices and the conversion of the low spatial frequencies to high spatial frequencies.

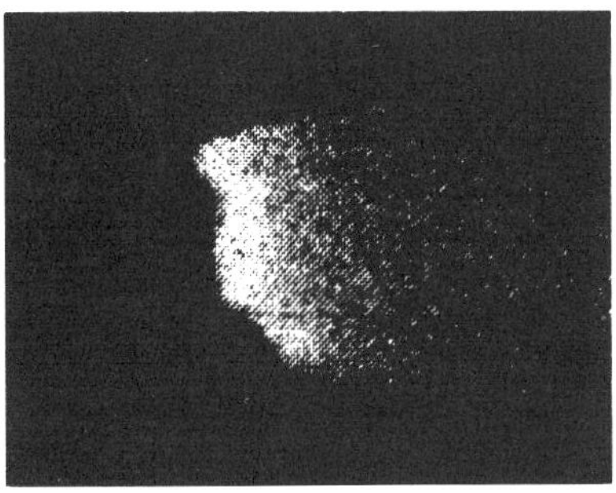

figure1a

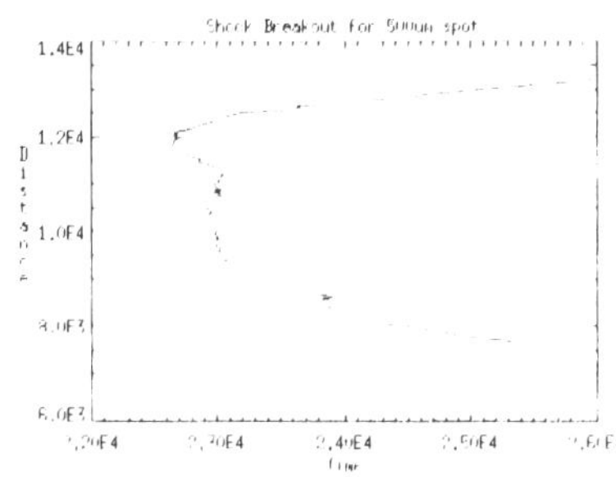

figure 1b

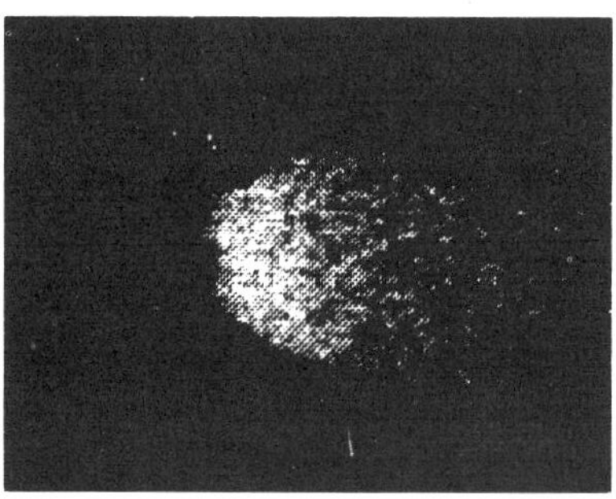

figure 2a

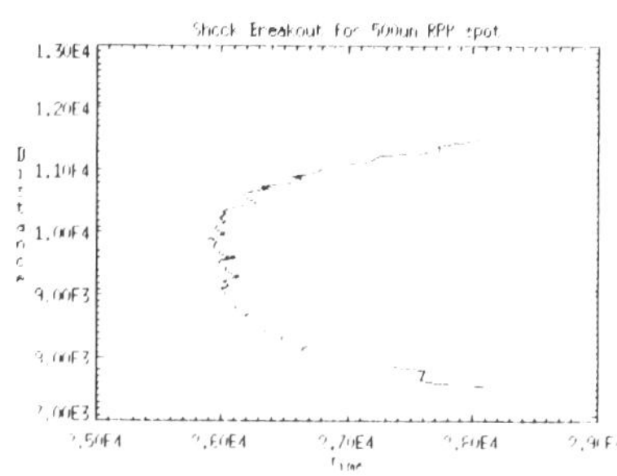

figure 2b

Since a Phased Zone Plate array to produce a 500μm laser spot was unavailable, the data for the PZP shots relates to an 850μm laser spot. The performance of a 850μm defocussed laser spot was determined and is shown in figures 3a & 3b. This shows the same general characteristics that were obtained for the 500μm defocused spot. However, the shock breakout that was produced using the PZP (figures 4a & 4b) shows an almost uniform shock front with no low spatial frequency modulations. This shock breakout using a PZP should be compared to that obtained using an RPP (figure 2).

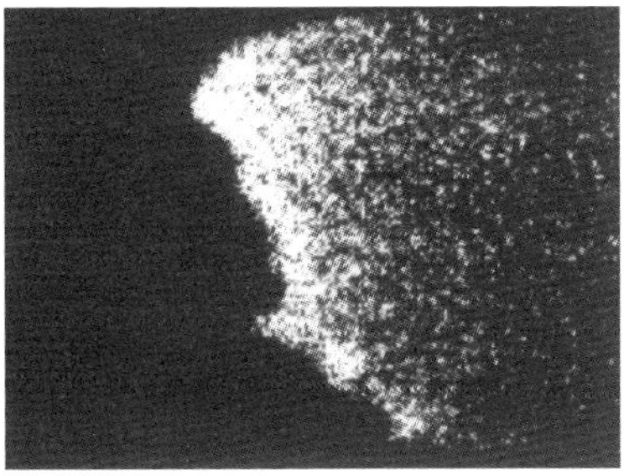

figure 3a

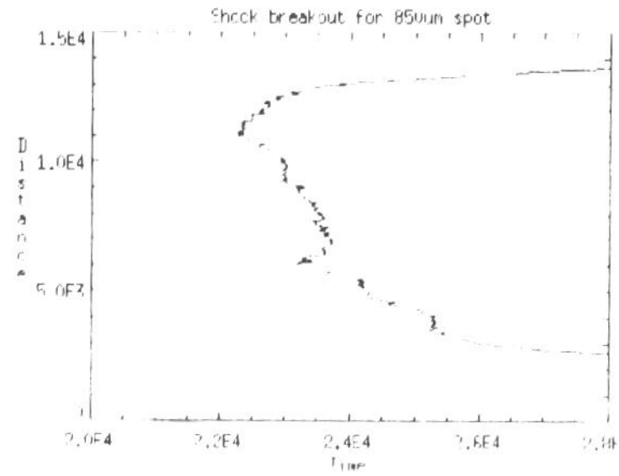

figure 3b

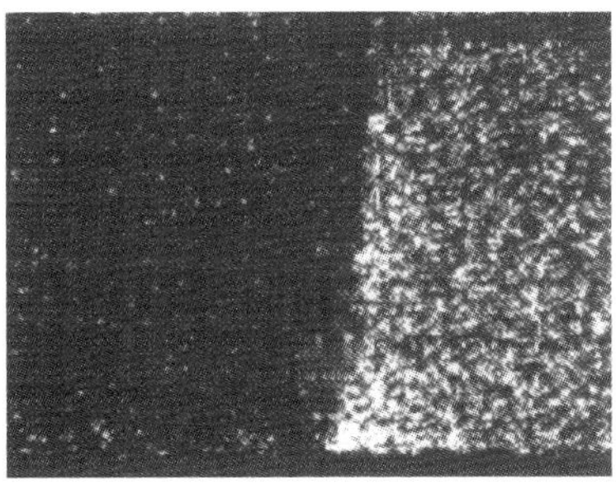

figure 4a

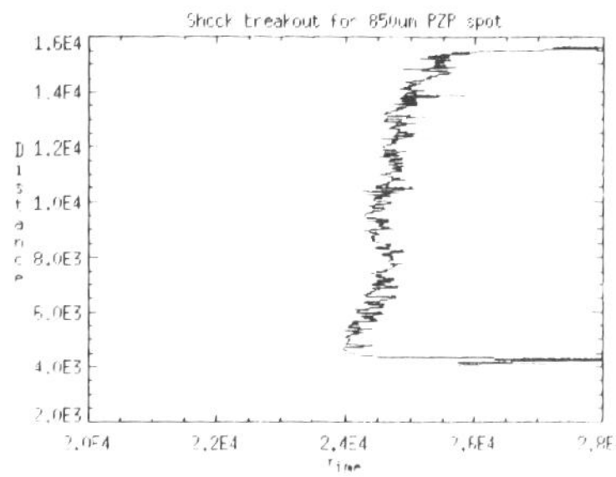

figure 4b

4. Conclusions

The use of phase plates is a well established technique for focal spot conditioning. This paper has demonstrated the production of a uniform shock breakout using a PZP array which has resulted from a four fold improvement in the incident intensity distribution. The conversion of low spatial frequencies to high spatial frequencies can be clearly seen and the improvement in performance for the production of a top-hat intensity profile using a PZP compared to an RPP is demonstrated.

References

[1] Y. Kato *et al* 1984 *Phys. Rev. Letters* **53**, 1057-1060

[2] R M Stevenson *et al* 1994 *Optics Letters* **19**, 363-365

Inst. Phys. Conf. Ser. No 140: Section 9
Paper presented at Laser Interact. with Matter Conf., Oxford, 19–23 September 1994
© 1995 IOP Publishing Ltd

Development of Low Tritium Targets for Inertial Fusion Reactor Systems

N A Tahir and D H H Hoffmann

Gesellschaft für Schwerionenforschung, Postfach 11 05 52, 64220 Darmstadt, Germany

Abstract. This paper shows how one can design high gain inertial fusion targets that use sustantially reduced tritium fraction . Use of such targets may lead to an overall reduction of tritium inventory in the reactor system which is highly desirable from the safety point of view .

1. Introduction

It is very likely that the first generation of fusion reactors may be based on a deuterium-tritium (DT) reaction scheme because this reaction is much more easy to bring about as compared to other fusion reactions including a deuterium- deuterium (DD) reaction and a deuterium-helium3 (DHe^3) reaction . This is because the former reaction has an appreciable fusion crossection at a temperature of 5 keV while the latter two reactions require temperatures of tens of keV which is too difficult to achieve with the present day technology . However two important problems are associated with a DT reactor system . First such a system would require a large tritium inventory for the daily use of tritium . Tritium is difficult to contain as it diffuses through metals very easily . Since it is a radioactive material, any relese of this substance from the system may lead to serious environmental problems.Moreover tritium does not exist in nature and has to be bred artificially which makes it expensive . It is therefore highly desirable that the use of tritium is minimized which in turn would allow us to reduce the overall tritium inventory . Second problem is the activation caused by the neutrons generated in DT reactions . This problem may be treated by designing and developing new low activation materials and will not be further discussed any further . A solution to the first roblem will be discussed in this paper .

2. Method of solution

It has been shown (Tahir and Hoffmann 1994) that it is not necessary to use equimolar DT mixture in inertial fusion targets but instead one may use a much higher percentage of deuterium than that of tritium as summarized below. The number of DT reactions taking place per unit time, per unit volume is given by

$$R_{DT} = (\sigma v)_{DT} f_D f_T n_i^2 \qquad (1)$$

where $(\sigma v)_{DT}$ is the DT reaction rate which is a function of ion temperature only. Also f_D, f_T and n_i denote fraction of deuterium ions, fraction of tritium ions and ion number density respectively. In Table 1 we give values of the product $f_D f_T$ for various combinations of f_D and f_T. It is seen that in the case of $f_D = 0.6$ and $f_T = 0.4$ the product of these two parameters is smaller by 4 % as compared to the equmolar DT case. This implies that a 20 % reduction of the tritium content in the initial design would lead to just a 4 % reduction in the output energy, provided the fuel density and temperature is kept constant. Similarly one may consider other combinations of the fractions of the two isotopes.

Table I Reduction in Tritium Inventory as a function of Deuterium-Tritium Fractions.

No.	f_D	f_T	$f_D f_T$	Reduction in $f_D \cdot f_T$	Reduction in Tritium
1	0.5	0.5	0.25		
2	0.6	0.4	0.24	4 %	20 %
3	0.7	0.3	0.21	16 %	40 %
4	0.8	0.2	0.16	36 %	60 %
5	0.9	0.1	0.09	64 %	80 %

3. Simulation results

The mathematical model described in the previous section has been implemented in our one-dimensional implosion simulations of a reactor-size radiation driven inertial fusion target (Tahir et al 1992). The results are presented in Table 2 and Fig.1. It is seen that the energy output obtained from the computer simulations is higher than that predicted by Eq.(1) for all low tritium cases. This is an interesting effect and can be explained as follows. As we reduce the tritium content in the initial design, the rate of energy production in the hot spot at ignition is reduced in accordance to Eq.(1). Cosequently the rate of temperature increase is lower than in case of equimolar DT fuel. This leads to a slower increase in the hot spot pressure. When the ignition starts in the hot spot, the bulk of the fuel around the hot spot and the tamper material is still imploding with high velocities . Because of the lower pressure in the hot spot the fuel in compressed to higher density which makes n_i higher . This increase in n_i compensates for the decrease in the value of the product $f_D f_T$ in Eq.(1) .

Table 2 Energy Output as a Function of Tritium Fraction f_T

f_D	f_T	E_{OS} [MJ] Simulation Results	E_{OA} [MJ] by Eq. 1
0.50	0.50	700	
0.60	0.40	699	672
0.65	0.35	691	637
0.70	0.30	686	588
0.75	0.25	606	525
0.80	0.20	467	448

In Table 3 we give various burn parameters which shows that smaller the tritium mass in the initial design , the higher is the tritium fractional burn . This means that a much smaller tritium mass will have to be evacuated with the target debris from the reactor chamber after each shot which will make the evacuation process much cleaner .

Table 3 Tritium Burn Parameters [Initial Fuel Mass 4.45 mg]

No.	Tritium Fraction f_T	Initial Tritium Mass (mg)	Final Tritium Mass (mg)	Tritium Fractional Burn (%)
1	0.50	2.67	1.43	46
2	0.40	2.22	0.99	55
3	0.35	1.99	0.78	61
4	0.30	1.74	0.52	70
5	0.25	1.49	0.42	72

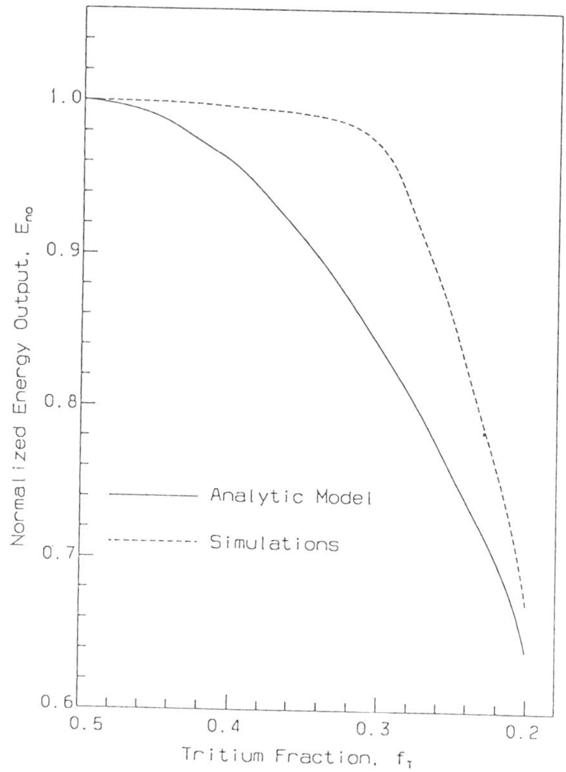

Figure 1. Normalized energy output vs tritium fraction

4. Conclusions

We have shown that it may be possible to design targets for inertial fusion reactors with tritium content as low as 25 % . This may enable us to reduce the overall tritium inventory by a factor of 2 which is an important conclusion from the safety and environmental acceptibility point of view . Moreover in such low tritium targets the tritium fractional burn is very haigh (70 %). This means that the evacuation process of the target debris from the reactor chamber will be much more cleaner as compared to the case of equimolar DT targets .

References

Tahir et al 1992 *Phys. Lett.* **172 A** 162

Tahir N A and Hoffmann D H H 1994 *Fusion Eng. Design* **24** 413

Tahir N A and Hoffmann D H H 1994 *Fusion Tech.* submitted

Inst. Phys. Conf. Ser. No 140: Section 9
Paper presented at Laser Interact. with Matter Conf., Oxford, 19–23 September 1994

© *1995 IOP Publishing Ltd*

Energy transport in plasmas produced by ~ ps krypton flouride laser pulses focused onto solid targets

G J Tallents[1], M G Holden[1], Y Al-Hadithi[1], P Fews[2], K Nazir[3], S Rose[3,4], P A Norreys[4], M H Key[4,5] and J Zhang[5]

[1]Department of Physics, Univcrsity of Essex, Colchester CO4 3SQ, UK.
[2]Department of Physics, University of Bristol, Bristol BS8 1TL, UK.
[3]Department of Physics, University of Birmingham, Birmingham B15 2TT, UK.
[4]Central Laser Facility, Rutherford Appleton Laboratory, Didcot OX11 0QX, UK.
[5]Clarendon Laboratory, University of Oxford, Oxford OX1 3PU, UK.

Abstract. The transport of energy from the target surface focal region for high irradiance (~ 10^{18} Wcm^{-2}), high contrast (~ 10^{10}), short pulse (300 fs and 12 ps) krypton flouride laser pulses focused onto solid targets has been studied. Diagnostic buried layers of sodium flouride (NaF) or iron (Fe) in the otherwise plastic (CH) targets are used to deduce axial heat penetration by recording the characteristic NaF or Fe spectra for varying overlay thicknesses, while a pinhole camera/penumbral image technique is used to search for lateral transport of heat.

1. Introduction

The transport of energy from the focal region when high power lasers are focused onto solid targets is often two-dimensional ('axially' in the direction of the laser and 'laterally' in the directions along the target surface perpendicular to the laser direction) and can occur due to several different mechanisms. For example, thermal electron conduction axially into solid targets can be significantly modified by superthermal electron and X-ray penetration ('preheat'). Superthermal electrons are produced by processes such as resonance absorption at the critical density [1] and Raman scatter at the quarter-critical density [2].

Radiation transport and heating in solid targets is now an active research area [3], but the processes can be effectively switched-off by use of low atomic number (Z) targets such as carbon for which the X-ray emission is small. Superthermal electrons are typically produced when the target irradiance I for a laser wavelength λ is such that $I\lambda^2 > 10^{14}$ - 10^{15} $Wcm^{-2}\mu m^2$ [4]. For this paper, high irradiance (~ 10^{18} Wcm^{-2}) krypton flouride laser pulses (wavelength $\approx$ 0.25 μm) were used to generate plasmas from solid plastic (CH) targets with buried diagnostic layers of either sodium fluoride or iron.

2. Method

The Rutherford Appleton Laboratory krypton flouride laser system has been used to pump a Raman process in methane gas and so produce $\approx$ 0.3 TW, very high contrast (~ 10^{10}), low divergence (3 X diffraction limited = 20 μrad) pulses of 12 ps duration at a wavelength of 0.268 μm [5] and has also been modified for chirped pulse amplification so that $\approx$ 0.8 TW pulses of 300 fs duration at a wavelength of 0.249 μm can be created [6]. With both laser systems, focused irradiances around 10^{18} Wcm^{-2} in a spot focus of $\approx$ 10 μm diameter (12 ps experiments) or ≈ 20 μm (300 fs experiment) are produced. In the 12 ps experiments a refracting lens is used to create the spot focus, while for the 300 fs experiments, the spot focus is generated using an off-axis parabolic mirror.

Energy transport from the focal spot has been studied by recording X-ray emission from tracer layers of sodium flouride (0.25 μm thick) or iron (0.2 μm thick) buried below overlay plastic of variable thickness. The laser is focused onto the overcoat and the penetration of the heat front is measured by recording the depth of overcoat for which the tracer layer can be observed to emit characteristic X-ray spectral lines or simply more emission due to its higher Z.

Flat crystal spectrometers with 10 μm thick beryllium filters (for light tightness) were used to record emission from the buried layers using crystals of beryl or KAP for the sodium and flourine emission and TLAP for the iron emission. The spatial distribution of the X-ray emission was measured using a pinhole camera with either a small pinhole (≈ 10 μm) and film recording (12 ps experiments) or a large pinhole (80 μm) and CCD recording (300 fs experiments). The shorter pulse measurements produce less overall X-ray flux and so a penumbral X-ray imaging technique [7] is used to deconvolve the X-ray pinhole images to give a final image with spatial resolution better than 5 μm. In the 300 fs experiments, two X-ray photodiodes with filtering of 10 μm and 250 μm Beryllium were used to measure X-ray emission around energies of 1 and 3 keV respectively.

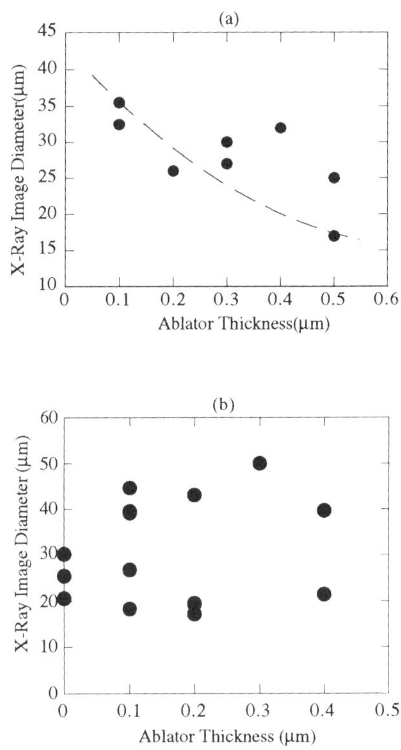

Figure 1. The measured X-ray emission diameter as a function of the depth of the signature layer of (a) NaF for the 12 ps experiments and (b) Fe for the 300 fs experiments. In (a), the focal spot diameter is ≈ 10 μm, while in (b) it is ≈ 20 μm.

3. Results

The X-ray image diameter as a function of the plastic overlay thickness is shown in figures 1 for 12 ps and 300 fs irradiation. For the 12 ps experiments, the depth of penetration of the heat front ≈ 0.5μm and there is evidence for lateral transport. For the 300 fs experiment, the depth of penetration of the heat front is ≈ 0.4 μm without significant lateral transport (the scatter in the diameters of figure 1(b) is believed to be mainly due to inadvertent target positioning outside the best focal position). The electron temperatures in the sodium flouride spectator layer were deduced using the intensity ratio of the NaXI 1s - 2p and NaX $1s^2$ - 1s2p spectral lines as outlined by Al-Hadithi et al [8] for the 12 ps experiments (see figure 2). The ratio of flourine-like to neon-like iron emission indicates a faster decrease of temperature with depth for the 300 fs experiments (figure 3).

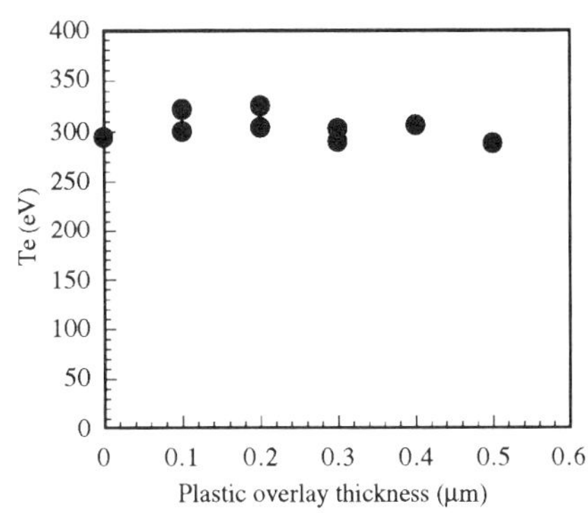

Figure 2 Electron temperature as a function of depth deduced from the intensity ratio of NaXI 1s - 2p to NaX $1s^2$ - 1s2p spectral lines emitted by a NaF signature layer for the 12 ps experiment.

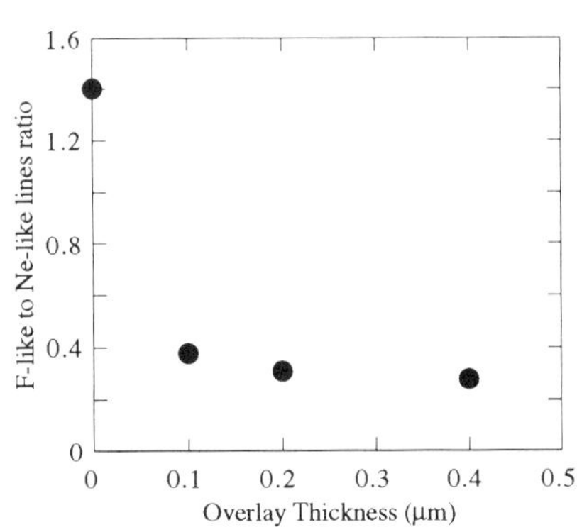

Figure 3 The ratio of the intensity of an FeXVIII (F-like) line (1.526 nm) and an FeXVII (Ne-like) line (1.2264 nm) as a function of depth of Fe signature layer for the 300 fs experiments.

4. Discussion

One-dimensional Medusa [9,10] simulations without superthermal electron production predict heat penetration into plastic of ≈ 1.6 μm ps^{-1} (assuming an absorption of 10^{17} Wcm^{-2}, a flux limiter of 0.1 and no lateral transport). For the 12 ps experiments in the absence of lateral transport, we could expect axial heat penetration to greater than 3 μm, but experimentally the heat penetration is reduced to ≈ 0.5 μm because of lateral transport. For the 300 fs experiments, the heat penetration is expected to be ≈ 0.5 μm (for a rate of 1.6 μm ps^{-1}) which is close to the measured value of ≈ 0.4μm.

The intensity contrast ratio between pre-pulse and main pulse of both the 12 ps and 300 fs duration pulses has been measured to be at least 10^{10} [5,6]. Using published mass ablation rate data [11], it can be readily shown that this level of pre-pulse will only ablate ~ 0.01 μm of the target plastic overlay and so not affect the heat penetration measurements.

Forslund, Kindel and Lee [4] have shown that a superthermal electron component of temperature (eV) ≈ $2 \times 10^{-3} (I\lambda^2)^{0.33} T_c^{0.5}$ (where T_c is the background, thermal temperature in eV) is generated by resonance absorption. For the experimental conditions presented here, we can expect a superthermal component of ~ 20 keV based on this scaling. In the 300 fs experiment, the photodiodes measuring photon energies around 1 and 3 keV produced signal levels consistent with such a temperature. Superthermal electrons of this temperature can be expected to penetrate ~ microns of plastic [12] and so modify thermal electron energy transport to enhance heat penetration. However, classical flux-limited thermal conductivity as modelled by the Medusa code [10] appears to describe the observed axial heat penetration adequately both with lateral transport (12 ps experiments) and without (300 fs experiments).

References

[1] Ginzburg V L 1970 *The propogation of EM waves in plasmas* (Pergamon: New York). p260.

[2] Barr H C, Boyd T J M and Coutts G A 1988 Phys. Rev. Lett. **60**. 1950.

[3] Edwards J, Dunne M, Riley D, Taylor R, Willi O and Rose S 1991 Phys. Rev. Lett. **67**, 3780.

[4] Forslund D W, Kindel J M and Lee K 1979 Phys. Rev. Lett. **39**, 284.

[5] Shaw M J et al 1993 Optics Lett. **18**, 1320.

[6] Ross I N et al 1994 Optics Commun.

[7] Nugent K A and Luther-Davies B 1984 Optics Comm. **49**, 393.

[8] Al-Hadithi Y, Tallents G J, Zhang J, Key M H, Norreys P A and Kodama R 1994 Phys. Plasmas **1**, 1279.

[9] Christiansen J P, Ashby D E T F and Roberts K V 1974 Comput. Phys. Commun. **7**, 271.

[10] Tallents G J, Key M H, Norreys P A, Jacoby J, Kodama R, Baldis H, Dunn J and Brown D 1992 Optics Commun. **89**, 410.

[11] Goldsack T J et al 1981 J. Phys. **D14**, L47.

[12] Hares J D, Kilkenny J D, Key M H and Lunney J G 1979 Phys. Rev. Lett. **42**, 1216.

Inst. Phys. Conf. Ser. No 140: Section 9
Paper presented at Laser Interact. with Matter Conf., Oxford, 19–23 September 1994
© *1995 IOP Publishing Ltd*

High contrast measurement of startup imprinting and induced Rayleigh-Taylor growth

R J Taylor, A Iwase, M Desselberger, O Willi

Imperial College, London, SW7 2BZ, UK

J P Dahlburg, D E Fyfe, J H Gardner

Naval Research Laboratory, Washington DC, USA

Abstract. High resolution, high contrast images have been obtained of thin plastic foils irradiated with optically smoothed laser irradiation in the first demonstration of a novel face-on XUV radiography technique. The areal density modulation induced in the foil creates a corresponding modulation in the contrast of the transmitted probe radiation, and time-resolved images hence provide a measure of the areal density perturbation corresponding to different levels of irradiation non-uniformity. The novel foam-buffered direct drive (FDD) scheme is also evaluated and found to provide superior smoothing when compared to RPP, ISI and hybrid ISI+RPP optical smoothing techniques.

1. Introduction

Laser irradiation can accelerate a thin foil target through ablation of a hot plasma from the target surface, and has received considerable attention as a possible driver for Inertial Confinement Fusion (ICF). A perfectly smooth irradiation at high intensity has proven very difficult to obtain, however, and experimental [1] and computational [2] studies have demonstrated that the most sophisticated smoothing techniques available are not sufficient to eliminate the imprinting of the laser non-uniformity during the early phases of illumination. So-called "start-up imprinting" is thought to occur when the laser is incident on the initially cold target surface, before sufficient separation has developed between critical density and the ablation surface (D_{ac}). Once D_{ac} is established, electron thermal conduction helps to smooth the drive non-uniformities [3].

Since a high acceleration is associated with an ICF implosion, and the ablation surface is unstable to the Rayleigh-Taylor (RT) instability [4], even very small perturbations induced in the early phase can grow rapidly, leading to shell breakup and an ineffective implosion. To design a successful direct-drive implosion pellet, it is necessary that the sophisticated hydrocodes already in existence [5] for this purpose are benchmarked against detailed experimental data on the imprint phenomenon. Obtaining such data requires a diagnostic technique which is sensitive to very small areal density perturbations on small spatial scales, such as are associated with the speckle pattern present in the focal spot of a typical high power laser. Conventional probing techniques are hampered by a combination of low spatial resolution (~10μm) and low contrast, making then unsuitable for the study of very small perturbations. We present here the first demonstration of a technique which at once possesses the resolution and the contrast requisite for a study of startup imprinting.

2. XUV Transmission Radiography

2.1. Experimental setup

XUV radiation (in this case, hυ=248eV) posses characteristics which make it ideal for the study outlined above. Multi-layer coatings are available with reflectivity of a few percent at this photon energy, hence allowing the use of reflective optical components for image formation. This at once has the advantage that a large entrance pupil to the imaging system can be used, hence increasing the x-ray flux relayed to the detector, and also provides us with the ability to generate high-magnification images at very high spatial resolution through the use of high-quality optical substrates, coated with the multilayer stack.

Figure 1 is a schematic of the experimental layout used for the experiment reported here, where a single, spherical multilayer mirror was used to image the plane of the target foil onto a gated XUV camera with a magnification of 70. The limiting resolution in the system was that of the detector, as demonstrated in shots where the target was replaced with a finely spaced copper mesh, leading to an overall resolution of ~1μm.

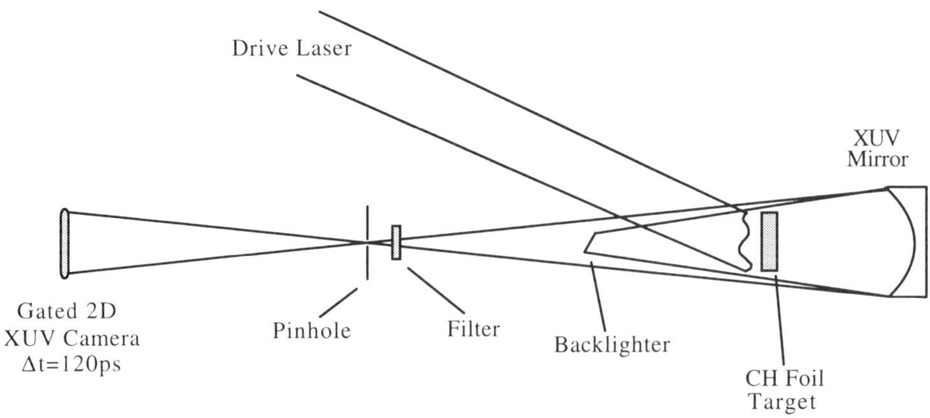

Figure 1 Experimental layout for face-on XUV radiography

The multi-layer coating is designed such that significant reflectivity is produced only in a narrow band (+/- ~5%) around the central photon energy of 248eV, with higher order reflections rejected due to the absorption and reflection coefficients of the layer materials (Platinum and Carbon). Some residual reflectivity is retained in the optical band, however, so a 1000Å Aluminium foil is used to optically isolate the detector enclosure from the target chamber itself.

The XUV backlighting is provided by a separate high-Z (in this case Gold) plasma, separated from the target by a sufficient distance such that the level of radiation incident on the target itself is negligible with respect to the hydrodynamic evolution of the system. The O-band radiation of gold, centred around ~300eV, is spectrally filtered by the multi-layer mirror to provide a monochromatic image at 248eV.

Target self-emission does not significantly affect the absorption image because the backlighter material produces a significantly higher level of conversion to x-rays in the spectral region sampled. This coupled with the spatial filtering in place at the backlighter focal plane and the time-resolution provided by the detector results in a backlighter:self-emission ratio estimated at better than 100. This was supported by null shots without a backlighter present, which failed to generate any images at all.

2.2. Diagnostic Sensitivity

The modulated areal density profile generated by startup imprinting and the subsequent RT growth produces a variation across the transmitted backlighter signal which is related to the amount of material in a given line of sight of the probe radiation, and the opacity of that material. The opacity itself is a function both of material temperature and density, and hence for a useful measurement of the areal density profile induced in the foil, it is requisite that we probe at a photon energy for which the opacity is only weakly dependent on the target condition for the experimental environment we wish to investigate.

The experiment reported here was designed to study the early phase of laser imprinting by using low irradiance levels ($\sim 5.10^{12}$W/cm^2) as would typically be used in the low-intensity foot of a direct-drive ICF pulse. CH target foils were used to represent the low-Z outer layer of a direct-drive ICF capsule.

Study of hydrocode predictions and analytical models suggest that such irradiance conditions will generate a target compression of ~3x solid density, and minimal shock heating. The accelerated, shock compressed target region is expected to remain below 5eV in temperature, whereas the ablated plasma is expected to be heated to several hundred eV at a density of less than 0.002 g/cc. The intermediate region will be small due to the localised deposition of laser energy around critical density and the consequently steep density gradient up to the ablation surface.

The ablated plasma is expected to be virtually transparent to the probe radiation, predominantly due to the low material density and high temperature, whereas the opacity of the compressed target lies around 2700cm^2/g, with the exact level dependent on the prevailing plasma conditions. Opacities predicted by coupling the IMP [6] opacity code to hydrocode simulations suggest that the opacity varies by up to 10% through the accelerated portion of the target, and so to gain detailed information as to the areal density profile in the target we must compare the measured optical depth with that predicted by various opacity codes coupled to hydrocodes.

To a first approximation, however, the sensitivity of the XUV probe can be estimated by assuming a constant target opacity at K_{cold}=2763cm^2/g (from Henke's tabulated data) and then evaluating the areal density perturbation which will generate the minimum detectable signal fluctuation. The XUV framing camera is capable of recording a ripple amplitude of ~2.5% without the signal being lost in noise, and hence through Equation 1 the minimum detectable areal density ripple amplitude ($\Delta\rho r_{min}$) is 0.09 μm g/cm^3.

$$\Delta\rho r_{min} \approx \ln((I+\Delta I_{min}) / I) / K_{cold} \qquad (1)$$

3. XUV Radiographs of startup imprinting

The technique detailed above has been implemented to obtain XUV radiographs of plastic foils (3-8μm thick CH) irradiated with the VULCAN laser (λ=0.527μm) at irradiances around 5.10^{12}W/cm^2. Images were recorded both during the shock transit (prior to the onset of acceleration and RT growth) and at later times once the imprinted seed had grown by the RT instability. The coherent beam and beams smoothed with Random Phase Plate (RPP), Induced Spatial Incoherence (ISI) and hybrid RPP+ISI techniques were used. Figure 2 shows typical radiographs for each of these irradiation conditions, and it can be seen that none of these techniques were capable of generating a uniform acceleration, even for the modest conditions prevailing here.

In contrast, however, the novel FDD [7] smoothing scheme was found to generate radiographs which were smooth to within the sensitivity of the XUV probe (see Figure 3), highlighting the promise FDD holds for direct drive ICF. The residual fractional ripple in the transmitted signal is within the noise:signal ratio of the camera, and was never seen to grow, even after the prolonged (>2ns) acceleration that was sufficient to destroy foils driven with optically smoothed beams.

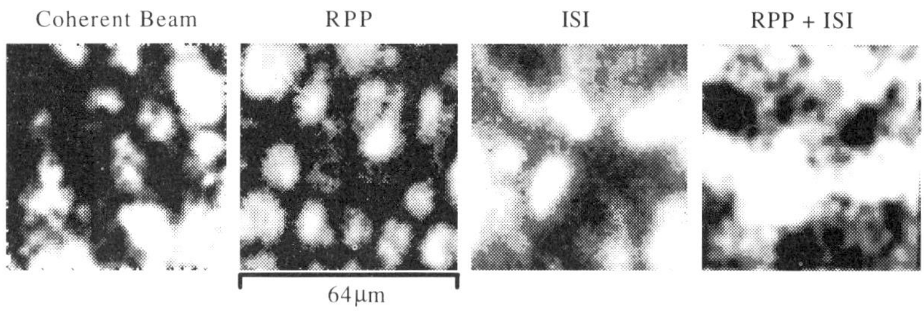

Figure 2 XUV Radiographs of CH foils driven by optically smoothed and unsmoothed laser irradiation.

FDD smoothing is achieved by placing a layer of low-density foam before the target, overcoated with a high-Z element to provide a flash of soft x-rays at early time. This supersonically pre-forms a plasma from the foam [8], providing rapid thermal smoothing of the subsequent drive, and demonstrably suppressing the laser imprint.

Figure 3 XUV Radiograph of a CH foil with FDD smoothed drive.

4. Conclusion

Detailed analysis of the imprinting data is underway, involving comparison with sophisticated, multi-dimensional hydrocodes. The extent to which conventional hydrocode physics can reproduce the data acquired will provide a valuable insight into the physical processes involved, and benchmarking these codes on actual data such as that presented here will bring confidence in our ability to successfully design a high-gain direct-drive ICF pellet. Further experimental investigation is planned to investigate the extent of imprinting produced when the target is optically isolated from the low-intensity rising edge of the laser pulse (by utilising a thin Aluminium overcoat) and to further evaluate the performance of the FDD scheme.

5. References

[1] M.Desselberger *et al.,* Phys. Rev. Lett. **68**, 1539 (1992)
[2] M.H.Emery,J.H.Gardner,R.H.Lehmberg,S.P.Obenschain, Phys. Fluids B, **3**, 2640 (1992)
[3] J.H.Gardner and S.E.Bodner, Phys. Rev. Letts. **47**, 1137 (1981)
[4] H.Takabe, K.Mima, L.Montierth and R.L.Morse, Phys. Fluids **28**, 3676 (1985)
[5] J.P.Dahlburg, J.H.Gardner, S.W.Haan & G.D.Doolen, Phys. Fluids B, **5**, 571 (1993)
[6] S.Rose, J.Phys. B, **25**, 1667 (1992)
[7] M.Desselberger *et al.,* Submitted to Phys. Rev. Lett.
[8] T.Afshar-rad *et al.,* Phys. Rev. Lett. **73**, 74 (1994)

Inst. Phys. Conf. Ser. No 140: Section 9
Paper presented at Laser Interact. with Matter Conf., Oxford, 19–23 September 1994
© *1995 IOP Publishing Ltd*

Photoablated carbon plasma in the presence of an ambient gas

R. K. Thareja, Abhilasha and R. K. Dwivedi

Department of Physics and Centre for Laser Technology, Indian Institute of Technology, Kanpur 208016, India

Abstract. A detailed investigation of photoablated carbon plasmas at low and moderate irradiances in the presence of ambient gases (Air, He, Ar) is reported. Practical aspects of pulsed laser deposition of C_{60} and DLC films are discussed. The deposited films have been characterized by Scanning Electron Microscopy, X-ray Diffraction and Raman spectroscopy.

1. Introduction

Pulsed laser ablated plasma of a material is characterized by laser parameters (laser irradiance, pulse duration and wavelength) and temperature-dependent optical (reflectivity, absorption coefficient, penetration depth) and thermophysical (heat capacity, diffusivity, thermal conductivity etc.) properties of the target material [1]. Characteristic features of the laser ablated plasma such as temperature, velocity, forward-directed nature, pressure, presence of atomic, ionic and molecular species mainly depend on laser irradiance [2]. There are many reports on studies at various laser irradiances but a few reports [3,4] are available on the comparative study of low and high laser irradiance produced plasmas. Laqua [3] explained theoretically two different cases of vaporization depending on the irradiance; lower than $10^8 W/cm^2$ and higher than $10^8 W/cm^2$.

Laser ablated plasma at low irradiance $\sim 10^8 - 10^9 W/cm^2$ is mostly used for pulsed laser deposition of thin films e.g. high T_c[5], diamond like carbon [6] and polymeric thin films [7], for microelectronic fabrication [8]; for studies of molecular species [9]; for production of other clusters [10-13]; for surface temperature studies [14] etc.. Recent applications of the laser ablated plasma at moderate irradiance $\sim 10^9 - 10^{12} W/cm^2$, include source of X-rays [15], VUV continuum [16] and highly charged ions, for generation of laser oscillations [17,18].

Here we present the optical emission diagnostics of laser ablated carbon plasma in the presence of ambient gas at low and moderate irradiance [19 - 22]. Pulsed laser deposited carbon films have been characterized by Scanning Electron Microscopy (SEM), X-ray Diffraction (XRD) and Raman spectroscopy.

2. Experiment and Results

The experimental setup used is similar to one described elsewhere [19,23].

2.1. Laser Produced Plasmas at Moderate Intensity

There has been great interest in studying the interaction between charged ions emanating from expanding laser ablated plasmas with a background gas medium. Various interaction processes such as recombination, collisional interactions, micro- instabilities etc. play an important role in line emission [24 - 26] in understanding the hydrodynamics of the expanding plasma and in exploring the possibility of laser generated controlled thermonuclear reaction.

The emission spectra of carbon plasma were recorded in the wavelength range $190 - 800nm$ at various distances away and parallel to the target surface ($z = 0 - 20mm$) at different laser irradiances and wavelengths (1.06, 0.532, 0.355 and $0.266\mu m$). Variation of electron temperature, estimated using the relative intensities of spectral lines, with intensity of laser radiation shows $I^{0.4}$ dependence consistent with theory. The electron density was estimated from the stark broadened profile of C II transition $3p^2p^O - 4s^2S$ at $392.0nm$ at various distances away from the target in vacuum and was found to be in the range $10^{16} - 10^{17}cm^{-3}$. Temporal profiles of C II transition $3d^2D - 4f^2F^O$ at $426.7nm$, C III transition $3s^3S - 3p^3P^O$ at $465.0nm$ and C IV transition $3s^2S - 3p^2P^O$ at $580.1nm$ were recorded to estimate the velocity of the plasma front. The expansion velocities of C II, C III and C IV species were found to be about $4x10^6, 7x10^6$ and $10x10^6 cm/\sec$, respectively. Studies of laser ablated carbon in the presence of helium and argon showed enhancement of line emission for the transitions $2p^2\ ^1D - 3s^1P^O$ of CI at $193.1nm$, $3p^2P^O - 5d^2D$ of C II at $213.7nm$, $3d^3D - 4f^3F^O$ of C III at $192.3nm$ and $3d^1D - 4f^1F^O$ of C III 216.3 nm. It was observed that line intensity was more in the presence of helium and argon gases than that in vacuum. An interesting feature observed is the appearance of a hump in the temporal profile of the C II transition $3d^2D - 4f^2F^O$ at $426.7nm$ at distances far away from the target viz. 12, 7 and $8mm$ for helium, argon and air, respectively. It was found that the onset distance for the double peak structure decreases with the increase in pressure and atomic mass of the ambient gas. It is proposed that at the interface region due to density fluctuations Rayleigh-Taylor instability occurs in the decelerating laser plasma front and leads to stratification of the plasma into fast and slow ion components. We found that at 0.1 Torr, mutual penetration of the laser plasma ions and the ambient gas decreases and interaction region becomes an interface. The experimental value of 0.1 Torr compares well with theoretically evaluated pressure of the ambient gases at which Rayleigh-Taylor instability occurs at the interface i.e. 0.16 Torr of air, 0.19 Torr of argon and 0.36 Torr of helium gas.

2.2. Laser Produced Plasmas at Low Intensity

At moderate laser intensities ($\sim 10^9 - 10^{12}W/cm^2$) atomic/ionic species from CV to CI which recombine through atomic to molecular carbon are observed. However, at low irradiance ($\sim 10^8 - 10^9W/cm^2$) molecular C_2 emission was found to dominate the plasma emission. C_2 emission studies are helpful to optimize the parameters for diamond like carbon film deposition, to correlate carbon clusters with the plasma dynamics and to investigate the spectroscopy of small carbon clusters. We observed the spectrum to be dominated by C_2 emission in the $\Delta v = -2, -1, 0, +1, +2$ sequences of the Swan $(d^3\pi_g - a^3\pi_u)$ bands and the $\Delta v = 0$ and $+1$ sequences of the Deslandres-d'Azambuja $(C^1\pi_g - A^1\pi_g)$ bands. The presence of helium gas enhances the intensity of emitted spectrum and of all Swan band heads. The presence of 'high pressure' band of the Swan system at $589.9nm$ originating from $v' = 6$ level suggests that the plasma is in fast recombining phase. Vibrational temperature estimated from the measured emission intensities of the Swan band heads of the sequence ($\Delta v = -1$) at various laser energies and wavelengths and under various helium pressures was found to be in the range of 6000 - 12000 K. Vibrational temperatures also increased with the decrease in laser wavelength.

2.3. Pulsed Laser Deposition of Thin Carbon Films

We have deposited the carbon films on silicon and glass substrates at $1cm$ away and parallel to the target surface at low irradiance under various helium and argon gas pressures. The deposited films were characterized using X-ray Diffraction and Scanning electron microscopy. The $X - ray$ diffraction pattern showed peaks at $10.3°, 11°, 17.9°$ and $20°$ which corresponds to (100), (002), (110) and (112) crystalline planes of C_{60}. SEM photographs of the films deposited under various helium pressures showed that the density of the microcrystalline clusters increased with the increase in helium pressure. The ablated carbon powder 'soot' collected from the chamber at different helium pressures was analyzed for the presence of fullerenes, C_{60} etc.. The carbon clusters were characterized using UV-visible absorption and IR spectroscopic technique.

The carbon films deposited in the presence of argon showed larger crystallite size on the substrate at 1 Torr of pressure. The micro Raman spectra showed two well defined characteristic peaks at $1350cm^{-1}$(D-line) and a broad peak at $1580cm^{-1}$(G-line). The relative intensity of the peaks I_D/I_G varies inversely with the size of graphitic crystallites [30]. The intensity ratio is minimum at 1 torr and increased on either side with change of argon pressure. The X-ray diffraction pattern gave the peaks at $43.5°, 76°, 91.5°$and $120°$ showing the presence of (111), (220), (311) and (400) crystalline planes of cubic diamond at 1 Torr of argon pressure.

3. Conclusions

An interesting feature observed at moderate intensity is the appearance of a hump in the temporal profile of C II transition at $426.7nm$ at distances far away from the target for He, Ar and air, respectively, which is attributed to Rayleigh - Taylor instability. The deposited films morphology and characteristics are strongly dependent on the choice and pressure of the ambient gas.

4. Acknowledgements

The work reported is partly supported by the Department of Science and Technology, New Delhi and Council of Scientific and Industrial Research, New Delhi.

References

[1] T.P. Hughes, *Plasmas and Laser Light* (Wiley, New York, (1975).
[2] R.K. Singh and J. Narayan, Phys. Rev. *B***41**, 8843(1990).
[3] K. Laqua, *Analytical Laser Spectroscopy*, N. Omenetto Ed., (Wiley, Interscience, New York, 1979).
[4] K. Dittrich and R. Wennrich, Prog. Analyt. Atom. Spect. **7**, 139 (1984).
[5] C. Champeaux, P. Marchet, J. Aubreton, J.P. Mercurio and A. Catherinot, Appl. Surf. Sci. **69**, 335 (1993).
[6] J. Seth, R. Padiyath, D.H. Rasmussen and S.V. Babu, Appl. Phys. Lett. **63**, 473 (1993).
[7] E. Dyer and R.J. Farley, J. Appl. Phys. **74**, 1442 (1993).
[8] L.J. Radziemski and D.A. Cremers, Ed. *Laser Induced Plasmas and Applications*, (Marcel Dekker, Inc., New York, 1989).
[9] D.L. Pappas, K.L. Saenger, J. Bruley, W. Krakow, J.J. Cuomo, T. Gu and R.W. Collions, J. Appl. Phys. **71**, 5675 (1992).
[10] E.A. Rohlfing, D.M. Cox, and A. Kaldor, J. Chem. Phys. **81**, 3322 (1984).
[11] U. Naher, H. Gohlich, T. Lange and T.P. Martin, Phys. Rev. Lett. **68**, 3416 (1992).
[12] R.E. Leuchtner, A.C. Harms and A.W. Castleman, Jr., J. Chem. Phys. **94**, 1093 (1991).
[13] D.E. Powers, S.G. Hansen, M.E. Geusic, D.L. Michalopopulos and R.E. Smalley, J. Chem. Phys. **78**, 2866 (1983).
[14] R.W. Dreyfus, R. Kelly, R.E. Walkup and R. Srinivasan, SPIE **710**, 46 (1986).
[15] B. Soom, H. Chen, Y. Fischer and D.D. Meyerhofer, J. Appl. Phys. **74**, 5372 (1993).
[16] G. Mehlman, D.B. Chrisey, P.G. Burkhalter, J.S. Horwitz and D.A. Newmann, J. Appl. Phys. **74**, 53 (1993).
[17] J.F. Seely and W.B. McKnight, J. Appl. Phys. **48**, 3691 (1977).
[18] C. Steden and H.J. Kunze, Phys. Lett. A **151**, 534 (1990).
[19] Abhilasha, R.K. Dwivedi and R.K. Thareja, J. Appl. Phys. **75**, 8237 (1994).
[20] Abhilasha and R.K. Thareja, Phys. Lett. A **184**, 99 (1993).
[21] R.K. Thareja and Abhilasha, J. Chem. Phys. **100,** 4019 (1994).
[22] P.S.R. Prasad, Abhilasha and R.K. Thareja, Phys. Stat. Sol.(a) **139,** *K*1(1993).
[23] Abhilasha, P.S.R. Prasad and R.K. Thareja, Phys. Rev. *E***48**, 2929(1993).
[24] J.N. Broughton and R. Fedosejevs, J. Appl. Phys. **74**, 3712 (1993).
[25] C. Timmer, S.K. Srivastava, T.E. Hall and A.F. Fucaloro, J. Appl. Phys. **70**, 1888 (1991).
[26] H.P. Gu, Q.H. Lou, N.H. Cheung, S.C. Chen, Z.Y. Wang and P.K. Lin, Appl. Phys. B **58**, 143 (1994).
[27] D. S. Knight and W. B. White, J. Mater. Res. **4**, 385(1989).

Inst. Phys. Conf. Ser. No 140: Section 9
Paper presented at Laser Interact. with Matter Conf., Oxford, 19–23 September 1994

© *1995 IOP Publishing Ltd*

Soft X-ray Emissions From Line-Focus Short-Pulse-Driven Recombining Laser Produced Plasmas

J.O. Uhomoibhi[1], C.L.S. Lewis[1], D. Neely[1], P. Norreys[2], R. Kodama[2], M.H. Key[2], G.J. Tallents[3], L. Dwivedi[3], J. Krishnan[3], J. Zhang[4], C. Smith[4], S. Ramsden[5], G.J. Pert[6], Y. Kato[7], H. Shiraga[7], K. Murai[7], G. Yuan[7]

1 Dept. of Pure and Applied Physics, Queen's University, Belfast, BT7 1NN, UK
2 Central Laser Facility, SERC Rutherford Appleton Laboratory, UK
3 Dept. of Physics, University of Essex, Colchester, CO4 3SX, UK
4 Clarendon Laboratory, University of Oxford, Oxford, OX1 3EU, UK
5 Dept. of Physics, University of Hull, Hull, HU6 7RX, UK
6 Dept. of Physics, University of York, York, YO1 5DD, UK
7 Institute of Laser Engineering, Osaka University, 2-6 Yamada-oka, Suita, Japan

Abstract The 0.268μm KrF Raman pumped SPRITE laser facility at the Rutherford Appleton Laboratory has been used to produce recombining line focus plasmas. SPRITE delivers up to 6 J in 12 psec in a single beam. Experimental studies of the sub-keV x-rays emitted from these short-pulse-driven recombining plasmas have been undertaken. Carbon fibre targets of up to 6.0 mm length have been used to produce plasmas which emitted soft x-rays in the range 50 - 124 eV. Time-integrated, space-resolved and time-resolved, space-integrated measurements of these emissions have been made with grazing incidence flat field spectrometers set at different angles of observation relative to the surface of the carbon fibre targets. The spatial and temporal intensity distributions of the carbon hydrogen-like Balmer-α and Balmer-β transitions, 182.17Å and 134.95Å respectively were investigated as a function of target length. Although modelling predicts peak gain-length products for the current conditions of as high as 10 we have not been able to unambiguously observe amplification on the Balmer-α line at 182.17Å with the data recorded.

1. Introduction

The availability of high power, short pulse laser systems has made possible the extension of picosecond pulse generation into the x-ray region of the spectrum [1-5]. A laser is focused onto the surface of a solid to produce a high temperature x-ray emitting plasma. For a short laser pulse with a rapid rising edge, the energy can be absorbed by the electrons at solid densities in an optical skin depth, before significant ablation of the solid occurs [1]. Rapid cooling of these hot electrons occurs as a result of rapid thermal conduction into the underlying cold material, electron energy loss to ions, and expansion into the surrounding vacuum [1,2,6,7,8]. This leads to rapid recombination in the plasma, thus terminating the x-ray emission. X-ray generation and the spectrum depend on many parameters of both plasmas and lasers. Those parameters which may significantly affect the x-ray generation efficiency include laser wavelength, laser intensity, pulse width, plasma density scale-length, and target materials [9,10,11,12,13] .

In the present work we review extensive studies of soft x-rays in the range of 50 - 124eV generated from line-focus short-pulse-driven recombining laser produced plasmas performed at the SPRITE laser facility at the Rutherford Appleton Laboratory.

2. Experiment

The KrF Raman laser facility, SPRITE, gives the possibility of very high focused intensity in a very narrow line foci by combining its short wavelength with low beam divergence and high beam intensity. The Sprite system provides a beam combining Raman amplifier which generates a final beam of 8cm × 8cm aperture. This generates 0.4 TW, 12 psec pulses with 20μrad beam divergence. The current performance of SPRITE (up to 6J in 10ps) with output beam intensity ~7GW/cm^2 leads to a two photon absorption 50% in 2cm of fused silica or 5% in 2cm of CaF_2. In the absence of CaF_2 focusing lenses, reflective focusing optics incorporating an off-axis paraboloid (OAP) mirror and an off-axis spherical mirror (OAS) were used to generate a line focus in the existing target chamber. The OAP mirror was F/3 with Ø110 aperture and an off-axis angle of 15°. Its surface accuracy was λ/4 in the visible and was coated for 100% reflection at both 249 and 268nm. The alignment procedure established the correct tilt for the OAP and thus provided an aberration-free focus. A HeNe beam expanded and aligned to the KrF beam was used for alignment and focusing onto target. The design of the line focus optics scheme utilised the OAP in series with a tilted off-axis spherical mirror (OAS) and with minor adjustments to the geometry and input beam diameter the line focus length could be varied from 4 to 10mm. A diagrammatic representation of the experimental arrangement of the target, the laser irradiation geometry, the laser and target alignment set up and the diagnostics/observation systems is shown in Fig. (1). In Fig. (1), M_1, M_2, M_3, M_4 and M_5 are alignment mirrors, I_1 and I_2 are iris/apertures, O_M is a microscope objective lens, L_c is a collimator lens, OAP is the off-axis paraboloid mirror, OAS is the tilted off-axis spherical mirror designed for obtaining the desired length of line focus at the centre of the target chamber, SSC is a single slit imaging camera, XSC is a double or cross slit imaging camera and all the other terms have usual meaning as defined in the text. The pointing of the carbon fibre targets was aligned to an accuracy better than ± 1 mrad relative to the spectrometer axis and positioning of the targets and focal line to better than ± 2μm was achieved using a split field microscope system. To increase the x-ray flux into the axial spectrometer, a toroidally shaped grazing incidence mirror was used to image the fibre end onto the entrance slit of the spectrometer in the dispersion direction and onto the spectrometer focal plane in the orthogonal direction. With this arrangement, the angular detection range from the fibre ends was ~ ± 2.3 mrad. Separate measurements have been made at angles directly in line with the fibre and at 4mrad to the fibre axis.

The heated plasma was monitored using active single slit and double slit cameras with a phosphor screen-image intensifier assembly. All of the data was recorded on individually calibrated HP5 films. The near- and far-field patterns of the SPRITE laser beam were monitored using two CCD cameras. Flat-field VUV spectrometers were used as spectroscopic diagnostics and were operated in both time integrated and time resolving modes. One of these (the axial spectrometer) observed axial emission from the free end of the fibres and used a 1200l/mm grating. The other (the transverse spectrometer) observed transverse emission of the resonance lines at 45° relative to the target axis and used a 2400l/mm grating. The axial spectral emission detection system comprised the grazing incidence spectrometer coupled to a streak camera (HAMAMATSU C4575) and a CCD read-out system (HAMAMATSU C3610) and was used to record x-ray streak images with a sweep speed at 35.2 ps/mm. The schematic of the grazing incidence spectrometer is shown is Fig. (2).

The toroidal mirror had curvatures of 5000mm and 26mm in the tangential plane and the sagittal plane, respectively. The incidence angle to the mirror was 88°. Its tangential curvature made a focus on the 300μm wide entrance slit along the vertical direction. Its sagittal curvature enabled us to form a space resolved image along the horizontal direction

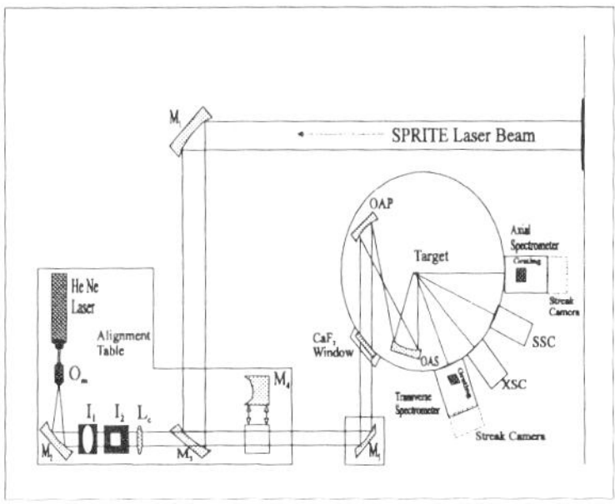

Fig. (1a) Experimental arrangement.

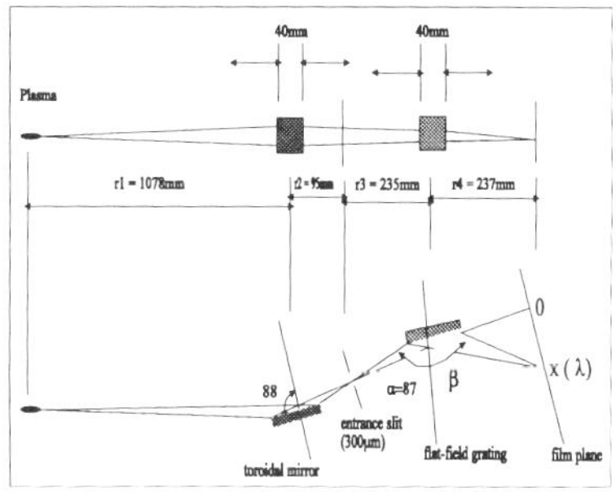

Fig. (1b) Diagram of set-up showing top view (upper) and side view (lower) of the axial spectrometer.

on the spectrometer focus plane. The magnification of the system due to this toroidal mirror was 0.53. The acceptance angle of the spectrometer along the horizontal direction and along the vertical direction were determined by the toroidal mirror and by the grating, respectively. The horizontal acceptance angle was 4.6mrad because of the effective width of the toroidal mirror which was about 5mm. The vertical acceptance angle of the grating was 0.8mrad, which is smaller than 1.3mrad, as determined by the mirror. The streak camera was attached to the spectrometer with its slit cathode parallel to the dispersion plane (vertical direction). During the experiment Kodak 10107 film was used to record space-resolved time-integrated spectra and for time resolved measurements, the streak camera was used instead of the film. The cathode material was CsI and the streak data were obtained with a 650μm wide and 10mm long cathode, evaporated on 50Å aluminium coated on a 1000Å CH substrate. From a time integrated shot obtained with the streak camera still attached to the axial spectrometer, the spatial width, (FWHM) for the carbon VI Hα, Hβ lines and continuum emission indicated a temporal resolution of ≈ 45 psec.

3. Results and Discussion

All photographic recordings were digitised using Applied Imaging (Joyce Loebl) MK6 microdensitometer and the images processed using Visual Numeric' PV-WAVE data analysis software on a DEC-Alpha Ultrix workstation. The optical densities above fog of the film were related back to the intensities of the radiation producing the exposure and the distances on the film in the dispersing direction were converted to wavelength or photon energy using an analysis program. Film calibrations for Kodak 10107 XUV film were incorporated in the program following Henke et al [14,15]. For Kodak 10107, the optical density, D, is related to, I, the incident flux, I, and is given by the thin emulsion formula [16],

$$D = A[1 - \exp(-BI)] \qquad (1\text{-}1)$$

where, A, is the saturation density of the film and B is a constant. Values of A and B were obtained from reference [17]. Typically A ~ 2.63 and B ~ 0.22, where I is in photons/μm^2.

Fig. (2) shows typical observed soft x-ray emission lines obtained from measurements made with the axial spectrometer operated in both the time-integrated and time-resolving modes.

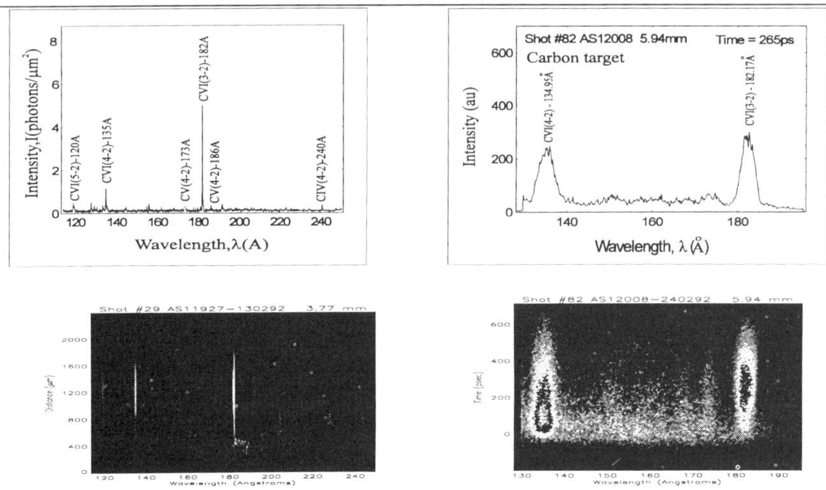

(2a)Time integrated (3.77mm plasma)   (2b)Time resolved (5.94mm plasma)

Fig.(2) Axial spectra of (a) 3.77mm and (b) 5.94mm carbon plasmas, showing identified emission lines assigned after refs. [17,18]. In (a) Spatial distances are referred to fibre (object) plane. In (b) Time is referenced to peak of continuum.

Intensity variations of the identified transition lines for shots constituting the time integrated and time resolved data sets of varying lengths of plasmas utilising carbon fibre targets are shown in Figs. 3(a and b).

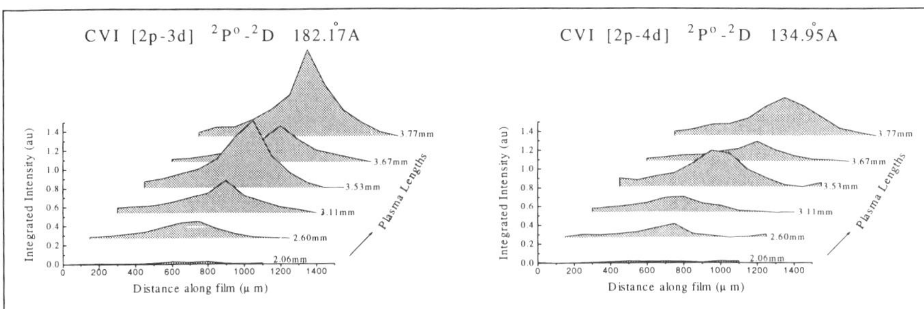

Fig (3a) Intensity variation with distance along film for time integrated CVI 182.17Å and 134.95Å transitions for different carbon plasma lengths. The axial spectrometer was pointing at 2mrad relative to the target axis for these shots

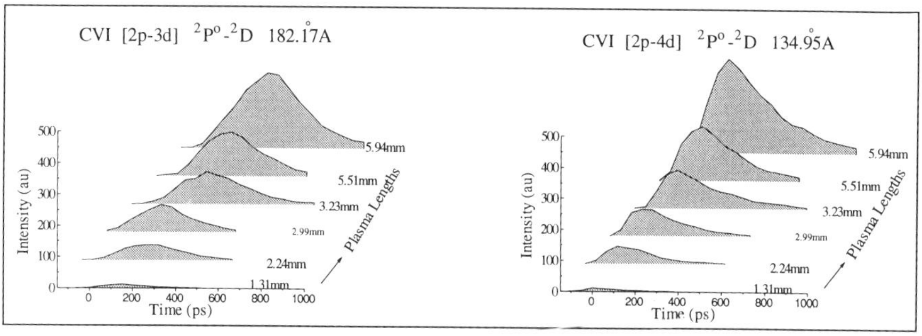

Fig (3b) Intensity variation with time for time resolved CVI 182.17Å and 134.95Å transitions for different carbon plasma lengths. The axial spectrometer was set at 4mrad relative to the target axis.

Intensity versus length growth curves for CVI Balmer-α and Balmer-β are plotted in Fig. 4a) and 4b). Here we have integrated the total signals in each data set for comparison and consistency checks between the two data sets. That is, in each case the "integrated" signal is integrated over time, space and spectral line profile.

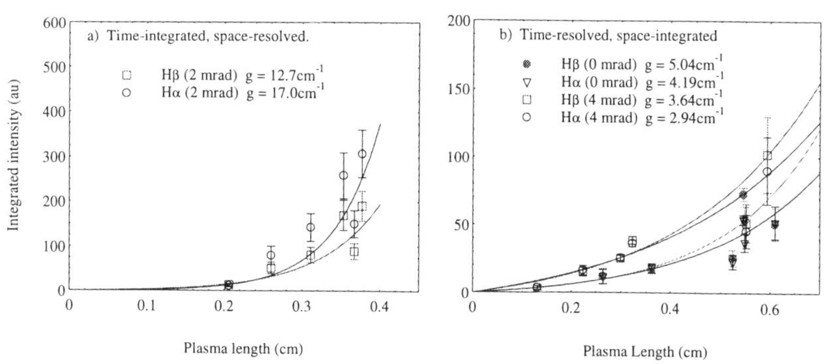

Fig 4 Integrated intensity growth data as a function of plasma length for (a) space and (b) time-resolved data sets and fits to exponential growth functions of the form $I = (\varepsilon/g)\{\exp(gl)-1\}$ where ε is emissivity and g is a gain coefficient.

From fits to exponential growth curves, apparent "gain" values of 17.0cm^{-1} (2mrad), 2.94cm^{-1} (4mrad) and 4.19cm^{-1} (0mrad) were deduced for the Hα line and apparent "gain" values of 12.7cm^{-1} (2mrad), 3.64cm^{-1} (4mrad) and 5.04cm^{-1} (0mrad) were deduced for the Hβ line. Broadly the two data sets are consistent but clearly both are suggesting non-physical effects in that the Hβ line cannot exhibit such gain.

To improve the observations/analysis, space and/or time resolved measurements are necessary to enhance the signal-to-noise ratio of the detected signal. Such measurements provide localised space and time intensity information about lasing transitions (the signal) against a relatively bright background emission of fluorescent radiation from the plasma on the same transitions (the noise). For example, typical growth curves of localised intensity in time at 150 ± 25ps for the Hα, Hβ and nearby continuum emission for the on-axis (0mrad) and off-axis(4 mrad) data sets are shown in Fig. (5). In these cases apparent "gain" values of

2.99 cm^{-1}, 5.26 cm^{-1} and 3.10 cm^{-1} for the on-axis case, and 2.06 cm^{-1}, 4.16 cm^{-1} and 3.30 cm^{-1} for the off-axis case were obtained for the Hα, Hβ and nearby continuum emission respectively. Similar conclusions are made at any other time window. Thus, even growth curves at fixed relative times manifest no strong differences between apparent growth of the Hα, Hβ and nearby continuum emissions.

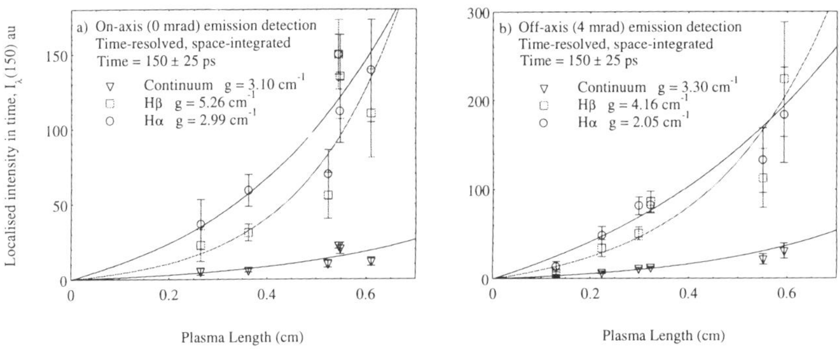

Fig.5 Time resolved intensity growth curves as a function of plasma length for emission lines obtained from (a) on-axis and (b) off-axis spectrometer measurements for the Hα, Hβ and nearby continuum emission.

We observe similar contradictions when using growth curves at fixed relative positions from the spatially resolved data sets. The overall conclusion is that we cannot claim to have unambiguously observe gain on the Hα line of CVI under these conditions.

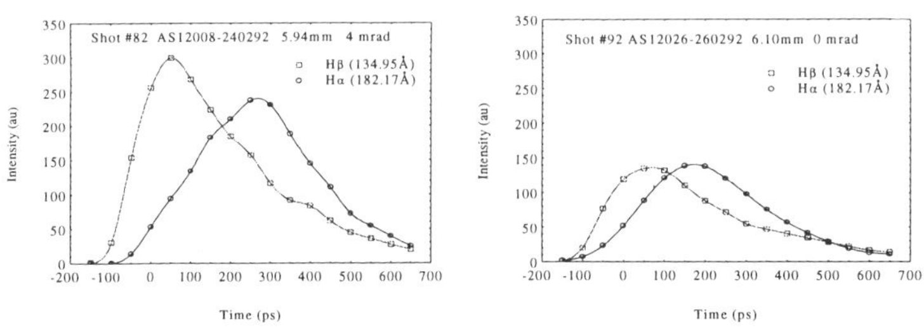

6a) Off-axis (4 mrad) measurements 6b) On-axis (0 mrad) measurements

Fig. (6) CVI Hα and Hβ intensity variation with time for axial spectrometer on- and off-axis measurements.

Intensity plots for the CVI Hα and Hβ transitions versus time for 5.94 mm and 6.10 mm plasmas obtained for spectrometer off-axis (4 mrad) and on-axis (0 mrad) settings are displayed in Figs. (6a) and (6b) respectively. Such plots provide information on the relative angular dependence of the intensity of emissions from the Hα and Hβ lines. In these shots (that is, in Fig. 6) the Hα emission appears to become relatively brighter at larger angles as time progresses and it also has a different temporal signature than the Hβ line, peaking typically 150 ± 50 psec later. Nevertheless we were not able to observe any real systematic directional effects for the growth curves of the Hα line. Again we conclude that no strong evidence for Hα amplification has been observed and that neither refraction nor very polar outputs are responsible for not having observed exponential growth.

4. Conclusion

We have observed Balmer series emissions from carbon fibre targets irradiated with 12 psec KrF laser pulses at an average incident intensity of ~5 x 10^{14} W/cm^2. We have measured line strengths as a function of time, space and emission angle and plotted curves of growth for up to 6 mm long fibres. Curves of growth for the Hα and Hβ lines as well as for continuum radiation in the same spectral region indicate in each case a gain coefficient of similar positive magnitude. This must indicate that the data is statistically unreliable and reinforces our belief that all such experiments with effectively low gain-length products require large numbers of shots to clearly show any true exponential growth trends. We have also attempted to couple radiation back into the plasma with an x-ray mirror but without any observed relative enhancement of the Hα line. Therefore we must conclude that from the data sets available we have not unambiguously demonstrated amplification on the Hα transition under the conditions of the present experiment.

REFERENCES

[1] R.W. Falcone, et al, in Short Wavelength Coherent Radiation: Generation and Applications, D.T. Attwood and J. Bokor, eds. (AIP 147, New York, 1986).
[2] O.L. Louden, et al, Opt. Comm. 63, 253 (1987).
[3] D.G. Stearns, et al, Phys. Rev. A 37, 1684 (1988).
[4] H.M. Milchberg, et al, in Advances in Laser Science II, M. Lapp,et al, eds. (AIP New York, 1987).
[5] C.H. Nam, et al, Phys. Rev. Lett. 59, 2427 (1987).
[6] S.E. Harris, et al, Phys. Rev. Lett.
[7] A. Caruso, et al, Plasma Phys. 11, 839 (1969).
[8] D.J. Bradley, et al, Opt. Comm. 15, 231 (1975).
[9] T. Mochizuki, et al, Phys. Rev., A33, 525 (1986).
[10] K. Eidmann, T. Kishimoto, Appl. Phys. Lett., 49, 377 (1986).
[11] P. Alaterre, et al, Phys. Rev., A34, 4184 (1986).
[12] R. Fedosejevs, et al, Proc. SPIE, 831, 54 (1987).
[13] M. Chaker, et al, Proc. SPIE, 831, 237 (1987).
[14] B.L. Henke, et al, J. Opt. Soc. Am. B3 (1986) pp. 1540.
[15] B.L. Henke, et al, J. Opt. Soc. Am. B1 (1984) pp. 828.
[16] J. Krishnan et al, Ann. Rept. to the LFC, SERC, RAL-92-020, 1992, pp.22-23.
[17] R.L. Kelly, Atomic and Ionic Spectrum Lines below 2000Å, J Phys. Chem. Ref. Data 16,Suppl. No. 1-3 (1987)
[18] S. Bashkin et al, Atomic Energy Levels and Grotrian Diagrams, Vol.1 [Hydrogen I-Phorphorus XV], North Holland Publishing Company, Amsterdam (1975), pp. 56-99.

Inst. Phys. Conf. Ser. No 140: Section 9
Paper presented at Laser Interact. with Matter Conf., Oxford, 19–23 September 1994
© *1995 IOP Publishing Ltd*

Calculation of target compression with account for the plasma self-radiation

G A Vergunova and V B Rozanov

Lebedev Physics Institute, Leninsky prospect, 53, 117924 Moscow, Russia

Abstract. Influense of self x-ray radiation on compression of complex spherical laser target was investigated. Protected layers were made for variation of plasma parameters and increasing of neutron yield.

We investigate the compression of a composite target under the action of the laser radiation penetrating into a small-density foamlike medium between an inner shell of a target containing a deuterium-tritium ice and an external heavy shell. A typical scheme of such a target is shown in Fig.1.

The target is subjected to a linearly growing radiation pulse of a Nd-laser, pulse duration $3\,ns$, total energy $10^5\,J$, and the wavelength $1.06\,mcm$.

We have studied a possibility of the radiation control in different parts of the complex target, and especially, the preheating of the DT layers.

The laser radiation penetrates into region V (we do not discuss here the way of the energy input into the target in region V). The energy is transferred into a polyethylene layer (region IV) by the heat conductivity wave and the radiative wave. The polyethylene layer plays the role of an ablator, and at the same time it protects the inner layers of the target from the soft X-radiation. A copper interlayer (region III) is placed in the interior of the DT matter (regions I and II). This interlayer protects an inner part of the target against a preheating by the hard X-ray emission. It is also possible to add high Z material into polyethylene layer (region IV). On the outside, this construction is covered by a heavy copper shell (region VII) with a deposited CH layer (region VI).

The target compression was calculated with taking into account x-ray radiation of the plasma by making use of the RADIAN code [1]. The code is based on a physical-mathematical model which includes the absorption of an external laser radiation, a system of equations describing the two-temperature gas dynamics, thermonuclear burning and the α-particle transfer. The radiation field is determined by a quasistationary transfer equation.

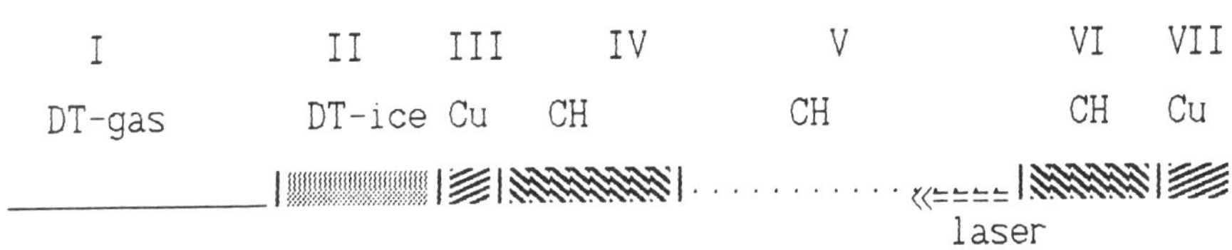

Figure 1.

Much computer time is required for making numerical calculation of the radiative gas problem with the calculation of the transfer equation. In order to make these expenses less we have employed the well known methods of the radiating gas dynamics calculation, that is the multi-group approximation, the characterization method, the photon energy averaging, the quasidiffusion method.

We used different quantitites of the spectral groups and widths of the studied spectrum, different values of the foam density, and the size of the target in order to obtain maximum neutron yield. Some parameters of one of the studied targets are given below: $R_{IV} = 0.10153\,cm$, $R_{VII} = 0.15373\,cm$.

One of the complex problems is the determination of the spectral optical coefficients. The optical coefficients for DT, CH and Cu were calculated preliminarily [2], and then interpolated with due regard for the given spatial-time profiles of the density and temperature.

Figure 2 illustrates r-t diagram of the target compression, where the collapse occurs at the moment $t \simeq 5.3\,ns$. The temperature and density of DT-gas and DT-ice are increasing in the process of compression. The maximum electron temperature of DT-ice is $10\,keV$ ($t = 5.3\,ns$); the maximum density of DT-ice is $\rho = 78\,g/cm^3 (t = 5.2\,ns)$; the neutron yield is 10^{17}, which corresponds to the gain of 2.4 in respect to the energy absorption.

At the beginning of the laser radiation interaction the density of region V is below critical, and the laser radiation is absorbed by the foam and the narrow internal part of the polyethylene which adjoins to the foam. One ns after the extenal shell started to expand and the inner target started to implode the foam temperature increases up to $1.5\,keV$ and the density up to $5\,10^{-3}\,g/cm^3$. Within subsequent moments of time the foam temperature is set at $\simeq 1.5 - 2.5\,keV$, and the density at $(1-3)\,10^{-2}\,g/cm^3$. The laser radiation will occur not throughout the entire region IV, but in its external part.

The energy is transferred from the laser absorption region to the inner target (IV) and to the CH external protection layer (VI) by means of the electron heat conductivity wave (see Fig.3). The reemission regions are formed (one region is in the internal compressive target, and another in the external shell). In the process of compression the temperature is varied in the range $T_e \simeq 1 - 5\,keV$ and the density in the range $(3-10)\rho_V$, which is higher than those of the laser absorption region V.

The radiative wave goes from the reemission region into the deeper layers of matter. At an initial moment, at $t = 1\,ns$ (Fig.3), this radiative wave is lower than the thermal wave, but it heats up the deeper layers of matter. As soon as the input energy increases, for example, by the time $t = 4\,ns$, the radiative wave becomes more intensive (Fig.4). The region which is heated up by the thermal wave becomes considerably larger, but it still does not penetrate into the inner layer of copper. The flow of the electron heat conductivity is considerably decreased, whereas the flow of the x-ray radiation increases and penetrates inside the copper layer.

The radiative spectra for different spatial patterns by the moment $t = 1\,ns$ and $t = 4\,ns$ are illustrated in Fig.4. At $1^{st}\,ns$ the soft x-ray radiation of region V heats internal layer of region IV and the harder x-ray radiation goes from region IV to the region V (curve IV-V). X-ray radiation from polyethylene penetrates into copper layer (curve III-IV). This radiation is absorbed by copper and does not penetrate in the DT. At $4^{th}\,ns$ x-ray radiation is more intensive, and copper layer protects DT from this radiation.

The calculations made for different thicknesses of the external shell (regions VI,VII)

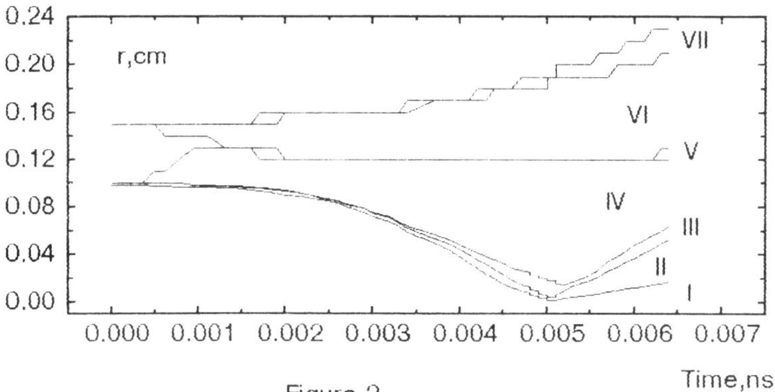

Figure 2

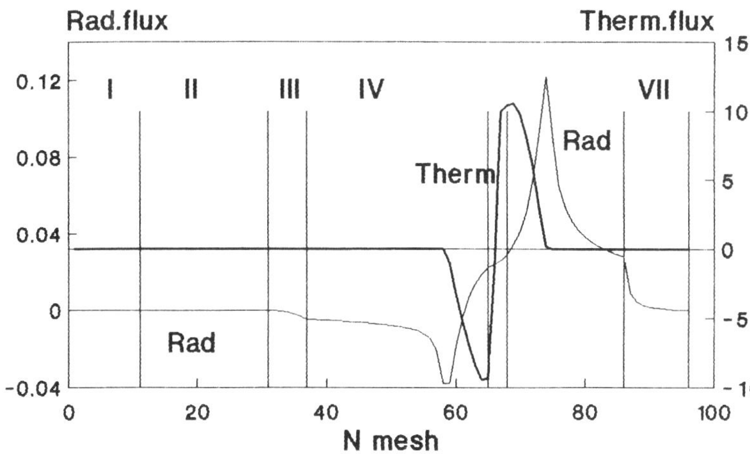

Figure 3.

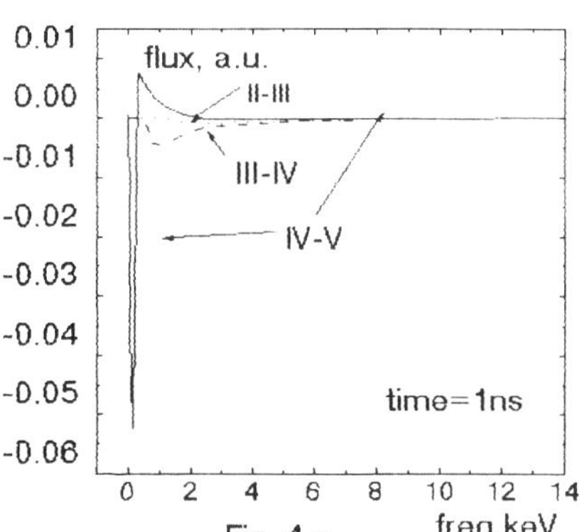

Fig. 4,a.

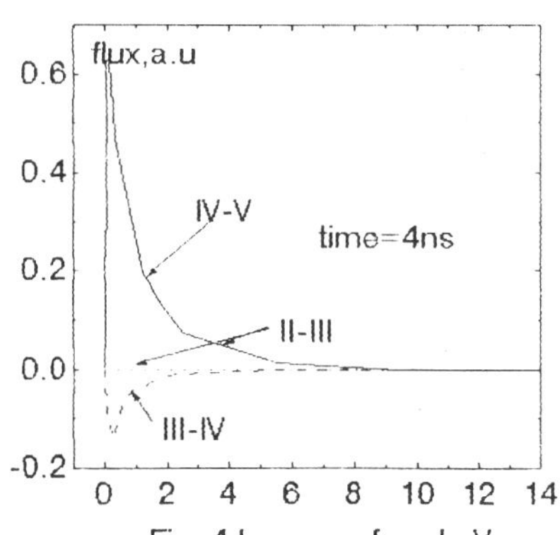

Fig. 4,b.

have shown that the shell should be rather thick in order to keep the internal target from fast expansion. But starting from some quantity of about $4\,mcm$ the thickness does not influence essentially the neutron yield.

We can follow the influence of the copper layer thickness (region III) and the ablator thickness (region IV) on the level of the radiation and the level of entropy in DT layers (regions I and II). The neutron yield has maximum at $\delta_{III} = 1.6\,mcm$ of the copper layer in this target. An increase in the neutron yield can be explained by a decrease in the entropy of DT layers due to a decrease in the radiation level. Let us note that the radiation in our case is substantially nonequilibrium. The plasma temperature in the absorption region is rather high, it achieves $2\,keV$, and the effective temperature of the radiation achieves $\simeq 50 - 100\,keV$.

References

[1] Govorun T K et al 1986 Preprint Inst. of Appl. Math N176
[2] Novikov V G 1990 Plasma physics 16 77-79 (in Russian)

Inst. Phys. Conf. Ser. No 140: Section 9
Paper presented at Laser Interact. with Matter Conf., Oxford, 19–23 September 1994
© 1995 IOP Publishing Ltd

> 10^{19} Wcm^{-2} irradiances for laser/plasma interaction studies

F N Walsh[1], J R M Barr[2], L J Barzanti[1], C N Danson[1], M D Ebbage[1], C B Edwards[1], M J Gander[1], D C Hanna[2], D W Hughes[2], M H R Hutchinson[3], M H Key[1], A A Majdabadi[2], I P Mercer[3], D Neely[1], P A Norreys[1], D A Pepler[1], S A Rivers[1], P F Taday[1], W T Toner[1], T B Winstone[1] and F Zhou[3]

[1] *Central Laser Facility, Rutherford Appleton Laboratory, Chilton, Didcot, Oxon. OX11 0QX*
[2] *Opto-electronics Research Centre, University of Southampton, Southampton, Hants. SO9 5NN*
[3] *Blackett Laboratory, Imperial College, Prince Consort Road, London. SW7 2BZ*

Abstract. Peak irradiances in excess of 10^{19} Wcm^{-2} have been delivered to target with the VULCAN Nd:glass laser using chirped pulse amplification (CPA) [1]. Recent developments have achieved sub picosecond pulses with compressed output powers of 35 TW in a single large aperture beam, with focused intensities on target in excess of 10^{19} Wcm^{-2}.

1. Laser system

The input pulse for the system was generated from a novel diode-pumped oscillator [2] providing pulses of 420 fs duration. The oscillator is based on additive pulse modelocking [3] (APM) of Nd:LMA. Nd:LMA was chosen as the active medium since: it has a large bandwidth (4.4 nm centred at 1055 nm); is suitable for amplification in phosphate glass amplifiers; and its absorption bands are at 800 nm enabling the device to be end pumped by a 3W laser diode. The shortest pulse duration achieved by the oscillator was 420 fs (assuming a sech^2 pulse shape), with a time-bandwidth product of 0.32 indicating that the pulses were near transform limited. The average output power from the modelocked oscillator was 115 mW at a repetition rate of 87 MHz , giving a single pulse energy of 1.3 nJ.

The output of the oscillator was injected into a double passed grating pair to produce a stretched pulse duration of ~200 ps. The gratings used were blazed holographic with 1740 lines mm^{-1}, gold coated and operated at an input angle of 73 degrees.

A schematic of the laser amplification system used is shown in figure 1. Three double passed 9 mm rod preamplifiers were used to enhance the oscillator pulse energy from 1 nJ to 1 mJ. It was then injected into the standard VULCAN amplfier chain consisting of rod and disc amplifiers separated by Faraday and Pockels cell optical isolators and air and vacuum spatial filters (ASFs and VSFs respectively) to maintain beam quality.

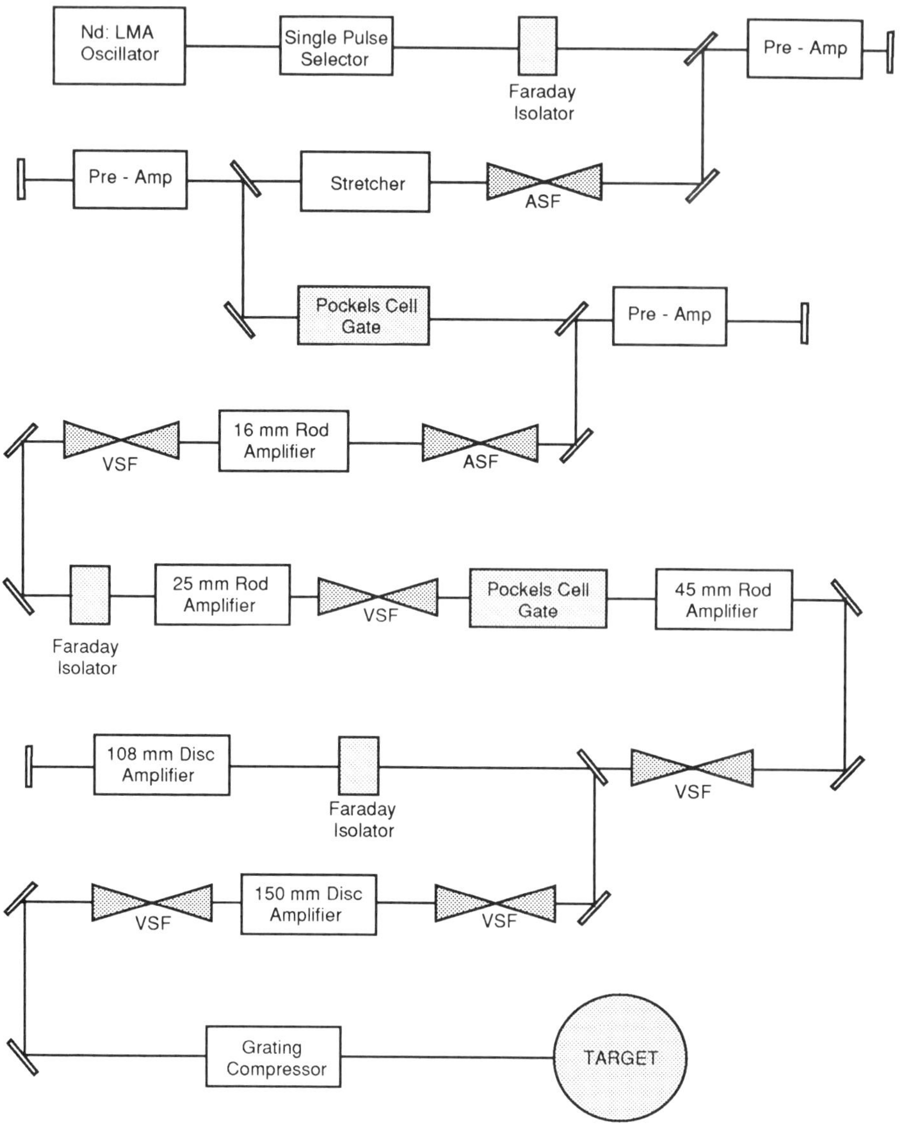

Figure 1. Schematic of the VULCAN Nd:glass laser chain configured for CPA operations.

Previous work [3] indicated that the fidelity of the recompression of the pulse was limited by self-phase modulation (effectively measured by the total B-integral) generated in non-linear media as the pulse passes through the system. It was essential to reduce these effects in order to produce the shortest possible pulse. The greater bandwidth associated with the 420 fs input pulse duration, compared to the original system, gave an increase in the stretched pulse duration to 200 ps, reducing the total B-integral by a factor of 3. The total amplification required in the amplifier chain to produce a given energy was minimised by ensuring good quality anti-reflection coatings on all transmissive optics. This together with limiting the thickness of chamber windows etc limited the B-integral to <1.

2. Compressed pulse diagnostics

The alignment of the large aperture diffraction gratings was optimised using a dual laser wavelength alignment technique [4] which produced a high quality focal spot of less than 3 times the diffraction limit. The focal spot size was measured both before and after grating alignment with low energy shots using a long focal length (10 m) equivalent plane monitor. Full energy shots verified the performance of the compressor, with a focal spot of less than 3 x the diffraction limit (see figure 2) in both the vertical and horizontal planes.

The grating separation was optimised by taking a series of test shots at low energy, and measuring the compressed pulse duration with a single shot second order auto-correlator. The gratings were set to the separation corresponding to the minimum pulse duration.

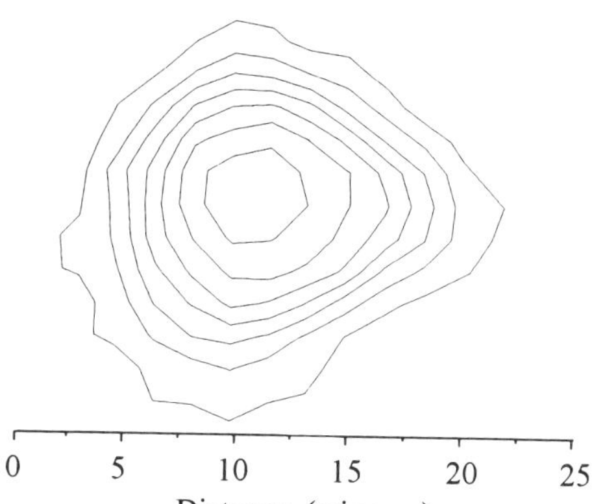

Figure 2. Iso density plot of the far-field image in an equivalent plane monitor. The axis has been scaled to show the expected distribution that would be obtained on target.

The pulse width and spectral data for high power shots are shown in figure 3. The data has been sampled from various experiments performed using the CPA system during the past year. The large scatter is due to slight changes in alignment between the stretcher and compressor gratings. Despite this scatter the data is consistent with an assumed $sech^2$ pulse shape (ie. a time-bandwidth product of 0.315). The incident energy onto the gratings for one particular shot (shot number 21281093) was 44 J, with losses due to diffraction efficiency and mirror reflectivity at ~ 50%, giving an energy on target of 22 J. The pulse width was measured at 620+/-100 fs ($sech^2$ profile), and the bandwidth of this pulse was 1.53 nm , giving a delivered laser power to target of 35 TW. The reduced bandwidth to target is mainly due to gain narrowing in the phosphate glass amplifiers. Using the measured focal spot size of ~15 μm diameter for the shot, obtained from an equivalent plane monitor, gives an irradiance on target of $>10^{19}$ Wcm^{-2}.

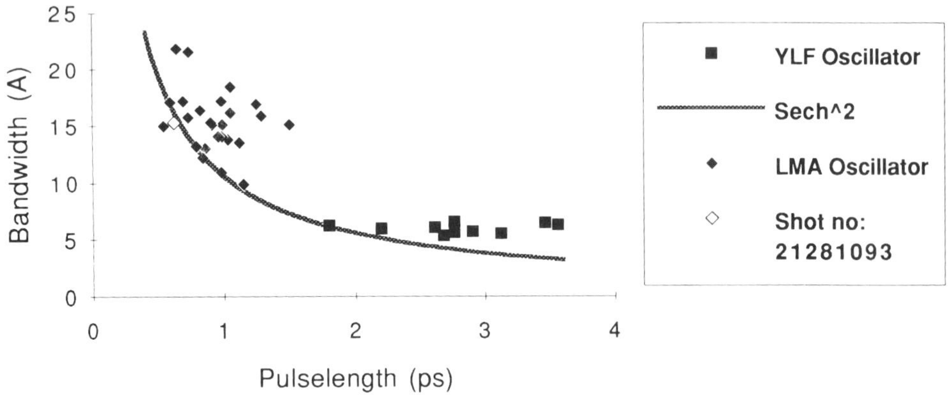

Figure 3. Plot of measured bandwidth and pulselength of the compressed pulse on full energy target shots

3. Laser / plasma interaction experiments

These pulses have been used in a series of laser/plasma interaction experiments to study: recombination x-ray lasers [5]; solid target interactions [6][7]; gas target interactions [8]; and particle acceleration techniques [9]. These facilities will open up new areas of investigations into areas such as short pulse ignitors for Inertial Confinement Fusion [10], particle acceleration by the laser wake field [11], relativistic self focussing [12], filamentation and optically field ionised plasmas and large amplitude plasma waves [13].

References

[1] P Maine et al, 1988 SPIE Vol 913, 140
[2] DW Hughes et al, 1993 OSA Topical Meeting on Advanced Solid State Lasers, paper AMC3, New Orleans, USA
[3] J M Liu and J K Chee, 1990 Optics Lett. **15** 685
[4] C N Danson et al, in press
[5] J Zhang et al, 1994 accepted to Phys Rev Lett.
[6] AR Bell et al, 1993 Phys Rev E **48** 2087-2093
[7] AP Fews et al, 1994 Phys Rev Lett **73** 1801-1804
[8] A MacKinnon et al, 1994 Proc. of ECLIM, Oxford
[9] Z Nadjmudin et al, 1994 Presented at ECLIM, Oxford
[10] M Tabak et al, 1994 Phys Plasmas 1 (5), 1626-1633
[11] P Chen et al, 1985 Phys. Rev. Lett. **54** 693
[12] P Sprangle et al, 1987 IEEE Trans on Plasma Science **15**(2), 145
[13] W L Kruer and S C Wilks, 1992 Plasma Physics and Controlled Fusion **34**(13), 2061

Inst. Phys. Conf. Ser. No 140: Section 9
Paper presented at Laser Interact. with Matter Conf., Oxford, 19–23 September 1994
© *1995 US Government*

Enhanced-Brightness, Short-Pulse X-rays Lasers

A. S. Wan, R. C. Cauble, L. B. Da Silva, J. C. Moreno, J. Nilsen

Lawrence Livermore National Laboratory, Livermore, California 94550, U. S. A.

Abstract. We are developing short-pulsed, enhanced-brightness, and coherent x-ray lasers (XRLs) for applications in areas such as plasma imaging. In a traveling wave pump setup the optical laser creating the XRL plasma sweeps along the lasant axis at the same speed as the x-rays. This technique becomes increasingly important as the target length increases and the gain duration shortens. An order of magnitude increase in output energy has been measured usinf the traveling wave pump. Using multiple pulse techniques and multilayer mirrors to inject the output of one pulse back into the plasma formed by a later pulse we have begun to develop the x-ray analog of a multi-pass amplifier. The use of multiple pulses separated by as much as 1.6 ns reduces multilayer mirror damage. This injection technique is demonstrated by imaging the near-field emission profiles of the XRL. The addition of multilayer beamsplitter will allow us to effectively produce a soft XRL cavity.

1. Introduction

With its short wavelength, controllable pulse duration, high brightness and coherence, the x-ray laser (XRL) is ideally suited as a diagnostic probe to image rapidly evolving (<1 ns) laser-driven plasmas with high electron densities (10^{21} cm^{-3} < n_e < 10^{24} cm^{-3}). [1] Over the past several years, we have developed a number of XRL applications to probe laser-produced plasmas on the Nova 2-beam facility at Lawrence Livermore National Laboratory (LLNL), such as radiography, [1] Moiré deflectometry, [1] and 2-D interferometry. [2] Experimentally we use one beam to set up the target plasma to be imaged, which leaves us one beam to use to generate the XRL. For current plasma imaging applications we typically use a 3-cm-long exploding-foil yttrium XRL [3] with an output energy of ~8 mJ, 200-ps FWHM pulse, ~100 μm diameter source size, 10-mrad divergence, and bandwidth ($\lambda/\Delta\lambda$) of 10^{-4}, which corresponds to a brightness of ~10^{17} W/sr-Å-cm^2. This brightness is equivalent to a 6-GeV blackbody, which overwhelms the self-emission of the target plasma.

Although the yttrium XRL has sufficient coherence and brightness for most current imaging applications, we still need to pursue a shorter-pulse, enhanced-coherence, and higher-brightness XRL to achieve better spatial resolution and to move toward 3-D holographic applications. Typical laser-produced plasmas have a blowoff velocity of ~10^7 cm/s. Thus, an XRL with a 200-ps FWHM pulse duration yields a spatial resolution of ~20 μm. To obtain higher spatial resolution images of the plasmas, we need to shorten the XRL pulse by another order of magnitude. At LLNL we are developing a more suitable XRL for plasma imaging using two approaches: traveling wave excitation and a short-pulse XRL cavity.

2. Traveling Wave Excitation

Traveling wave excitation is configured by tilting the wave front of the driving laser to match the propagation velocity of the XRL photons. This technique for exciting the plasma can be an effective method of increasing the effective gain length to extract the maximum energy out of the XRL. This technique has been employed in various lasers to shorten the pulse duration and increase the amplification [4-5]. Here we have demonstrated the application of the traveling wave technique [6] to neon-like germanium XRLs using 3-cm-long slab targets.

The laser pulse consisted of three 100-ps gaussian pulses separated by 400 ps peak to peak. Each pulse had an intensity on target of 1.1×10^{14} W/cm^2 at a laser wavelength of 0.53 μm (herein denoted as 2ω). Our diagnostics consisted of a 1-meter grazing incidence spectrograph (McPigs) with a microchannel plate detector viewing one end of the XRL axis and a streaked flat field spectrograph viewing the opposite end of the XRL.

We compared laser irradiation of germanium slab targets for three different wave front angles: first with the traveling wave going toward the detector; second with no traveling wave; and third with the traveling wave going away from the detector. We observe a factor of ten increase in XRL line intensity compared to no traveling wave and a factor of seventy increase compared to having the traveling wave go the opposite direction. A simple numerical model, including transit time effects, was developed to simulate the effect of the transit time of the XRL photons on the integrated output line intensity. In Fig. 1 we compare the measured integrated intensities of the J=0–1 line of neon-like germanium at 196 Å, taken with the McPigs spectrograph, to this numerical model. The points show the measured data while the lines are from the numerical model. There is good agreement between the numerical model and measurements.

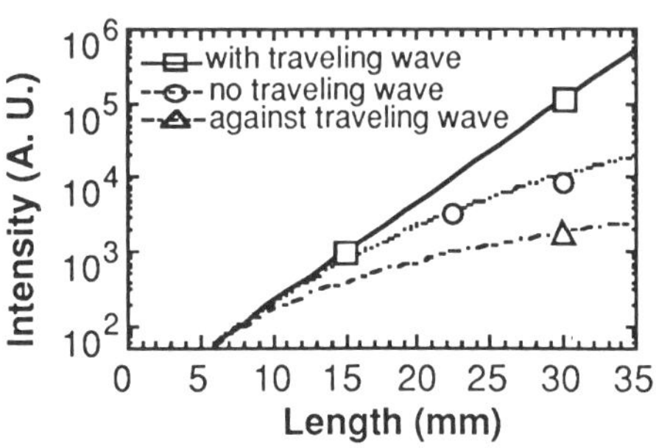

FIGURE 1. Intensity of the J=0–1 line plotted vs length. The curves are from the numerical model while the points are experimental data.

3. Development of an X-Ray Laser Cavity

Our approach on cavity development is to place the injection mirror far away from the XRL to minimize multilayer damage and to use the multiple-pulse configuration on Nova, timed such that the XRL photons produced in earlier pulses will be reinjected and propagate through the gain medium created by latter pulses. We verify the successful locking by measuring the near-field emission profiles of the XRL plasma, from which we can also estimate the spatial gain and n_e profiles.

Our primary diagnostic is the XUV Imaging Diagnostic, [2] with a magnification of 20 resulting in a spatial resolution of order of 3 μm. We measure the near-field emission profile of the J = 2–1 yttrium laser line at 155 Å, for a 1-cm-long, 100-μm-thick yttrium-slab target. The driving Nova pulse shape is a multiple-pulse configuration with four 600-ps gaussian pulses at ~150 TW/cm^2 peak incident intensity at 2ω. The temporal peak-to-peak separation between these four pulses is 1.6 ns, which allows the photons generated by the first pulse to reinject into the gain medium created by the third pulse, and the second pulse to inject into the fourth pulse. We observed little evidence of injection mirror damage during our experiments.

We used both available beams to drive two sides (designated as east and west sides) of the yttrium-slab XRL. The injection mirror is aligned such that the east-side XRL can reinject while the west-side XRL will not couple. By taking the west-side XRL as a reference, the output intensity of the east-side XRL is a factor of six or more stronger. This enhanced intensity agrees well with our estimates and demonstrates the successful coupling between the pulses due to reinjection. Closer examination of the images of the near-field emission profiles reveal several clues to the properties of this multiple-pulse XRL configuration. The peak emission regions for all four pulses peak at ~100 μm off the target surface, and they stay at approximately the same positions for all four pulses with no appreciable shifts in position. We also observe some residual emission at 155 Å after the pulses, moving away from the target surface with increasing time.

For collisionally dominated plasmas we can model the XRL plasma in stages and decouple the detailed level populations and line transfer physics from the rest of the problem. We use LASNEX [7] to carry out the laser-deposition and hydrodynamics simulations. From LASNEX we obtain plasma characteristics, such as T_e and n_e, and variables such as mesh positions and velocities. Using our atomic kinetic post-processing code, GLF, [8] we calculate the level populations (and therefore the small signal gains) and line transfer. With the spatial gain and n_e profiles, we can perform ray propagation calculations to match the simulations with actual observable quantities such as laser output intensities. For this purpose we use SPECTRE, which calculates time-dependent ray trajectories, and BEAM3, [9] which is a 3-D Monte-Carlo ray propagation code to study the effect of refraction and gain saturation.

We use a 1-D wedge geometry to represent the 2-D hydrodynamic expansion of the laser plasma. An on-axis image of the XRL plasma [10] shows a wedge-shaped expansion of order of 20 degrees, which is consistent with 2-D LASNEX simulations. Each driving laser pulses heats the plasma to a T_e of ~1.5–1.6 keV and burns through slightly beyond an averaged ionization state of neon-like. Between pulses the plasma expands and cools to ~200-300 eV. Given the long time (1.6 ns) between pulses, by the time the next pulse hits the target, we observe a large n_e reduction. Therefore each pulse behaves as if it incidents a new slab target, without significant perturbation by the pre-formed plasma in front of the slab, and the plasma behavior reproduces pulse after pulse, which is in agreement with our experimental observation with spatially fixed emission profiles.

The GLF calculations show strongly peaked gain profiles, with gain >20 cm^{-1} near the slab surface. The effect of radiation trapping, obtained by GLF line transfer calculation, reduces the peak gain to ~13 cm^{-1}. With the reproducible plasma parameters for every pulse, the gain profiles also stay fixed at these positions, in agreement with observation. We then use SPECTRE and BEAM3 to study the role of refractive ray propagation on the near-field emission profiles. Rays near the slab traverse across large n_e gradients and are refracted out of the gain region, while rays launched further away from the slab surface face smaller n_e gradients and they refract by about 100 μm as they traverse across a 1-cm region. This 100-μm displacement is in excellent agreement with our experimental observations.

With the low T_e between each driving pulse, the plasma recombines past the neon-like configuration. At these temperatures, the collisional excitation rate is low and we observe a factor-of-2 increase in recombination flux from fluorine-like ions to the $3p_{3/2}$ upper laser state. The calculated gains between pulses are weak, by about an order of magnitude, and the spatial gain profiles move away from the slab surface with increasing time due to changes in the ionization balance as the plasma cools and expands. These features are also in agreement with our experimental observation and further strengthen our model assumptions.

To produce a short-pulse, enhanced-coherence XRL cavity, we need to further reduce the driving pulse width. Using multiple 100-ps laser pulses and traveling wave technique we have previously been able to generate 45-ps yttrium XRL pulses at 155 Å. [1] We will focus our effort on the continual development of the short-pulse XRL cavity in the future.

4. Summary

This paper describes our experimental and theoretical studies of the development of short-pulse, enhanced-brightness XRLs. We observed an order of magnitude enhancement of the neon-like germanium XRL intensity when the traveling wave technique is used to compensate for the short gain duration of the XRL plasma and long target length. We have also begun to develop a short-pulse XRL cavity, and have obtained good agreement between the measured near-field emission profiles and the simulations using a combination of LASNEX/GLF /SPECTRE-BEAM3. In our simulations we have demonstrated the importance of properly modeling the effect of hydrodynamic expansion, atomic physics, and refractive ray propagation. Beyond the 2-D interferometer, a diffraction-limited XRL can yield 3-D holographic images of ICF capsules. To improve laser coherence, for a fixed wavelength, we need to either increase the XRL length [11,12] or to reduce the XRL aperture. [13]

Acknowledgment

Work performed under the auspices of the U. S. DOE by LLNL under contract number W-7405-ENG-48 and is partially supported by the Institute Sponsored Research Program.

References

[1] Da Silva L B *et al* 1993 *Proc. Ultrashort Wavelength II, SPIE* **2012**, 158, Suckewer S, Ed.

[2] Da Silva L B *et al* 1994 to be published in *X-Ray Lasers 1994*, Eder D C and Matthews D L, Eds.

[3] Da Silva L B *et al* 1993 *Opt. Lett.* **18**, 1174–1176

[4] Barty C P J *et al* 1988 *Phys. Rev. Lett.* **61**, 2201-2203

[5] Sher M H, Macklin J J, Young J F, and Harris S E 1987 *Opt. Lett.* **12**, 891-893

[6] Moreno J C *et al* 1994 to be published in *X-Ray Lasers 1994*, Eder D C and Matthews D L, Eds.

[7] Zimmerman G B *et al* 1975 *Com. Plasma Phys. and Cont. Fus.* **2**, 51

[8] Scott H A and Mayle R W 1994 *Applied Physics B* **58**, 35–43

[9] Maenchen G private communications

[10] Seely J F *et al* 1994 *Physics of Plasmas* **1** 6, 19972002

[11] Kato Y *et al* 1993 *Proc. Ultrashort Wavelength II, SPIE* **2012**, 12, Suckewer S, Ed.

[12] Lewis C L S *et al* 1992 *Opt. Commun.* **91**, 71–76

[13] Wan A S *et al* 1994 *Optical Engineering* **33** 7, 2434–2441

Inst. Phys. Conf. Ser. No 140: Section 9
Paper presented at Laser Interact. with Matter Conf., Oxford, 19–23 September 1994

© 1995 IOP Publishing Ltd

Indirectly Driven Targets for Ignition

D. C. Wilson, and William J. Krauser

Los Alamos National Laboratory, Los Alamos, New Mexico 87545 USA

Abstract. Both Los Alamos and Lawrence Livermore Laboratories have studied capsule and laser driven target designs for the National Ignition Facility. Our current hohlraum design is a 2.76mm radius, 9.5mm long gold cylinder with 1.39mm radius laser entrance holes covered by 1μm thick plastic foils. Laser beams strike the inside cylinder wall from two separate cones with a peak power less than 400 TW. The problem with a pressure pulse caused by wall plasma stagnating on axis has been overcome by filling the hohlraum with gas. Currently this is equi-molar hydrogen-helium gas at 0.83 mg/cc density. One capsule uses a 160μm plastic ablator doped with oxygen and bromine surrounding an 80μm thick DT ice layer with an inner radius of 0.87 mm. Los Alamos integrated calculations of the hohlraum and this capsule using 1.4 MJ of laser energy achieve yields of 4.9 MJ using LTE atomic physics, and 3.5 MJ with non-LTE. This confirms Livermore calculations of ignition. For radiation driven implosions, a beryllium ablator offers a viable alternative to plastic. It is strong enough to contain high DT pressures. Copper, soluble at required levels, is an excellent dopant to add opacity. A beryllium capsule with a 155 μm thick ablator doped with 0.9 atom % copper, and the same inner dimensions as the plastic capsule, placed in a similar hohlraum , yields 6.9 MJ with LTE. Although these calculations show the designs are sensitive, they add to our confidence that NIF can achieve ignition. Using our best integrated calculations which are not yet fully optimized, we confirm Livermore calculations of ignition with a plastic capsule, and have added an alternate capsule design with a beryllium ablator.

1. Introduction

The near-term goal of the U.S. Inertial Confinement Fusion (ICF) Program is to demonstrate ignition and moderate gain in a proposed next-generation laser, the National Ignition Facility (NIF). Of the many possible target designs, we have focussed our work on two different capsules, one using a plastic ablator, and the other a beryllium ablator, placed into a common hohlraum with a few variations. We have integrated the hohlraum and capsule calculations into a single two dimensional radiation hydrodynamics calculation, using the laser light input. We calculate the laser absorption, x-ray radiation, material motion, energy transport to the capsule, the capsule implosion, and fusion burn. This comprehensive and computationally intensive technique was developed over the last two years. Previously a hohlraum might be calculated alone, the angularly and temporally dependent radiation fluxes tabulated, and then linked to a second calculation of the capsule implosion.

These integrated calculations use our best numerical packages for hydrodynamics, radiation transport, atomic physics, and thermonuclear burn. They include effects such as the refraction, reflection and absorption of laser light passing through the spatially and temporally varying laser entrance hole and hohlraum plasma. They include the filling of the hohlraum with plasma, the motion of the laser absorption region, the time dependent emission of X-rays from the laser heated plasma, their absorption and reradiation, and final absorption onto the capsule, the spatial and temporal variation of the capsule ablation, its implosion, and thermonuclear burn. But our integrated calculations are incomplete. The calculation of colliding low density plasmas is lagrangian, and does not model plasma interpenetration. Particle-In-Cell simulations may help understand this phenomenon. Laser plasma instabilities such as Brillouin and Raman scattering are not modeled. Capsule mixing and yield degradation due to short scale length hydrodynamic instability growth must be modeled separately.

To date our integrated calculations of one plastic capsule and one beryllium capsule in NIF hohlraums have ignited. They have also shown us how to modify the laser target for better performance.

2. The Hohlraum

Our colleagues at Lawrence Livermore Laboratory designed the basic hohlraum, a 9.5 mm long by 2.76 mm radius gold cylinder, 40 μm thick, with 1.39 mm radius laser entrance holes. The NIF lasers were simulated by two cones of beams. We added a 10 μm thick plastic liner around the lip of the entrance hole to help keep it open. Our first integrated calculations used a 2.5μm thick polystyrene (CH) lining on the gold hohlraum walls with the expectation that it would slow the expansion of the gold wall into the hohlraum center. However we found that the plastic stagnated on axis and created a pressure pulse, whose added impulse caused the capsule axis to implode early, and the capsule to fail to ignite.

Our solution was to remove the plastic liner and add gas with a density of approximately 0.001 g/cc. Since the capsule must be maintained at approximately 18 K, the gas could be only hydrogen or helium, or a mixture of the two. Our most recent design uses 0.000833 g/cc of an equi-molar mix. This mixture was chosen to reduce Brillouin scattering based on theory and experiments presented in other papers at this conference. To contain the gas, 1 μm thick plastic windows were added over the entrance holes. With these changes stagnation pressures were reduced to the point where they no longer degraded the capsule implosion.

3. Plastic Capsule

In the center of this hohlraum we placed a 1.11 mm radius spherical capsule. In its center, out to 0.87 mm is DT gas at an equilibrium density of 0.0003 g/cc. From 0.87 mm to 0.95 mm is solid DT, 0.25 g/cc. From 0.95 mm to 1.11 mm is a doped plastic ablator of density 1.05 g/cc with an atomic composition of 0.475 C12, 0.475 H1, 0.05 O16, and 0.00025 Br. A 1D calculation of this capsule implosion with the radiation temperature drive of figure 1 gives a yield of 14 MJ.

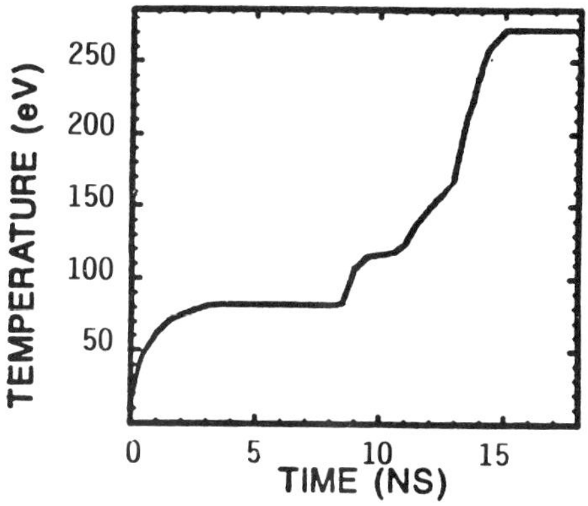

Figure 1. Temperature drive - plastic capsule

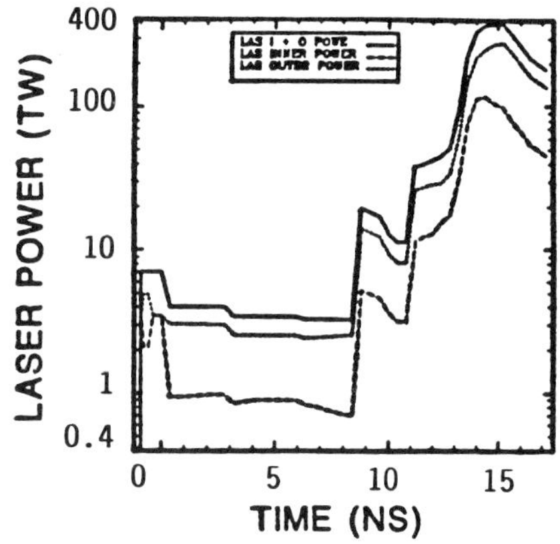

Figure 2. Laser power drive - plastic capsule

For the 2D integrated calculation the time dependent power in two laser beam cones and the spatial pointing of those beams had to be specified. Figure 2 shows a typical power history of the inner and outer beam cones, and the total. The total laser power was adjusted to drive a capsule implosion similar to the 1D calculation. Finally the time dependent split between the inner and outer cone powers was adjusted to minimize the time dependent P2 asymmetry in the radiation flux. The pointing of the cones was adjusted to minimize the time averaged P4 asymmetry. Our best calculation using an LTE assumption in the atomic physics calculations of the opacity and EOS gives 4.9 MJ yield with an input laser energy of 1.37 MJ.

Our non-LTE atomic physics package attempts to model the time dependent level populations, particularly in the laser irradiated plasma. This change causes less radiation emission from the laser deposition regions, and more plasma motion. The radiation source for the capsule moves, and therefore a new beam balance is needed. Our best non-LTE calculation to date produces a yield of 3.5 MJ with 1.32 MJ of laser energy. In no calculation is the full 1.8 MJ of expected laser energy used. We allow for unexpected losses of laser light and some expected loss from Brillouin scattering.

Throughout this iteration process the capsule imploded asymmetrically, but produced some yield. Table 1 shows the yield and asymmetry of the capsule as the power balance, beam pointing, and total energy were changed. It shows that repointing the beams can reduce asymmetry, and with increased laser energy, can increase the yield. The yield degradation seems due to the asymmetry causing the hot spot temperature to rise later in the implosion. Adequate fuel ρr is always achieved. Adding extra laser energy compensates by producing a higher implosion velocity, and an earlier rise in the hot spot temperature.

Table 1. The effects of asymmetry, energy, and pointing on yield

Yield (MJ)	Asymmetry (Pole/Equator)	Laser Energy (MJ)	Outer Beam Pointing (μm)
0.013	7	1.12	0
0.060	6	1.12	0
0.20	5	1.12	+250
0.47	3	1.12	+250
1.3	6	1.37	+250
4.9	3.5	1.37	+250

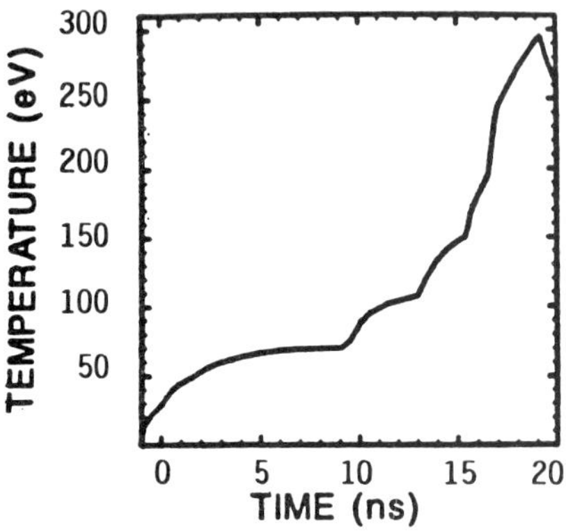

Figure 3. Temperature drive - beryllium capsule

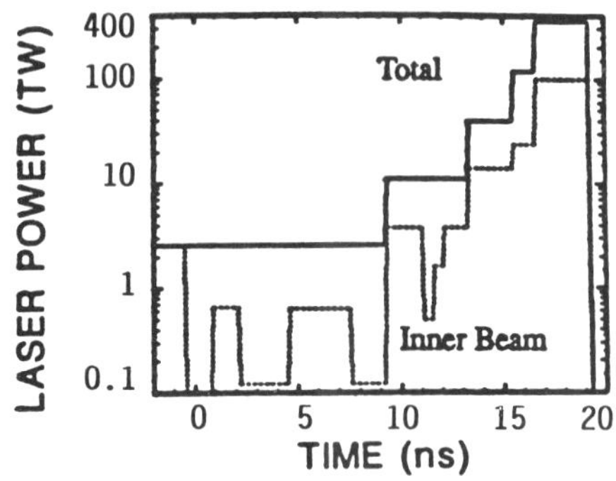

Figure 4. Laser power - beryllium capsule

4. Beryllium Capsule

Beryllium capsules are viable alternatives to plastic. Pure beryllium is a better ablator. It absorbs more hohlraum energy, and it produces a higher pressure than plastic at the same temperature. The technology of beryllium capsule fabrication is well developed – surface finishes of 20 nm have been demonstrated by lapping. It is strong enough to contain high DT pressures. Copper is an excellent dopant to add opacity and is soluble up to required levels. The fraction of dopant can be chosen to provide adequate opacity at the final drive temperature.

Our beryllium capsule is similar to the plastic but with an outer radius of 1.105 mm. The atomic composition of the ablator is 0.991 beryllium and 0.009 copper. A one dimensional calculation of this capsule with the radiation temperature drive of figure 3 gives a yield of 14 MJ. It can also be driven with five flat steps in the radiation pulse between 70 and 300 eV to give a yield of 8.8 MJ. As with the plastic capsule, the timing of the steps is critical.

Figure 4 shows the time history of the laser beam power driving the two dimensional integrated calculations. The highest yield achieved to date is 6.9 MJ in an LTE calculation. We used the same beam pointing as in the plastic capsule calculation, but different power history and balance.

In summary, using our best integrated calculations we confirm Livermore calculations of ignition with a plastic capsule, and have added an alternate capsule design with a beryllium ablator. The problem with a pressure pulse caused by stagnation on axis has been overcome by filling the hohlraum with gas. Although these calculations show the designs are sensitive, they add to our confidence that NIF can achieve ignition.

Acknowledgments

We wish to thank our colleagues at Lawrence Livermore Laboratory, particulary Steve Haan and Steve Pollaine, who have collaborated with us and shared their hohlraum and capsule designs. This work is supported by the USDOE under contracts W-7405-ENG-36 and W-7405-ENG-48.

Inst. Phys. Conf. Ser. No 140: Section 9
Paper presented at Laser Interact. with Matter Conf., Oxford, 19–23 September 1994
© 1995 IOP Publishing Ltd

Ta^{53+} ions with energies up to 4.8 MeV registered by Ion Electrostatic Analyser and Thomson Parabola Spectrograph from a laser produced plasma.

E. Woryna, P. Parys, J. Wołowski, J.Farny [a] and W. Mróz [b]
Institute of Plasma Physics and Laser Microfusion, 00-908 Warsaw, 23 Hery St., Poland
[a] 95 445 Bayreuth, Meranierring 28 / 7, Germany
[b] Institute of Optoelectronics, Military Univ. of Technology, 01-489 Warsaw, Poland

B. Králiková, J. Krása, L. Láska, K. Mašek, M. Pfeifer, K. Rohlena, J. Skála, P. Trenda
Inst. Physics, Acad. Sci. Czech Rep., 180-40 Prague 8, Na Slovance 2, Czech Republic

H. Haseroth, J. Collier
European Organization for Nuclear Research (CERN), 1211 Geneva 23, Switzerland

B. Yu. Sharkov, A. V. Shumshurov
Inst. for Theoretical and Experim. Phys., B. Cheremushkinskaja 25, 117 259 Moscow, Russia

ABSTRACT : An iodine laser PERUN (E<50 J, τ~350 ps, λ=1.315 μm, I<10^{15} Wcm^{-2}) was used for the production of highly ionized tantalum ions. In the experiment, the following set of complementary corpuscular diagnostics were used : an ion collector, an ion electrostatic analyser (EIA) and a Thomson parabola spectrograph (TS). The maximum registered charge state of the Ta ions (registered by IEA) was z=53 and maximum recorded energy of the tantalum ions with z=53 was 4.8 MeV. The Ta ion energy distributions are presented, in addition to the average charge state vs energy. A comparison of the experimental results from the EIA and TS is made.

1. INTRODUCTION

Investigations of ion energy and composition of a laser-produced plasma in large distances from a target started probably simultaneously with the beginning of laser-matter interaction experiments [1]. Soon the fact of registration of ionization states z>20 was reported, for example Co^{27+} [2]. Detailed measurements made with the EIA [3] in CO_2 laser system with $(CH_2)_n$, Al and Cu targets shown composed structure of ion energy distributions. The authors registered maximum charge state Cu^{25+} at the distances L=179 cm from the target. Specially interesting estimation of ion energy and angular distributions were made by spectral measurements (Nd:YAG laser, I~(5-7)×10^{14} W/cm^2). Maximum P^{13+} ion energy was estimated to be about 4 MeV [4]. The CERN project to investigate the feasibility of application of laser ion sources for heavy ion production accelerated investigations in this matter. The main experimental investigations were performed with CO_2 lasers [5,6] and an iodine laser [7]. This paper presents the last experimental results registered by the EIA and the TS with iodine laser installation PERUN.

2. EXPERIMENTAL ARRANGEMENT

The laser radiation from the laser system PERUN [8] was focused by an f/2 lens onto a massive, flat tantalum target at an angle of 30° to the target normal (Figure 1). The target was housed in a holder that was capable of being moved in the three linear directions with a resolution of 5 μm and provide target rotation. The spot diameter containing 80% of the incident radiation was 80 μm. The laser energy was below 50J. The FWHM laser pulse length was variable from 350 ps to 600 ps. The maximum intensity attainable was 10^{15} W/cm^2.

The electrostatic ion analyser with a deflection angle of 90° [9] was located 207.4 cm from the plasma in the plane of the laser beam and target normal at 10° to the target normal, opposing the laser beam. The ions transmitted through the analyser were detected by means of a windowless electron multiplier, type EMI 9643/2B with CuBe dynodes. The signal from the detector was recorded by a two channel Hewlett-Packard digitizing oscilloscope type HP 54510B. The signal from an ion collector located 20° degrees below the laser-target plane underneath the EIA was recorded by the second channel of the oscilloscope. The flight path of the ion collector was 94 cm.

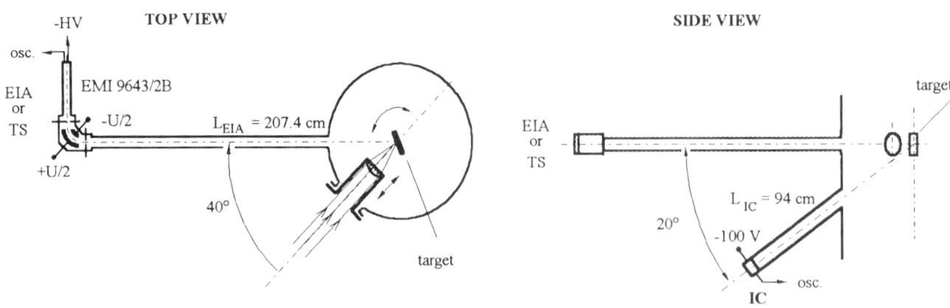

Figure 1. Experimental arrangement.

When installed, the Thomson parabola ion spectrograph [10] directly replaced the EIA at a distance from target to screen of 125 cm. The ions from this device were detected by a MCP - Phosphor screen assembly and recorded by a Tektronix CCD camera C1002 under the control of an IBM PC.

3. EXPERIMENTAL RESULTS

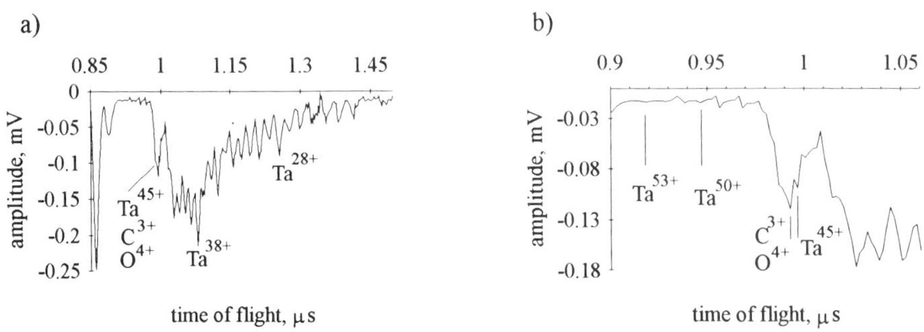

Figure 2. Ta ion spectra from EIA with long timebase - (a), early portion of the EIA spectrum containing the highest charge states and their theoretical positions - (b).

Figure 2a shows the output of the EIA with long time base. Figure 2b illustrates early portion of the EIA trace in Figure 2a containing the highest charge states. The maximum charge state present on this trace is found to be z=53.

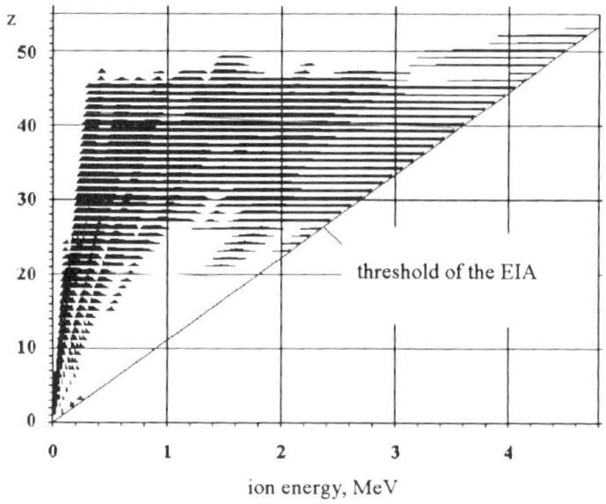

Figure 3. Charge state - energy diagram of recorded Ta ions.

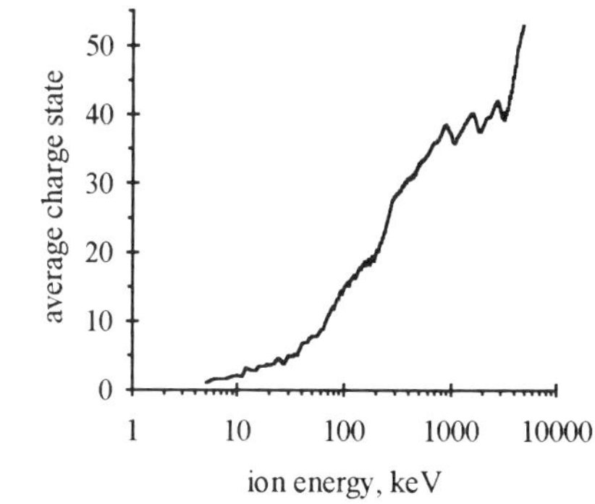

Figure 4. Average charge state of Ta plasma as a function of ion energy.

Figure 3 shows the charge state - energy diagram of recorded Ta ions, derived by assimilating more than 40 traces similar to those in Figure 2a taken under different analyser conditions. The laser energy over these 40 shots was 29 J with jitters of 10%. The thickness of the lines is logarithmically proportional to the magnitude of energy distribution. The variation of the secondary ion-electron emission coefficient with charge state and energy -$\gamma(z,E)$ has not been taken into account due to lack of data. The diagonal "cut-offs" running through Figure 3 are due to the lower limit caused by analysing procedure and the upper physical limit of the analyser. It clearly shows two distinct regions. For ions up to charge state z=30, the distributions are considerably narrower than those above z=30, terminating at approximately E/z=38 keV. Ions with charge state greater than 30 exhibit a substantially larger range of energies, the upper limit being unknown due to the analyser limitation, but in excess of this limit of E/z=90 keV, i.e. at least 4.8 MeV for ions with z=53.

Figure 4 shows the average charge state of Ta ions as a function of energy, again without consideration of $\gamma(z,E)$.

The difficulty with the above methods for ion analysis is the need for many laser shots in order to reconstruct the spectra with an acceptable statistical confidence. This is a particular problem for large lasers. On possibility to circumvent this problem would be to use a Thomson spectrograph, which can record simultaneously all ion charge states and energies over a large range. Figure 5 illustrates the a typical output of such a spectrograph when used in place of the EIA. The various charges states of tantalum are clearly shown for the lower charge states but are unresolved for the high charge states. Light ion surface contaminants are also displayed. Charge states up to 17+ are just resolved and one could derive information from the recorded parabolas, but above this it is not possible. The spectrograph is still however very useful for providing a general observation of the plasma, for example the difference in the spectra obtained from a clean surface (by laser pre-pulse) as in Figure 5 and that obtained for an unclean surface. For the unclean surface there is an absence of the highest charge states, and a greater proportion of lower charge states prevalent. Furthermore, the maximum registered energy of those ions appears to be independent of charge state for the unclean surface and was estimated to be about 400 keV. For clean surface, their maximum energy appears to be proportional to charge (the Ta parabola ends lying on a line parallel to the Av/z axis) with E/z=18 keV. Similar effect was observed before for carbon ions [11]. These energies are much lower than those registered by the EIA. This is due to the fact that the used Thomson detector is much less sensitive than the EIA detector by about 3 orders of magnitude.

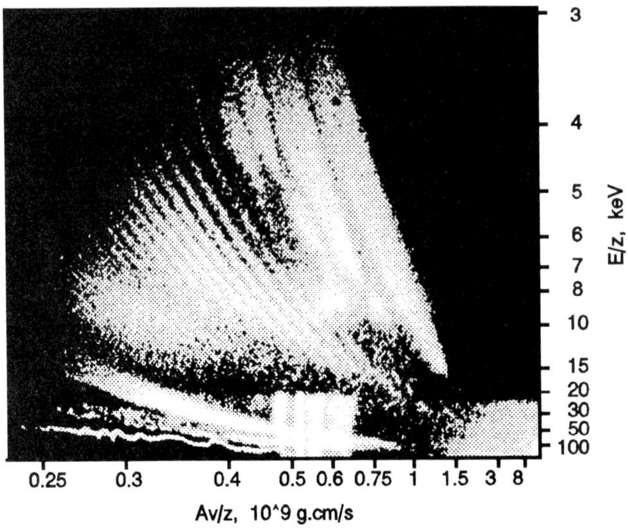

Figure 5. Thomson parabola picture from clean target .

4. CONCLUSION

The use of the techniques described above has enabled us to show the presence of ionized tantalum up to Ta^{53+}, at energies at least 4.8 MeV emitted from a laser plasma produce at $I \sim 10^{15}$ Wcm^{-2} at long distances from the target. Higher charge states could be present in the plasma itself. Further inferences with regard to absolute values of relative composition and energy distributions cannot be reliably made due the lack of knowledge of the secondary ion-electron emission coefficient $\gamma(z,E)$ of the detecting surfaces. Furthermore, an upper detection limit of the EIA and the lower sensitivity of the Thomson spectrograph mean it is not possible to investigate the highest energy part of the distributions. Work is continuing to alleviate these problems.

References:

[1] Linlor, W. I. 1963 Appl. Phys. Lett. **3**, 210.
[2] Bykovskij, Yu. A. et al. 1971 Sov. Phys. JETP **33**, 706.
[3] Denus, S. et al. 1978 J.Tech. Phys..**19,** 503.
[4] Basov, N. G. et al. 1987 Sov. Phys. JETP **65**, 727.
[5] Amidouche, Y. et al. 1992 Rev. Sci. Instrum. **63**, 2838.
[6] Sherwood, T. R. 1992 Rev. Sci. Instrum. **63**, 2798.
[7] Láska, L. 1994 Appl. Phys. Lett. **65**, 691.
[8] Chvojka, M. et al. 1988 Czech J. Phys. **B38**, 1337.
[9] Denus, S. et al. 1977 J.Tech. Phys.**18**, 25.
[10] Wołowski, J. et al. 1982 Zh. T. Ph. (in Russian) **52**, 366.
[11] Ehler, W. et al. 1980 J. Phys. D: Appl. Phys. **13**, 29.

Inst. Phys. Conf. Ser. No 140: Section 9
Paper presented at Laser Interact. with Matter Conf., Oxford, 19–23 September 1994

© 1995 IOP Publishing Ltd

High Harmonic Generation with a CPA KrF Laser

M. Zepf,[1] W. Blyth,[1] M.H. Key,[1,2] M. Nakai,[2,*] S.G. Preston,[1] and J.S. Wark[1]

[1]Department of Physics, Clarendon Laboratory, University of Oxford, Parks Road, Oxford, OX1 3PU, UK
[2]Rutherford Appleton Laboratory, Chilton, Didcot, Oxon, OX11 OQX, UK
*on leave from Institute of Laser Engineering, Osaka University 2-6 Yamada-oka, Suita-Osaka 565, Japan

Abstract. High harmonics have been observed from a 0.8 TW, 350-fs KrF CPA laser focused to 8×10^{17} W cm $^{-2}$ in gas jet targets. Estimates of the absolute conversion efficiencies have been made. The efficiency and relative strength of the harmonics was found to be largely independent of the position of the laser focus within the gas jet, and of the density, suggesting a coherence length limited by dephasing effects due to plasma electrons. The highest harmonics are attributed to ions rather than neutrals.

The considerable progress that has been made in the development of high-power short-pulse lasers has led to a wealth of interesting non-linear phenomena, including the generation of extremely high odd harmonics of the laser radiation by the interaction of the laser field with a gaseous target. The intensities of these harmonics have been found to be slowly varying from approximately the seventh up to very high order (the plateau) after which they decrease rapidly (the cut-off). Most of the past work has concentrated on the use of relatively long wavelength short-pulse lasers[1-8]. This is because the energy of the highest harmonic, at the cut off E_m has been shown to be approximately $E_m = E_i + 3U_p$, where E_i is the ionisation energy of the neutral atom, and U_p the ponderomotive energy [9] so that, because U_p scales quadratically with wavelength, a long wavelength fundamental gives the highest harmonics. At present the highest harmonic photon energy produced is 166 eV in the 141st harmonic of 1.053-µm light.[10]

The generation of harmonic radiation from shorter wavelength lasers is still of great interest however for two reasons. Firstly, it has been shown by Perry *et al*, that the efficiency of harmonic generation at a given photon energy can be up to 30x greater for 0.53-µm fundamental wavelength relative to 1.05-µm [8]. Part of the reason for this is attributable to the smaller dephasing effect that a given density of free electrons has at shorter fundamental wavelength. In addition there is also some evidence that the single atom response may be greater. The possibility of still higher relative efficiency under optimised conditions with 248-nm wavelength remains to be investigated.

Secondly the few experiments that have been performed at 248 nm [13-15] together with theoretical analysis by Krause, Schafer and Kulander [9] suggest that, at least in part, the generation of harmonics arises from ions as well as neutral atoms, while recent calculations by Sanpera *et al* . predict similar single atom response from singly ionised and neutral helium for 248-nm radiation [14]. Even though ionic response is not well understood, there appears to be an interesting possibility of generating high order harmonics of high photon energy from ions using a short wavelength laser.

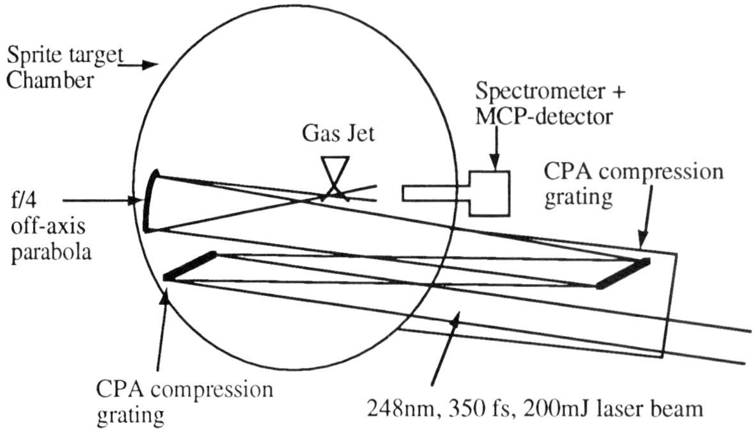

Fig 1: Schematic of experimental set-up

We report here on the generation of high harmonics of 248-nm radiation from noble gas targets. The experiment was performed using the SPRITE chirped pulse amplification KrF laser at the Central Laser Facility of the Rutherford Appleton Laboratory. A schematic diagram of the experiment is shown in Fig.1. Laser pulses of 200-mJ energy at 248-nm radiation in a pulse-length of 300-fs (FWHM) were focused into a gaseous target by an f/4 off-axis parabolic mirror. The focal spot diameter was ~ 10 μm, resulting in an intensity at focus of 8×10^{17} W cm^{-2}. The gases investigated - He, Ne, Ar and Kr were injected into the vacuum chamber through the 1-mm diameter nozzle of a gas jet, with atomic number densities in the focal region varying from approximately 5×10^{17} to 2×10^{19} cm^{-3}. The intensity at the edges of the gas jet (+/- 1-mm from the focus) was only 10^{15} Wcm^{-2}, much lower than at the focus. After passing through the gas target, the fundamental beam and associated harmonics were incident on the entrance slit of a flat-field XUV spectrometer, used with a Hitachi 1200-l/mm grating. A zero-order beam block intercepted the 248-nm radiation, the signal from which was further reduced by a filter pack placed over the front of the Galileo micro-channel-plate (MCP) detecting system. The filter pack consisted of two separate filters, arranged such that the full spectrum from the harmonics and plasma radiation was recorded through each filter. The first filter was 1000 Å of aluminum, and the second 1600 Å of boron, both having a substrate of 1000 Å of CH polymer. The output from the channel plate was recorded on a CCD detecting system.

A lineout from a typical MCP image of harmonics generated by focusing into Neon at an atomic density of $1.3\ 10^{19}$ cm^{-3} is shown in Fig.2. This spectrum was recorded through the boron filter. When the filter and instrument response are taken into account we deduce relative harmonic intensities as a function of harmonic order as shown in Fig. 3. It can be seen that, over the range of harmonics observed, the brightest harmonic from Neon is the 13th, with the highest harmonic readily detectable being the 19th, although the 21st was faintly visible on some laser shots. Stray light from the zero order image masked the spectrum in the short wavelength limit. For comparison the results are shown alongside those of Sarakura *et al* .[12], where we have normalised the two sets of Neon results to the intensity of the brightest (13th) harmonic. It can be seen that the relative intensity of the harmonics recorded in our experiment is very similar to that observed by Sarakura *et al*, and the fact that the brightest harmonic is the 13th is also consistent with the earlier results reported by McPherson *et al*.[11]. Furthermore, the reduction in harmonic intensity of two orders of magnitude from the 13th to about the 19th is also in agreement with these previous results. A rough estimate of the efficiency of harmonic production of 10^{-8} has been made by calibrating the response of the spectrometer with the MCP against that of the same spectrometer, but using Ilford Q-plates and Kodak 101-07 film as the recording medium. By using published calibration curves for the film [15] and the efficiency of diffraction of the grating, the total number of photons incident on the spectrometer in each harmonic was estimated. We estimate that this calibration procedure is only accurate to about a factor of 4. The efficiencies for our data (not applicable to those of Sarakura *et al)* [12] are also shown in Fig. 3.

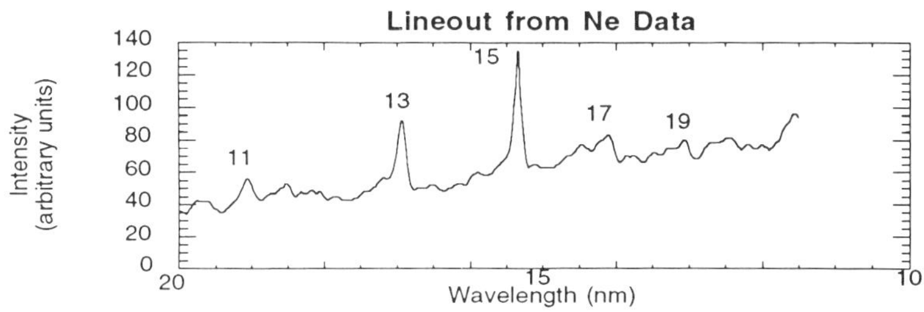

Fig 2: Lineout from Neon spectrum, harmonic order denoted in the diagram

During the experiment the position of the gas jet with respect to the laser focus was varied by ± 0.5 mm, i.e. data was recorded when the focal spot was located at both the front and rear of the column of gas emitted from the gas-jet nozzle. The harmonic yields and relative intensities were found to be largely independent of the position. Similarly, when the density of the gas was varied from approximately 5 10^{17} to 2 10^{19} cm^{-3} there was little effect on the harmonic production. This invariance of the harmonic yield and relative intensity with position and density suggests that the conversion is limited by a short coherence length due to the de-phasing effect introduced by the free electrons.

For a 300-fs 248-nm pulse we predict an approximate saturation intensity of neutral Neon at an irradiance of 6 x 10^{14} W cm^{-2}, which corresponds to a value of $3U_p$ of 10.3 eV; which when added to the ionisation potential for neutral Neon (21.6 eV) gives a simplistic estimate of the maximum photon energy for harmonics from neutrals of 32 eV corresponding to the 7th harmonic of 248-nm radiation. Corresponding estimates for singly ionised Neon give 3 x 10^{15} W cm^{-2} with $3U_p$ of 51.6 eV i.e. the 17th harmonic . Our observations of harmonics up to the 21st are therefore consistent with generation from at least singly ionised ions.

In the high intensity at focus in our experiment it is obvious that ionisation stages far beyond singly-ionised are produced. Indeed, the peak intensity of 8 x 10^{17} W cm^{-2} implies optical ionisation up to helium-like Neon. At present, to our knowledge, no calculations have been performed of the harmonic response of such high ion stages. We cannot therefore conclude whether singly ionised or more highly ionised ions are responsible for the harmonics of highest photon energy in our experiments though it is expected that the single atom response will be weaker for higher ionisation stages.

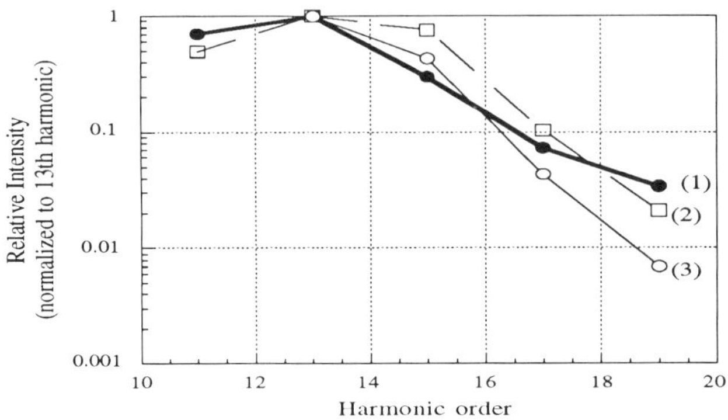

Fig 3: Neon results at 1.3 10^{19} cm^{-3}, 140 mJ (3); 1.9 10^{19}, 180 mJ (2) and results by Sarakura et al. (1)

In summary we have observed up to the 19th harmonic of 248-nm radiation in interactions with Neon. The relative intensity of the harmonics as a function of harmonic number is in good agreement with those reported by Sarakura *et al* [12]under similar experimental conditions. In agreement with the interpretation of Krause *et al*, [9] we believe that the higher harmonics are due to the response of the ions.

We are happy to acknowledge the able assistance of the staff of the SPRITE laser facility in the performance of these experiments. This work was supported by the Science and Engineering Research Council under grant number GR/J 02560. One of us (M.Z.) was supported by Konrad Adenauer Stiftung.

References:

[1] M. Ferray, A. L'Huiller, X.F. Li, L.A. Lompre, G. Mainfray, and C. Manus, J. Phys. B **21**, L31 (1988).

[2] X.F. Li, A. L'Huiller, M. Ferray, L.A. Lompre, and G. Mainfray, Phys. Rev. A **39,** 5751 (1989).

[3] K.C. Kulander and B.W. Shore, Phys. Rev. Lett. **62**, 524 (1989).

[4] Anne L'Huillier and Ph. Balcou, Phys. Rev. Lett. **70**, 774 (1993).

[5] J.J. Macklin, J.D. Kmetic, and C.L. Gordon III, Phys. Rev. Lett. **70**, 766 (1993).

[6] Anne L'Huiller, Phillipe Balcou, and L.A. Lompre, Phys. Rev. Lett. **68**, 166 (1992).

[7] Phillipe Balcou and Anne L'Huiller, Phys. Rev. A **47,** 1447 (1993).

[8] John K. Crane, Michael D. Perry, Donna Strickland, Steve Herman, and Roger W. Falcone, IEEE Transactions on Plasma Science, Vol. **21**, 82 (1993).

[9] J.L. Krause, K.J. Schafer, and K.C. Kulander, Phys. Rev. Lett. **68**, 3535 (1992).

[10] Michael D. Perry and Gerard Mourou, Science **264**, 917 (1994).

[11] A. McPherson, G. Gibson, H. Jara, U. Johann, T.S. Luk, I. A. McIntyre, K. Boyer, and C. K. Rhodes, J. Opt. Soc. Am. B **4**, 595 (1987).

[12] Nobuhiko Sarakura, Kiyoshi Hata, Takashi Adachi, Ryoichi Nodomi, Masayoshi Watanabe, and Shuntaro Watanabe, Phys. Rev. A **43**, 1669 (1991).

[13] K. Kondo, N. Sarukura, K. Sajiki, and S. Watanabe, Phys. Rev. A **47**, R2480 (1993).

[14] A. Sanpera, P. Jönsson, J.B. Watson, K. Burnett, submitted to Phys. Rev. A (1994)

[15] J. Krshnan, D. Neely, C.Danson, L. Dwivedi, C.L.S. Lewis, G.J. Tallents, Rutherford Appleton Laboratory Report RAL-92-020, 22 (1992)

Index